Quasars and Active Galactic Nuclei
An introduction

This book provides an up-to-date and comprehensive account of
quasars and active galactic nuclei (AGN). The latest observations
and theoretical models are combined in this clear, pedagogic textbook
for advanced undergraduates and graduate students. Researchers will
also find this wide-ranging and coherent review invaluable.

Throughout, detailed derivations of important results are provided
to ensure the book is self-contained. And theories and models are
critically compared with detailed and often puzzling observations from
across the spectrum. After an introduction to the discovery and early
models of quasars and AGN, we are led through all the key topics,
including quasar surveys and statistics, continuum radiation, time
variability, relativistic beaming and superluminal motions, accretion
disks, jet sidedness, gravitational lensing, unification and detailed,
multi-wavelength studies of individual objects. Particular emphasis
is placed on radio, X- and gamma-ray observations – not covered
in depth in any previous book – and the technical challenges of
making such observations.

All those entering into this exciting and dynamic area of astronomy
research will find this book an ideal introduction.

Quasars and Active Galactic Nuclei
An introduction

Ajit K. Kembhavi and Jayant V. Narlikar

Inter-University Centre for Astronomy and Astrophysics
Pune 411 007, India

CAMBRIDGE
UNIVERSITY PRESS

PUBLISHED BY THE PRESS SYNDICATE OF THE UNIVERSITY OF CAMBRIDGE
The Pitt Building, Trumpington Street, Cambridge CB2 1RP, United Kingdom

CAMBRIDGE UNIVERSITY PRESS
The Edinburgh Building, Cambridge CB2 2RU, UK   http://www.cup.cam.ac.uk
40 West 20th Street, New York, NY 10011-4211, USA   http://www.cup.org
10 Stamford Road, Oakleigh, Melbourne 3166, Australia

First published 1999

Printed in the United Kingdom at the University Press, Cambridge

Typeset in 10/13pt Monotype Times [EPC]

A catalogue record for this book is available from the British Library

Library of Congress Cataloguing in Publication data

Kembhavi, A. K. (Ajit K.)
    Quasars and active galactic nuclei : an introduction / Ajit K. Kembhavi,
Jayant V. Narlikar.
        p.   cm.
    Includes bibliographical references.
    ISBN 0 521 47477 9
    1. Quasars. 2. Active galactic nuclei. I. Narlikar, Jayant Vishnu, 1938–   .
    II. Title.
    QB860.K46   1999
    523.1′15–dc21   97-28655CIP

ISBN 0 521 47477 9 hardback
ISBN 0 521 47989 4 paperback

Contents

Preface

The year 1963 marks a watershed in extragalactic astronomy. The optical identification of radio sources 3C 273 and 3C 48 and the measurement of their redshifts demonstrated to astronomers the existence of a new class of energy sources that have a star-like appearance, yet produce luminous energy at a rate comparable to a galaxy of a hundred billion (10^{11}) stars.

The quasi-stellar objects (QSOs) or 'quasars' as these sources came to be called, arrived on astrophysicists' plates just about when they had digested the long-standing mystery of stellar energy. By the 1960s, the problems of stellar structure and evolution built on the pillars erected by Eddington, Milne, Chandrasekhar, Bethe, Lyttleton, Schwarzschild and Hoyle had been tackled successfully, thanks to the advent of fast electronic computers. The quasars, however, presented challenges of an altogether different nature. How could so much energy come with such rapid variability out of such a compact region and be distributed over such a wide range of wavelengths?

The classic book *Quasi-Stellar Objects* by Geoffrey and Margaret Burbidge, published in 1967, captured this early excitement and posed the numerous challenges of quasar astronomy very succinctly. Now, three decades later, we have the benefits of vast progress in the techniques of observational extragalactic astronomy and the intricate sophistication of ideas in high energy astrophysics. Yet it is fair to say that the understanding of quasars and the related field of active galactic nuclei (AGN) has not reached the same level of success that stellar studies had attained thirty years ago.

In this book we have attempted the admittedly difficult task of putting together the astronomical information on quasars and AGN and the interpretations given to it by theorists. Although a 'Standard Model' is beginning to emerge, there are numerous details that still remain to be fitted. Many of the interpretations are tentative, either because the data is scarce or because there is so much of it that theorists have not yet been able to make much headway. However, there are some established paradigms, like accreting massive black holes as the central engine, accretion disks, relativistic beaming and, at a weaker level, the idea of unification. Much of the theoretical interpretation makes use of these basic ideas, which at least provide a convenient framework within which one may judge the successes and failures of the present limited understanding. We have also drawn the reader's attention to a series of anomalous findings by a minority of workers in the field, findings which perhaps deserve greater consideration than the dismissal usually accorded to them as being 'accidental', 'insignificant' or 'artifacts'.

Although we have individually worked on quasars and AGN over a number of years, neither of us can claim to be a pioneer in the field. In writing this book, therefore, we have not been motivated by the desire to champion a specific theory, scenario or paradigm, which perhaps gives us the advantage of being neutral and critical. Over the last decade, there has been a phenomenal increase in the range and quantity of data on quasars and AGN. Digesting, or even developing a superficial understanding of, these data has become a rather difficult task, especially for those who are away from the centres of activity and lack ready access to experts. Our motivation in writing the book has been to remedy this situation somewhat, by providing an overview with a strong pedagogic content.

The text itself is aimed at a typical reader who is familiar with astronomy and astrophysics up to the undergraduate level. It could serve as a teaching text at the advanced undergraduate or graduate level, or as a source book for the graduate student embarking on research in high energy astrophysics. The content of some of the chapters is more detailed than would be required in a graduate course, as we have also aimed at providing a review of the field for the research worker who is not an expert in the general area we address. The book is largely self-contained, but the reader is occasionally referred to a few well-known texts for details of derivations of some formulae, especially where these are a part of the general physics curriculum. We have tried to represent different points of view and cover many important areas of which we are aware, but, owing to the sheer magnitude of the work, we may have missed important facts and arguments. We have omitted detailed discussion on the physics of emission line regions, as this would make a book in itself, and several excellent treatments are available. Because of the vast range of the subject covered, and our own lack of expertise with many of the areas, we have had to depend heavily on published reviews and research articles by leaders in the field, as citations given in the text will testify. We may have inadvertently omitted to cite some important references, especially where the material has become a part of our teaching notes over the years, or is simply a part of the folklore. We gratefully acknowledge our debt to all these sources.

IUCAA   Ajit Kembhavi
Pune   Jayant Narlikar

Acknowledgments

It is a pleasure to thank the many well-informed colleagues who have helped us in understanding some of the more difficult concepts related to quasars and AGN. In this connection we wish to mention D.J. Saikia, with whom we have had several discussions about radio astronomy, R. Srianand and N. Dadhich. K.P. Singh and V.R. Venugopal kindly provided comments on portions of the manuscript. Several young persons have helped us with their comments, have drawn some figures for us and helped with digitization, typesetting and proofreading. The list is long, but we must particularly mention Vidyullata and Ashish Mahabal, Yogesh Wadadekar, Ramana Athreya, Shyamal Banerjee, Ranjeev Misra, Soma Mukherjee, K. Srinivasan, Santosh Khadilkar, Archana Kamanapure, Susan Mathew and S.R. Tarphe. One of us (AKK) wrote some chapters at the Astronomical Institute 'Anton Pannekoek' in Amsterdam where he was a guest of E.P.J. van den Heuvel and at the Institute of Astronomy, Cambridge. We are grateful to Martin Rees who encouraged us to write the book, to Simon Mitton of CUP who took us through the early steps, to Adam Black for his unbounded patience in extending the deadline again and again, and to Susan Parkinson who has marvellously copy-edited the manuscript. Finally we thank our families who, during the writing, were deprived even of the rather scant attention that they normally receive from us.

Acknowledgments for using copyrighted material

We wish to thank the Annual Review Inc., the Astronomical Society of the Pacific and Kluwer Academic Publishers for permission to reproduce figures from material published by them. We have also reproduced figures from the *Astronomical Journal*, the *Astrophysical Journal* and *Monthly Notices of the Royal Astronomical Society*, as well as several publications of Cambridge University Press. We wish to thank all publishers and authors for use of their material.

Acronyms and abbreviations

Acronyms and abbreviations used in the text are listed here. Further information can be found through the index.

AGN, active galactic nucleus
ARIEL, X-ray astronomy satellite
ASCA, advanced satellite for cosmology and astrophysics
   (the Japanese ASTRO-D X-ray satellite)
AXAF, advanced X-ray astronomy facility
BAL, broad absorption line quasar
BATSE, burst and transient source experiment
BLRG, broad line radio galaxy *123*, 185
BQS, Palomar Bright Quasar Survey
COMPTEL, Compton telescope
COMPTON, Compton gamma-ray observatory
EGRET, energetic gamma-ray experiment telescope
EINSTEIN, high energy astronomy observatory II
EMSS, Extended Medium Sensitivity Survey
EUVE, extreme ultraviolet explorer
EXOSAT, European X-ray astronomy satellite
FIRST, Faint Images of the Sky at Twenty Centimetres survey
FR-I, Fanaroff–Riley class I
FR-II, Fanaroff–Riley class II
FSRQ, flat spectrum radio quasar
GINGA, the Japanese ASTRO-C X-ray satellite
GRANAT, Russian X-ray and gamma-ray satellite
HALCA, highly advanced laboratory for communications and
   astronomy (Japanese orbiting radio telescope)
HEAO, high energy astronomical observatory
HRI, high resolution imager
IPC, imaging proportional counter
IRAS, infrared astronomy satellite
IUE, international ultraviolet explorer
LAC, large area counter

LBQS, Large Bright Quasar Survey
LINER, low ionization emission line region
MERLIN, multi-element radio linked interferometer network
NLRG, narrow line radio galaxy
OSO, orbiting Solar observatory
OSSE, oriented scintillation spectroscopy experiment
OVV, optically violent variable quasars
POSS, Palomar Observatory Sky Survey
PSPC, position sensitive proportional counter
PTGS, Palomar Transit Grism Survey
QSO, quasi-stellar object
RBL, radio selected BL Lac object
ROSAT, Röntgen-Satellit
SAS, small astronomy satellite
TENMA, the Japanese ASTRO-B X-ray satellite
UHURU, X-ray astronomy satellite
VLA, very large array
VLBA, very long baseline array
VLBI, very long baseline interferometry
XBL, X-ray selected BL Lac object

1 Historical background

1.1 Introduction: the energy problem

The discovery of quasi-stellar objects (QSOs or quasars) in 1963 represents a landmark in observational astronomy. Thanks to a coordination between optical and radio astronomers, it was possible to discover a new and important class of astronomical objects. Because this text book is all about quasars and related phenomena, it will not be out of place to begin at the beginning of the subject and to review briefly how these remarkable objects were first discovered.

The science of radio astronomy really began after the end of World War II, when some of the scientists and engineers engaged in wartime radar projects used their know-how to follow up the pioneering works of Karl Jansky in the 1930s and Grote Reber in the early 1940s. Thus radio dishes and interferometers appeared in England and Australia, at Jodrell Bank, Cambridge, Sydney and Parkes.

The early observations revealed the existence of cosmic radio sources and by the mid-1950s it became an accepted fact that radio galaxies exist. The nature of their radiation was non-thermal, and its polarization properties indicated that its origin lay in the synchrotron process. As we will discuss in Chapter 3, in this process radiation comes from electrons accelerated by a magnetic field. Thus a typical radio source has as its energy reservoir the dynamical energy of relativistic particles and magnetic field energy.

In 1958 Geoffrey Burbidge drew attention to the enormous size of this energy reservoir. Since we will go through his argument in Chapter 3, we simply state the result here. He found that the *minimum* energy available is of the order of 10^{60} erg! This estimate would go up further if one included the energy of the protons as well as that of the electrons. Indeed, so large was this estimate that it was the last nail in the coffin of the collision hypothesis, which sought to explain the radio source phenomenon as the outcome of colliding galaxies. The typical gravitational potential energy of a pair of colliding galaxies of masses M_1 and M_2 separated by a distance R is $-GM_1M_2/R$. For typical galactic mass $\sim 10^{11} M_\odot$, where $M_\odot$ is the mass of the Sun, and $R \sim 10$ kpc, we get an answer $\sim 10^{59}$ erg. This calculation also assumes an optimistic situation where one energy reservoir can be converted to another with perfect efficiency, whereas astrophysical processes seldom exhibit high efficiency. The mass–energy conversion in the thermonuclear fusion of hydrogen to helium operates with an efficiency of 0.007. Thus even a 10 per cent efficiency would require the primary source of energy in a radio source to be of the order of 10^{62} erg. What can such a source be like?

Burbidge himself had suggested that the source could lie in a chain reaction that triggers off one supernova after another in the nuclear region of a galaxy. Since the nuclear region is expected to be much denser than average, and supernovae do pour out a lot of energy in an explosive fashion, this idea seemed plausible. However, it was not pursued in detail.

Instead, in 1962, Fred Hoyle and William A. Fowler proposed that the nuclear region of a galaxy may permit the formation of a supermassive star having, say, a million or more solar masses which would evolve to become a 'super-supernova' thus generating the requisite energy from a thermonuclear source. However, by this time, Hoyle and Fowler had come to realize that a thermonuclear source of energy will not be as efficient as a gravitational one for masses of this order. This is clear from a qualitative argument to start with: the nuclear energy increases in proportion with the mass whereas the gravitational potential energy increases as the *square* of the mass. Thus while at the solar-mass level the former wins over the latter (*vide* the historical argument that dethroned the Kelvin–Helmholtz contraction hypothesis in favour of thermonuclear energy), the situation is the exact opposite for supermassive stars. But how can one tap the gravitational energy?

In a pioneering paper in *Nature* in early 1963, Hoyle and Fowler proposed that the energy for such sources was of gravitational origin, being derived from the collapse of very massive objects under their own strong gravitational fields. They pointed out that for very massive objects the internal pressures are inadequate to withstand this force of gravity. The objects therefore start contracting and as they contract the force of gravity grows, thus widening the gap between the inward and outward pressures. The situation becomes unstable, leading to the phenomenon of *gravitational collapse*. Thus the stage was set on theoretical grounds for highly collapsed supermassive objects as sources of high energy.

1.2   The discovery

We now return to the observational aspects once again. Even in the early days of radio astronomy, the importance of optical identification of a radio source was appreciated. The identification of Cygnus A had been the key observation to underscore the extragalactic nature of radio sources. Following the optical identification of a suspected extragalactic source, one can hope to do spectroscopy of the optical object and, if possible, determine its redshift. Then, by Hubble's law (see Chapter 2) one can infer the source's distance and hence its radio and optical luminosity.

One of the early catalogues of radio sources was the *Third Cambridge Catalogue* prepared by the Mullard Radio Astronomy Observatory at the University of Cambridge. Its sources were picked on the basis of their falling within a declination band and being brighter than 9 Jy.[1] The sources were listed in increasing order of right ascension

----

[1]   1 Jy (jansky) $= 10^{-26}$ W m^{-2} Hz.

with the prefix 3C. Of these, two sources, 3C 273 and 3C 48, were to play a major role in the developments of the early 1960s.

Radio astronomers at Jodrell Bank were interested in looking at the angular sizes of radio sources. In their first survey they looked at about 300 sources and found that their average size was around 30 arcsec. The size of most sources was in the range of 5 arcsec to a few arcminutes. However, a set of some 10 sources were extremely small, less than 1 arcsec in size. What could they be?

One of them was 3C 48, a source that was optically identified with a star-like object. Allan Sandage found its spectrum to be very unusual, one peculiarity being that it had strong emission lines. Also, Matthews and Sandage (1963) found the light from the object to be variable. These considerations led, in 1962, to this object's being labelled a radio star.

Meanwhile, Cyril Hazard was trying out a new method of fixing the position of a radio source very accurately in the sky. The method, which involved observing the occultation of the source by the Moon, when used on the source 3C 212 seemed to be very promising. And so he wished to try it for the compact source 3C 273, which was also in the Moon's path. Hazard proposed to observe it from Australia along with M.B. Mackey and A.J. Shimmins. The observation was successfully carried out in 1962 and the position of 3C 273 (a source with two components separated by about 20 arcseconds) was obtained with an accuracy of $\sim$ 1 arcsec (Hazard, Mackay and Shimmins 1963).

This positional accuracy was sufficient for astronomers to identify the optical counterpart of the source. It turned out to be a star-like object of some thirteenth magnitude. However, its spectrum, taken by Maarten Schmidt, was peculiar: it had four emission lines. It turned out that they were one of the doublets of [O III], as well as the three hydrogen lines H α, H β and H γ. However, all four were appearing with their wavelengths increased by about 16 per cent. Later, using a spectrum scanner, Beverley Oke also found the expected line H α with a similar shift.

Thus the object had a redshift of 16 per cent (see the next chapter for the interpretation of redshift); and as such its original classification as a star in our galaxy went overboard! Instead of being one of the 100 billion or so stars in our galaxy, it had to be an extragalactic object whose redshift-related distance was so large that it was a hundred times brighter than an entire galaxy. Yet, it had to be compact enough to be mistaken for a star. Further, examination of old plates showed that the light from this object had also fluctuated significantly with time. Figure 1.1 illustrates the optical object with the optical jet clearly seen at its lower right-hand side.

The radio source structure of 3C 273 determined by lunar occultation showed it to be a two-component system with the components separated by 19.5 arcsec. Of the two components A and B, the optically identified object sits on component B and shows a jet directed towards the other component, A. The jet is another indication that the object is not a star but is probably a much more violent system. The radio structure on various angular scales can be seen in Figure 9.13.

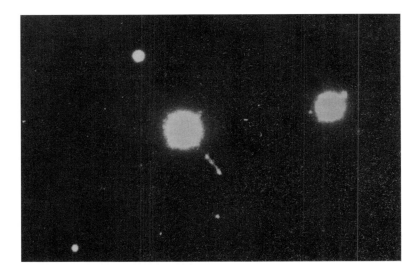

Fig. 1.1.   The quasar 3C 273 in the optical band. Reproduced from Narlikar (N93).

Alerted by these unusual features in 3C 273, astronomers took a second look at the object 3C 48 and found that its spectrum had looked unusual (or, unfamiliar) because it too was redshifted, and by an even larger degree, around 36.7 per cent.

It should be recalled that in the early 1960s the measured redshifts of galaxies usually ranged up to about 0.2, i.e., 20 per cent. The galaxy identified with the radio source 3C 295 had a redshift of 0.46, which was then the record! Thus the large redshifts of these two unusually compact objects immediately drew the attention of theoreticians. Further, this discovery, coming as it did in early 1963, coincided with the theoretical expectations of Hoyle and Fowler that highly compact massive objects could serve as high energy reservoirs.

The role of theoretical inputs was considered so important that an international symposium was convened in Dallas, Texas in December 1963 to bring together general relativists, theoretical astrophysicists and observers to discuss the implications of the discovery of these remarkable objects. This was to be the first of a continuing series of biennial symposia under the name 'Texas Symposia'. The objects became known as *quasi-stellar objects (QSO)* or *quasars*. We will use either of these names in this text.[1]

1.3   The nature of redshift

The quasars 3C 48 and 3C 273 had large redshifts in the context of the galactic redshifts known in the early 1960s. Subsequently, however, quasars with even higher redshifts began to be discovered. The source 3C 9 was the first very large redshift object to

---

[1] See the discussion in Section 6.1.

be found, with redshift $z = 2$. In fact, a large number of quasars with redshifts in this neighbourhood became known during the first decade of the discovery of quasars. These discoveries were also accompanied by theoretical discussions on their nature.

In the 1960s the astronomer was familiar with three kinds of redshift: (1) the Doppler shift arising from the relative motion between the source of light and the observer, (2) the gravitational redshift of light travelling from a strong to a weak gravitational field and (3) the redshift arising from the expansion of the universe. In the early days of quasar astronomy, all three interpretations were tried, before the consensus settled in favour of the last of these. It may be worth looking at the pros and cons of these different hypotheses briefly. We will return to this issue in a more detailed fashion in Chapter 15.

1.3.1 *Doppler shift*

If a source of light moves with velocity **v** relative to the observer so that the velocity vector makes an angle θ with respect to the radial vector from the observer to the source, the spectral shift of the light from the source measured by the observer will be given by

$$1 + z = \frac{1 + (v/c)\cos\theta}{\sqrt{1 - (v/c)^2}},$$ (1.1)

where $v = |\mathbf{v}|$ and c is the speed of light. For small velocities, this formula reduces to the Newtonian limit $z = v/c$ for radially outward motion.

For stellar motion in our galaxy it is common to find small Doppler shifts both positive and negative and these are interpreted in terms of stars moving away from or towards us. James Terrell in 1966 interpreted the quasar redshifts as arising from fast ejection of quasars from the galactic centre. The problem with this concept was that it endowed the galactic centre with violent activity, which was not consistent with the undisrupted motions of stars in the area. This idea was taken further by Hoyle and Burbidge (1966), who suggested that, unlike the case for our galaxy, ejection from a galactic nucleus that shows violent activity would appear normal. In particular, they suggested NGC 5128 as a likely site for ejection.

The Doppler hypothesis thus delinks redshift from distance; that is, a large redshift does not imply that the source is very distant. This eases the energy budget of a typical source. However, if several such sources are to be ejected from a single site (such as NGC 5128) the energetics of that site becomes problematical. Moreover, there is the problem of blueshifts.

The problem was first highlighted by P. Strittmatter in 1966 (unpublished). A site of explosion will eject quasars in all directions. Thus, unless the explosion occurred a long time ago, some quasars would still be travelling *towards* the observer and these should show blueshifts. Moreover, these quasars would appear brighter due to the blueshift effect and would tend to dominate in a sample that is complete with respect to all quasars brighter than a specified flux level. Detailed calculations based on isotropic ejection of all such quasars show that the ratio of blueshifted (N^-) quasars to redshifted (N^+) quasars in a flux-limited sample is

$$\frac{N^-}{N^+} = (1 + z_{\mathrm{m}})^{2.5+1.5\alpha}, \tag{1.2}$$

where z_{m} is the maximum redshift in the sample, and α is the spectral index of a typical source. Thus, for a maximum redshift of 2, say, and a spectral index unity, we may expect 81 times as many blueshifted sources as redshifted ones!

We will return to this problem in Chapter 15. For the time being we leave the topic of Doppler shifts as it was perceived in the early sixties, and move on to the next alternative.

1.3.2   *Gravitational redshift*

Einstein's general theory of relativity drew the attention of astronomers to this source of redshift: indeed, the early demonstrations of this effect were for the white dwarf stars Sirius B and δ Eridani B. In either case the effect was very small, being less than 0.001. Theoretically one could, on the basis of Schwarzschild's solution, define the surface redshift of a mass M with radius R by the formula

$$1 + z = (1 - 2GM/Rc^2)^{-1/2}. \tag{1.3}$$

Thus, by having R arbitrarily close to the critical value $2GM/c^2$, it may seem possible to have large redshifts. This is not, however, the case.

In 1964 Hermann Bondi showed that whatever equation of state is chosen, provided it is physically realistic (i.e., with sound speed in the material not exceeding the speed of light), the surface redshift cannot exceed 0.62. Thus surface gravitational redshift proved inadequate to explain the large redshifts of quasars. There was also an observational objection put forward by J. Greenstein and M. Schmidt (1964). We briefly reproduce their elegant argument as applied to the quasar 3C 273.

For small redshifts we get from the above formula

$$z \simeq \frac{GM}{Rc^2} \simeq 1.47 \times 10^5 \frac{M}{M_\odot} \frac{R}{1\,\mathrm{cm}}. \tag{1.4}$$

Consider a thin shell of radius R and thickness ΔR emitting radiation in a line of width w. For line widths in 3C 273 arising from the slight change of redshift over ΔR we get

$$\frac{\Delta R}{R} = \frac{w}{\Delta \lambda} = 0.07. \tag{1.5}$$

Now suppose that the object is of stellar type with mass $\sim M_\odot$. For such an object a redshift 0.158 as in 3C 273 gives a radius of $\sim 10^6$ cm, and so $\Delta R \sim 7 \times 10^4$ cm.

Next we determine the volume of the shell and use the volume emissivity of the Hβ line to estimate the flux of radiation in that line. For the source 3C 273 the volume emissivity at a temperature $\sim 10^4$ K is $\sim 10^{-25} n_{\mathrm{e}}^2$ erg sec^{-1} cm^{-3}, where n_{e} is the electron number density. Multiplying this with the volume of the shell and equating the result to the observed luminosity we get

$$10^{-25} R^2 \Delta R \, n_e^2 = 3.4 \times 10^{-12} d^2, \tag{1.6}$$

where d is the expected distance of the source measured in centimetres. This gives

$$n_e^2 = 5.10 \times 10^{-4} d^2. \tag{1.7}$$

Since the searches for proper motion of 3C 273 proved negative, Greenstein and Schmidt assumed that the source could not be nearer than 100 pc. This gave

$$n_e > 6 \times 10^{18} \, \mathrm{cm}^{-3}. \tag{1.8}$$

This density is too high, and is precluded by the appearance of the forbidden line [O III]λ 5007 in the source spectrum. Thus the stellar possibility is ruled out. Similarly Greenstein and Schmidt were able to reject the alternative hypothesis that the source is much more massive, extragalactic and located within, say, 25 Mpc.

To get round this argument and the Bondi limit, Hoyle and Fowler (1967) proposed a rather unusual model of a quasar in which the redshift arose from light coming from the interior. In this model the gravitational potential well at the centre of the object is provided by a distribution of a large number of very compact objects (such as, say, neutron stars). Any gas present in the system would descend to the centre and settle down there and form the emission line region. It is not difficult to see that the gravitating compact objects would leave a sufficient gap for the central radiation to emerge with redshifts much higher than the Bondi limit. The interior Schwarzschild solution used by Hoyle and Fowler to demonstrate this effect gives arbitrarily high central redshifts. However, more realistic models discussed by P.K. Das (1977) yielded redshifts of the order of 2–3. These models also get around the Greenstein–Schmidt criticism provided the masses are of galactic order.

We will take a second look at the gravitational redshift option in Chapter 15, when considering other evidence that has a bearing on the nature of redshifts.

1.3.3 *Cosmological redshift*

The third alternative, which is often referred to as the *cosmological hypothesis*, has occupied the central place in this field. In this hypothesis our Galaxy, along with other galaxies and any other extragalactic objects, is located in an expanding universe and the redshift of any such object is explained as arising from the time-dilation produced in the curved space–time of such a universe. This hypothesis explains the Hubble law (see Chapter 2) observed for galaxies, namely that the redshift of a source is proportional to its distance from us. Right from the early days this has been the favourite hypothesis for quasars also, for the following reasons.

- The redshifts of quasars, although by and large greater than those of galaxies, seem to arise as a natural consequence of the expanding universe hypothesis.

If this is true, no detailed structural scenarios are needed (as in the other two options discussed above) to understand the redshift.

- Quasar properties are in many respects similar to those of the nuclei of Seyfert galaxies, and of active galactic nuclei in general. Since the redshifts of these galaxies and nuclei are believed to be cosmological, the redshifts of quasars can likewise assumed to be so.

All the same there still persist some doubts about the universality of the cosmological hypothesis and we will return to the contentious issues in Chapter 15. For the time being, and in the following chapters, we will assume that this hypothesis holds for the quasars.

1.4   Physical characteristics: quasars

Keeping in mind the serendipitous discovery of the first two quasars, we need to specify the broad criteria that would help distinguish a quasar from a star or a galaxy. In the classic book *Quasi-Stellar Objects* by Geoffrey and Margaret Burbidge, written in the early years of quasar searches, the following criteria specified by Maarten Schmidt were given:

- star-like object identified with a radio source;
- variable light;
- large ultraviolet flux of radiation;
- broad emission lines in the spectra with absorption lines in some cases;
- large redshift.

Later searches showed that the radio source property is not generic and only some 10 per cent of quasars may be radio-loud. We shall discuss this property of quasars in Chapter 9. In fact, X-ray emission seems more common with quasars, as became clear with the studies of the *EINSTEIN* observatory in 1980. Time variability has been found in several wavebands, e.g. visible, radio, X-ray etc. In the 1970s very long baseline interferometry revealed several cases of apparent superluminal separation of the components of radio-loud quasars on the scale of parsecs. We shall consider again in Section 6.1 the defining characteristics of quasars and other related objects.

The spectra of quasars have been used to get information on the physical environment in the universe at large redshifts. This environment can be in the quasar itself or in its vicinity or in the intermediate region between it and us (the observers). In particular, the absorption lines may arise from absorbing material in the quasar itself or in the intervening medium. With the discovery of high redshift galaxies this property is no longer the prerogative of quasars.

Finally, in 1979 the first case of the gravitational lensing of quasars came to light. Two quasar images named PHL 0957+561 A and B seemed so similar in physical properties that it became plausible to argue that they were virtual images of only one real object. The multiple imaging could be caused by the bending of light from

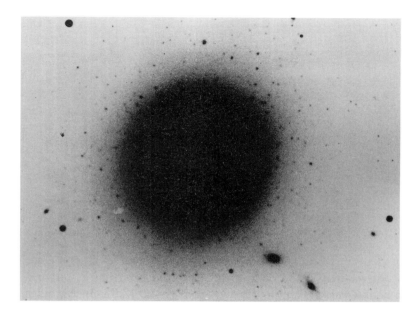

Fig. 1.2. The galaxy M87 at optical wavelengths. Reproduced from *The Cambridge Atlas of Astronomy*, second edition (1988).

the source to the observer by an intervening galaxy (or a cluster of galaxies, or some massive lump of dark matter), allowing more than one path. Gravitational lensing has since been proposed for several multiple quasar systems with members having very close angular separations ($\lesssim 5$–6 arcsec).

We will discuss the details and implications of all these physical characteristics in later chapters.

1.5   Physical characteristics: active galactic nuclei

In Figure 1.2 we see an optical photograph of the galaxy M87. This galaxy is of Hubble type E0, with the peculiar feature that again a jet appears to come out of it. The emergence of the jet is an indication of some violent activity in the nucleus, which, as mentioned above, is of an exceptional nature in the context of galaxies. Normally one would expect that more stars would congregate towards the galactic centre, which might lead to greater velocity dispersion and greater brightness compared to the outer regions. However, when in the late 1970s, with one of the early uses of charge coupled device (CCD) cameras, the nucleus of M87 was looked at more closely, it showed a sharp rise in the luminosity towards the centre (see Figure 1.3). This rise could be explained by supposing that there is a massive attractor at the centre, which pulls a large population of stars into a compact region, thereby pushing up the luminosity of the region.

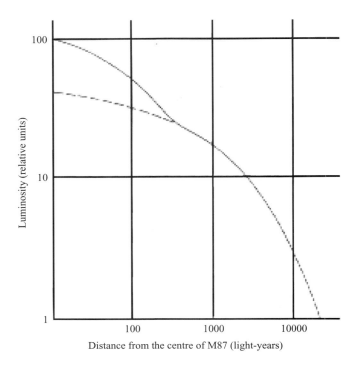

Fig. 1.3.   The luminosity profile of M87. Reproduced from Narlikar (N96).

Around the same time the velocity dispersion of stars in the nuclear region was also measured and it also seemed to rise rapidly inwards. This too suggested the presence of a massive object in the nucleus. The conclusion from both these studies was that this central attractor may well be a massive black hole of $\sim 5\times10^9 M_\odot$. The galaxy M87 is also a strong emitter of X-rays and thus requires an energy production mechanism that goes beyond the typical stellar energy production found in normal galaxies.

There is thus a similarity between the nucleus of M87 and a quasar-like 3C 273, in that both are seats of intense energy production, both require a collapsed massive object, both are X-ray sources and, in this particular case, both have jets issuing from the central region. As has become clearer over the years, there is a whole class of galactic nuclei of this kind that show signs of energetic activity. These are collectively referred to as *active galactic nuclei* (AGN).

Indeed, a class of galaxies originally studied by C. Seyfert show the unusual quasar-like feature of emission lines from their centres. Seyfert galaxies, as these galaxies are called, probably form no more than ~ 2 per cent of the class of spiral galaxies. The Seyferts are further subdivided into two classes, Seyfert 1 and Seyfert 2, depending on the width of the emission lines found in them. The distinction made in the literature between a quasar and a Seyfert 1 nucleus is largely morphological: in the latter a galactic envelope is seen whereas a quasar is generally star-like, without a nebulosity.

Indeed, the strength of the argument that the redshift of a quasar is of cosmological origin rests mainly on the continuity of properties between Seyfert nuclei and quasars. It should be remembered that most Seyferts are known within the nearby region and therefore have low redshifts, whereas quasars are, by and large, high redshift objects. Arguments could be made either way on this distinction. Thus one could argue that a quasar has manifestly different morphology from a Seyfert and this generates an extra intrinsic redshift. Or, one could argue that if we look at a very distant Seyfert, its galactic envelope would be very faint and would be outshone by its bright nucleus, which is what a quasar is.

The conventional view is the latter of the above two and regards quasars and AGN as different manifestations of the same basic phenomena. In the following chapters we shall subscribe to this view and develop the astrophysics of these remarkable objects using the cosmological hypothesis. Only when we get to Chapter 15 shall we take a critical look at the cosmological hypothesis and the evidence claimed for and against it.

As the starting point of the hypothesis we outline the expanding universe paradigm in the following chapter.

2 The cosmological framework

2.1 Introduction

The third decade of the twentieth century brought about a significant advance in our perception of the universe. In particular, it became clear to the astronomer that our Milky Way galaxy is but one amongst many such galaxies. And, what is more, the vast collection of galaxies appeared to be boundless (even as it does today). Towards the end of the decade Edwin Hubble came up with the remarkable discovery that the spectra of galaxies appear to be shifted towards the red end of the spectrum, and that the shift in a given galaxy is proportional to the distance of the galaxy from us. Hubble's actual observations showed that redshift increases with increasing faintness of the galaxies. Distance is inferred from the inverse square law of illumination. If we interpret the spectral shift as due to the Doppler effect, the corresponding radial velocity is then proportional to distance. The constant of proportionality is known as Hubble's constant and its value is denoted by H_0.

Since the discovery of Hubble's law in 1929, evidence has steadily grown that, barring very few exceptions, the phenomenon of redshift is universally found in all extragalactic objects. The theoreticians have had no difficulty in explaining the phenomenon; in fact, seven years *before* Hubble, A. Friedmann had found world models as solutions of Einstein's equations of general relativity wherein the property of redshift arose naturally. In 1922 there was no systematic observational evidence for redshifts of extragalactic objects, and Friedmann's models did not receive any significant attention. The majority of astronomers took it for granted that the universe is static. Even Einstein, who wished to depart from the traditional concepts of Euclidean geometry and Newtonian dynamics for constructing cosmological models, did not contemplate large scale motions of galaxies.

Thus the observational impetus of 1929 was primarily responsible for theoretical cosmologists' turning to Friedmann's models as the simplest framework for extragalactic redshifts. Gradually therefore the *expanding universe* concept took root and the extragalactic universe came to be described by the non-Euclidean geometry of Friedmann's models.

So, while galactic astronomy is carried on as before within the framework of Euclid and Newton, its extragalactic counterpart must take note of relativistic cosmology. In this chapter we will summarize the essential features of relativistic cosmology that are needed for extragalactic astronomy. For details the reader is referred to standard texts in cosmology (e.g. Narlikar N93), whose notation we will follow here).

2.2   The Robertson–Walker line element

The essential feature of general relativity is that the large scale distribution of matter has large scale gravitational effects that show up through the geometry of space–time: this geometry is non-Euclidean. The fact that the universe appears to have galaxies, clusters and superclusters distributed out to unlimited distances implies that the geometry will reflect ambient gravity on a large scale. As a starting point (and historically as well) the simplest models should be considered first. These models are based on the following set of assumptions:

1 Weyl's postulate   The world lines of matter form a highly ordered non-interacting bundle of geodesics[1] which can be parameterized by three space-like coordinates x^μ, $\mu = 1, 2, 3$. Thus

$$x^\mu = \text{constant} \tag{2.1}$$

along a world line. Also, there exists a set of space-like hypersurfaces given by $t = $ constant, orthogonal to this set of world lines. We may identify t with the time coordinate that monotonically increases along a typical world line.

The time t may therefore be given a global status and called the *cosmic time*. The observers whose world lines follow Equation 2.1 are sometimes called *fundamental observers*.

Figure 2.1 illustrates the Weyl scenario and contrasts it with a more general (and more chaotic) scenario. Notice that the time t is also the proper time of a fundamental observer. This allows each such observer to relate his time measurements to a unique cosmological epoch.

2 Cosmological principle   The Weyl postulate tells us that the universe is orderly in terms of its large scale motions. The cosmological principle assumes it to be homogeneous and isotropic in terms of the distribution of its matter. Observationally it implies that the universe at a given time t looks the same to all fundamental observers and that each such observer sees the same vista in all directions. Since, in Einstein's theory of general relativity, matter contents are intimately related to the geometry of space–time, the cosmological principle restricts the possible forms of geometry to only those wherein the three-space at any given time t is an invariant under spatial translation and local rotation.

Although such geometries can be intuitively guessed, and were used by Einstein and de Sitter in 1917, by Friedmann in 1922–4 and by Abbé Lemaitre in 1927, their

---

[1] A geodesic is the name given to the path of stationary distance between any two space–time points. It is therefore the analogue of 'uniform motion in a straight line' of Newtonian–Euclidean physics.

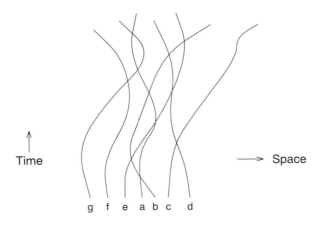

Time → Space

g f e a b c   d

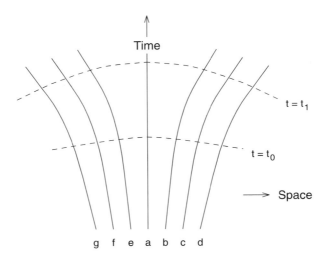

Time

t = t₁

t = t₀

→ Space

g f e a b c d

Fig. 2.1. Upper panel: an arbitrary bundle of world lines. Lower panel: particles moving along non-intersecting world lines.

rigorous derivation had to await the work of H.P. Robertson in 1935 and A.G. Walker in 1936. Independently these authors showed that there are only *three* kinds of such space–time, denoted below by the parametric values $k = 0, 1, -1$. The line element is given by

$$ds^2 = c^2 dt^2 - S^2(t) \left[\frac{dr^2}{1 - kr^2} + r^2(d\theta^2 + \sin^2\theta \, d\phi^2) \right].$$ (2.2)

Here r, θ, ϕ are the three coordinates x^μ. The function $S(t)$ sets the scale of the space-like sections spanned by r, θ, ϕ. That $S(t)$ is an increasing function of time expresses the notion of the expansion of the universe.

For $k = 0$ the expression in the brackets is simply the line element for three-dimensional Euclidean geometry, i.e., for flat space. For this reason this case is often referred to as the *flat* Robertson–Walker model. For $k = 1$, the space is finite but unbounded with $0 \leq r < 1$. This is the *closed* model, while for $k = -1$ the model is said to be *open*. For both $k = -1$ and $k = 0$, the r coordinate goes over the range $0 \leq r < \infty$. The θ, ϕ coordinates range over the intervals $-\pi \leq \theta \leq \pi$ and $0 \leq \phi < 2\pi$ in all three cases.

As a first approximation we assume that galaxies (treated as point particles!) follow the Weyl geodesics and that each galaxy has fixed comoving coordinates r, θ, ϕ. This frame is often referred to as the *cosmological rest frame*. A particle travelling from one galaxy to another moves relative to this frame since its r, θ, ϕ coordinates keep on changing. In practice, galaxies themselves are not at rest with respect to this frame and have peculiar motions relative to it.

2.3   The redshift

Consider light communication between two galaxies of which the receiving galaxy G_o has the radial coordinate $r = 0$ while the sending galaxy G_1 is at $r = r_1$. (Since the universe is assumed homogeneous there is no loss of generality in assigning the status of origin, $r = 0$, to any arbitrarily chosen fundamental observer.) Relativity theory tells us that light travels along a null geodesic $ds = 0$, which in this case turns out to be a path along which θ and ϕ are constant. From Equation 2.2 we get

$$\frac{dr}{\sqrt{1 - kr^2}} = -\frac{cdt}{S(t)}, \tag{2.3}$$

where the minus sign on the right hand side indicates that along the path of the light ray r decreases as t increases.

Let galaxy G_1 emit monochromatic light of wavelength λ_1. Consider two epochs at $t_1, t_1 + \delta t_1$ when two successive wave crests leave G_1, reaching G_0 at t_0 and $t_0 + \delta t_0$ respectively. Then integrating Equation 2.3 we get

$$\int_0^{r_1} \frac{dr}{\sqrt{1 - kr^2}} = \int_{t_1}^{t_0} \frac{cdt}{S(t)} = \int_{t_1 + \delta t_1}^{t_0 + \delta t_0} \frac{cdt}{S(t)}. \tag{2.4}$$

Assuming that the function $S(t)$ varies slowly so that it does not change significantly over intervals like δt_0 and δt_1, we get from the above

$$\frac{\delta t_0}{S(t_0)} = \frac{\delta t_1}{S(t_1)}. \tag{2.5}$$

Now recall that δt_1 is the period during which the wave travels a distance λ_1 at G_1.

Let the wavelength perceived at G_0 be λ_0. Then $\lambda_0 = c\delta t_0$ and $\lambda_1 = c\delta t_1$ so that

$$\frac{\lambda_0}{\lambda_1} = \frac{S(t_0)}{S(t_1)} = 1 + z \quad \text{(say)}. \tag{2.6}$$

The parameter z, called *redshift* in the familiar astronomical terminology, is nothing other than the spectral shift, i.e., the fractional change in the wavelength in terms of the original wavelength. Clearly, for a redwards shift of wavelength $z > 0$ and hence $S(t_0) > S(t_1)$.

So, to be able to explain the observed extragalactic redshift it is necessary to have $S(t)$ increasing with t. However, Hubble's law goes further than that! We need a redshift–magnitude relation. To obtain it we next consider how magnitudes are calculated in extragalactic astronomy.

2.4   Extragalactic radiation fluxes

Since astronomers use magnitudes, flux densities etc. at different wavelengths we shall give a general derivation that can be adapted to specific situations.

Suppose in our previous example the galaxy G_1 is isotropically emitting radiation at the rate of $L(\lambda)d\lambda$ energy units per unit time in the wavelength range $(\lambda, \lambda + d\lambda)$. Then

$$L = \int_0^\infty L(\lambda)d\lambda \tag{2.7}$$

is the total (bolometric) luminosity of G_1. Let us calculate the flux of radiation received by G_0 in the wavelength range $(\lambda_0, \lambda_0 + d\lambda_0)$ per unit time per unit area held normal to the direction $\overrightarrow{G_1G_0}$.

First notice that radiation leaving G_1 at $t = t_1$ crosses the surface of a sphere containing G_0 at time $t = t_0$. A glance at the angular part of the line element in Equation 2.2 tells us that this surface area at $t = t_0$ is $4\pi r_1^2 S^2(t_0)$. So one would expect our desired answer to be

$$F(\lambda_0)d\lambda_0 = \frac{L(\lambda_0)d\lambda_0}{4\pi r_1^2 S^2(t_0)}. \tag{2.8}$$

We shall refer to $r_1 S(t_0)$ as the *metric distance* d_{m} of G_1.

Two corrections are, however, needed in the above expression. First consider the amount of radiation leaving G_1 in the wavelength range $(\lambda_1, \lambda_1 + d\lambda_1)$ in time δt_1. By definition, this is $L(\lambda_1)d\lambda_1\delta t_1$. Taking each photon to contribute energy hc/λ_1 we find that this flux of radiation contains

$$L(\lambda_1)d\lambda_1 \left(\frac{\lambda_1}{hc}\right) \delta t_1 \tag{2.9}$$

photons. Since these photons will cross our spherical surface between t_0 and $t_0 + \delta t_0$, the number crossing per unit area per unit time at G_0 is

$$L(\lambda_1)d\lambda_1 \left(\frac{\lambda_1}{hc}\right)\left(\frac{\delta t_1}{\delta t_0}\right)\left[\frac{1}{4\pi r_1^2 S^2(t_0)}\right]. \tag{2.10}$$

However, at G_0 each photon because of redshift has wavelength λ_0 and energy hc/λ_0. Therefore, the required answer is, instead of Equation 2.8

$$F(\lambda_0)d\lambda_0 = \frac{L(\lambda_1)d\lambda_1}{4\pi r_1^2 S^2(t_0)}\left(\frac{\lambda_1}{\lambda_0}\right)\left(\frac{\delta t_1}{\delta t_0}\right) \tag{2.11}$$

$$= \frac{L(\lambda_0/(1+z))d\lambda_0/(1+z)}{4\pi r_1^2 S^2(t_0)(1+z)^2}. \tag{2.12}$$

The two $1+z$ factors in the denominator come from Equations 2.5 and 2.6. Thus we have

$$F(\lambda_0) = \frac{L(\lambda_0/(1+z))}{4\pi r_1^2 S^2(t_0)(1+z)^3}. \tag{2.13}$$

If we express these results in in terms of the frequency v by writing $L(\lambda)\,d\lambda = L(v)\,dv$, with $\lambda v = c$, we get

$$F(v_0) = \frac{L(v_0(1+z))}{4\pi r_1^2 S^2(t_0)(1+z)}, \tag{2.14}$$

with $F(v_0)\,dv_0 = F(\lambda_0)\,d\lambda_0$. We are using the same symbols F and L to denote flux and luminosity as functions of wavelength or frequency. The functional form depends on which independent variable is being used. If we integrate over all λ or v we get the apparent bolometric flux as

$$F = \frac{L}{4\pi r_1^2 S^2(t_0)(1+z)^2} = \frac{L}{4\pi D_L^2} \tag{2.15}$$

where we may call $D_L \equiv r_1 S(t_0)(1+z)$ the *luminosity distance* of G_1. Note that the luminosity distance $D_L = (1+z)d_m$, where d_m is the metric distance.

2.5   Hubble's law

From Equation 2.4 we see that the larger the value of r_1, the smaller is t_1 compared to t_0 and hence larger the ratio $S(t_0)/S(t_1)$. Thus we expect z to increase with r_1. Similarly, from Equation 2.8 we see that for galaxies of the same luminosity, with increasing r_1 (and hence, with z also increasing) $F(\lambda_0)$ will fall, thereby increasing their apparent magnitude. So the model predicts fainter magnitudes for increasing redshifts.

All this is in qualitative agreement with Hubble's law. For a detailed quantitative prediction we need to know $S(t)$. We shall consider the dynamical behaviour of our models in the next section. Here we show that, *whatever* the functional form of $S(t)$, provided $S(t)$ is an increasing function a *velocity–distance relation* à la Hubble follows for nearby galaxies.

If $r \ll 1$, we can approximate the integrals in Equation 2.4 to write

$$r_1 \simeq \frac{c(t_0 - t_1)}{S(t_0)}. \tag{2.16}$$

Also, by Taylor expansion near t_0 we get

$$S(t_1) \simeq S(t_0) - (t_0 - t_1)\dot{S}(t_0), \tag{2.17}$$

i.e.,

$$\frac{S(t_1)}{S(t_0)} \simeq 1 - \frac{r_1 S(t_0)}{c}\left[\frac{\dot{S}(t_0)}{S(t_0)}\right]. \tag{2.18}$$

Using Equation 2.6 in the approximation $z \ll 1$, we can write the left hand side of Equation 2.18 as $1 - z$. Further, for small r_1, the expression for the metric distance $r_1 S(t_0)$ is equal to the luminosity distance D_L of G_1. We therefore obtain the above expression as

$$cz = \left(\frac{\dot{S}}{S}\right)_{t_0} D_L, \tag{2.19}$$

i.e., with the Doppler velocity $v = cz$,

$$v = H_0 D_L; \quad H_0 \equiv \left(\frac{\dot{S}}{S}\right)_{t_0}. \tag{2.20}$$

This is Hubble's velocity–distance relation, the Hubble constant being identified as $\dot{S}/S$.

2.6   The Friedmann models

Einstein's field equations provide the missing information, viz. the dynamical behaviour of the function $S(t)$. In keeping with the general philosophy of relativity theory, the parameters of the space–time geometry, viz. $S(t)$ and k, turn out to be related to the matter and radiation contents of the universe. The field equations are a set of ten partial differential equations in the four space–time variables. The simplifying assumptions of Section 2.2 reduce their number to two ordinary differential equations with t as the independent variable:

$$\frac{2\ddot{S}}{S} + \frac{\dot{S}^2 + kc^2}{S^2} = -\frac{8\pi G}{c^2}\left(p + \frac{u}{3}\right), \tag{2.21}$$

$$\frac{\dot{S}^2 + kc^2}{S^2} = \frac{8\pi G}{3c^2}(\rho c^2 + u). \tag{2.22}$$

Here ρ is the density of matter treated as a fluid, p its pressure and u the energy density of radiation. From these equations simple calculus leads to the conservation law

$$\frac{d}{dS}\left[(\rho c^2 + u)S^3\right] + 3S^2\left(p + \frac{u}{3}\right) = 0. \tag{2.23}$$

This is, in fact a restatement of the first law of thermodynamics. What are the values of ρ, p, u at present?

If one restricts discussion to luminous matter in the form of galaxies, intergalactic medium etc. then $\rho \sim 3 \times 10^{-31}\,\mathrm{g\,cm^{-3}}$. The pressure term may be estimated as a fraction $\sim v^2/c^2$ of ρ, where v is the average peculiar velocity of galaxies in clusters. With $v \sim 300\,\mathrm{km\,sec^{-1}}$, this fraction is $\sim 10^{-6}$, i.e., quite negligible. Thus if we set $p = 0$ we arrive at the so-called *dust* approximation.

What about radiation? Of the various radiation backgrounds in different wavelength bands, that in the microwave is the most important, with an energy density $u \sim 4 \times 10^{-13}\,\mathrm{erg\,cm^{-3}}$. Compared to ρc^2 therefore we find that the ratio $u/(\rho c^2) \sim 10^{-3}$. Again, radiation density can be neglected in comparison with matter density at the present epoch.

However, it is instructive to note that both p and u were more important in the past than now. For example, peculiar motions drop off in an expanding universe as S^{-1}. Thus for S smaller by $\sim 10^3$ the pressure term would be comparable to ρ. Although other factors (relating to structure formation) may have intervened to make the S^{-1} law invalid early on, this illustrates the importance of the pressure term in the early epochs.

Similarly, if we consider matter and radiation as basically decoupled, then Equation 2.23 with $p = 0$ splits into two:

$$\frac{d}{dS}(\rho S^3) = 0, \quad \frac{d}{dS}(uS^3) + uS^2 = 0, \tag{2.24}$$

giving

$$\rho \propto S^{-3}, \quad u \propto S^{-4}. \tag{2.25}$$

It therefore follows that in the past the ratio $u/(pc^2)$ was higher, increasing as S^{-1}. Thus at an epoch when S was approximately 10^{-3} of its present value u and ρc^2 were comparable. Prior to this epoch u was more dominant than matter. More correctly, even matter would have been relativistic (and thus simulate the $p = \rho c^2/3$ law for radiation) if we go sufficiently far back into the past.

The redshift corresponding to the epoch when the switchover occurred from radiation to matter domination was thus $\sim 10^3$, far higher than the redshift of any discrete source such as a quasar. For doing the astronomy of discrete sources therefore it is sufficient to neglect the u-term.

Accepting the dust approximation for the present epoch, Equations 2.21 and 2.22 are further simplified to

$$\frac{2\ddot{S}}{S} + \frac{\dot{S}^2 + kc^2}{S^2} = 0, \tag{2.26}$$

$$\frac{\dot{S}^2 + kc^2}{S^2} = \frac{8\pi G\rho}{3}, \tag{2.27}$$

with

$$\rho = \rho_0 \frac{S_0^3}{S^3}. \tag{2.28}$$

We are thus back to the models considered by Friedmann.

Cosmological parameters Before considering solutions of Equations 2.21 and 2.22 we use them to define certain well-known parameters. We will adopt the convention of labelling the *present* value of a parameter by the subscript 0. Thus H_0 is the value of the Hubble constant at the present epoch t_0.

Define the density parameter Ω and the deceleration parameter q by

$$\Omega = \frac{8\pi G\rho}{3H^2}, \quad q = -\frac{1}{H^2}\frac{\ddot{S}}{S}, \tag{2.29}$$

where, following Equation 2.20, we define

$$H = \frac{\dot{S}}{S}. \tag{2.30}$$

Using these values, Equations 2.26 and 2.27 are rewritten as

$$-2qH^2 + H^2 + \frac{kc^2}{S^2} = 0, \tag{2.31}$$

$$H^2 + \frac{kc^2}{S^2} = H^2\Omega. \tag{2.32}$$

From these equations we get

$$\Omega = 2q. \tag{2.33}$$

Also, $q = 1/2$ for $k = 0$, $q > 1/2$ for $k = 1$ and $q < 1/2$ for $k = -1$. It is clear that the dynamical properties of the universe, e.g. the behaviour of $S(t)$, are linked with its geometry as distinguished by the parameter k. We will treat the three cases $k = 0$, $k = 1$, $k = -1$ separately.

The flat model For this case Equations 2.26 and 2.27 together give

$$S = \left(\frac{t}{t_0}\right)^{2/3}, \tag{2.34}$$

where we have, without loss of generality, set $S_0 = 1$. We have

$$H = \frac{2}{3t}, \quad \Omega = 1, \quad q = \frac{1}{2}. \tag{2.35}$$

This model expands from $S = 0$ at $t = 0$ to $S = \infty$ at $t = \infty$. For a galaxy of redshift z, the expansion law along with Equations 2.4 and 2.6 give

$$t_1 = t_0(1+z)^{-3/2}, \quad r_1 = \frac{2c}{H_0}\left(1 - \frac{1}{\sqrt{1+z}}\right). \tag{2.36}$$

The present age of the universe is

$$t_0 = \frac{2}{3H_0}.$$ (2.37)

Since $S_0 = 1$, the metric distance d_m of G_1 is also r_1. The density of matter is given by

$$\rho = \frac{3H_0^2}{8\pi G} = \rho_c \quad \text{(say)},$$ (2.38)

where ρ_c is called the *critical density* or the *closure density*, for reasons to be made clear soon. This model is also known as the Einstein–de Sitter model as it was jointly proposed by the two theorists.

The closed model   The $k = 1$ case reduces to the solution of the differential equation

$$\dot{S}^2 = c^2 \left(\frac{S_m}{S} - 1 \right)$$ (2.39)

where S_m, the maximum value attained by S during expansion, is given by

$$S_m = \frac{2q_0}{(2q_0 - 1)^{3/2}} \left(\frac{c}{H_0} \right).$$ (2.40)

This model expands from $S = 0$ to $S = S_m$ and then contracts back to $S = 0$.
It is convenient to express both S and t in terms of a parameter χ:

$$S = \frac{S_m}{2}(1 - \cos \chi), \quad ct = \frac{S_m}{2}(\chi - \sin \chi).$$ (2.41)

The complete cycle of expansion from $S = 0$ to contraction to $S = 0$ takes a time $\pi S_m/c$, while the maximum in the scale factor occurs at $t_m = \pi S_m/(2c)$. The value of q_0 is related to χ_0 by

$$q_0 = (1 + \cos \chi_0)^{-1}.$$ (2.42)

The present age of the universe in the closed model ($q_0 > 1/2$) is

$$t_0 = \frac{q_0}{(2q_0 - 1)^{3/2}} \left[\arccos \left(\frac{1 - q_0}{q_0} \right) - \frac{(2q_0 - 1)^{1/2}}{q_0} \right] H_0^{-1}.$$ (2.43)

For a galaxy G_1 with redshift z, we have from Equations 2.4 and 2.6

$$r_1 = \sin(\chi_0 - \chi_1), \quad 1 + z = \frac{1 - \cos \chi_0}{1 - \cos \chi_1}.$$ (2.44)

It is possible to eliminate χ_0 and χ_1 from Equation 2.44 to arrive at the final answer for r_1:

$$r_1 = \frac{(2q_0 - 1)^{1/2}}{q_0^2(1 + z)} \left[q_0 z + (q_0 - 1)(\sqrt{1 + 2zq_0} - 1) \right].$$ (2.45)

The corresponding metric distance is

$$d_{\mathrm{m}} = r_1 S(t_0) = \left(\frac{c}{H_0 q_0^2}\right) \frac{q_0 z + (q_0 - 1)(\sqrt{1 + 2z q_0} - 1)}{1 + z} \tag{2.46}$$

while the luminosity distance is

$$D_{\mathrm{L}} = d_{\mathrm{m}}(1 + z) = \frac{c}{H_0 q_0^2} \left[q_0 z + (q_0 - 1)(\sqrt{1 + 2z q_0} - 1) \right]. \tag{2.47}$$

The open model For $k = -1$, $q_0 < 1/2$ we have a solution given by the parametric representation

$$S = \frac{S_{\mathrm{m}}}{2}(\cosh \psi - 1), \quad ct = \frac{S_{\mathrm{m}}}{2}(\sinh \psi - \psi) \tag{2.48}$$

where

$$S_{\mathrm{m}} = \frac{2q_0}{(1 - 2q_0)^{3/2}} \left(\frac{c}{H_0}\right). \tag{2.49}$$

Unlike the $k = 1$ case here the universe expands from $S = 0$ all the way to infinity. Thus S_{m} is *not* the maximum value of S. The present age of the universe in this model $(q_0 < 1/2)$ is

$$t_0 = \frac{q_0}{(2q_0 - 1)^{3/2}} \left\{ \frac{(1 - 2q_0)^{1/2}}{q_0} - \ln \left[\frac{1 - q_0 + (1 - 2q_0)^{1/2}}{q_0} \right] \right\} H_0^{-1}. \tag{2.50}$$

The behaviour of $S(t)$ for different values of q_0 is shown in Figure 2.2.

We have corresponding to Equations 2.42 and 2.44

$$q_0 = (1 + \cosh \psi_0)^{-1}, \tag{2.51}$$

$$r_1 = \sinh(\psi_0 - \psi_1), \quad 1 + z = \frac{\cosh \psi_0 - 1}{\cosh \psi_1 - 1}. \tag{2.52}$$

The expressions for d_{m} and D_{L}, however, remain the same as in Equations 2.46 and 2.47.

Notice from Equation 2.33 that $\Omega > 1$ for the closed model ($k = 1$), while $\Omega < 1$ for the open ones ($k = -1$) with the flat model ($k = 0, \Omega = 1$) lying on the border between the two. Thus for a density exceeding that given by Equation 2.38, the universe will be closed. This is the reason why this density is called the critical or closure density.

2.7   The Λ-cosmologies

Soon after writing down his equations of general relativity Einstein felt the need to modify them by adding an extra term, the so-called Λ-term, whose interpretation in Newtonian parlance was that of a repulsive force that varied in proportion to the spatial separation between two bodies. Einstein needed such a force to balance the attractive force of gravity in order to obtain a static model for the universe. (It is easy to verify that Equations 2.21 and 2.22 do not have a solution with $S = $ constant.)

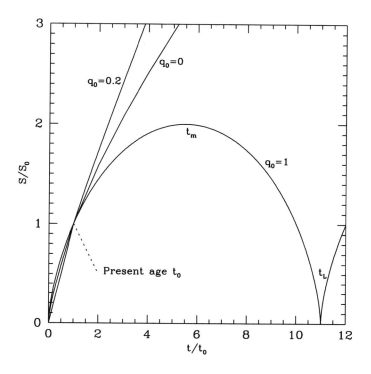

Fig. 2.2. The cosmological scale factor $S(t)$ as a function of the cosmic time for an open model ($q_0 = 0.2$), a closed model ($q_0 = 1$) and the Einstein–de Sitter model ($q_0 = 0$). The scale factor and time are plotted in units of their present values. For the closed model $S(t)$ passes through a maximum at $t = t_m$ and reduces to zero at $t = t_L$. The reexpansion shown beyond t_L is only notional.

In the dust approximation the introduction of the Λ-term modifies Equations 2.21 and 2.22 to

$$\frac{2\ddot{S}}{S} + \frac{\dot{S}^2 + kc^2}{S^2} - \Lambda c^2 = 0, \tag{2.53}$$

and

$$\frac{\dot{S}^2 + kc^2}{S^2} - \frac{\Lambda c^2}{3} = \frac{8\pi G \rho}{3}. \tag{2.54}$$

The solution of these equations, which Einstein obtained in 1917, was

$$S = \text{constant} \equiv S_0 \tag{2.55}$$

with the density ρ_0 and cosmological constant Λ_c given by

$$\Lambda_c = \frac{1}{S_0^2}, \qquad \rho_0 = \frac{\Lambda_c c^2}{4\pi G}. \tag{2.56}$$

This solution requires $k = 1$. Thus the Einstein universe is closed.

In 1917, shortly after Einstein's solution, de Sitter obtained another solution with different assumptions: de Sitter's universe was not static but it was empty. For this model $k = 0$ and $\rho = 0$, with

$$S \propto \exp(H_0 t), \qquad \Lambda c^2 = 3H_0^2. \tag{2.57}$$

The de Sitter solution resurfaced in 1948 as the solution for the steady state cosmology and in 1981 as the solution for inflationary cosmology.

In addition to the de Sitter universe, Equations 2.53 and 2.54 along with Equation 2.28 offer a variety of other dynamical models that have been considered by cosmologists from time to time. Apart from the parameters Ω, q and H we also define

$$\Omega_\Lambda = \frac{\Lambda c^2}{3H^2}. \tag{2.58}$$

It is easy to verify that for the Λ-cosmologies the relation 2.33 is modified to

$$\Omega = 2(\Omega_\Lambda + q), \tag{2.59}$$

and for the flat case ($k = 0$) we have

$$\Omega_\Lambda + \Omega = 1. \tag{2.60}$$

The Einstein–de Sitter models play certain limiting roles in Λ-cosmologies. For example, by choosing Λ arbitrarily close to Λ_c we can have a model that expands from $t = 0$, $S = 0$, then loiters around in a quasistatic state near $S = S_0$ and subsequently expands exponentially in the de Sitter fashion (for $\Lambda > \Lambda_c$) or contracts to $S = 0$ (for $\Lambda < \Lambda_c$).

2.8   Some useful formulae for cosmography

We end this chapter by deriving a few formulae that are needed by astronomers in the data reduction pertaining to extragalactic sources.

Apparent magnitude and K-correction   We have seen in Section 2.4 how the observed radiation flux from a source is related to its luminosity. Using the definition of luminosity distance D_L given in Section 2.4 and Equations 2.14 and 2.47 we can write

$$F(v_0) = \frac{(1 + z)L(v_0(1 + z))}{4\pi D_L^2}, \tag{2.61}$$

with

$$D_L = (c/H_0 q_0^2) \left[q_0 z + (q_0 - 1)(\sqrt{1 + 2q_0 z} - 1) \right]. \tag{2.62}$$

Optical astronomers conventionally express the flux F in terms of apparent magnitude as $m = -2.5 \log F + $ constant, with the constant depending on the filter used in obtaining the flux. The luminosity is then given in terms of the absolute magnitude M, which is defined as the apparent magnitude corresponding to the flux that the source would have at a distance of 10 pc. Using these definitions and Equation 2.14 we get

$$m = M + 5 \log \left\{ \frac{1}{q_0^2} \left[q_0 z + (q_0 - 1)(\sqrt{1 + 2q_0 z} - 1) \right] \right\}$$
$$- 2.5 \log(1 + z) + K(z) + 42.39 - 5 \log h_{100}, \tag{2.63}$$

where Hubble's constant has been expressed as

$$H_0 = h_{100} \times 100 \, \text{km sec}^{-1} \, \text{Mpc}^{-1} \tag{2.64}$$

and $K(z)$ is the *K-correction* term, defined by

$$K(z) = -2.5 \log \left[\frac{L(v_0(1 + z))}{L(v_0)} \right]. \tag{2.65}$$

The flux received in a given filter band-pass comes from a spectral region that depends on the redshift of the emitter. The K-correction term allows for this effect so that the relevant absolute magnitude corresponding to zero redshift is obtained. In the case of AGN and quasars, the continuum spectrum in the optical region can be at least crudely approximated by a power law $L(v) \propto v^{-\alpha_{op}}$, where α_{op} is the spectral index in the optical window. In this case

$$K(z) = 2.5 \alpha_{op} \log(1 + z). \tag{2.66}$$

When the spectrum is more complex, as in the case of galaxies, an average spectrum is obtained by studying a number of similar objects (say galaxies of the same Hubble type) at very low redshift. The K-correction can then be numerically evaluated by transforming the spectrum to the required redshift (see e.g. Rocca-Volmerange and Guiderdoni 1988; the correct expression to be used when the magnitude is defined by integration over a frequency range may also be found there). The apparent magnitude as a function of redshift for a power law optical continuum is shown in Figure 2.3.

Angular size A source of linear size l at a distance d in a Euclidean space subtends an angle $\theta = l/d$ at the observer, if held perpendicular to the line of sight. How is this formula modified in the expanding universe?

Suppose the galaxy G_1 is spherical with diameter l at the redshift z. If it subtends an angle θ at the observer, where $\theta \ll 1$, we can use the Robertson–Walker line element to relate l to the emission epoch t_1:

$$l = r_1 S(t_1)\theta = \frac{d_m \theta}{1 + z}, \tag{2.67}$$

where d_m is the metric distance. Thus we get

$$\theta = \frac{(1 + z)l}{d_m} = \frac{(1 + z)^2 l}{D_L} \equiv \frac{l}{d_A}. \tag{2.68}$$

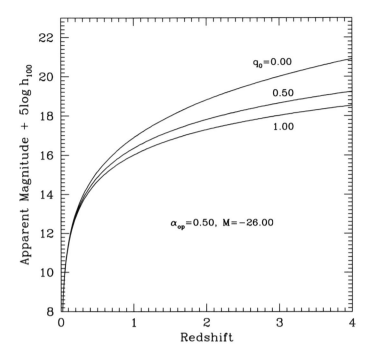

Fig. 2.3. Apparent magnitude as a function of redshift for different values of q_0. The scale on the magnitude axis corresponds to $H_0 = 100\,\mathrm{km\,sec^{-1}\,Mpc^{-1}}$.

$d_A = D_L/(1+z)^2 = d_m/(1+z)$ is known as the *angular diameter distance*. Using the expressions for D_L in Section 2.6 we can write

$$\theta = \frac{q_0^2(1+z)^2\,l}{\left(\frac{c}{H_0}\right)\left[q_0 z + (q_0 - 1)(\sqrt{1 + 2q_0 z} - 1)\right]}, \quad q_0 \neq 0, \tag{2.69}$$

$$\theta = \frac{(1+z)^2\,l}{\left(\frac{c}{H_0}\right)z(1 + \frac{1}{2}z)} \qquad q_0 = 0. \tag{2.70}$$

Notice that the $1 + z$ factor in the numerator produces a peculiar effect in that it does not automatically follow that the more distant a source is, the smaller it appears. If we use Equation 2.46 for $q_0 = 1$, say, we find that $\theta \propto (1+z)^2/z$, which has a *minimum* at $z = 1$. The linear diameter as a function of redshift for a source with angular size 1 arcsecond, for various values of q_0, is shown in Figure 2.4.

Surface brightness   Suppose the galaxy G_1 above has the bolometric luminosity L. The solid angle subtended by G_1 at the observer is $\pi\theta^2/4$ where θ is given by

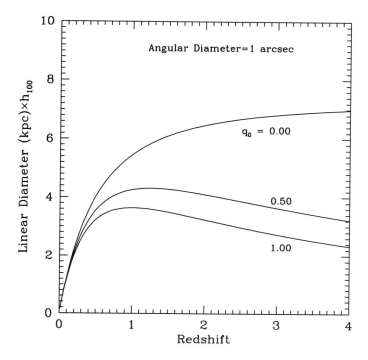

Fig. 2.4. Linear diameter as a function of redshift, for a source with angular diameter 1 arcsecond, for different values of q_0. The scale on the diameter axis corresponds to $H_0 = 100 \, \mathrm{km \, sec^{-1} \, Mpc^{-1}}$.

Equation 2.68. The surface brightness is therefore

$$\sigma = \frac{F}{(\pi \theta^2 / 4)} = \frac{L}{\pi^2 l^2} (1 + z)^{-4}. \tag{2.71}$$

This dependence of σ on $1 + z$ is a hallmark of the expanding universe hypothesis. Other interpretations of redshift may not give this dependence.

Volume–redshift dependence What is the comoving volume of a spherical shell sandwiched between redshifts z and $z + dz$? This question often arises when counting sources at different redshifts for a comparison between theory and observation.

From the relation 2.4 we get r_1 as a function of t_1 and then using Equation 2.6 we can translate it into a relationship between r_1 and z. Of course we need to know k and $S(t)$ explicitly in order to arrive at the answer. We illustrate this with the example of the $k = 1$ Friedmann model. Using the relations 2.41–2.44 we have the comoving volume element as

$$dV = \frac{4\pi r_1^2 dr_1}{\sqrt{1 - r_1^2}} = 4\pi r_1^2 |d\chi_1|, \tag{2.72}$$

$$\frac{dz}{1+z} = \frac{\sin \chi_1}{1 - \cos \chi_1} d\chi_1 = \left(\frac{1 + 2q_0 z}{2q_0 - 1}\right)^{1/2} d\chi_1. \tag{2.73}$$

Therefore

$$dV = \frac{4\pi (2q_0 - 1)^{3/2}}{q_0^4 (1+z)^3} \frac{[q_0 z + (q_0 - 1)(\sqrt{1 + 2q_0 z} - 1)]^2}{\sqrt{1 + 2q_0 z}} dz. \tag{2.74}$$

What is the corresponding proper volume? This is obtained by multiplying dV by $S_1^3 \equiv S_0^3 (1+z)^{-3}$. Since from Equation 2.32 $S_0 = (c/H_0)(2q_0 - 1)^{-1/2}$, we get the proper volume element

$$dV_{\mathrm{P}} = 4\pi \left(\frac{c}{H_0}\right)^3 \frac{[q_0 z + (q_0 - 1)(\sqrt{1 + 2q_0 z} - 1)]^2 \, dz}{q_0^4 (1+z)^6 \sqrt{1 + 2q_0 z}}. \tag{2.75}$$

Radiation background from sources   Using the above expressions we can compute the background produced by radiating sources all over the universe. From Equation 2.14 we have the flux $F(\nu_0) d\nu_0$ arising from a typical source at redshift z. Suppose the comoving number density of such sources is $n(z) dz$ in the redshift range $(z, z + dz)$. Then the total background from all sources has the spectrum $B(\nu_0) d\nu_0$ where

$$B(\nu_0) = \int_0^z F(\nu_0) \frac{dV(z)}{dz} n(z) \, dz$$

$$= \int_0^z \frac{L(\nu_0(1+z))}{4\pi r_1^2 S_0^2(t)(1+z)} \left[\frac{4\pi r_1^2 \sqrt{(2q_0 - 1)}}{(1+z)\sqrt{1 + 2q_0 z}}\right] n(z) dz$$

$$= \frac{c}{4\pi H_0} \int_0^z \frac{L(\nu_0(1+z))}{(1+z)^2 \sqrt{1 + 2q_0 z}} n(z) dz. \tag{2.76}$$

This formula in fact works for all q_0. Using it, the background in some spectral region due to a specific set of sources, say the X-ray background due to quasars, can be calculated once the spectral form of their luminosity and the comoving density are specified. The background due to a diffuse emitter, say the intergalactic medium, can be obtained by combining the luminosity and number density into a single term specifying the energy emitted per unit volume.

Look-back time   The look-back time to an epoch with redshift z is defined by

$$\tau(z) = t_0 - t, \tag{2.77}$$

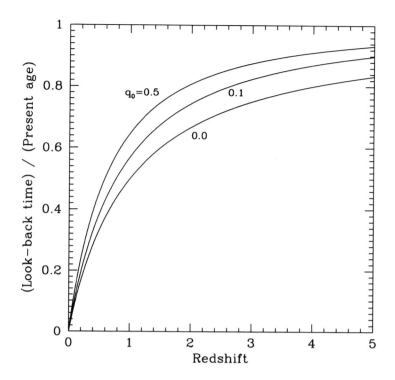

Fig. 2.5. Look-back time in units of the present age of the universe t_0 as a function of redshift for different values of q_0.

where t_0 is the present age of the universe and t is the age (cosmological epoch) corresponding to redshift z. $\tau(z)$ measures the time elapsed since a given epoch and is often used to parameterize evolution in some quantity (like quasar number density) as a function of the epoch. Given a cosmological model, $\tau(z)$ can be calculated as the time taken by a light ray to travel from a fundamental observer at epoch z to one at the present time.

The calculation of look-back time is particularly easy in the case of the Einstein–de Sitter universe ($q_0 = 0.5$) because of the simple time dependence of the scale factor: $S(t) \propto t^{2/3}$. This gives

$$1 + z = \frac{S(t_0)}{S(t)} = \left(\frac{t_0}{t}\right)^{2/3}, \tag{2.78}$$

i.e.,

$$\tau(z) = t_0 \left(1 - \frac{t}{t_0}\right) = \frac{2}{3H_0}\left[1 - \frac{1}{(1+z)^{3/2}}\right]. \tag{2.79}$$

In the case of an open universe ($0 < q_0 < 0.5$) a useful expression has been provided by Schmidt and Green (1983):

$$\tau(z) = \left(1 - \frac{g(z)}{g(0)}\right) t_0,$$ (2.80)

$$g(z) = a(z) - \frac{1}{a(z)} - 2 \ln a(z),$$ (2.81)

$$a(z) = \frac{q_0 - 1 + z}{1 + z} + \left[\left(\frac{q_0 - 1 + z}{1 + z}\right)^2 - 1\right]^{1/2},$$ (2.82)

where t_0 is the age of the open universe as given in Equation 2.50. The case $q_0 = 0$ corresponds to an empty universe with $S(t) \propto t$, age $t_0 = H_0^{-1}$ and

$$\tau(z) = \frac{z}{H_0(1 + z)}.$$ (2.83)

This simple expression provides a good approximation to the look-back time for values of q_0 greater than but close to 0. The look-back time as a function of redshift is shown in Figure 2.5 for various values of q_0.

3 Radiative processes – I

3.1 Introduction

The continuum radiation from AGN stretches over the entire range of the electromagnetic spectrum, from the radio to the high energy γ-ray region, where pair production by photons becomes important. The continuum spectrum has an overall complex shape, but it can often be approximated by a simple power law form over fairly wide wavelength intervals. The radiation is produced in elementary processes like synchrotron emission and bremsstrahlung, and is modified by scattering, absorption and reemission. In this chapter and the next we shall consider some aspects of radiation processes which are important to the basic understanding of the continuum spectrum. The discussion will be brief, and the emphasis will be on developing concepts, summarizing important results and providing them in such a form that they can be directly applied to situations pertinent to quasars and AGN. The subject has been treated in detail in a pedagogic manner by Jackson [J75] and Rybicki and Lightman [RL79]. The more advanced and formal aspects have been covered by Blumenthal and Gould (1970) and an excellent summary, especially of the synchrotron process, with application to AGN, may be found in Moffett (1968). Our treatment and notation owe much to these sources.

In the present chapter we will consider mainly synchrotron radiation and some consequences of relativistic radiation. The other processes important to AGN and quasars will be discussed in Chapter 4.

3.2 Synchrotron emission from a single particle

Synchrotron radiation is emitted when a relativistic electron is accelerated by a magnetic field and for this reason it is sometimes referred to as *magnetic bremsstrahlung*. It is believed that the radio emission received from radio galaxies and radio quasars is synchrotron emission, sometimes modified by other emission and absorption processes. There are situations in which some of the flux received in other wavelength regions is also of the synchrotron type.

The motion of a charged particle in a magnetic field $\mathbf{B}$ is described in classical electrodynamics by the Lorentz force equation

$$\frac{d}{dt}(\gamma m \mathbf{v}) = \frac{e}{c}(\mathbf{v} \times \mathbf{B}), \qquad (3.1)$$

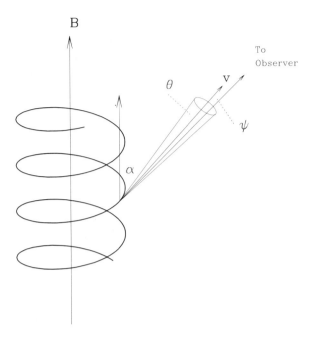

Fig. 3.1. The helical motion of a charged particle in a magnetic field. When the particle is relativistic, the radiation due to accelerated motion is emitted in a narrow cone with the instantaneous velocity vector as the axis and opening angle $1/\gamma$.

where e and m are the charge and rest mass of the particle, $\mathbf{v}$ is the velocity and $\gamma = (1 - v^2/c^2)^{-1/2}$ the Lorentz factor, with v the magnitude of the velocity. The acceleration $d\mathbf{v}/dt$ is normal to the velocity and therefore v and the Lorentz factor γ are constant. There is no force acting on the charged particle in the direction of the magnetic field $\mathbf{B}$, and therefore the component of the velocity parallel to it, $\mathbf{v}_\parallel$, is constant. Since $v =$ constant, the magnitude of the velocity component normal to the magnetic field is also constant: $v_\perp = \sqrt{v^2 - v_\parallel^2} =$ constant. The net effect is that the particle moves in a helix with its axis parallel to $\mathbf{B}$, as is seen in Figure 3.1.

The *gyration frequency*, i.e., the frequency of the projected orbit on a plane normal to $\mathbf{B}$, is

$$v_g = \frac{eB}{2\pi\gamma mc}. \tag{3.2}$$

The total power emitted by an accelerated relativistic charged particle is

$$P = \frac{2e^2}{3c^3}\gamma^4 \left(a_\perp^2 + \gamma^2 a_\parallel^2 \right), \tag{3.3}$$

where $a_\perp$ and $a_\parallel$ are the components of the acceleration perpendicular and parallel to the magnetic field respectively. For synchrotron motion it follows from Equation 3.1

and $a_\parallel = 0$ that $a_\perp = 2\pi v_\mathrm{g} v \sin\alpha$, where α is the angle between the velocity and the magnetic field and is known as the *pitch angle*. The emitted power then is

$$P = \frac{2}{3} r_0^2 c \beta^2 \gamma^2 B^2 \sin^2\alpha, \quad r_0 = \frac{e^2}{mc^2}. \tag{3.4}$$

As shown in Figure 3.1, this power is emitted mainly in a narrow cone of opening angle $\sim 1/\gamma$ around an axis coinciding with the instantaneous velocity vector. As we shall see later, the radiation is polarized.

The emitted power $P \propto m^{-2}$, and therefore synchrotron radiation from an electrically neutral plasma is overwhelmingly from the electrons. We will deal exclusively with electrons in the following, and e and m will be taken to mean the magnitude of the electron charge and mass respectively. In this case r_0 is the classical electron radius.

A radiating electron loses energy at the rate $dE/dt = -P \propto E^2 B_\perp^2$. Given a constant magnetic field, it is straightforward to integrate this expression to obtain E as a function of t and to find that the time taken by the electron to lose half its energy is

$$t_{1/2} = \left(\frac{3m^4 c^7}{2e^4 \sin^2\alpha} \right) \frac{1}{B^2 E} \tag{3.5}$$

$$= 8.5 \times 10^9 \left(\frac{B_\perp}{1\,\mu\mathrm{G}} \right)^{-2} \left(\frac{E}{1\,\mathrm{GeV}} \right)^{-1} \mathrm{yr}. \tag{3.6}$$

In terms of the critical frequency v_c to be introduced in Equation 3.13,

$$t_{1/2} = \left(\frac{27m^5 c^9}{16\pi e^7 \sin^3\alpha} \right)^{1/2} \frac{1}{B^{3/2} v_\mathrm{c}^{1/2}} \tag{3.7}$$

$$= 3.7 \times 10^8 \left(\frac{B_\perp}{1\,\mu\mathrm{G}} \right)^{-3/2} \left(\frac{v_\mathrm{c}}{1\,\mathrm{GHz}} \right)^{-1/2} \mathrm{yr}. \tag{3.8}$$

When particle velocities are distributed isotropically, the distribution of α is

$$p(\alpha)\,d\alpha = \tfrac{1}{2} \sin\alpha\,d\alpha, \tag{3.9}$$

with α ranging over $[0, \pi]$. Averaging over the pitch angle α gives

$$P = \tfrac{4}{3} \sigma_\mathrm{T} c \beta^2 \gamma^2 U_B, \tag{3.10}$$

where σ_T is the Thomson scattering cross-section (see Section 4.1) and $U_B = B^2/(8\pi)$ is the energy density of the magnetic field. For highly relativistic electrons it is usual to set $\beta = 1$ in these expressions.

If the electron were non-relativistic, i.e., $\gamma \simeq 1$, it would emit *cyclotron radiation* at the frequency $v_\mathrm{g} = eB/(2\pi mc)$. As the speed increases, the higher harmonics of v_g begin to contribute to the spectrum with a strength that depends on powers of v/c. As the velocity becomes relativistic, γ increases and the gyration frequency decreases because $v_\mathrm{g} \propto \gamma^{-1}$. An important effect of the relativistic motion is that the radiation is concentrated in a narrow cone with axis in the direction of $\mathbf{v}$ and half-angle $\theta \propto 1/\gamma$ (see Figure 3.1). As the electron goes round the magnetic field, an observer whose line of sight happens to intersect the cone sees a sequence of pulses with a period equal to the Doppler shifted gyration frequency

$$\nu'_g = \frac{\nu_g}{1 - (v/c)\cos^2\alpha} \simeq \frac{\nu_g}{\sin^2\alpha}. \tag{3.11}$$

The width of the pulse Δt is given by the time taken by the cone to sweep across the observer's line of sight and, in the highly relativistic case,

$$\Delta t = \frac{1}{2\pi\gamma^3\nu_g\sin\alpha}. \tag{3.12}$$

The frequency spectrum of this radiation consists of a series of spikes at ν'_g and its harmonics, with cutoff at $\sim 1/(2\pi\Delta t)$. For large values of γ the harmonics are closely spaced, and in addition, the frequency of each harmonic is broadened when radiation from an ensemble of particles is considered, because of the distribution of γ and pitch angle. The spectrum thus appears to be continuous, and can be shown to have a maximum at the *critical frequency*

$$\nu_c = \frac{3eB\sin\alpha}{4\pi mc}\left(\frac{E}{mc^2}\right)^2 = 16.1 \times \left(\frac{B_\perp}{1\,\mu\text{G}}\right)\left(\frac{E}{1\,\text{GeV}}\right)^2 \text{MHz}. \tag{3.13}$$

The emitted power, i.e., the energy emitted per unit time per unit frequency interval, as a function of frequency can be obtained from the Fourier transform of the electric field of the synchrotron pulses, and is given by (see RL79)

$$P(E,\nu) = \frac{\sqrt{3}e^3B\sin\alpha}{mc^2}F\left(\frac{\nu}{\nu_c}\right), \tag{3.14}$$

with the function F defined by

$$F(x) = x\int_x^\infty K_{5/3}(\xi)\,d(\xi), \tag{3.15}$$

where $K_{5/3}(\xi)$ is the modified Bessel function of order 5/3. A related function, which is useful in considerations about the polarization of the synchrotron radiation (see Section 3.4), is

$$G(x) = xK_{2/3}(x), \tag{3.16}$$

where $K_{2/3}(x)$ is the modified Bessel function of order 2/3. The functions F and G are shown in Figure 3.2.

The total power emitted at all frequencies is obtained by integrating $P(\nu,E)$ over ν and is given in Equation 3.4, where it was more directly obtained by considering the total energy output from an accelerated charge.

3.3 Synchrotron emission from an ensemble of electrons

Consider an ensemble of electrons with energy in the range (E_1, E_2). Let $n(E)\,dE$ be the number density of electrons with energy between E and $E + dE$. The power emitted by

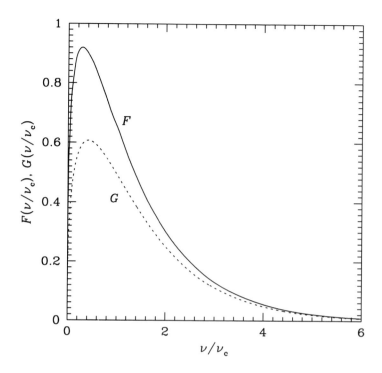

Fig. 3.2. The function $F(v/v_c)$, which provides the frequency dependence of the synchrotron radiation emitted by an electron, and the related function $G(v/v_c)$.

the electrons as a function of the frequency of the emitted radiation is given by

$$P(v) = \int_{E_1}^{E_2} P(E, v) \, n(E) \, dE, \tag{3.17}$$

We will see below that observed radio spectra from emission regions which are transparent to the radiation are of the power law form $P(v) \propto v^{-\alpha}$, where α is a constant. Such spectra are also encountered in the X-ray domain, where Compton scattering from highly relativistic electrons is central to the production of the spectrum. In the synchrotron as well as the Compton case, power law spectra are naturally produced if the electrons have a power law distribution of energy. Such power law electrons can be produced in a variety of ways, such as acceleration through shocks (see e.g. Blandford 1990). For such a distribution the number density of electrons as a function of energy is given by

$$n(E) \, dE = CE^{-p} dE, \tag{3.18}$$

where C and p are constants. The radiation spectrum produced by electrons having energy E peaks at the critical frequency v_c given in Equation 3.13. Using this to define the new variable $x = v/v_c$, we have

$$P(v) = \frac{\sqrt{3}e^3}{2mc^2} \left(\frac{3e}{4\pi m^3 c^5} \right)^{(p-1)/2}$$
$$\times C(B \sin \alpha)^{(p+1)/2} v^{-(p-1)/2} G(v/v_1, v/v_2, p), \qquad (3.19)$$

where v_1 and v_2 are the critical frequencies corresponding to the energies E_1 and E_2 respectively and

$$G(x_1, x_2, p) = \int_{x_2}^{x_1} x^{(p-3)/2} F(x) \, dx, \qquad (3.20)$$

with the function F defined as in Equation 3.15. When $v_1 \ll v \ll v_2$, the function G reduces to

$$g(p) \equiv G(0, \infty, p) = \frac{2^{(p-3)/2}}{3} \left(\frac{3p+7}{p+1} \right) \Gamma \left(\frac{3p-1}{12} \right) \Gamma \left(\frac{3p+7}{12} \right), \qquad (3.21)$$

where Γ is the usual gamma function. In this case the emitted spectrum has the very simple power law form

$$P(v) \propto v^{-\alpha}, \quad \alpha = \frac{p-1}{2}. \qquad (3.22)$$

In the case of extended radio sources, which are transparent to the synchrotron emission, $\alpha \sim 0.5 - 1$, so that $2 \lesssim p \lesssim 3$. For $G(v/v_1, v/v_2, p)$ to reduce exactly to $g(p)$, it would be necessary for the electron energies to extend over the interval $(0, \infty)$. This is of course not possible in a power law function, because of the divergence that would be encountered at the endpoints. The energy spectrum must therefore depart from the pure power law form at least at one of the ends. The frequency spectrum shows a corresponding departure from a pure power law, but the power law approximation is valid when the frequency v is far from v_1 and v_2, so that $v/v_1 \gg 1$ and $v/v_2 \ll 1$. A table listing the range of v over which the approximation is valid for various values of p has been provided by Moffett (1968). The energy density u_e, and the pressure p_e, due to the electrons are given by

$$u_e = \int_{E_1}^{E_2} n(E)E \, dE = \frac{C}{2-p}(E_2^{2-p} - E_1^{2-p}), \quad p_e = \frac{u_e}{3}, \qquad (3.23)$$

for $p \neq 2$. When $p = 2$ the integral reduces to $\log(E_2/E_1)$. The energy density and pressure are dominated by the upper cutoff energy E_2 for $p < 2$ and the lower cutoff E_1 for $p > 2$.

For an isotropic distribution of particles, the average emitted power is obtained by integration, using the distribution of α given in Equation 3.9:

$$\frac{1}{\sqrt{\pi}} \int_0^{\pi} (\sin \alpha)^k \sin \alpha \, d\alpha = \frac{\Gamma\left(\frac{k+2}{2}\right)}{\Gamma\left(\frac{k+3}{2}\right)}. \qquad (3.24)$$

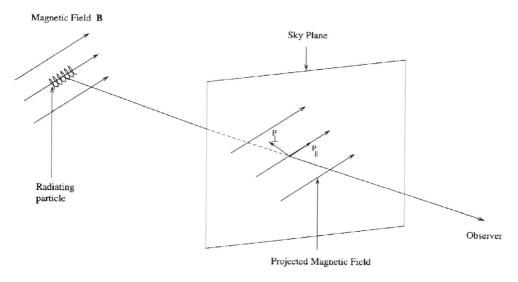

Fig. 3.3. Projection of the vectors relevant to synchrotron radiation on the plane of the sky.

This results in the replacement of the factor $(\sin \alpha)^{(p+1)/2}$ in Equation 3.19 by

$$\frac{\sqrt{\pi}}{2} \frac{\Gamma\left(\frac{p+5}{4}\right)}{\Gamma\left(\frac{p+7}{4}\right)}. \tag{3.25}$$

Using the function $G(x)$ in Equation 3.16, we now define a function which is useful in the treatment of polarization:

$$G'(x_1, x_2, p) \equiv \int_{x_2}^{x_1} x^{(p-3)/2} G(x)\, dx. \tag{3.26}$$

When $v_1 \ll v \ll v_2$, the function G' reduces to

$$g'(p) = G'(0, \infty, p) = 2^{(p-3)/2} \Gamma\left(\frac{3p-1}{12}\right) \Gamma\left(\frac{3p+7}{12}\right). \tag{3.27}$$

3.4 Polarization of synchrotron radiation

The synchrotron radiation from a single electron is elliptically polarized (Westfold 1959). Let $\mathbf{B}_\perp$ be the projection of the magnetic field $\mathbf{B}$ on the plane perpendicular to the line of sight (see Figure 3.3), i.e., the plane of the sky. It can be shown that the maximum in the polarized electric field $\mathbf{E}$ is perpendicular to the direction of $\mathbf{B}_\perp$, while the sense of the polarization depends on the frequency of the radiation and the viewing angle.

As described in Section 3.2, since the motion of the electron is highly relativistic the radiation is beamed in the forward direction and is mainly concentrated in a cone with half-angle $\theta \propto 1/\gamma \ll 1$ around the direction of the velocity **v**. The observed elliptical polarization of this radiation depends upon the angle ψ between the direction of **v** and the line of sight. When $\psi = 0$, the line of sight is located on the cone swept out by **v** during the helical motion of the electron around **B** and the polarization is observed to be linear. When $\psi \neq 0$, the rotation of the electric field around the ellipse of polarization is opposite for $\psi < 0$ and $\psi > 0$, i.e., for the situations when the line of sight lies inside or outside the cone generated by **v**. For an ensemble of electrons, it is expected that the number density is only a slowly varying function of the pitch angle. The radiation received by an observer therefore arises equally from electrons with $\psi < 0$ and for $\psi > 0$, and the elliptical component cancels out. The result is that the radiation is partially linearly polarized, with the degree of polarization given by

$$\Pi(E,v) = \frac{P_\perp(E,v) - P_\|(E,v)}{P_\perp(E,v) + P_\|(E,v)}, \tag{3.28}$$

where $P_\perp(E,v)$ is the synchrotron power emitted with polarization perpendicular to $\mathbf{B}_\perp$ and $P_\|(E,v)$ is the power with polarization parallel to $\mathbf{B}_\|$. It can be shown that in terms of the functions F and G introduced in Equations 3.15 and 3.16 the two components are

$$P_\perp(E,v) = \frac{\sqrt{3}e^3 B \sin \alpha}{2mc^2}[F(x) + G(x)], \tag{3.29}$$

and

$$P_\|(E,v) = \frac{\sqrt{3}e^3 B \sin \alpha}{2mc^2}[F(x) - G(x)], \tag{3.30}$$

where $x = v/v_c$, so that

$$\Pi(E,v) = \frac{G(x)}{F(x)}. \tag{3.31}$$

For a power law distribution of electron energy as in Equation 3.18, the power emitted by all the electrons can be obtained by multiplying $F(x) + G(x)$ and $F(x) - G(x)$ by $x^{(p-3)/2}$ and integrating as in Equation 3.20. Extending the integration over $(0, \infty)$ and using Equations 3.21 and 3.27 gives the simple result

$$\Pi(v) = \frac{3p + 3}{3p + 7}. \tag{3.32}$$

The partial linear polarization of synchrotron radiation from power law electrons is therefore independent of the frequency, and depends only on the power law index. Π is conventionally expressed as a percentage. For $p = 3$, which corresponds to the radio spectral index $\alpha = 1$, $\Pi = 75$ per cent. The polarization here has been obtained by using the direction of the projected magnetic field. If the magnetic field in a given region is well tangled, its projection on the plane of the sky does not favour any direction and the net polarization over the region vanishes. The net polarization of radiation coming from an extended source can also be weakened if the magnetic field direction changes from one region to another.

3.4.1 Faraday rotation

The interstellar medium (ISM) in a galaxy contains ionized gas as well as a magnetic field. Consider a linearly polarized beam of radiation travelling through a screen (slab) of such a magneto-ionic medium. In general, linearly polarized radiation can be considered to be made up of two circularly polarized components, one left handed and the other right handed. In the ISM, consisting of plasma in magnetic fields, each of these components sees a different refractive index in the plasma and therefore the two travel with different phase velocity, which leads to an alteration in the relative phases of the two components. When the two beams recombine in a detector, a linearly polarized wave is again seen, but with the plane of polarization now rotated relative to its orientation when it first entered the plasma. This is known as the *Faraday effect*.[1] The angle of rotation of the plane of polarization is given by

$$\Theta = \lambda^2 \times \frac{e^3}{2\pi m^2 c^4} \int n_e(s) B_\parallel \, ds, \tag{3.33}$$

where λ is the wavelength of the radiation, $n_e(s)$ is the electron density of the plasma at distance s along the beam, $B_\parallel$ is the component of the magnetic field tangent to the path and the integration extends over the entire path taken by the beam through the plasma. The term multiplying λ^2 in the above equation is called the *rotation measure* (*RM*) and is conventionally expressed in units of radians per square metre. In practical units

$$\Theta = RM \times \left(\frac{\lambda}{1\,\text{m}}\right)^2 \text{ radians}, \tag{3.34}$$

where

$$RM = 8.11 \times 10^2 \int \left(\frac{n_e}{1\,\text{cm}^{-3}}\right)\left(\frac{B_\parallel}{1\,\mu\text{G}}\right)\left(\frac{ds}{1\,\text{kpc}}\right) \text{ radians m}^{-2}. \tag{3.35}$$

The integral in Equation 3.33 is called the *Faraday depth*.

The characteristic λ^2 dependence of Faraday rotation allows it to be measured and used as an important diagnostic tool. *RM* can be deduced from multiwavelength observations, using Equation 3.34. When observations are available only at two frequencies, as is quite often the case, and the plane of polarization is found to be rotated from one frequency to the other, one can attribute the observed rotation to the Faraday effect and obtain *RM*. However, one cannot distinguish between rotations of the plane of polarization through angles θ and $\theta + n\pi$, where n is an integer. There is therefore an ambiguity in *RM* at least to the extent of $\pi/(\lambda_2^2 - \lambda_1^2)$ radians m^{-2}, where λ_1 and λ_2 are the two wavelengths at which measurements are made.

This ambiguity can be resolved when polarization measurements are available at several wavelengths. One first estimates *RM* from two closely spaced wavelengths λ_1 and λ_2, assuming that the rotation angle is less than $\pi/2$. Using this *RM* the angle

---

[1] When an elliptically polarized beam of radiation travels through a magnetized plasma, the Faraday effect results in a rotation of the polarization ellipse.

expected at some other wavelength λ_3 is estimated. The ambiguity from the measured angle at λ_3 can now be removed by adding or subtracting from the estimated value multiples of π until the difference between the estimate and the measured value is less than $\pi/2$. The angle determined in this manner can be used in a fit to Equation 3.34 with three points, and the process continued in a similar manner to include all the observed values. The original plane of polarization can be recovered by extrapolating the fit to $\lambda = 0$. The projection of the magnetic field onto the plane of the sky at any point in the image is perpendicular to the deduced direction of the polarization. The direction of the magnetic field in many extended radio sources has been mapped in this manner.

Consider polarized radiation travelling through a medium that is in the form of a foreground screen between the source of radiation and the observer. For a screen in the form of a slab with constant thickness L, uniform electron density n_e and a constant magnetic field B, the plane of polarization of radiation with wavelength λ is rotated through an angle $\theta \propto \lambda^2 n_e B_\parallel L$. Since this rotation is the same for all the radiation of a given wavelength, there is no change in the polarized intensity of the radiation (see Dreher, Carilli and Perley 1987 for a useful review of polarized radiation propagation). However, due to transverse variations in the properties of the screen, the plane of polarization can rotate through different angles for different rays. When such structure is present on a scale smaller than the beam size used in observation, at any wavelength there is vector addition of the radiation propagating through screen areas with different Faraday depth. The polarization angle of the observed radiation therefore shows a departure from the λ^2 law and a reduction in the polarized flux density. For a given medium this *beam depolarization* changes with beam size, which allows it to be distinguished from other depolarizing effects.

In the case of a depolarizing screen, the thermal electrons, which rotate the plane of polarization, are located in the foreground, separated from the relativistic electrons which emit the synchrotron radiation. A situation in which the two kinds of electron are mixed together, called internal depolarization, is more complicated to deal with. Here the plane of polarization of the radiation coming from different parts of the source is rotated through different angles, so that the polarized intensity is affected. A simple example of internal depolarization is provided by a uniform slab containing relativistic and thermal electrons. The degree of polarization, which is the ratio of the polarized intensity to the total intensity, is given by

$$m = \left(\frac{3p+3}{3p+7} \right) \left| \frac{\sin(k_0 n_e B_\parallel L)}{\lambda^2 k_0 n_e B_\parallel L} \right|, \tag{3.36}$$

and the observed position angle is

$$\chi = \frac{1}{2} \arctan \left| \frac{\sin(k_0 n_e B_\parallel L)}{\cos(k_0 n_e B_\parallel L)} \right|. \tag{3.37}$$

The radiation produced at different depths in the slab undergoes different amounts of Faraday rotation. The vector addition of these parts gives rise to the sinusoidal dependence. The rotation angle again has a λ^2 dependence, as in the case of external depolarization due to a screen, but the rate of rotation is half as much. Moreover, the position angle is limited to the range $(0, \pi/2)$, since it changes by $\pi/2$ whenever the polarized intensity passes through a zero. This upper limit on the rotation also applies to more realistic models of internal polarization, though there are cases where the limit can be exceeded. A rotation that exceeds $\pi/2$ is therefore taken to be indicative of external polarization due to a screen. A derivation of results in the case of a uniform slab model has been given by Dreher, Carilli and Perley (1987), difficulties encountered in the interpretation of polarization data have been discussed by Laing (1984), while detailed derivations, on which much of the later work has been based, have been provided by Burn (1966). A detailed review of polarization structures in radio galaxies and quasars can be found in Saikia and Salter (1988).

The observed, integrated value of RM for a majority of the sources is $\lesssim 50$ radians m^{-2}, so that the Faraday rotation is $\lesssim 10$ deg for $\lambda \lesssim 6$ cm. For high frequency measurements, therefore, one can assume that there is negligible rotation and map the magnetic field from the observed direction of polarization. When the magnetic field is highly tangled the intrinsic polarization will of course be greatly reduced.

3.5   Absorption of synchrotron radiation

We have seen in Section 3.3 that the synchrotron spectrum emitted by an ensemble of electrons with a power law distribution of energy has the simple power law form $P(v) \propto v^{-\alpha}$, $\alpha = $ constant. An observer will see such a spectrum only if there is no absorption of photons by the emitting region, or by intervening matter such as the host galaxy of the AGN, or an intervening galaxy or our own Galaxy. It is observed on the one hand that the spectrum of radio emission from the lobes and large scale jets of giant radio sources often has a power law form with $\alpha \gtrsim 0.5$. This corresponds to an electron spectral index $p \gtrsim 2$, if the regions are taken to be optically thin. On the other hand the radiation from compact regions often has flat, inverted or complex spectra. If the basic radiation process is the same in the extended and compact regions, the different shapes observed in the latter must be the result of absorption or scattering. Even the spectra from optically thin emission regions sometimes show signs of low frequency absorption, which can be traced to thermal absorption in the galactic interstellar medium (ISM). We shall now review some important absorption processes. A discussion of the observational effects produced by these processes will be left to later chapters.

3.5.1   *Synchrotron self-absorption*

We have so far considered only the emission of photons by relativistic electrons moving in a magnetic field, but there are two other effects to be taken into account: (1) in the presence of a magnetic field an electron can absorb a photon and make a transition to a higher energy state; (2) photons in a particular quantum mechanical state can stimulate further emissions of photons into the same state. Both these processes are proportional to the intensity of the radiation and are important in highly luminous compact sources, that have a high radiation density. The results of the two processes can be combined together into a single effective absorption coefficient, to which stimulated emission makes a negative contribution.

In the theory of the emission and absorption of radiation by matter (see for example RL79), the coefficient for spontaneous emission j_v by an electron that passes from a higher energy state to a lower energy state, even in the absence of any ambient radiation, is given by

$$j_v = \frac{hv}{4\pi} \phi(v) \int_{E_1}^{E_2} A_{21} n(E) \, dE, \tag{3.38}$$

where E_1 and E_2 are the limits of the electron energy distribution, and A_{21} is Einstein's A-coefficient. A_{21} gives the probability per unit time of making a spontaneous transition from state 2 with energy E to state 1 with energy $E - hv$, emitting a photon of energy hv. $\phi(v)$ is a line profile function that is dimensionless and normalized to unity. It takes into account the line broadening produced by the broadening of electron energy levels; this occurs for various physical reasons such as the uncertainty principle and the Doppler effect due to motion of the emitters and absorbers. $\phi(v)$ is sharply peaked around the central frequency under consideration and may be taken to be a Dirac delta function for our present purposes. Since j_v is the amount of energy emitted per unit time, per unit frequency interval and per unit solid angle, it follows from Equation 3.17 that

$$A_{21} = \frac{1}{hv} \frac{P(E,v)}{\phi(v)}. \tag{3.39}$$

We have assumed here that the distribution of electron pitch angles is isotropic. If the electrons moved in some preferred direction relative to the magnetic field, the electron number density would be a function of the pitch angle, and so would be A_{21}. The emission coefficient in general also depends on the polarization of the radiation.

We have mentioned above that electrons spiralling in magnetic fields can absorb photons of the ambient radiation to pass to a higher energy state. The number of such absorptions taking place is proportional to the intensity of the radiation, with an absorption coefficient at frequency v given by

$$\frac{hv\phi(v)}{4\pi} \int_{E_1}^{E_2} [n(E - hv)B_{12}] \, dE, \tag{3.40}$$

where the Einstein coefficient B_{12} is the probability per unit time and per unit radiation intensity for a photon of energy hv to be absorbed, with the electron now making a transition from energy $E - hv$ to energy E.

Simulated emission causes electrons to make a transition from a higher energy state to a lower energy state with the emission of a photon, with a probability proportional to the intensity of the radiation. It therefore constitutes negative absorption and can be conveniently described in terms of an absorption coefficient

$$-\frac{h\nu\phi(\nu)}{4\pi} \int_{E_1}^{E_2} [n(E)B_{21}] \, dE. \tag{3.41}$$

The coefficient B_{21} gives the probability of transition from the higher energy state to the lower energy state, i.e., of the emission of a photon in the presence of a radiation field. An effective absorption coefficient including true absorption as well as stimulated emission is now defined by

$$\alpha_\nu = \frac{h\nu\phi(\nu)}{4\pi} \int_{E_1}^{E_2} [n(E-h\nu)B_{12} - n(E)B_{21}] \, dE. \tag{3.42}$$

The change in the intensity of radiation $I(\nu)$ in traversing a distance ds is given by

$$dI(\nu) = j_\nu \, ds - \alpha_\nu I(\nu). \tag{3.43}$$

Spontaneous and stimulated emission add to the intensity while true absorption decreases it.

The three Einstein coefficients are connected by the simple relations (see RL79)

$$g_1 B_{12} = g_2 B_{21}, \qquad \frac{A_{21}}{B_{21}} = \frac{2h\nu^3}{c^2}, \tag{3.44}$$

where g_1 and g_2 denote the statistical weights in the states 1 and 2 respectively. For a state of energy E the weight is proportional to the allowed phase space volume, which is $\propto 4\pi p^2 \, dp \propto 4\pi E^2 \, dE$ for highly relativistic particles with energy $E \simeq pc$, where p is the momentum of the particle. We therefore have

$$\frac{g_2}{g_1} = \left(\frac{E}{E-h\nu}\right)^2, \tag{3.45}$$

and the absorption coefficient reduces to

$$\alpha_\nu = \frac{c^2}{8\pi h\nu^3} \int_{E_1}^{E_2} \left[\left(\frac{E}{E-h\nu}\right)^2 n(E-h\nu) - n(E)\right] P(E,\nu) \, dE. \tag{3.46}$$

For $h\nu \ll E$, which is pertinent to radio astronomical situations, we have

$$n(E-h\nu) = n(E) - h\nu\frac{dn}{dE}, \qquad \left(\frac{E}{E-h\nu}\right)^2 \simeq 1 + \frac{2h\nu}{E} \tag{3.47}$$

and

$$\alpha_\nu = -\frac{c^2}{8\pi\nu^2} \int_{E_1}^{E_2} E^2 \frac{d}{dE}\left[\frac{n(E)}{E^2}\right] P(E,\nu) \, dE. \tag{3.48}$$

We now consider a power law distribution of electrons as in Equation 3.18, use Equation 3.14 for $P(E,\nu)$ and change from E to the variable $x = \nu/\nu_c$ as in Section 3.3. The integrand is then $\propto x^{-(p-2)/2}F(x)$. The limits of integration can again be stretched so that $x_2 \to 0$ and $x_1 \to \infty$ and the integral obtained using Equation 3.21, with p in that equation replaced by $p + 1$. For an isotropic distribution of electrons we average over the angle using Equations 3.9 and 3.24. The final result is that

$$\alpha_v = A v^{-(p+4)/2}, \tag{3.49}$$

with the proportionality constant

$$A = \frac{\sqrt{3}e^3}{8\pi m}\left(\frac{3e}{4\pi m^3 c^5}\right)^{p/2} 2^{p/2}\frac{\sqrt{\pi}}{2}CB^{(p+2)/2}\frac{\Gamma\left(\frac{3p+2}{12}\right)\Gamma\left(\frac{3p+22}{12}\right)\Gamma\left(\frac{p+6}{4}\right)}{\Gamma\left(\frac{p+8}{4}\right)}. \tag{3.50}$$

If the emitting region is homogeneous and has size $\sim l$, it follows from the usual radiation transfer relation (see RL79) that

$$I(v) = \frac{j_v}{\alpha_v}\left[1 - \exp\left(-\alpha_v l\right)\right]. \tag{3.51}$$

When the source is optically thin, i.e., $\tau_v = \alpha_v l \ll 1$, the spectrum of the radiation leaving the medium is $I(v) \propto j_v l \sim v^{-\alpha}$. However, for an optically thick medium with $\tau_v \geq 1$ it follows from Equations 3.22 and 3.49 that

$$I(v) \sim \frac{j_v}{\alpha_v} \sim v^{5/2}, \tag{3.52}$$

Notice that $I(v)$ now has a positive exponent for v.

Since the synchrotron self-absorption coefficient is inversely dependent on the frequency, a given source will be optically thick below the frequency v_a at which $\tau_v \simeq 1$ is first satisfied, and may be considered to be optically thin at all higher frequencies. If the source subtends a solid angle $\Delta\Omega$ at the observer, the flux density received is $F(v) = I(v)\Delta\Omega$, and its frequency dependence is given by

$$F(v) \propto v^{-\alpha}, \quad v > v_a, \tag{3.53}$$
$$F(v) \propto v^{5/2}, \quad v < v_a.$$

Synchrotron self-absorption therefore produces an inverted spectrum of power law form, with spectral index 5/2 regardless of the value of the power law index in the unabsorbed part. The shape of the synchrotron spectrum in the presence of self-absorption is shown in Figure 3.4.

A slope of 5/2 has never so far been observed, and what is found in compact sources is a spectral index considerably flatter than is found in extended sources or in spectra that are complex and undulating. The reason for this is that different parts of the source become optically thick at different frequencies, and therefore the observed spectrum is a superposition of spectra that peak there. Some typical radio spectra, including examples with shapes that can be attributed to synchrotron self-absorption, are shown in Figure 3.5. In the figure, 3C 123 is a radio galaxy whose radio flux is dominated by power law emission from the extended structure, which is transparent at the observed frequencies; some steepening is seen at the higher frequencies. 3C 48 is a radio quasar; its spectrum is self-absorbed for frequencies $\lesssim 100\,\mathrm{MHz}$ and a smooth power law at higher frequencies. 3C 84 is a radio galaxy with a complex spectrum

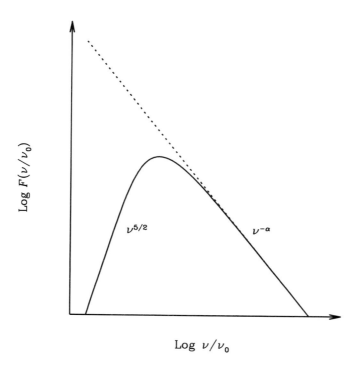

Fig. 3.4. The turnover in the synchrotron spectrum due to synchrotron self-absorption. ν is the observed frequency while ν_0 is the frequency at which the self-absorption optical depth becomes unity. The dotted line shows how the optically thin $\nu^{-\alpha}$ spectrum would extend to low frequencies in the absence of self-absorption.

dominated at low frequencies by a power law component arising in an extended structure, an intermediate scale component that becomes self-absorbed at a few GHZ and a compact component that becomes self-absorbed at $\sim 20\,\text{GHz}$. 3C 454.3 is a quasar that again has a complex spectrum with multiple components, which become self-absorbed at different frequencies. The different fluxes noticed at the same or closely placed frequencies in some sources are due to variability.

It appears from the above discussion that the power law slope of self-absorbed synchrotron emission can never exceed 5/2. This is, however, true only for a single power law, while for a two-power-law distribution that is concave, i.e., becomes steeper at low energies, it is possible to have a slope exceeding 5/2 over 1–1.5 decades of energy, for a suitable choice of parameters (de Kool and Begelman 1989). When the effect of self-absorption on the absorbing electrons is taken into account, low energy electrons are found to acquire a relativistic Maxwell–Boltzmann distribution that connects to the power law, producing the concavity (Ghisellini, Guilbert and Svensson 1988) required for a steep slope. Detailed models of spectra produced in these circumstances have been considered by Schlickeiser, Biermann and Crusius-Wätzel (1991), who take into

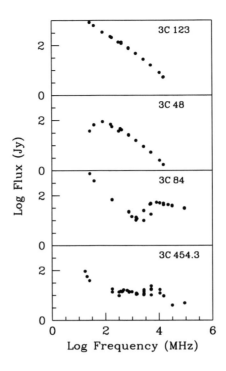

Fig. 3.5. Typical radio spectra. See text for detailed description. The data were kindly provided by D.J. Saikia.

account various non-thermal processes that can provide energy to electrons and show that slopes $\gtrsim 4$ can be produced. We will see in Section 8.6 that radio-quiet quasars and AGN have very steep slopes in the submillimetre region, which can in principle be self-absorbed synchrotron spectra of the kind discussed here.

3.5.2   *Source size and synchrotron self-absorption*

A simple relation between source size and synchrotron self-absorption frequency can be obtained as follows. Consider a source that first becomes synchrotron self-absorbed at the frequency ν_a. Since this is the marginal frequency for reaching unit optical depth, the flux here can be obtained in two ways: (1) using the relation $F(\nu) \sim I(\nu) \sim j_\nu / \alpha_\nu$ in Equation 3.51 and (2) by extrapolating the unabsorbed power law flux to ν_a. Denoting this by $F(\nu_a)$ and equating the two expressions, we get

$$\frac{j_{\nu_a}}{\alpha_{\nu_a}} = \frac{F(\nu_a)}{\Delta\Omega}. \tag{3.54}$$

Using Equations 3.19, 3.24 and 3.49 and the relations $j_\nu = P(\nu)/(4\pi)$ and $\Delta\Omega \simeq \theta^2$, where θ is the angle subtended by the source, we get

$$\theta = c \left(\frac{3e}{4\pi m^3 c^5} \right)^{1/4} B^{1/4} \chi(p) v_a^{-5/4} F^{1/2}(v_a). \tag{3.55}$$

Here all the p-dependence in the relation is lumped together into a function $\chi(p)$. Using the expressions provided in the previous sections, it can be shown that $\chi(p) \simeq (2p-1)^{1/2}$ (see Moffett 1968). In practical units

$$\theta \simeq 0.17 \left(\frac{B}{1\,\mu G} \right)^{1/4} \left[\frac{F(v_a)}{1\,\text{Jy}} \right]^{1/2} \left(\frac{v_a}{1\,\text{MHz}} \right)^{-5/4} \chi(p) \, \text{arcsec}, \tag{3.56}$$

i.e.,

$$v_a \simeq 0.23 \left(\frac{B}{1\,\mu G} \right)^{1/5} \left[\frac{F(v_a)}{1\,\text{Jy}} \right]^{1/2} \left(\frac{\theta}{1\,\text{arcsec}} \right)^{-4/5} \text{MHz}. \tag{3.57}$$

The flux and frequency in this expression are in the rest frame of the emitter. When it is at a cosmological distance and moves relativistically in the local rest frame, the expression has to be suitably modified. This results in an additional factor

$$(1+z)^{-3/4} \mathscr{D}^{-1/4}, \quad \mathscr{D} = \frac{1}{\gamma(1 - \beta \cos \psi)}, \tag{3.58}$$

on the right hand side of Equation 3.56. Here $\beta = v/c$, v is the bulk velocity of the emitter relative to the local rest frame and ψ is the angle that the velocity vector makes with the line of sight. The origin of the additional factors will become clear after the discussion on relativistic bulk motion in Section 3.8.

The functional relationship between source parameters in Equation 3.56 can be derived in a straightforward manner by invoking thermodynamics (Williams 1963). For this, it is convenient to express the intensity of radiation $I(v)$ in terms of the *brightness temperature* T_b, which is defined by the relation

$$I(v) = \frac{2v^2 k T_b}{c^2} = F(v)/\theta^2. \tag{3.59}$$

For black body radiation in the Rayleigh–Jeans approximation, T_b is the usual thermodynamic temperature and has the same value for all $I(v)$ and v for which the Rayleigh–Jeans approximation is valid. As defined in Equation 3.59, however, T_b provides a measure of the intensity and is in general different at different frequencies (see RL79).

Let the radio source become self-absorbed at the frequency v_a. The synchrotron radiation at this frequency is mostly emitted by electrons with some energy E which is related to v_a through Equation 3.13, if we identify v_a with the critical frequency. Even though the electrons do not have a thermal spectrum, one can formally associate with them a kinetic temperature $T = E/k$. It then follows from thermodynamics that the brightness temperature of the radiation cannot exceed the kinetic temperature of the emitting electrons, $kT_b \lesssim kT$, and the radiation becomes self-absorbed when

$$kT_b \simeq kT = E \simeq \left(\frac{3e}{4\pi m^3 c^5}\right)^{-1/2} B^{-1/2} v_a^{1/2}. \qquad (3.60)$$

The condition for avoiding synchrotron self-absorption is then

$$T_b \lesssim 9 \times 10^{13} \left(\frac{B}{1\,\mu G}\right)^{-1/2} \left(\frac{v}{1\,\mathrm{GHz}}\right)^{1/2} \mathrm{K}. \qquad (3.61)$$

Using Equations 3.59 and 3.60, the angular size of a source that becomes self-absorbed at frequency v_a is

$$\theta \simeq c \left(\frac{3e}{4\pi m^3 c^5}\right)^{1/2} B^{1/4} F^{1/2}(v_a) v_a^{-5/4}, \qquad (3.62)$$

which is consistent with Equation 3.55. The spectral form of the self-absorbed radiation is

$$F(v) \sim I(v) \sim v^2 k T_b \sim v^2 \times v^{1/2} \sim v^{5/2}. \qquad (3.63)$$

In the above expressions relating angular size to self-absorption, all the quantities except the magnetic field are directly observable and therefore B can in principle be determined. However, B depends on high powers of the measured values and any uncertainties in these are highly magnified in B. Nevertheless, if a value of B can be guessed then an idea of source size can be obtained for unresolved sources.

The distortions produced in radio spectra can be related to the nature of the sources which are found in radio surveys at different frequencies. In a low frequency survey, carried out at a few hundred MHz say, the dominating sources are those that have extended structure and relatively steep, unabsorbed radio spectra with spectral index $\alpha \gtrsim 0.5$. At high frequencies, say a few GHz, these sources become faint, while compact self-absorbed sources, with flat or inverted spectra, have higher flux. High frequency surveys, which are often flux limited, are therefore dominated by compact sources.

3.5.3   *Thermal absorption*

As a synchrotron photon travels through plasma, it can be absorbed by the bremsstrahlung process, also known as free–free absorption. In this process an electron absorbs a photon and so accelerates in the field of an ion, the latter helping with momentum conservation. When the electrons and ions have a thermal distribution, the process is known as thermal bremsstrahlung absorption. The absorption coefficient for the process is given by (see RL79)

$$\alpha_v^{ff} = \frac{4e^6}{3mhc} \left(\frac{2\pi}{3km}\right)^{1/2} Z^2 n_e n_i T^{-1/2} v^{-3} (1 - e^{-hv/kT}) \bar{g}_{ff}(T, v), \qquad (3.64)$$

where n_e and n_i are the number density of electrons and ions respectively, Z is the ionic charge, and T the temperature; e and m are the electron charge and mass respectively.

When there is more than one ionic species present, a summation over Zn_i for all the species is made. The *Gaunt factor* $\bar{g}_{ff}(T,v)$ depends on the approximations made in deriving the expression for the absorption cross-section. It is available in the form of analytic formulae valid over specific regions of frequency and temperature, or as extensive tables (see references in RL79).

In the radio frequency region of the spectrum the Gaunt factor is given by (see Scheuer 1960 and Oster 1961)

$$\bar{g}_{ff}(T,v) = \frac{\sqrt{3}}{\pi} \ln \left[\left(\frac{2kT}{\gamma^* m} \right)^{3/2} \frac{m}{\pi \gamma^* Z e^2 v} \right],$$

(3.65)

where $\gamma^* = e^\gamma = 1.781\ldots$, and $\gamma = 0.5772\ldots$ is Euler's constant. This expression is valid for $T \leq 8.92 \times 10^5\,$K. At very low frequencies and temperatures ($T \lesssim 10^2\,$K) the approximations used in deriving the above expression are not valid and numerical values provided by Oster (1961) may be used. For $T \geq 8.92 \times 10^5\,$K the Gaunt factor is

$$\bar{g}_{ff}(T,v) = \frac{\sqrt{3}}{\pi} \ln \left[\frac{2kT}{\pi v \hbar \gamma^*} \right].$$

(3.66)

The higher temperature case is useful in calculating the thermal radio emission from hot gases, as in an H II region. For thermal absorption in the galaxy the form in Equation 3.65 is applicable and in practical units this reduces to

$$\bar{g}_{ff}(T,v) = 11.69 + 0.83 \ln \left(\frac{T}{10^4\,\text{K}} \right) - 0.55 \ln \left(\frac{v}{100\,\text{MHz}} \right).$$

(3.67)

For $T \simeq 10^4\,$K and $100\,\text{MHz} \leq v \leq 10\,\text{GHz}$, $\bar{g}_{ff} \simeq 10$ and with this value the absorption coefficient is

$$\alpha_v^{ff} = 5.47 \times 10^{-2} \left(\frac{n_e}{1\,\text{cm}^{-3}} \right)^2 \left(\frac{T}{10^4\,\text{K}} \right)^{-3/2} \left(\frac{v}{100\,\text{MHz}} \right)^{-2} \text{kpc}^{-1}.$$

(3.68)

An absorber of size l becomes opaque at the frequency v_a at which the optical depth $\alpha_{v_a}^{ff} l$ roughly equals unity. Using Equation 3.68, v_a is given by

$$v_a = 23.4 \left(\frac{n_e}{1\,\text{cm}^{-3}} \right) \left(\frac{T}{10^4\,\text{K}} \right)^{-3/4} \left(\frac{l}{1\,\text{kpc}} \right)^{1/2} \text{MHz}.$$

(3.69)

Above this frequency the absorption can be neglected, while below it the intensity falls off exponentially. When a synchrotron-radiation-emitting region also contains a thermal plasma, then for $v < v_a$ the intensity of the radiation is given by

$$I(v) \sim \frac{j_v}{\alpha_v} \sim \frac{v^{-\alpha}}{v^{-2}} \sim v^{-(\alpha-2)}.$$

(3.70)

In most AGN the radio spectral index $\alpha < 2$ and the spectrum becomes inverted at low frequencies owing to thermal absorption. Even for steeper values of α there is conspicuous flattening of the spectrum.

3.6   Radio source energetics

Consider a source of synchrotron radiation in which the electrons have a power law distribution of energy, as in Equation 3.18. If the volume of the source is V, the *total energy* in the electrons is

$$U_e = V \int_{E_1}^{E_2} n(E)E \, dE = \frac{VC\left(E_2^{2-p} - E_1^{2-p}\right)}{2-p}. \tag{3.71}$$

From Equation 3.10 with $\beta \simeq 1$ for the power radiated by an electron with energy E, the total synchrotron luminosity of the source is given by

$$L = V \int_{E_1}^{E_2} n(E)P(E) \, dE = \frac{4\sigma_T U_B VC}{3m^2c^4}\left(\frac{E_2^{3-p} - E_1^{3-p}}{3-p}\right). \tag{3.72}$$

Using Equation 3.13 we can replace the energy limits E_1 and E_2 by their critical synchrotron frequencies ν_1 and ν_2 and use $p = 2\alpha + 1$ to obtain for the ratio of the total energy of the electrons to the synchrotron luminosity,

$$\frac{U_e}{L} = \left(\frac{27m^5c^9}{16\pi e^7}\right)^{1/2} \frac{1}{B^{3/2}} \left(\frac{2-2\alpha}{1-2\alpha}\right) \frac{\nu_2^{1/2-\alpha} - \nu_1^{1/2-\alpha}}{\nu_2^{1-\alpha} - \nu_1^{1-\alpha}} \equiv \frac{A}{B^{3/2}}, \tag{3.73}$$

where A is a constant that depends only on the spectral index. Since there are other species of particle besides the electrons present in the source, the total particle energy $U_p = aU_e$, where $a > 1$. The total energy of the source in the particles and the magnetic field is therefore

$$U_{\text{tot}} = U_p + U_B = \frac{aAL}{B^{3/2}} + \frac{VB^2}{8\pi}. \tag{3.74}$$

When regarded as a function of B, the total energy is a minimum at

$$B_{\text{min}} = \left(\frac{6\pi aAL}{V}\right)^{2/7}, \tag{3.75}$$

which is within ~ 10 per cent of the *equipartition field*

$$B_{\text{eq}} = \left(\frac{8\pi aAL}{V}\right)^{2/7} \tag{3.76}$$

at which the energy density in the particles is equal to the energy density in the magnetic field. The total energy for the equipartition value of the magnetic field is

$$U_{\text{tot}}(\text{eq}) = 2V\left(\frac{B_{\text{eq}}^2}{8\pi}\right) = 0.50(aAL)^{4/7}V^{3/7} = 1.01\,U_{\text{tot}}(\text{min}), \tag{3.77}$$

where $U_{\text{tot}}(\text{min})$ is the minimum total energy. It was G. Burbidge (1959) who pointed out that $U_{\text{tot}}(\text{eq})$ is very nearly equal to $U_{\text{tot}}(\text{min})$, and it is customary to use the equipartition value of the magnetic field to assess the total energy in a radio source. Uncertainties are introduced because of the ignorance regarding the source size, the volume occupied by the emitting electrons, the cutoff energies for the electron distribution and the energy in the different kinds of particles. But it is clear from the above equation that the dependence on any of these parameters is not very strong. U_{tot}(min) for radio galaxies is found to be in the range $\sim 10^{57}$–10^{61} erg, with $B_{eq} \sim 10^{-6}$–10^{-4} G.

The pressure terms due to the relativistic particles and the magnetic field are given by $u_e/3$ and $B^2/(8\pi)$ respectively, where u_e is the particle energy density. The minimum pressure in the radio source therefore can be obtained from the minimum energy condition and is given by

$$p_{min} = \frac{U_{tot}}{3V} = 0.16 \left(\frac{aAL}{V}\right)^{4/7} = \frac{B_{eq}^2}{12\pi}. \tag{3.78}$$

To express the minimized quantities in terms of observable parameters, we express the luminosity as $L \simeq L(v_p)v_p/(1-\alpha)$, where v_p is the upper or lower cutoff frequency of the power law spectrum, depending on the value of the radio spectral index (see comments in Section 3.3). Using Equation 2.61 we get $L \simeq 4\pi D_L^2 F(v_{p0})$, where $v_{p0} = v_p(1+z)$ is the cutoff frequency in the observer's frame, $F(v_{p0})$ the observed flux density and D_L the luminosity distance. Assuming that the source is approximately spherically symmetric with diameter d, we have $V = \pi d^3/6 = \pi \theta^3 d_m^3/6$, where $d_m = d_L/(1+z)^2$ is the angular diameter distance defined in Section 2.8. We then have $L/V = 24F(v_{p0})v_{p0}(1+z)^6/(\theta^3 D_L)$ and

$$B_{eq} = (192\pi a)^{2/7} \left(\frac{27m^5 c^9}{16\pi e^7}\right)^{1/7} F^{2/7}(v_{p0})v_{p0}^{1/7} D_L^{-2/7} \theta^{-6/7}(1+z)^{11/7}. \tag{3.79}$$

Notice that cosmological effects enter this calculation through the redshift factors.

3.7 Energy loss and the electron spectrum

Electrons spiralling in a magnetic field lose their energy by synchrotron emission as well as by inverse Compton scattering of the synchrotron photons and other ambient radiation (see Section 4.2). In either case the electron energy decreases at a rate $dE/dt \propto E^2$. There can also be other processes causing energy loss such as (1) bremsstrahlung and adiabatic expansion of the source, in which case $dE/dt \propto E$, and (2) ionization of atoms and molecules in the plasma, for which dE/dt has a logarithmic energy dependence and can be approximately taken to be constant. The net energy loss rate can therefore be written as

$$\frac{dE}{dt} = a_0 + b_0 E + c_0 E^2, \tag{3.80}$$

where a_0, b_0, c_0 are constants. Given an initial power law spectrum of electrons $n(E) \propto E^{-p_0}$, $p_0 = $ constant, the more energetic electrons lose their energy more rapidly and the spectrum is distorted from its initial power law. The initial spectrum can be replenished by the injection of fresh electrons or reacceleration, but this does not in general restore the original power law.

For some energy loss mechanisms such as synchrotron emission or Compton scattering with the Thomson limit applicable in the electron rest frame (see Section 4.2), the energy lost in the emission of a single photon is only a small fraction of the electron energy. In such cases the evolution of the time-dependent electron density can be found by considering a small energy bin $(E, E + dE)$ and counting the electrons gained and lost from it in a small interval of time. Electrons are lost from the bin as their energy decreases due to emission, while electrons are gained from higher energies, as well as from fresh injection or reacceleration. The result can be expressed as the partial differential equation (see Longair L86 for a derivation)

$$\frac{\partial n(E,t)}{\partial t} = -\frac{\partial}{\partial E}\left[\frac{dE}{dt} n(E,t)\right] + q(E,t), \tag{3.81}$$

where $q(E, t)$ is the rate at which electrons with energy E are supplied per unit volume. When the number density of electrons is not constant across the source, electrons can be gained and lost from a volume element due to diffusion. In this case the term $D\nabla^2 n$, where D is the diffusion coefficient, has to be added to the right hand side of the evolution equation. When the energy lost by an electron in a single event is a significant fraction of its energy, as in the cases of Compton scattering in the Klein–Nishina limit or bremsstrahlung emission, the evolution is described by an integro-differential equation. While a general solution of the evolution equation can be obtained in terms of Green's functions,[1] there are several cases of practical importance that can be treated in a simple manner.

Consider a continuous injection of electrons at the constant rate $q(E)$ and let an equilibrium between the injection and energy loss be reached. Then $\partial n/\partial t = 0$ and the steady state number density of electrons of energy E is

$$n(E) = \left(\frac{1}{a_0 + b_0 E + c_0 E^2}\right)\int_E^\infty q(E')\,dE'. \tag{3.82}$$

The lower limit of integration is taken to be E because electrons at energy E are either produced at that energy, or result from the energy loss by electrons with higher energy. We shall consider the situation where the energy loss is restricted to the synchrotron or Compton processes, i.e., $a_0 = b_0 = 0$. When the injection is at a single energy E_0, $q(E)$ is the Dirac delta function $\delta(E - E_0)$ and

$$n(E) \propto E^{-2}, \quad E \le E_0, \tag{3.83}$$
$$n(E) = 0, \quad E > E_0. \tag{3.84}$$

For a power law injection spectrum with $n(E) \propto E^{-p_0}$, $p_0 > 1$, we get

$$n(E) \propto E^{-(p_0+1)}, \quad p_0 > 1, \tag{3.85}$$

i.e., the electron power law index is steepened by one power of E, and thus the

[1] See GS64 and Bleumenthal and Gould (1970) for detailed discussions of the equations and solutions.

spectral index of the synchrotron radiation is steepened by $1/2$. This result can also be obtained in a simple manner as follows. An injected electron with energy E loses its energy in time $\sim t_{1/2} \propto E^{-1}$ (see Equation 3.6). In the steady state, therefore, $n(E) \sim q(E)t_{1/2} \propto E^{-p_0}E^{-1} \propto E^{-(p_0+1)}$, which explains the steepening of the spectrum. For $p_0 < 0$ the integration has to be cut off at some $E_{max} \gg E$, in which case again $n(E) \propto E^{-2}$.

We now consider a one-time injection of electrons that takes place at $t = 0$. In this case $q(E) = C_0 E^{-p_0}\delta(t)$, where $\delta(t)$ is the Dirac delta function and the solution of the diffusion equation is

$$n(E,t) = C_0 E^{-p_0}(1 - C_0 Et)^{p_0-2}, \quad C_0 Et < 1. \tag{3.86}$$

For other values of $C_0 Et$, $n(E,t) = 0$. At a given time t, electrons with energy exceeding $1/(C_0 t)$ would have lost a considerable fraction of their energy, while lower energy electrons would not have sufficient time for energy loss. For $p_0 > 2$ there is smooth steepening with energy to $n(E,t) = 0$ at $E = 1/(C_0 t)$, while for $p < 2$ there is a minimum in $n(E,t)$ at $E = p/(2C_0 t)$ and therefore a cusp at $E = 1/(C_0 t)$.

A related case is an injection that starts at $t = 0$ and ceases after a time t_0. For $t > t_0$ the energy loss of electrons that have a half-life $t_{1/2} > t - t_0$, i.e., $E < 8.5[(t - t_0)/10^9 \text{ yr}]^{-1}(B/1\,\mu\text{G})^{-2}$, may be neglected so that they effectively preserve the injection power law and $n(E) \propto E^{-p_0}$. However, electrons with $t_{1/2} < t - t_0$ lose a considerable fraction of their energy by time t, which leads to steepening of the spectrum as described above and $n(E) \propto E^{-(p_0+1)}$. The result is that the radio spectrum can be approximately described by a power law with the spectral index steepening from $\alpha = (p_0 - 1)/2$ at the low energies to $\alpha = p_0/2$ where the energy loss is significant.

3.8 Relativistic bulk motion

With the advent of very large baseline interferometry (VLBI) in radio astronomy in the late 1960s and early 1970s, it became possible to measure the structure of radio sources on the angular scale of milliarcseconds. One of the early observational results of VLBI was that some compact radio sources were made up of more than one component, and these components were seen to separate at an apparent speed exceeding that of light. A number of possible explanations have been found for such *superluminal motion* after the discovery (see Blandford McKee and Rees 1977 for a review) but the most straightforward and useful model is the one proposed by Rees (1966) *before* the first observations. Rees invoked bulk motion at relativistic speed to explain the very rapid radio variability found in some quasars; a straightforward consequence was the apparent expansion of the source at a speed exceeding that of light. We will now derive a few important consequences of relativistic bulk motion and discuss in Chapter 9 the present observational status.

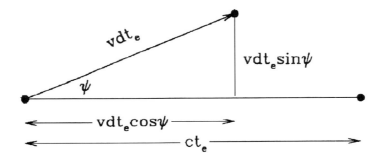

Fig. 3.6. A blob of radiating particles ejected from a stationary source, with speed v at an angle ψ to the line of sight.

3.8.1   *Superluminal motion*

Consider a blob of matter ejected from a stationary source at an angle ψ to the line of sight of a distant observer (see Figure 3.6). The observer sees only the component of the motion normal to the line of sight. Let the angular displacement of the blob observed in a time dt_0 be $d\theta$. In the proper reference frame of the stationary source the angular displacement corresponds to a linear displacement $dl = r_1 S(t_e) d\theta$. Here r_1 is the radial coordinate of the source used in the Robertson–Walker line element (see Section 2.2), and $S(t_e)$ is the cosmological scale factor corresponding to the time of ejection of the blob, t_e. The observed interval dt_0 corresponds to a time interval $dt_e = dt_0/(1 + z)$ in the reference frame of the source. The magnitude of the proper velocity normal to the line of sight in the source frame, as determined by the observer, is

$$v_a = \frac{dl}{dt_e} = r_1 S(t_e) \frac{d\theta}{dt_e} = \mu r_1 S(t_0), \tag{3.87}$$

where $\mu = d\theta/dt_0$ is the *apparent proper motion*, i.e., the observed rate of change of angular separation, and we have used Equation 2.5. Using Equation 2.46 the apparent velocity becomes

$$v_a = \mu \left(\frac{c}{H_0}\right) \frac{1}{q_0^2} \left[\frac{q_0 z + (q_0 - 1)(\sqrt{1 + 2zq_0} - 1)}{1 + z}\right]. \tag{3.88}$$

Introducing $\beta_a = v_a/c$ and practical units this can be written in the useful form

$$\beta_a = \frac{47.2}{h_{100} q_0^2} \left(\frac{\mu}{1\ \mathrm{marcsec\ yr^{-1}}}\right) \left[\frac{q_0 z + (q_0 - 1)(\sqrt{1 + 2q_0 z} - 1)}{1 + z}\right]. \tag{3.89}$$

Here 'marcsec' abbreviates 'milliarcsec'.

A compact component in the quasar 3C 216 has a rate of angular separation 0.11 milliarcsec yr^{-1} from a stationary component (the optical counterpart). At the known quasar redshift of $z = 0.669$ this corresponds to $\beta_a = 2.35/h_{100}$ for an assumed value of $q_0 = 0.5$. Since $\beta_a > 1$, this is a case of superluminal motion. Even larger values of β_a have been observed in many cases (see subsection 9.5.2)

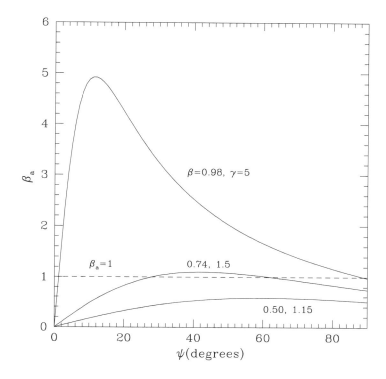

Fig. 3.7. Apparent transverse speed as a function of the angle between blob velocity **v** and the line of sight. Values of β and the Lorentz factor γ are indicated against each curve.

The observed superluminal motion can be explained very simply by considering the consequence of the motion of the radiating blob on the arrival time of signals at the observer. We will first assume that the observer is close enough to the source for cosmological considerations to be neglected. As shown in Figure 3.6, the blob moves at an angle ψ to the line of sight with speed v. Consider signals emitted by the blob in time dt_e in the frame of the stationary source. During this time the blob moves a distance $v dt_e \cos \psi$ in the direction of the observer. Because of the reduction in the distance, the signals emitted over dt_e arrive in a time interval

$$dt_0 = \left(1 - \frac{v}{c} \cos \psi\right) dt_e. \tag{3.90}$$

If the observed angular displacement of the moving component in time dt_0 is $d\theta$, the observer concludes that the transverse speed v_a is given by

$$\beta_a = \frac{v_a}{c} = \frac{\beta \sin \psi}{1 - \beta \cos \psi}. \tag{3.91}$$

For motion of the blob close to the line of sight the apparent transverse speed can exceed the speed of light c, as is shown in Figure 3.7. It is seen that for sufficiently large values of γ superluminal motion occurs even for large inclination angles.

The apparent speed β_a has many interesting properties, some of which are as follows.

- Consider a given value of β. β_a is now a function ψ, and there is a maximum in β_a at $\cos \psi = \beta$, i.e., $\sin \psi = 1/\gamma$, with $\beta_a^{max} = \gamma \beta$. Thus for large superluminal motions the angle ψ is small, i.e., the motion has to be more or less beamed at the observer for it to be seen. When $\beta \to 1$, i.e., γ is very large, the maximum occurs at $\psi \sim 1/\gamma$. The condition for superluminal motion to be obtained at some angle is $\beta_a^{max} > 1$, i.e., $\beta > 1/\sqrt{2}$. The condition for $\beta_a > \beta$ is $0 < \cos \psi < 2\beta/(1+\beta^2)$.
- Consider a given value of β_a. The minimum value of β required to produce this value of β_a occurs at $\psi = \cot^{-1} \beta_a$ and $\beta_{min} = \beta_a/\sqrt{1+\beta_a^2}$. The minimum value of the Lorentz factor is $\gamma_{min} = \sqrt{1+\beta_a^2}$. For fixed β_a, as the speed β increases, the angle ψ at which the given value of β is reached becomes larger. The maximum angle is obtained when $\beta \to 1$ and is $\psi_{max} = 2 \cot^{-1} \beta_a$. ψ_{max} is twice the angle that minimizes β.
- Consider ψ as a fixed parameter. Then β_a increases with β and $\beta_a^{max} = \sin \psi/(1 - \cos \psi)$, which $\to \infty$ as $\beta \to 1$.

When the observer is at a cosmological distance from the source, it is only necessary to multiply the right hand side of Equation 3.90 by a factor $1 + z$ to allow for the cosmological redshift. The condition for observing superluminal motion therefore remains unchanged.

3.8.2   *The distribution of* β_a

Consider a source that emits a blob of matter with a speed β. The direction of motion of the blob will make some angle ψ with the line of sight to an observer, who will measure an apparent transverse speed β_a, as given in Equation 3.91. Let a number of sources emit blobs with the same speed β in random directions relative to the observer. The probability distribution of the angles made with the line of sight to the observer is

$$p(\psi)\, d\psi = \tfrac{1}{2} \sin \psi \, d\psi, \quad 0 < \psi < \pi. \tag{3.92}$$

Therefore the probability of obtaining an angle ψ lying in a range of angles (ψ_1, ψ_2) is given by

$$p(\psi_1, \psi_2) = \tfrac{1}{2}(\cos \psi_2 - \cos \psi_1). \tag{3.93}$$

Now given a value of β the probability $p(> \beta_a)$ of obtaining an apparent transverse speed that exceeds some value β_a is given by Equation 3.93, ψ_1 and ψ_2 being the two values of the angle of inclination between which the apparent transverse speed exceeds β_a. Equation 3.91 can be expressed as a quadratic in $\cos \psi$, using which it is very simple to show (Cawthorne *et al.* 1986a, see the *Erratum* in Cawthorne *et al.* 1986b) that

$$p(> \beta_a) = \frac{1}{2}(\cos \psi_2 - \cos \psi_1) = \frac{1}{1 + \beta_a^2} \sqrt{1 - \frac{\beta_a^2}{\gamma^2 - 1}}. \tag{3.94}$$

The probability is maximized when $\beta \to 1$, with

$$p_{\max}(\beta_a) = \frac{1}{1 + \beta_a^2}, \tag{3.95}$$

and we can obtain an upper limit on $p(> \beta_a)$ even when the true speed β is not known.

3.8.3 Flux amplification

When a source consisting of radiating particles moves at relativistic speed towards the observer, the observed flux is Doppler boosted to levels that can be far higher than the flux observed when the source is stationary. However, when the source moves away the observed flux can be greatly reduced, so that features that are intrinsically symmetric, such as a two-sided radio jet, can appear to be one-sided.

Consider a source of radiation which moves towards the observer as in Figure 3.6. To obtain an expression for the Doppler boosting of the flux we need to perform a Lorentz transformation between the rest frame of the source and that of the observer (cosmological factors will be introduced later). If v and v' are the observer-frame and rest frame frequencies respectively, we have (see RL79, p. 121)

$$v = \mathscr{D}v', \tag{3.96}$$

where the Doppler boosting factor $\mathscr{D}$ is given by

$$\mathscr{D} = \frac{1}{\gamma(1 - \beta \cos \psi)}. \tag{3.97}$$

If $I(v)$ is the intensity of radiation, it can be shown that $I(v)/v^3$ is a Lorentz invariant (see RL79, p. 146). It follows that the transformation law for the intensity is

$$I(v) = \mathscr{D}^3 I'(v'). \tag{3.98}$$

For an unresolved source that is optically thin, or optically thick and spherically symmetric, the flux $F(v) \propto I(v)$ and transforms in the same way as the intensity. For a source with a power law spectrum $F(v) \propto v^{-\alpha}$ the transformation can be written as

$$F(v) = \mathscr{D}^{3+\alpha} F'(v), \tag{3.99}$$

which shows the boosting of the observed flux over the rest frame flux. The luminosities in the observer frame and rest frame are related in the same manner as the corresponding fluxes.

When the source is at a redshift z the frequency transformation in Equation 3.96 is replaced by

$$v = \frac{\mathscr{D}v'}{1 + z}. \tag{3.100}$$

Consequently it follows from Equations 3.98 and 3.99 that

$$F(v) = \left(\frac{\mathscr{D}}{1 + z}\right)^{3+\alpha} F'(v). \tag{3.101}$$

Using this transformation and Equation 2.61, the intrinsic luminosity of a relativistically moving source at redshift z is given in terms of the observed flux by

$$L'(v') = \left(\frac{\mathscr{D}}{1+z}\right)^{(3+\alpha)} \frac{4\pi D_{\mathrm{L}}^2}{1+z} F(v). \tag{3.102}$$

Now consider two identical sources that move in opposite directions with the same speed v and with line-of-sight angles ψ and $\pi + \psi$ respectively. If F_{in} is the observed flux from the source moving towards the observer and F_{out} is the flux from the source moving away, we have

$$\frac{F_{\mathrm{in}}}{F_{\mathrm{out}}} = \left(\frac{1+\beta\cos\psi}{1-\beta\cos\psi}\right)^{3+\alpha}. \tag{3.103}$$

The boosting formula is somewhat different when the source is a jet-like feature (which may be considered to be made up of a number of unresolved components). In such a case the observed flux is obtained by integrating over the area of the feature. Following Blandford (1990) we can write

$$F(v) = \int I(v)\,d\Omega\,dl = \int j_v \frac{da\,dl}{d_{\mathrm{A}}^2}, \tag{3.104}$$

where j_v is the emissivity, $d\Omega$ is an element of solid angle, da is an area element normal to the line of sight and d_{A} is the distance to the source.[1] The emissivity transforms as $j_v = \mathscr{D}^2 j'(v')$, which, for a power law spectrum, becomes $j_v = \mathscr{D}^{2+\alpha} j'(v)$. Introducing the volume element $dV = dA\,dl$, the transformation law for the flux becomes

$$F(v) = \frac{\mathscr{D}^{2+\alpha}}{d_{\mathrm{A}}^2} \int j'(v)\,dV = \mathscr{D}^{2+\alpha} F'(v), \tag{3.105}$$

and the ratio of the flux from an advancing jet to an identical receding jet is

$$\frac{F_{\mathrm{in}}}{F_{\mathrm{out}}} = \left(\frac{1+\beta\cos\psi}{1-\beta\cos\psi}\right)^{2+\alpha}. \tag{3.106}$$

The flux ratio for the point source and jet cases is shown in Figure 3.8. Given an observed flux ratio, it is not possible to determine the value of β as well as the inclination angle. However, assuming the extreme relativistic case $\beta = 1$, an upper limit to the angle of inclination can be obtained:

$$\psi \le \frac{R^{1/(2+\alpha)} - 1}{R^{1/(2+\alpha)} + 1}, \quad R = \frac{F_{\mathrm{in}}}{F_{\mathrm{out}}}. \tag{3.107}$$

The Doppler boosting factor $\mathscr{D}$ increases monotonically as the angle of inclination decreases from 90 deg to 0 deg, with $\mathscr{D}_{\mathrm{min}} = \mathscr{D}(90\,\mathrm{deg}) = 1/\gamma$ and $\mathscr{D}_{\mathrm{max}} = \mathscr{D}(0\,\mathrm{deg}) =$

---

[1] The relation between solid angle and area element can be preserved over cosmological distances by interpreting d_{A} as the angular diameter distance, as explained in Section 2.8.

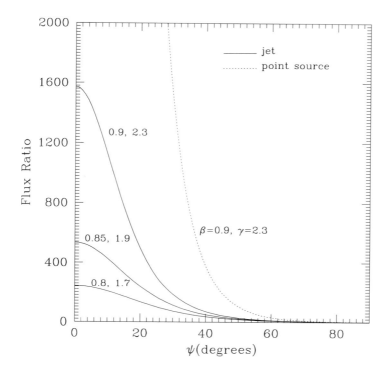

Fig. 3.8. The observed flux ratio of advancing and receding jets for $\beta = 0.8$–0.9 and for the point source case with $\beta = 0.9$. A power law spectrum with $\alpha = 0.5$ is assumed.

$\gamma(1 + \beta)$. Some useful relations which connect γ, $\mathscr{D}$, ψ and the apparent speed β_a are

$$\gamma = \frac{\beta_a^2 + \mathscr{D}^2 + 1}{2\mathscr{D}}, \tag{3.108}$$

$$\tan \psi = \frac{2\beta_a}{\beta_a^2 + \mathscr{D}^2 - 1}, \tag{3.109}$$

$$\frac{1 + \beta \cos \psi}{1 - \beta \cos \psi} = \beta_a^2 + \mathscr{D}^2. \tag{3.110}$$

These relations have been used by Ghisellini *et al.* (1993) to put limits on the Lorentz factor and the angle of inclination using observable quantities.

3.8.4 Variability

Consider a source in which the flux density changes over a time $\sim \Delta t_{\mathrm{var}}$. By the usual argument that relates variability to source size, the linear size of the source is $\sim c\Delta t_{\mathrm{var}}$. If the source is at a distance D from the observer, the angular size is $\theta \sim c\Delta t_{\mathrm{var}}/D$. If $F(\nu)$ is the flux density of the source, using Equation 3.59 the *variability brightness temperature* is defined by

$$T_{\text{var}} = \frac{F(v)D^2}{2kv^2\Delta t_{\text{var}}^2}. \tag{3.111}$$

In the case of many sources with observed rapid variability, the estimated brightness temperature exceeds the temperature at which catastrophic energy loss through the inverse Compton process begins to dominate (see Section 4.4). In fact it was this problem of rapid variability that led Rees (1966) to propose relativistic motion within the source. To see how this helps, one needs only to notice that if the source is in internal motion, it is not the observed time scale Δt_{var} that should go into determining the limit on the size of the emitting region. The appropriate quantity is the variability time scale in the rest frame of the source, given by

$$\Delta t_{\text{var}}' = \mathscr{D}\Delta t_{\text{var}} \tag{3.112}$$

where $\mathscr{D}$ is the Doppler boosting factor introduced in Equation 3.97. The source size is therefore $R \sim c\Delta t_{\text{var}}' \sim c\mathscr{D}\Delta t_{\text{var}}$, and the solid angle subtended at the source is $\sim (c\mathscr{D}\Delta t_{\text{var}}/D)^2$. The observed flux is then given by

$$F(v) = I(v)\left(\frac{c\mathscr{D}\Delta t_{\text{var}}}{D}\right)^2, \tag{3.113}$$

where $I(v)$ is the intensity in the observer's frame. Substituting this in Equation 3.111, using the transformations in Equations 3.96 and 3.98 and introducing the rest frame brightness temperature $T' = I'(v')c^2/(2k^2v'^2)$ we get

$$T_{\text{var}} = \mathscr{D}^3 T'. \tag{3.114}$$

Now for a source with superluminal motion we have for the angle of inclination $\sin\psi \sim 1/\gamma$ so that $\mathscr{D} = \gamma^{-1}(1 - \beta\cos\psi)^{-1} \sim \gamma$. A Doppler factor of ~ 10, which is consistent with the statistics of superluminal motion (see subsection 9.5.2), therefore implies a rest frame brightness temperature below the inverse Compton limit for $T_{\text{var}} \gtrsim 10^{15}$ K.

Qian *et al.* (1991) have observed variability with time scales of ~ 1 day at 5 GHz in the flux density of the quasar 0917+624. The amplitude of variation is ~ 20 per cent. Additional variability at shorter time scales of 0.3 day has been observed at higher frequencies. Such variability implies that $T_{\text{var}} \gtrsim 2\times10^{18}$ and a Doppler factor $\gtrsim 100$ is required to avoid the Compton catastrophe. The Lorentz factors needed are therefore very much larger than the estimates for more normal sources and unless the quasar is highly unusual (a few other cases of such rapid variability are also known), some other mechanism has to be invoked to explain the observations. Qian *et al.* have interpreted the variability in terms of the apparent superluminal motion of a relativistic shock through inhomogeneous structures along a jet beamed towards the observer (see Section 8.7 for further discussion).

3.8.5   Synchrotron self-Compton emission

Synchrotron photons emitted by energetic electrons can undergo inverse Compton scattering from the same set of electrons, acquiring in the process $\sim \gamma^2$ times the energy before the scattering, where γ is the Lorentz factor of the scattering electron. Depending on the value of γ, radio and IR photons can be scattered to X-ray and γ-ray energies through this process, which could therefore be responsible for emission of high energy photons by AGN and quasars. The ratio of the rates at which the electrons lose energy to the synchrotron and Compton processes is equal to the ratio of the energy density of the magnetic field to the energy density of the radiation (see subsection 4.2.1). The self-Compton emission is therefore important when the source is compact, which is the situation in which synchrotron self-absorption is observed. One therefore expects a simple relationship between the self-absorbed radio emission and inverse Compton emission. However, the relationship depends on the physical parameters and geometry of the emitting region in a complicated manner and is also affected by relativistic bulk motion.

The simplest situation is where the source is uniform and spherical, and the electrons have a power law distribution of energy with number density $n(E) \propto E^{-(2\alpha+1)}$, where E is in the range (E_1, E_2) and α is as usual the power law index of the synchrotron emission at frequencies where self-absorption can be neglected. The inverse Compton flux density emitted at energy E is then given by

$$F^{SC}(E) = f(\alpha) \left(\frac{\theta}{1\,\text{milliarcsec}}\right)^{-2(2\alpha+3)} \left(\frac{E}{1\,\text{keV}}\right)^{-\alpha} \left(\frac{\nu_a}{1\,\text{GHz}}\right)^{-(3\alpha+5)}$$

$$\times \left[\frac{F(\nu_a)}{1\,\text{Jy}}\right]^{2(\alpha+2)} \left(\frac{1+z}{\mathscr{D}}\right)^{2(\alpha+2)} \ln\left(\frac{\nu_2}{\nu_a}\right), \tag{3.115}$$

where θ is the angular size of the source, z is the redshift, $F(\nu_a)$ is the flux at the self-absorption frequency ν_a and $\mathscr{D}$ is the Doppler boosting factor given in Equation 3.97 (see Marscher 1987, Ghisellini et al. 1993 and references there; the derivation essentially uses the results of subsection 4.2.1 and subsection 3.5.1.) In this equation all quantities except $\mathscr{D}$ are in principle observable and therefore it can be inverted to obtain $\mathscr{D}$. Equation 3.115 has been derived assuming that the source is a moving sphere, but an analogous expression is available for other geometries. Ghisellini et al. (1993) have discussed the case of a continuous jet, and the Doppler factor $\mathscr{D}_j$ obtained in this case is related to $\mathscr{D}$ in Equation 3.115 through

$$\mathscr{D}_j = \mathscr{D}^{(2\alpha+4)/2\alpha+3)}. \tag{3.116}$$

3.8.6   Beaming and luminosity functions

Beaming amplifies the observed flux, i.e., the observed source luminosity, in an orientation-dependent manner. If a population of sources all having the same intrinsic luminosity is subject to beaming, the sources will appear to be distributed over a range of luminosity. The luminosity function, i.e., the comoving number density of sources as a function of luminosity, is therefore spread out from its original delta function form. The effect of beaming on luminosity functions was studied in detail by Urry and Shafer (1984) and Padovani and Urry (1992). We shall consider a somewhat simplified version of their treatment here and refer the reader to the original papers and a review by Urry and Padovani (1995) for further details.

Consider sources in the emitting region, all having the same luminosity $\mathscr{L}$ and in relativistic motion in directions distributed at random. The observed luminosity is given by

$$L = \mathscr{D}^m \mathscr{L}, \quad \mathscr{D} = \frac{1}{\gamma(1 - \beta \cos \psi)}, \tag{3.117}$$

in our usual notation, with $m = 2 + \alpha$ for a jet and $m = 3 + \alpha$ for a single emitting blob (see Equations 3.99 and 3.104). The range of the observed luminosity as ψ varies from 0 to $\pi/2$ is $(2\gamma)^{-m}\mathscr{L} < L < (2\gamma)^m\mathscr{L}$. Consider sources with a fixed value of γ. Let $P(\mathscr{D})\,d\mathscr{D}$ be the probability of obtaining a beaming factor in the range $(\mathscr{D}, \mathscr{D} + \mathscr{D})$. For randomly oriented sources this is given by

$$P(\mathscr{D})\,d\mathscr{D} = \sin \psi \, d\psi, \tag{3.118}$$

with $\mathscr{D}$ and ψ related through Equation 3.117. Let $P(L|\mathscr{L})\,dL$ be the probability of obtaining an observed luminosity in the range $(L, L + dL)$ due to beaming of the intrinsic luminosity $\mathscr{L}$. From Equation 3.117 these probabilities are related by

$$P(L|\mathscr{L})dL = P(\mathscr{D})d\mathscr{D}. \tag{3.119}$$

Using Equations 3.117 and 3.118 we then get

$$P(L|\mathscr{L}) = \left(\frac{1}{\beta\gamma m}\right) \mathscr{L}^{1/m} L^{-(m+1)/m}. \tag{3.120}$$

The distribution of the observed luminosity is therefore a power law, with index in the range ~ 1–1.5 for the usual values of m.

When the intrinsic luminosity $\mathscr{L}$ is distributed according to some function $\Phi_{\text{int}}(\mathscr{L})$ with $\mathscr{L}_{\min} < \mathscr{L} < \mathscr{L}_{\max}$, the distribution of observed luminosity is given by

$$\Phi_{\text{obs}}(L) = \int d\mathscr{L}\, P(L|\mathscr{L})\Phi_{\text{int}}(\mathscr{L}), \tag{3.121}$$

with the limits depending on L, γ, $\mathscr{L}_{\min}$ and $\mathscr{L}_{\max}$. To see how these are fixed, we will consider the simple case of a power law intrinsic luminosity function $\Phi_{\text{int}}(\mathscr{L}) = K\mathscr{L}^{-q}$, $q > 1$, $\mathscr{L}_{\min} < \mathscr{L} < \infty$. Given a small bin of observed luminosity around L, the range of $\mathscr{L}$ that can contribute to it is $L/(2\gamma)^m < \mathscr{L} < (2\gamma)^m L$. The observed luminosity function is therefore given by

$$\Phi_{\text{obs}}(L) = \left(\frac{K}{\beta\gamma\, m}\right) L^{-(m+1)/m} \int_{\mathscr{L}_1}^{(2\gamma)^m L} d\mathscr{L}\, \mathscr{L}^{-q+1/p}, \tag{3.122}$$

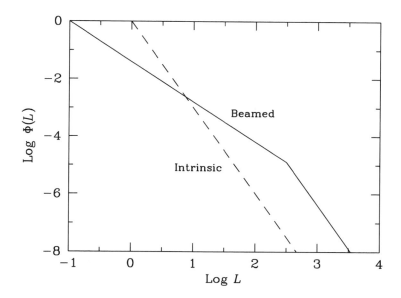

Fig. 3.9. The broken power law luminosity function (solid line) produced by relativistic beaming. The intrinsic luminosity function (broken line) is $\propto \mathscr{L}^{-3}$. The beamed emission is assumed to originate in a jet with $\gamma = 5$ and spectral index 0.5. Arbitrary units are used along the two axes.

with the lower limit $\mathscr{L}_1 = \max(\mathscr{L}_{\min}, L/(2\gamma)^m)$. The result of the integration is that $\Phi(L) \propto L^{-q}$ for $L > (2\gamma)^m \mathscr{L}_{\min}$ and $\Phi(L) \propto L^{-(m+1)/m}$ for $L < (2\gamma)^m \mathscr{L}_{\min}$. At high luminosities a whole range of $\mathscr{L}$ contributes to the observed luminosity function, and therefore one simply gets a scaling from $\mathscr{L}$ to L and the slope of the intrinsic luminosity function is preserved. For sufficiently small values of L the lower limit of integration becomes independent of L, and the main contribution comes from sources around $\mathscr{L}_{\min}$. The observed slope therefore takes on the same value as in Equation 3.120. When $L = (2\gamma)^m \mathscr{L}_{\min}$ the upper and lower limits of integration become equal and therefore there are no observed values smaller than this limit. The effect of the beaming is therefore to produce a power law luminosity function with a break, which is flattened at the low luminosity end and has the same slope as the intrinsic function at high luminosities. These features are found even when the intrinsic luminosity function is more complex than a power law. An example of the 'broken' power law produced from a single power law is shown in Figure 3.9. We shall consider the application of beamed luminosity functions to the unification of different kinds of AGN in Chapter 12.

When beamed and unbeamed components are both present, the observed luminosity is

$$L = \mathscr{L}_u + L_j = \mathscr{L}_u(1 + f\mathscr{D}^m), \qquad (3.123)$$

where $\mathscr{L}_u$ is the isotropic component, from extended lobes say, L_j is the beamed component and it is assumed that its intrinsic luminosity is a fraction f of $\mathscr{L}_u$ (Urry and Padovani 1995). The intrinsic luminosity function is taken to be a function of $\mathscr{L}_u$ and the observed function obtained from Equation 3.121.

The concept of bulk relativistic motion was first introduced in radio astronomy to account for problems associated with rapid variability and led to the prediction of apparent superluminal motion, which was subsequently observed. While there are other plausible explanations for this superluminal motion, relativistic motions provide the most natural explanation and remain valid in a variety of situations. In radio astronomy, relativistic motion manifests itself in the form of jet-like structures on the milliarcsecond scale, which corresponds to parsec scale linear sizes. But it can be argued that even the one-sided kiloparsec scale jets seen in radio quasars are relativistic and that beaming has an important role to play in determining the morphology of radio sources as perceived by the observer. In the X-ray and γ-ray domain too, observed properties imply the existence of bulk relativistic motion. We shall return to these issues elsewhere in the book.

4 Radiative processes – II

We shall consider in this chapter the scattering of photons by electrons and thermal and non-thermal pair plasmas. Some of the processes covered here are closely related to those in Chapter 3 and the division between the chapters is made only to keep each at a manageable length.

4.1 Thomson scattering

We shall review in this section some important facts about the scattering of electromagnetic radiation by electrons. As in the case of synchrotron radiation, we will only provide a brief summary of the process relevant to the study of AGN, and refer the reader to J88, BG70 and RL79 for a more detailed treatment.

Consider the scattering of an electromagnetic wave incident on an electron, with photon energy $h\nu \ll mc^2$. In this domain the process can be treated classically, and is known as Thomson scattering. The electromagnetic wave sets the electron oscillating, which then radiates as per the Larmor formula, which is applicable for electron velocity $v \ll c$. In Thomson scattering there is no change in the frequency of the scattered radiation, i.e., the scattering is elastic, which is its most important characteristic.

For incident radiation that is *linearly polarized*, the differential scattering cross-section is given by

$$\left(\frac{d\sigma_T}{d\Omega}\right)_{pol} = r_0^2 \sin^2 \Theta, \quad r_0 = \frac{e^2}{mc^2}, \tag{4.1}$$

where Θ is the angle between the direction of scattering and the polarization vector of the incident wave, and r_0 is the *classical electron radius*. The total scattering cross-section summed over all angles Θ is

$$\sigma_T = \frac{8\pi r_0^2}{3} = 6.65 \times 10^{-25} \, cm^2. \tag{4.2}$$

When the incident radiation is unpolarized, it can be considered to be made up of two linearly polarized components, with polarization vectors normal to each other. The differential scattering cross-section can then be obtained by treating each component separately and summing the two cross-sections, and is given by

$$\left(\frac{d\sigma_T}{d\Omega}\right)_{unpol} = \frac{r_0^2}{2}(1 + \cos^2 \theta), \tag{4.3}$$

where θ is the angle between the incident and the scattered radiation. Since the radiation is not polarized, there is now no reference to Θ. The total cross-section is given by integrating over all θ and again leads to the value on the right hand side of Equation 4.2. There is no difference in the total scattering cross-section for polarized and unpolarized radiation because classically the electron has no intrinsically defined direction.

When unpolarized radiation is Thomson scattered by an electron, it becomes partially linearly polarized, the degree of polarization being given by

$$\Pi = \frac{1 - \cos^2 \theta}{1 + \cos^2 \theta}. \tag{4.4}$$

If the incident radiation is polarized in the direction $\mathbf{e}$, then radiation scattered in the direction $\mathbf{n}$ is polarized in the direction $\mathbf{n} \times (\mathbf{n} \times \mathbf{e})$.

4.2   Compton scattering

When the scattering of electromagnetic radiation is viewed as the scattering of photons by electrons, it is evident that energy and momentum will be exchanged between the interacting particles. An electron at rest that scatters a photon will gain energy from the event, as it acquires a recoil velocity to satisfy momentum conservation. As a result, there is a decrease in the energy of the photon, i.e., its wavelength increases. This is known as *Compton scattering*.

A simple calculation using the conservation of relativistic four-momentum shows that

$$\epsilon_1 = \frac{\epsilon}{1 + \frac{\epsilon}{mc^2}(1 - \cos \theta)}, \tag{4.5}$$

where ϵ and ϵ_1 indicate the energies of the incident and scattered photon respectively, and θ as before is the angle between the incident and scattered directions (see Figure 4.1). Since we begin with an electron at rest, the photon always loses energy. When the incident photon energy $\epsilon \ll mc^2$, we get $\epsilon_1 \simeq \epsilon$, which is just Thomson scattering: it becomes applicable because the energy of the photon is so small that the recoil of the electron can be neglected.

The differential cross-section for Compton scattering of unpolarized radiation, obtained by using a quantum electrodynamic treatment, is given by the Klein–Nishina formula

$$\frac{d\sigma_{KN}}{d\Omega} = \frac{r_0^2}{2} \left(\frac{\epsilon_1}{\epsilon}\right)^2 \left(\frac{\epsilon}{\epsilon_1} + \frac{\epsilon_1}{\epsilon} - \sin^2 \theta\right), \tag{4.6}$$

while the total cross-section is

$$\sigma_{KN} = \frac{3\sigma_T}{4} \left\{ \frac{1+x}{x^3} \left[\frac{2x(1+x)}{1+2x} - \ln(1+2x) \right] + \frac{\ln(1+2x)}{2x} - \frac{1+3x}{(1+2x)^2} \right\}, \tag{4.7}$$

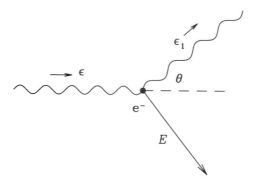

Fig. 4.1. A photon of energy ϵ is scattered to an energy ϵ_1 by an electron at rest. θ is the angle between the directions of the incident and the scattered photon. E is the energy acquired by the electron in the scattering.

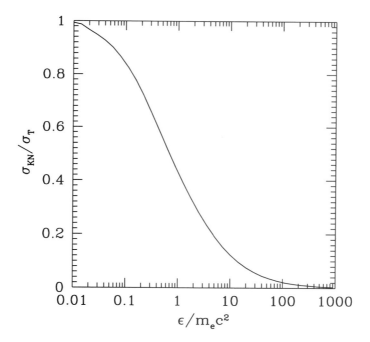

Fig. 4.2. The variation of the Compton scattering cross-section with energy.

with $x = \epsilon/(mc^2)$. For $x \ll 1$, Thomson scattering applies and $\epsilon \simeq \epsilon_1$. In this case the Klein–Nishina cross-section reduces to the Thomson cross-section. The variation of the scattering cross-section with energy is shown in Figure 4.2. Notice that σ_{KN} falls rapidly as the electron energy increases to $\gg m_e c^2$.

Inverse Compton scattering   When the electron that scatters the radiation is already in motion, energy can pass either from the electron to the photon or vice versa, depending upon the kinematical details of the collision. When the photon gains energy, the process is called *inverse Compton scattering*, though it is basically no different from Compton scattering. The general expression for the change in energy of the photon is rather complex (see Felten and Morrison 1966 for a detailed discussion), but it is possible to arrive at some very useful approximations using Lorentz transformations.

Consider a collision between a photon of energy ϵ and an electron with Lorentz factor γ in the laboratory frame S. Suppose the event is viewed in a Lorentz frame S' in which the electron is at rest before the collision. Energy transforms as the *time component* of the energy–momentum four-vector, so that the energy ϵ' of the photon, in the frame S', before scattering is

$$\epsilon' = \gamma\epsilon(1 - \beta\cos\theta), \tag{4.8}$$

where θ is the angle between the incident electron and photon directions in the laboratory frame S. For $\beta \simeq 1$, the range of photon energy in S' is $\epsilon/2\gamma \lesssim \epsilon' \lesssim 2\gamma\epsilon$, with the maximum occurring for head-on collisions. In frame S' the collision appears to be occurring between an electron at rest and a photon of energy $\epsilon' \simeq \gamma\epsilon$ for all but very small angles θ. Now suppose $\epsilon' \ll mc^2$, i.e., in the rest frame of the electron the photon has negligible energy. The event in S' can then be viewed as *Thomson scattering*, so that the energy of the scattered photon in S' is $\epsilon'_1 \simeq \epsilon'$. We can transform back to the laboratory frame using an expression similar to Equation 4.8 and the energy of the scattered photon in S is $\epsilon_1 \simeq \gamma^2\epsilon$. The energy of the inverse Compton scattered photon is therefore increased by a factor γ^2, which can be very large for a highly relativistic electron. The condition of applicability of the approximation used is that the incident photon energy $\epsilon \ll mc^2/\gamma$. This means that the fraction of energy lost by the electron in a single collision $\gamma^2\epsilon/(\gamma mc^2) \ll 1$. Since the total energy must be conserved in the collision, there is an upper limit to the increase in the photon energy: $\Delta\epsilon < \gamma mc^2$.

When the condition for Thomson scattering is not fulfilled in S', i.e., the incident photon has very high energy relative to the electron rest mass, as seen in S' it loses some of its energy after scattering. The transformation back to the laboratory frame therefore produces less energy than in the Thomson case for the same scattering angle. However, the energy lost by the electron can now be a large fraction of its energy before the collision. In the extreme Klein–Nishina limit $\epsilon' \simeq \gamma\epsilon \gg mc^2$ and it follows from Equation 4.5 that $\epsilon'_1 \simeq mc^2$. In the laboratory frame, therefore, the energy of the scattered photon is $\epsilon_1 \simeq \gamma mc^2$, i.e., $\Delta\epsilon/(\gamma mc^2) \simeq 1$ and the photon carries away a sizeable fraction of the electron energy. An electron undergoing such collisions therefore loses its energy in several discrete steps.

4.2.1   *Power from single Compton scattering*

When radiation is incident on an ensemble of relativistic electrons, each photon of energy ϵ that is scattered acquires on an average the energy $\sim \gamma^2\epsilon$, and the region in which the interaction takes place becomes a source of high energy radiation. The luminosity and spectrum produced depend on the nature and number of scattering events taking place per unit time per unit volume, which in turn depends on the spectrum and number of incident photons and electrons and the geometry of the source. The energy gained by the photons is balanced by the energy lost by the electrons. We will now discuss the characteristics of photons emerging from a medium which is optically thin, so that the photons undergo a single scattering event before escaping from the source.

A method for obtaining the Compton power generated by a single electron was developed by Blumenthal and Gould (1970) (see also the discussion in RL79). Suppose $n(\epsilon)$ is the number density of photons of energy ϵ in the laboratory frame S. In the rest frame S' of a given electron this appears as a function $n'(\epsilon')$ that is related to $n(\epsilon)$ through a Lorentz transformation. If we assume that $\gamma \epsilon \ll mc^2$, then in S' the energy of the photon remains almost unchanged after the scattering and the Thomson scattering cross-section can be used. The number of photons in the energy range $(\epsilon', \epsilon' + d\epsilon')$ that are scattered by the electron per unit time is then $c\sigma_T n'(\epsilon')d\epsilon'$. The power in the scattered photons, i.e., the energy carried away by them per unit time in the frame S' is

$$P'_{\text{scat}}(\epsilon'_1) = c\sigma_T \int \epsilon'_1 n'(\epsilon')\, d\epsilon', \qquad (4.9)$$

where ϵ'_1 is the energy of the scattered photon. Now the power and the quantity $n(\epsilon')d\epsilon'/\epsilon'$ are relativistically invariant (see RL79, p. 199) and $\epsilon'_1 \simeq \epsilon'$, so that in the laboratory frame the above equation becomes

$$P_{\text{scat}}(\epsilon) = c\sigma_T \gamma^2 \int (1 - \beta \cos\theta)^2 \epsilon n(\epsilon)\, d\epsilon, \qquad (4.10)$$

where we have used Equation 4.8. For an isotropic distribution of incident photons the average scattered power is obtained by integrating over the distribution in Equation 3.9. Using the same symbol as before for the power after averaging we get

$$P_{\text{scat}}(\epsilon) = c\sigma_T \gamma^2 \left(1 + \tfrac{1}{3}\beta^2\right) U_{\text{ph}}, \quad U_{\text{ph}} = \int \epsilon n(\epsilon) d\epsilon \qquad (4.11)$$

where U_{ph} is the total energy density of the electromagnetic radiation. The energy incident on the electron per unit time is $c\sigma_T U_{\text{ph}}$, so that the Compton power, which is the rate of the net energy output, is

$$P_{\text{Comp}} = \tfrac{4}{3}\sigma_T c \beta^2 \gamma^2 U_{\text{ph}}. \qquad (4.12)$$

The total number of incident photons per unit time is $c\sigma_T n_{\text{tot}}$, where $n_{\text{tot}} = U_{\text{ph}}/\langle \epsilon \rangle$ and $\langle \epsilon \rangle$ is the average incident photon energy. The average energy of the scattered photons for $\beta \simeq 1$ is therefore

$$\langle \epsilon_1 \rangle = \tfrac{4}{3}\gamma^2 \epsilon. \qquad (4.13)$$

The expression 4.12 for Compton power is strikingly similar to the synchrotron power in Equation 3.10. The similarity arises because synchrotron emission can be considered to be the scattering by an electron of the virtual photons of the magnetic field. An ensemble of electrons in a magnetic field can radiate by the synchrotron process, or lose energy through inverse Compton scattering of the ambient electromagnetic radiation, which can include the synchrotron photons. The ratio of the power generated in the two processes is

$$\frac{P_{\text{Comp}}}{P_{\text{sync}}} = \frac{U_{\text{ph}}}{U_B}.$$

(4.14)

When the Thomson approximation is not applicable in S', the calculation of the Compton power becomes rather involved because now the differential Klein–Nishina cross-section has to be used, and this is dependent on photon energy and the scattering angle in a complicated way. For an isotropic photon distribution it can be shown that (see Blumenthal and Gould 1970)

$$P_{\text{Comp}} = \tfrac{4}{3}\sigma_{\text{T}} c \beta^2 \gamma^2 U_{\text{ph}} \left(1 - \frac{63}{10} \frac{\gamma \langle \epsilon^2 \rangle}{mc^2 \langle \epsilon \rangle} + \cdots \right),$$

(4.15)

where the angle brackets indicate mean values. It should be noted that the luminosity can be negative, i.e., energy can pass from the photons to the electrons. When the incident radiation has the black body form, the number density of photons is

$$n(\epsilon) = \frac{\epsilon^2}{\pi^2 (\hbar c)^3 (e^{\epsilon/kT} - 1)},$$

(4.16)

and the correction term becomes equal to $152.1 \gamma kT/(mc^2)$.

In the extreme Klein–Nishina limit $\gamma \epsilon/(mc^2) \gg 1$, the expression for the power radiated by a single electron is (Blumenthal and Gould 1970)

$$P_{\text{Comp,KN}} \simeq \frac{3\sigma_{\text{T}} m^2 c^5}{8} \int \frac{n(\epsilon)}{\epsilon} \left(\ln \frac{4\epsilon\gamma}{mc^2} - \frac{11}{6} \right) d\epsilon,$$

(4.17)

which for the black body form of radiation gives

$$P_{\text{Comp,KN}} \simeq \frac{\sigma_{\text{T}} (mckT)^2}{16\hbar^3} \left(\ln \frac{4\gamma kT}{mc^2} - 1.981 \right).$$

(4.18)

4.2.2   *Spectrum from single Compton scattering*

When a photon of energy ϵ is scattered by an electron with Lorentz factor γ, the energy of the emergent photon depends on the incident and scattering angles, as does the differential Klein–Nishina cross-section, which gives the probability of scattering in a specific direction. It is therefore possible to produce a whole spectrum of photons starting with photons and electrons each with just one energy.

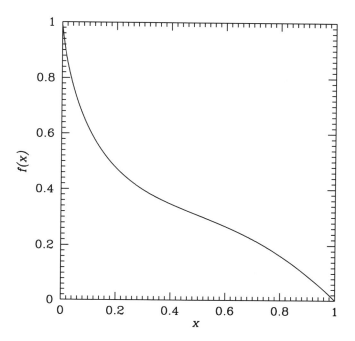

Fig. 4.3. The function $f(x)$, $x = \epsilon_1/(4\epsilon\gamma^2)$ which is used in describing the energy dependence of the spectrum of photons that have undergone a single Compton scattering (see Equation 4.20).

Consider the scattering of photons with energy in the range $(\epsilon, \epsilon + d\epsilon)$ by an electron with energy γmc^2. When $\gamma\epsilon \ll mc^2$ and the incident radiation is isotropic, the number of photons of energy ϵ_1 produced per unit range of ϵ_1 per unit time is

$$\frac{dN(\epsilon_1; \epsilon, \gamma)\, d\epsilon}{dt\, d\epsilon_1} = \frac{3\sigma_T c}{4\gamma^2\epsilon} n(\epsilon) d\epsilon f\left(\frac{\epsilon_1}{4\epsilon\gamma^2}\right), \qquad (4.19)$$

Here $n(\epsilon)$ is the number density of photons, and the function f is given by (Blumenthal and Gould 1970)

$$f(x) = 2x \ln x + x + 1 - 2x^2. \qquad (4.20)$$

$f(x)$ decreases monotonically from $f(0) = 1$ to $f(1) = 0$, with $x = 1$ corresponding to the maximum photon energy produced in the scattering. The function $f(x)$ is shown in Figure 4.3.

The Compton photon spectrum can be obtained by integrating the above expression for scattering from one electron over the entire electron energy distribution. A very useful form is the power law in Equation 3.18. In this case the integration extends over the range from

$$\gamma_{\min} = \max \left[\frac{1}{2} \left(\frac{\epsilon_1}{\epsilon} \right)^{1/2}, \gamma_1 \right] \tag{4.21}$$

to

$$\gamma_{\max} = \min \left[\gamma_2, \frac{1}{2} \left(\frac{\epsilon_1}{\epsilon} \right)^{1/2} \right], \tag{4.22}$$

where $\gamma_1 = E_1/(mc^2)$ and $\gamma_2 = E_2/(mc^2)$ are the ends of the electron distribution. If it is assumed that $\gamma_1 \ll \frac{1}{2}(\epsilon_1/\epsilon)^{1/2} \ll \gamma_2$, the limits of integration can be extended over the range $(0, \infty)$ of γ and the energy spectrum of the scattered photons is

$$P_{\text{Comp}}(\epsilon_1) = \frac{3\sigma_T cC}{8} \left[\frac{2^{p+3}(p^2 + 4p + 11)}{(p+3)^2(p+1)(p+5)} \right] \epsilon_1^{-(p-1)/2} \int_0^\infty \epsilon^{(p-1)/2} n(\epsilon) \, d\epsilon, \tag{4.23}$$

which is a power law, whatever be the spectrum of the incident photons. The spectrum obtained therefore has the same form as in the case of the synchrotron spectrum produced by power law electrons. The similarity arises because in both processes the energy of the photon produced is $\propto \gamma^2$. When the incident photon spectrum has the black body form of Equation 4.16, then

$$\int \epsilon^{(p-1)/2} n(\epsilon) \, d\epsilon = \frac{(kT)^{(p+5)/2}}{\pi^2(\hbar c)^3} \Gamma \left(\frac{p+5}{2} \right) \zeta \left(\frac{p+5}{2} \right), \tag{4.24}$$

where Γ and ζ are the gamma function and Riemann ζ function respectively.

The form of the spectrum produced by single scattering when the Thomson limit is not applicable has been provided by Blumenthal and Gould (1970). The spectrum again has an approximate power law form, with $P_{\text{Comp}}(\epsilon_1) \propto \epsilon_1^{-(p+2)}$, which is much steeper than the power law in Equation 4.23.

4.3   Multiple Compton scattering

A photon moving in a finite optically thick material medium can be scattered a number of times by electrons in the medium before it manages to escape. As the photon random-walks through the medium, the vector displacement $\mathbf{R}$ after a number of scatterings is given by

$$\mathbf{R} = \mathbf{r}_1 + \mathbf{r}_2 + \cdots + \mathbf{r}_i + \cdots, \tag{4.25}$$

where $\mathbf{r}_i$ indicates the vector displacement between scattering numbers $i-1$ and i. The square of the average net displacement L after a large number of scatterings is

$$L^2 = \langle \mathbf{R}^2 \rangle = \langle \mathbf{r}_1^2 \rangle + \langle \mathbf{r}_2^2 \rangle + \cdots + \langle \mathbf{r}_1 \cdot \mathbf{r}_2 \rangle + \langle \mathbf{r}_2 \cdot \mathbf{r}_3 \rangle + \cdots, \tag{4.26}$$

where the mean is taken over a large number of trials in each of which a photon is scattered a number of times. The mean over the scalar products vanishes and $L^2 = N l_v$, where N is the number of scatterings and

$$l_v^2 \equiv \langle \mathbf{r}_1^2 \rangle = \langle \mathbf{r}_2^2 \rangle = \cdots = \langle \mathbf{r}_N^2 \rangle \tag{4.27}$$

is the square of the *mean free path* l_v, which is defined as the average distance travelled between successive scatterings. It is easy to show (RL79, p. 14) that $l_v = 1/(n\sigma_v)$, where n is the number density of the particles from which scattering takes place and σ_v is the scattering cross-section, which is in general frequency dependent. For a photon to escape from a finite-sized medium, it will have to travel on average a distance approximately equal to the scale size of the medium. If we take this to be L, the number of scatterings before escape is given by

$$N_{\rm sc} \simeq \frac{L^2}{l_v^2} = \tau_v^2, \quad \tau_v \gg 1, \tag{4.28}$$

where $\tau_v = 1/\alpha_v^{\rm sc} \simeq 1/(n\sigma_v)$ is the *optical depth* to scattering of the medium and $\alpha_v^{\rm sc}$ is the *scattering coefficient*.

When a beam of radiation travels through a scattering medium, the attenuation of the beam due to photons scattered out of it is given by

$$dI(v) = -I(v)d\tau_v, \tag{4.29}$$

which for a homogeneous medium with constant α_v becomes

$$I(v) = I(v,0)\exp(-\tau_v), \quad \tau_v = n\sigma_v s, \tag{4.30}$$

where s is the distance travelled and $I(v,0)$ is the intensity incident on the medium.

The relation in Equation 4.28 applies when the optical depth, and therefore the number of scatterings that a photon undergoes, is large, so that the photon random-walks its way through the medium and the averaging is meaningful. When the optical depth is small, only a fraction of the photons produced in the medium or incident on it from the outside will undergo even a single scattering event. The attenuated intensity of the beam is still given by Equation 4.30, where $\exp(-\tau_v)$ is now the probability that there is no scattering, with $\tau_v \ll 1$. The probability of scattering is therefore $1 - \exp(-\tau_v) \simeq \tau_v$; this can also be looked upon as the fractional number of scatterings suffered on an average by a single photon, and

$$N_{\rm sc} \simeq \tau_v, \quad \tau_v \ll 1. \tag{4.31}$$

The expressions for large and small optical depths can be combined together to give

$$N_{\rm sc} \simeq \tau_v(1 + \tau_v), \tag{4.32}$$

which gives the correct values for τ_v large and small compared to unity.

For large optical depth, if a photon undergoes $N_{\rm sc}$ scatterings before escaping from a medium of size L, then the time spent by the photon in the medium is $N_{\rm sc}l_v/c$. If the medium were optically thin the photon would cross the medium in time L/c. The ratio of the times spent in the two circumstances is therefore given by

$$\frac{t_{sc}}{t_{nsc}} \sim N_{sc}\left(\frac{l_v}{L}\right) \sim 1 + \tau_v. \tag{4.33}$$

The average energy density of photons in the medium also increases by the same factor. The extra time spent in the medium means that the photon moves with a diffusion velocity $c/(1 + \tau_{sc})$ in it.

When absorption and scattering are both present, the *effective optical depth* is given by

$$\tau_* = \sqrt{\tau_{ab}(\tau_{ab} + \tau_{sc})}, \tag{4.34}$$

where the optical depth to absorption τ_{ab} is defined in terms of the absorption coefficient in the same manner as the scattering optical depth. τ_{ab} is a function of the frequency of the radiation. The behaviour of the medium towards the radiation depends on the magnitude of τ_* (RL79, p. 38). When $\tau_* \ll 1$ the medium is said to be *effectively thin* and most photons escape from it without being absorbed, but after a random walk. When $\tau_* \gg 1$ the medium is *effectively* thick and most photons are absorbed, the ones escaping being those emitted within one effective path length $l_* = L/\tau_*$ of the surface. For $\tau_{sc} \ll \tau_{ab}$, $\tau_* \simeq \tau_{ab}$, while for $\tau_{ab} \ll \tau_{sc}$, $\tau_* \simeq \sqrt{\tau_{ab}\tau_{sc}}$. When $\tau_* \ll 1$, the form of the emergent spectrum is the same as the form of the emitted spectrum. When $\tau_* \gg 1$ and $\tau_{sc} \ll \tau_{ab}$, the spectrum takes the black body form, but when the scattering depth is significant, the spectrum has a modified black body form with a Rayleigh–Jeans part at $h\nu/kT \ll 1$, a flat portion at intermediate frequencies and a Wien peak at high frequencies (see subsection 4.3.3).

There is exchange of energy between the photon and electron in each collision, and the magnitude of this exchange and the net loss or gain suffered by each particle after multiple collisions are considered below.

4.3.1   *Non-relativistic thermal electrons*

In a thermal scattering medium with temperature $T \ll mc^2/k$, the electrons are non-relativistic. It can be shown (see RL79 for a simple derivation) that in this case the average energy exchanged per scattering between a photon and an electron is

$$\langle \Delta\epsilon \rangle_{nr} = \frac{\epsilon}{mc^2}(4kT - \epsilon). \tag{4.35}$$

When $kT > \epsilon/4$, $\Delta\epsilon > 0$, so that on the average energy passes from the thermal electrons to the photons. When the temperature is not high enough for this condition to be satisfied the electrons end up gaining energy, and becoming hotter, until a *Comptonization temperature*

$$T_C = \frac{\epsilon}{4k} \tag{4.36}$$

is reached. At this temperature the energy gained by the electrons is on average equal
to the energy lost, so that they are not further energized.

The effect on the photon spectrum of repeated scatterings is conveniently measured
by the dimensionless *Comptonization parameter*

$$y \equiv \frac{\Delta\epsilon}{\epsilon} N_{\text{sc}} = \frac{4kT}{mc^2} \tau_{\text{sc}} (1 + \tau_{\text{sc}}), \tag{4.37}$$

where it is assumed that $\epsilon \ll 4kT$ and the expression for the number of scatterings is
valid at small and large optical depths. The photon energy after dN scatterings is

$$\epsilon_{dN} \simeq \epsilon \left(1 + \frac{4kT}{m_e c^2} \right)^{dN} \simeq \epsilon \left(1 + \frac{4kT\,dN}{m_e c^2} \right), \tag{4.38}$$

and the ratio of the total change in energy to the initial energy is

$$\frac{\Delta\epsilon}{\epsilon} \simeq \frac{4kT\,dN}{m_e c^2}. \tag{4.39}$$

Integrating this equation gives for the energy of the photon after N scatterings

$$\epsilon_{\text{f}} = \epsilon_{\text{i}} \exp\left(\frac{4kT}{m_e c^2} N \right), \tag{4.40}$$

where ϵ_{i} is the initial energy. In an optically thick medium the photon will have
undergone $\sim \tau^2$ scatterings until it diffuses out of the medium and, from the definition
of the Comptonization parameter, its final energy is $\epsilon_{\text{f}} \sim \epsilon_{\text{i}} e^y$. The repeated scattering
cannot increase the energy of the photon beyond $4kT$ since at this stage energy loss
due to Compton recoil of the electron becomes comparable to the energy gain.

4.3.2 *Relativistic electrons*

When the temperature is high enough to make the electrons highly relativistic, the
average energy transferred per collision for a fixed γ is given by Equation 4.13.
Averaging over a thermal distribution of electron energy, which in the relativistic case
is $n(E) \propto E^2 \exp(-E/kT)$, gives

$$\langle \Delta\epsilon \rangle_{\text{r}} \simeq 16\epsilon \left(\frac{kT}{mc^2} \right)^2. \tag{4.41}$$

The Comptonization parameter in the relativistic case is

$$y = 16 \left(\frac{kT}{mc^2} \right)^2 \tau_{\text{sc}} (1 + \tau_{\text{sc}}). \tag{4.42}$$

As in the non-relativistic case, if $y \gg 1$ the energy of a photon is changed significantly
owing to repeated scattering before it leaves the system.

4.3.3   *The Kompaneets equation*

Consider a non-relativistic thermal distribution of electrons at temperature T, and let photons with some initial distribution of energy interact with these electrons. There is exchange of energy between the particles and radiation, and the evolution of the photon phase space density is in general described by the Boltzmann equation, which is of the integro-differential type. In the situation where the electrons are non-relativistic, i.e., $kT \ll m_e c^2$, the energy transfer per Compton scattering is a small fraction of the electron energy,[1] $\Delta\epsilon/kT \ll 1$. The Boltzmann equation can now be expanded to second order in $\Delta\epsilon/kT$, which leads to an useful approximate form[2] that was first obtained by A. S. Kompaneets (1957).

The Kompaneets equation describes the evolution of the photon phase density $n(x)$,[3] which is expressed as a function of the variable $x = \epsilon/kT$. $n(x)$ is related to the number of photons $n_d(\epsilon)$ per unit volume of space per unit energy range through

$$n_d(\epsilon) = \frac{4\pi(kT)^3}{(hc)^3} n(x)x^2. \tag{4.43}$$

The mean free path of a photon between successive scatterings is $l = 1/(n_e\sigma_T)$ and the mean time between scatterings is $t_f = l/c = 1/(n_e\sigma_T c)$, where n_e is the number density of electrons. It is convenient to express the evolution in terms of the dimensionless parameter $t_c = t/t_f = (n_e\sigma_T c)t$, which is the time measured in units of the mean time between scatterings. With these dimensionless variables, the Kompaneets equation is given by

$$\frac{\partial n}{\partial t_c} = \left(\frac{kT}{mc^2}\right) \frac{1}{x^2} \frac{\partial}{\partial x} \left[x^4(n' + n + n^2)\right], \tag{4.44}$$

where $n' = \partial n/\partial x$.

This equation has the form

$$\frac{\partial n}{\partial t} = -\frac{1}{x^2} \frac{\partial x^2 j(x)}{\partial x}, \tag{4.45}$$

which is a continuity equation in energy (or momentum) space, with an appropriately determined radial current $j(x)$. The continuity equation applies because the number of photons is conserved in the scattering. Owing to the recoil of the electron in a scattering event, there is a change in the energy of the photon given by $\Delta\epsilon/\epsilon \simeq \epsilon/(m_e c^2)$. This leads to steady energy loss by the photon at the rate equal to the product of the energy lost per collision with the number of collisions per unit time, i.e., $d\epsilon/dt = -n_e\sigma_T\epsilon^2/(m_e c)$, and the contribution to the current is $n(\epsilon)d\epsilon/dt$. The second term on the right in Equation 4.44 corresponds to this Compton recoil term in dimensionless units.

---

[1] Note that as before we assume $\gamma\epsilon \leq m_e c^2$, with the electron Lorentz factor $\gamma \sim 1$ in the present case.
[2] This is in the general case called the Fokker–Planck approximation.
[3] We normally use the symbol n to indicate number density. In this section we use the same symbol to indicate phase space density, without putting an additional prefix, to avoid complicating the notation and maintain uniformity with the literature.

The first term on the right corresponds to the energy gain of the photons due to inverse Compton scattering, which can be treated as a diffusion process in energy space, with a current $j = -D_\epsilon \partial n / \partial \epsilon$ and diffusion coefficient $D_\epsilon = n_e \sigma_T c \langle (\Delta \epsilon)^2 \rangle / 3$ (see Blandford 1990); the mean square shift in photon energy due to scattering by an electron with speed v is $\langle (\Delta \epsilon)^2 \rangle = \epsilon^2 v^2 / c^2 = 3kT\epsilon^2/c^2$. The last two terms on the right hand side can together be written as $n(1 + n)$, which reduces to n for $n \ll 1$. For $n \gtrsim 1$ the factor $1 + n$ provides the enhancement due to induced scattering over the spontaneous rate. Detailed derivations of the Kompaneets equation may be found in RL79 and K87, and an heuristic discussion has been given by Blandford (1990).

Photons with some initial distribution of energy and with $\epsilon \ll kT$ are gradually Compton scattered to higher energies, and the distribution reaches an equilibrium, i.e., $\partial n / \partial t_c = 0$, when $n(x)$ attains the Bose–Einstein form

$$n(x) = \frac{1}{e^{\alpha + x} - 1}, \qquad (4.46)$$

where α is the chemical potential. When photons interact with matter through absorption and emission processes their number is not conserved and the equilibrium distribution is given by the Planck function, which is simply the Bose–Einstein distribution in Equation 4.46 with the chemical potential $\alpha = 0$. When equilibrium is attained through scattering, the number of photons is conserved and a non-zero α arises as an extra Lagrange multiplier in the derivation of the equilibrium distribution from statistical mechanics. When equilibrium is reached, the spectrum is said to be *saturated*. When $\alpha \gg 1$, the phase space density is small, i.e., $n \ll 1$, and

$$n(x) \propto e^{-x}, \quad n_d(x) \propto x^2 n(x). \qquad (4.47)$$

The intensity of the radiation is therefore of the form $I(v) \propto v^3 \exp(-hv/kT)$, which is of the same form as Wien's law of radiation and is known as the Wien spectrum. The average energy of photons with a saturated Wien spectrum is $3kT$.

While saturated Comptonization produces a Wien spectrum, *unsaturated Comptonization*, which we will now discuss, leads to a power law form over some energy range, making it a potentially useful device for the description of quasar and AGN spectra. In the unsaturated case one considers a finite medium in which there is a source of soft photons and escape of photons from the medium, with most photons not saturating to the Wien spectrum before escape. Following [RL79], we will assume that the probability for a photon to escape from the medium per Compton scattering time t_f is equal to the inverse of the mean number of scatterings, which, as we have seen at the beginning of this section, is given by $\tau_{sc}(1 + \tau_{sc})$. The Kompaneets equation modified for the presence of the source and escape becomes

$$\frac{\partial n}{\partial t} = \left(\frac{kT}{mc^2} \right) \frac{1}{x^2} \frac{\partial}{\partial x} \left[x^4 (n' + n) \right] + Q(x) - \frac{n}{\tau_{sc}(1 + \tau_{sc})}, \qquad (4.48)$$

where $Q(x)$ is the photon source. The induced emission term is not included, assuming that $n \ll 1$.

We are interested in the scattering of soft photons to higher energies, so it is assumed that $Q(x) \neq 0$ only for $x \leq x_s$, for some $x_s \ll 1$. The low photon energy means that the electron recoil too can be neglected, relative to the diffusive (i.e., random walk) energy gain, and $n \ll n'$. For $x > x_s$, $Q(x) = 0$ and Equation 4.48 reduces to

$$\frac{1}{x^2}\frac{d}{dx}\left(x^4\frac{dn}{dx}\right) = \frac{4n}{y}, \tag{4.49}$$

where y is the non-relativistic Comptonization parameter in Equation 4.37. This equation has a power law solution $n(x) \propto x^{-p}$, with the index satisfying the quadratic equation (Shapiro *et al.* 1976)

$$p(p-3) - 4/y = 0 \tag{4.50}$$

with solutions

$$p = \frac{3}{2} \pm \sqrt{\frac{9}{4} + \frac{4}{y}}. \tag{4.51}$$

For $y \gg 1$, the Comptonization is saturated and the negative sign is used as this leads to $p = 0$, which gives the correct approximation to Wien's law for $x \ll 1$. For $y \ll 1$, choosing the positive sign gives a power law for the intensity, $I(\nu) \propto x^3 x^{-p} \propto \nu^{-(p-3)}$, where $p = 3/2 + \sqrt{4/y}$.

Compton recoil can no longer be neglected when $n' \simeq n$, which for the power law solution occurs when $x = p$, i.e., when the photon energy $\epsilon = pkT$. Up to this limit the spectrum is of power law form. For $x \gg 1$ the approximate solution is $n \propto e^{-x}$, i.e., $I(\nu) \propto x^3 e^{-x} \propto \nu^3 e^{-h\nu/kT}$. Unsaturated Comptonization is therefore able to produce a power law spectrum over a wide energy (frequency) range, starting with a soft photon input and a non-relativistic thermal distribution of electrons.

4.4   Synchrotron self-Compton emission

Synchrotron photons can be Compton scattered by the ensemble of electrons that emits them in the first place, boosting photons of energy ϵ to energy $\sim \gamma^2 \epsilon$ (see Section 4.2). The electrons can therefore lose energy either through synchrotron emission or through Compton scattering of photons. The ratio of the luminosity generated by an ensemble of electrons in the Compton and synchrotron channels is proportional to the ratio of the energy density of the photons and that of the magnetic field, as shown in Equation 4.14. The radiation energy density is high in the case of luminous, compact sources and it is to be expected that they produce copious high energy photons through Compton scattering. These photons can again be Compton scattered, to still higher energies, and so on until the condition $\gamma \epsilon \ll mc^2$ is violated. If the energy density of the once-scattered radiation exceeds that of the magnetic field, the energy density of the twice-scattered photons exceeds that of the once-scattered photons. As a result of such multiple scattering the electrons will lose their energy very rapidly to high energy photons, thus quenching the source; this is known as the Compton catastrophe (Hoyle, Burbidge and Sargent 1966).

A catastrophic energy loss can be avoided if the energy density in the once-scattered photons is less than that in the magnetic field. A simple condition for its avoidance was derived by Kellermann and Pauliny-Toth (1969) along the following lines. The photons involved in the first Compton scattering are synchrotron photons (if it is assumed that there is no other ambient radiation present) whose energy density is $U_{ph} = L_s/(4\pi r^2 c)$, where L_s is the synchrotron luminosity and r is the radius of the source. If the synchrotron flux is F, we have $L_s = 4\pi D^2 F = 4\pi r^2 F/\theta^2$, where θ is the angular size of the source and D is its distance. If L_C is the luminosity in the once-Compton-scattered synchrotron photons, we have using Equation 4.14

$$\frac{L_C}{L_s} = \frac{U_{ph}}{U_B} = \frac{8\pi F}{B^2\theta^2 c}, \tag{4.52}$$

where $U_B = B^2/8\pi$ is the energy density of the magnetic field. It is clear that Compton scattering dominates in sources with high surface brightness, which are the sources that show synchrotron self-absorption. If ν_a is the frequency at which self-absorption sets in and the spectrum turns over, one has $F \sim F(\nu_a)\nu_a$ and

$$\frac{L_C}{L_s} = \frac{8\pi F(\nu_a)}{B^2\theta^2 c}\nu_a. \tag{4.53}$$

Now following the treatment in Equations 3.59 to 3.61, we can express the intensity (surface brightness) of radiation $I(\nu) = F(\nu)/\theta^2$ in terms of the brightness temperature T_b. The self-absorption frequency can be expressed in terms of the energy $E = kT$ of the emitting electrons using Equation 3.13. From the condition for self-absorption $T \simeq T_b$ we then get

$$\frac{L_C}{L_s} \simeq 16\pi \left(\frac{3e}{4\pi m^3 c^5}\right)^2 \frac{(kT_b)^5}{c^3}\nu_a \simeq \left(\frac{T_b}{10^{12.1}\,\text{K}}\right)^5 \frac{\nu_a}{100\,\text{GHz}}. \tag{4.54}$$

For $\nu_a \simeq 100\,\text{GHz}$, we have $L_C \ll L_s$ for $T < 10^{12}\,\text{K}$ and the synchrotron emission dominates. When $T \gtrsim 10^{12}\,\text{K}$ the luminosity in the Compton photons dominates that in the synchrotron photons and the second order scattering becomes important. Including this effect, the ratio of the Compton to the synchrotron luminosity becomes

$$\frac{L_C}{L_s} \simeq \left(\frac{T_b}{10^{12.1}\,\text{K}}\right)^5 \frac{\nu_a}{100\,\text{GHz}} \left[1 + \left(\frac{T_b}{10^{12.1}\,\text{K}}\right)^5\right]. \tag{4.55}$$

Since $T_b \gtrsim 10^{12}\,\text{K}$ the second order term dominates and

$$\frac{L_C}{L_s} \sim \left(\frac{T_b}{10^{12.1}\,\text{K}}\right)^{10}, \tag{4.56}$$

leading to the Compton catastrophe. It was first noticed by Kellermann and Pauliny-Toth (1969) that the lower limits on T_b for unresolved opaque radio sources were all $\lesssim 10^{12}\,\text{K}$. They attributed this to the inverse Compton cooling of sources in which the brightness temperature exceeded $\sim 10^{12}\,\text{K}$.

Sources with brightness temperatures exceeding the Compton limit by many orders of magnitude are known. The high brightness temperature here is inferred from the small angular size deduced from rapid variability in the observed flux on time scales as short as a day. The situation is saved by attributing the observed rapid variability to the apparent decrease in longer intrinsic variability time scales, owing to relativistic bulk motion (see Section 3.8). Rapid variability at low frequencies is attributed to refractive interstellar scintillations rather than to intrinsic changes in the source itself.

4.5   Thermal bremsstrahlung emission

When an electron is accelerated by the electric field of a positively charged ion, it emits radiation known as *bremsstrahlung* (braking radiation) or free–free radiation. The spectrum of the emission can be obtained using semi-classical approximations and depends on the velocity of the accelerating particle. In a hot gas at temperature T the electrons have a Maxwell–Boltzmann distribution of velocities, and the summed radiation from all the electrons can be obtained by averaging the expression for a single electron over the velocity distribution. The resulting expression for the emission coefficient, which is the energy emitted per unit volume of the gas per unit time, is (RL79)

$$\epsilon_v^{\mathrm{ff}} = \frac{32\pi e^6}{3mc^3}\left(\frac{2\pi}{3km}\right)^{1/2} Z^2 n_{\mathrm{e}} n_{\mathrm{i}} T^{-1/2} e^{-hv/kT} \bar{g}_{\mathrm{ff}}(T,v). \tag{4.57}$$

Here n_{e} and n_{i} are the electron and ion density respectively, Z is the ionic charge and the *Gaunt factor* $\bar{g}_{\mathrm{ff}}(T,v)$ is defined as in subsection 3.5.3, where approximate expressions applicable in the radio domain were discussed. Expressions over a variety of regimes were considered by Novikov and Thorne (1973) and may be found in RL79, where the run of numerical values is also shown. $\bar{g}_{\mathrm{ff}} \sim 1$ for $hv/kT \sim 1$ and is in the range 1–5 for $10^{-4} < hv/kT < 1$. The emission coefficient could also have been obtained from the absorption coefficient in Equation 3.64 and Kirchoff's law for the thermal gas.

Introducing numerical values for the constants and specializing to the case of an electrically neutral hydrogen plasma gives

$$\epsilon_v^{\mathrm{ff}} = 6.8 \times 10^{-38} n_{\mathrm{e}}^2 T^{-1/2} e^{-hv/kT} \bar{g}_{\mathrm{ff}}(T,v)\,\mathrm{erg\,sec^{-1}\,cm^{-3}\,Hz^{-1}}. \tag{4.58}$$

The total power emitted per unit volume by the hot gas is obtained by integrating ϵ_v^{ff} over all frequencies and is given by

$$\epsilon^{\mathrm{ff}} = \left(\frac{2\pi kT}{3m}\right)^{1/2} \frac{32\pi e^6}{3hmc^3} Z^2 n_{\mathrm{e}} n_{\mathrm{i}} \bar{g}_{\mathrm{B}}, \tag{4.59}$$

where $\bar{g}_{\mathrm{B}}$ is the frequency-average of the velocity-averaged Gaunt factor $\bar{g}_{\mathrm{ff}}(T,v)$ and its value is in the range 1.1–1.5. For the hydrogen plasma

$$\epsilon^{\mathrm{ff}} = 1.4 \times 10^{-27} T^{1/2} n_{\mathrm{e}}^2 \bar{g}_B\,\mathrm{erg\,sec^{-1}\,cm^{-3}}. \tag{4.60}$$

The radiation from the ions can be neglected because of their high mass. Thermal bremsstrahlung radiation is obtained from encounters between positrons and electrons at approximately the same rate as from encounters between protons and electrons. For non-relativistic speeds, electron–electron and positron–positron emission is of the quadrupole type and is much weaker than the dipole ion–electron radiation; however, such emission can become significant in the relativistic case.

A gas emitting thermal bremsstrahlung radiation loses energy at a rate given by Equation 4.59. Since the thermal energy of the gas per unit volume is $\sim n_e k T$, the time scale for cooling is

$$\tau_B \sim \frac{n_e k T}{\epsilon^{\text{ff}}} \sim \frac{1}{n_e \alpha_f \sigma_T c} \left(\frac{k T}{m c^2}\right)^{1/2}, \tag{4.61}$$

where $\alpha_f = e^2/(\hbar c) = 1/137.04$ is the fine structure constant.

4.6 Non-thermal pair models

Classical electrodynamics is a linear theory and in it the electromagnetic field does not interact with itself. In quantum electrodynamics, however, it is possible to have photon–photon interactions, through virtual pairs of charged particles, as a higher order effect. Consequently the scattering of photons by photons as well as photon–photon interactions that produce electron–positron pairs are possible, and the latter have been the basis of models that attempt to explain the characteristics of the continuum spectrum at X-ray and higher energies.

Consider an interaction between two photons, with energies ϵ_1 and ϵ_2 respectively, that leads to the formation of an electron–positron pair. Let the photons be moving in the directions $\mathbf{n}_1$ and $\mathbf{n}_2$ respectively. Using the conservation of the energy–momentum four-vector in the interaction, it is easy to show that the condition on the photon energies for the formation of a pair is that

$$\epsilon_1 \epsilon_2 \geq \frac{2(m_e c^2)^2}{1 - \mathbf{n}_1 \cdot \mathbf{n}_2}. \tag{4.62}$$

The right hand side is a minimum for a head-on collision with $\mathbf{n}_2 \cdot \mathbf{n}_2 = -1$. When the photon energies are equal, the condition simply states that in a head-on collision each photon must have at least the energy equivalent of the rest mass of the electron. However, for the production of a pair it is not always necessary for each photon to have energy exceeding $m_e c^2$ and the condition allows a high energy γ-ray photon to interact with a relatively low energy X-ray photon and produce a pair. To see the kind of photon energies that are involved, we write the condition in the normalized form

$$\left(\frac{\epsilon_1}{1\,\text{keV}}\right) \gtrsim 500 \left(\frac{\epsilon_2}{1\,\text{MeV}}\right)^{-1}. \tag{4.63}$$

If one of the pair-producing photons has energy $\epsilon_2 = 100\,\text{MeV}$, the lower limit on the energy ϵ_1 of the other photon is $5\,\text{keV}$.

For a γ-ray of a given energy, the cross-section for pair production $\sigma_{\gamma\gamma}$ is a function of the X-ray energy ϵ_x. Starting from zero at the threshold energy $\epsilon_{xs} = 2(m_ec^2)^2/\epsilon_\gamma$, the cross-section rises steeply to a maximum of $\sim 0.2\sigma_T$ at $2\epsilon_{xs}$, where σ_T is the Thomson scattering cross-section, and then falls off $\propto \epsilon_x^{-2}$ (see Herterich 1974 and references therein).

The optical depth to pair production for a γ-ray of energy ϵ_γ in a homogeneous region of size R is

$$\tau(\epsilon_\gamma) = R \int_{\epsilon_{xs}}^{\infty} n(\epsilon_x)\sigma_{\gamma\gamma}d\epsilon_x, \tag{4.64}$$

where $n\epsilon_x$ is the energy density of the X-ray photons per unit energy interval. Approximating the cross-section by a rectangular function with height $0.2\sigma_T$ and width $2\epsilon_{xs}$, we get

$$\tau(\epsilon_\gamma) \simeq 0.2\sigma_T n(2\epsilon_{xs})2\epsilon_{xs}R, \tag{4.65}$$

with the number density taken at the energy at which the maximum in the exact cross-section occurs. Expressing the number density of photons in terms of the luminosity through $n(2\epsilon_{ns})2\epsilon_{xs} = L(2\epsilon_{xs})/(4\pi R^2 c)$, the optical depth becomes

$$\tau(\epsilon_\gamma) \simeq \frac{0.2\sigma_T L(2\epsilon_{xs})}{4\pi Rc}. \tag{4.66}$$

4.6.1   *The compactness parameter*

If the power law spectra observed in AGN and quasars in the X-ray region extend to much higher energies, then pairs may be expected to be produced if the optical depth for the γ-ray photons for pair production exceeds unity. The condition under which this happens is conveniently stated in terms of the *compactness parameter* (see Guilbert, Fabian and Rees 1983)

$$l = \frac{L}{R}\frac{\sigma_T}{m_ec^3}, \tag{4.67}$$

where L is the luminosity produced in a region of radius R. If one assumes that the luminosity is produced by accretion onto a black hole of mass M, l can be expressed in terms of the Eddington luminosity $L_{Edd} = 4\pi cGMm_p/\sigma_T$ and the Schwarzschild radius $r_S = 2GM/c^2$ as (Svensson 1994)

$$l = \frac{2\pi}{3}\frac{m_p}{m_e}\left(\frac{L}{L_{Edd}}\right)\left(\frac{3r_S}{R}\right). \tag{4.68}$$

Since $L < L_{Edd}$ and it is expected that $R \geq r_S$, we have $l \lesssim 3600$.

The optical depth for pair production in photon–photon interactions is given by Equation 4.66. For γ-ray photons with energy $\sim m_ec^2$,

$$\tau \simeq \frac{0.2\sigma_T L}{4\pi Rm_ec^3} \simeq \frac{l}{60}, \tag{4.69}$$

where we have used $L \simeq L(m_e c^2) m_e c^2$ and Equation 4.67. The optical depth therefore exceeds unity when $l \gtrsim 60$ and copious pair production takes place. For a source where the size of the emitting region is not much bigger than a few times the Schwarzschild radius, l fulfills the required condition for $L \gtrsim 1.5 \times 10^{-2} L_{Edd}$.

To estimate the value of the compactness parameter requires knowledge of the luminosity of the source and its radius. The difficulty with the former is that though a source may produce significant γ-ray emission, this may be depleted by the formation of pairs. Therefore a source that is apparently not luminous in γ-rays could still be undergoing pair-creation-related processes, which, as we will see below, can significantly affect the shape of the spectrum that emerges from the source. The size of the source can be determined from variability time scales, but a difficulty can again arise when a source shows significant variation on different time scales.

Done and Fabian (1989) have determined the compactness parameter for a number of Seyferts for which good X-ray spectra in the 2–10 keV range and variability data are available. They assumed that the power law X-ray spectrum with $\alpha_x \simeq 0.7$ could be extrapolated to the higher energy region with a cutoff above 2 MeV, which is needed if the observed γ-ray background is not to be exceeded by the summed contribution from the Seyfert galaxies (the high energy spectra of AGN are discussed in Section 11.1). The luminosity that goes into the compactness parameter was determined using this spectrum. Done and Fabian determined the source size as $R = c\Delta_2$, where Δ_2 is the time scale over which the intensity varies by a factor of 2 and relativistic effects are ignored. There are other, shorter, variability time scales associated with AGN, and reprocessing by pairs can lengthen variability time scales relative to the light-crossing time $\sim R/c$. As a result $c\Delta_2$ is usually an overestimate of the source size, and the compactness parameter obtained from it is underestimated. Done and Fabian found from the data available to them that several of the Seyferts had $l \gtrsim 60$, while some had l as small as 0.1.

We have seen in Equation 4.12 that the Compton power radiated by a highly relativistic electron through the inverse Compton scattering of photons is $\propto \gamma^2 U$, where U is the energy density of the radiation field. Expressing this in terms of the luminosity we get for the rate of energy loss by the electron $dE/dt = \gamma^2 L \sigma_T/(3\pi R^2)$. The time scale for the energy loss due to the Compton loss is then given by

$$ t_{Comp} \sim \frac{E}{dE/dt} = \frac{3\pi}{\gamma l}\left(\frac{R}{c}\right) \sim \frac{1}{\gamma}\left(\frac{l}{10}\right)^{-1}\left(\frac{R}{c}\right). \tag{4.70} $$

Since the electrons are highly relativistic, the Lorentz factor γ is $\gg 1$. Therefore, for a compactness parameter $l \gtrsim 10$, the Compton cooling time is less than the escape time t_{esc}, which is $\sim R/c$. The energetic electrons and protons produced in photon–photon interactions, or injected into the region by other processes, cool before they can escape from the system. The particle–antiparticle pairs annihilate after cooling, since the annihilation time scale is long at high energy.

4.6.2   *Pair cascades*

When soft photons of energy ϵ_s are introduced into a system containing relativistic electrons with maximum Lorentz factor γ_{max}, the maximum energy after the scattering is $\epsilon_{1,max} = \gamma_{max} m_e c^2 + \epsilon_s \simeq \gamma_{max} m_e c^2$, as they cannot acquire more than the entire energy of the electron (see Section 4.2). In terms of the dimensionless energy variable $x = \epsilon/(m_e c^2)$ this means $x_{1,max} = \gamma_{max}$. These photons have sufficient energy to produce pairs in interaction with lower energy photons. The number of Compton-scattered photons with energies less than $x_{1,max}$ that are energetic enough to produce pairs will depend on the distribution of γ and the energy of the soft photons. If it is assumed that the energy of the high energy photon involved in the photon–photon interaction is shared equally by the electron and positron, the maximum energy of each particle, in units of $m_e c^2$, is $\gamma_{1,max} = x_{1,max}/2 = \gamma_{max}/2$.

The particles in the first pair generation are themselves capable of scattering the soft photons to higher energy. The maximum energy of these second-generation-scattered photons is $x_{2,max} = \gamma_{1,max} = \gamma_{max}/2$. If these photons are energetic enough they can again produce a second generation of pairs, with the maximum energy of a particle in a pair being half of the maximum energy in the first generation. As this process continues with the production of successive generations of pairs and scattered photons, the maximum energy of the produced particles halves in each generation. The photons scattered at some generation will no longer be capable of producing pairs, since their maximum energy will be $< m_e c^2$.

When an electron with energy γ Compton-scatters soft photons it loses energy at the rate $\dot{\gamma} \propto \gamma^2$. If the number of scattering electrons as a function of energy has the power law form $q(\gamma) \propto \gamma^{-\Gamma}$, the energy loss leads to a steepening of the spectrum. In the steady state the distribution of energy is given by $N(\gamma) \propto \gamma^{-p}$, where $p = \Gamma + 1$ for $\Gamma > 1$. However, if the injected electrons all have the same energy, the steady state distribution is given by $p = 2$ (see Section 3.7). The scattered photons, which end up gaining energy, develop a power law spectrum with number density $n(\epsilon) \propto \epsilon^{-\alpha_1}$, where $\alpha_1 = (p+1)/2$ (see subsection 4.2.2). If we begin with primary scattering electrons with constant energy γ_{max}, the scattered photons have $\alpha_1 = (2+1)/2 = 1.5$. When the energy of the scattered photons extends to $> m_e c^2$, and the optical depth to photon–photon scattering is high, it may be assumed that every γ-ray photon makes a pair, the energy of the photon being shared equally by the two particles. the energy distribution of these first-generation pair particles is given by $N_1(\gamma) \propto n(\epsilon) \propto \epsilon^{-\alpha_1} \propto \gamma^{-\alpha_1}$. The power law index of the injection spectrum of the first pair generation is therefore $\Gamma_1 = \alpha_1 = 1.5$. The spectrum of these particles is in turn steepened owing to energy loss to Compton scattering, with the power law index $p_1 = \Gamma_1 + 1 = 2.5$. The photons scattered by these particles acquire a power law form with index $\alpha_2 = (p_1 + 1)/2 = 1.75$. The next generation has $\Gamma_2 = 1.75$, $p_2 = 2.75$ and $\alpha_3 = 1.875$, the one after that has $\Gamma_3 = 1.875$, $p_3 = 2.875$ and $\alpha_4 = 1.9375$, and so on, with the limiting values $\Gamma = 2$, $p = 3$ and $\alpha = 2$, which correspond to a photon energy index of 1 (see Svensson 1987). In this case the cascade is said to be saturated and the injected electrons have equal power in every decade of energy, which is also the case with photons. When pair production does not continue beyond a few generations because the maximum photon energy falls below $m_e c^2$, the photon number-index is closer to 1.9 than 2.0 and one may take it to be about 1.9.

4.6.3 *Scattering by cool pairs*

Pairs produced in photon–photon interactions lose energy in scattering soft photons and cool on the time scale t_{Comp} given in Equation 4.70, which is much shorter than the annihilation time scale for relativistic particles. The cooled pairs annihilate at a rate $\sim 0.4\sigma_T c n_+ n_- = 0.4\sigma_T c n^2$ per unit volume, where n_+ and n_- represent the number density of positrons and electrons respectively and n that of the pairs. The reservoir of cool pairs is continuously replenished by the cooling of high energy pairs and a steady state is obtained when the rate of replenishment is equal to the annihilation rate. Following Svensson (1994), we will write for the pair production rate per unit volume

$$\dot{n}_{pp} = \frac{Y}{2 m_e c^2} \left(\frac{L_i}{4\pi R^3 / 3} \right), \tag{4.71}$$

where L_i is the injected power and Y, called the *pair yield*, is the fraction of this power which goes into the production of pair rest mass (more sophisticated expressions may be found in e.g. Lightman and Zdziarski 1987). In the steady state, therefore (Guilbert, Fabian and Rees 1983),

$$\tau_T \sim \sqrt{Y l_i}, \tag{4.72}$$

where l_i is the compactness parameter corresponding to L_i and $\tau_T = \sigma_T (n_+ + n_-) R = 2\sigma_T n R$ is the optical depth to scattering by cool pairs (Guilbert, Fabian and Rees 1983). The pair yield increases with l_i and reaches a constant value of ~ 0.1 for saturated pair cascades (a detailed discussion is given in Svensson 1987). In such a case

$$\tau_T \sim 2\sqrt{l_i / 40}. \tag{4.73}$$

When high energy photons are scattered by the cool pairs, the photons lose energy to the recoil of the electrons (or positrons). The fractional energy lost per collision is $\sim -\epsilon / m_e c^2$ for $\epsilon \ll m_e c^2$ (see Section 4.2) and the energy lost in dN collisions can be written as

$$\frac{d\epsilon}{\epsilon} = -\left(\frac{\epsilon}{m_e c^2} \right) dN. \tag{4.74}$$

Integrating this equation with the condition that $\epsilon = \epsilon_i$ before any collision takes place gives

$$\frac{\Delta \epsilon}{\epsilon_i} \equiv \frac{\epsilon - \epsilon_i}{\epsilon_i} = -\frac{\epsilon N}{m_e c^2}, \tag{4.75}$$

where N is the number of collisions. For the photon to be able to diffuse out, $N \simeq \tau_{\mathrm{T}}^2$, and the condition for $|\Delta\epsilon| \gtrsim \epsilon_i$ is $\epsilon \gtrsim m_e c^2 / \tau_{\mathrm{T}}^2$. Photons with energy greater than $m_e c^2 / \tau_{\mathrm{T}}^2$ will therefore undergo energy-losing collisions until they reach that value. If an initial photon power law spectrum with index α is injected uniformly into a homogeneous medium of cold pairs the effect of the energy loss is to steepen the spectrum to $\alpha + 1/2$ for $\epsilon > m_e c^2 / \tau_{\mathrm{T}}^2$. The break in the spectrum occurs at

$$\epsilon_b \sim \frac{m_e c^2}{\tau_{\mathrm{T}}^2} \sim 50 \left(\frac{l_i}{100} \right)^{-1} \text{keV}. \tag{4.76}$$

4.6.4 Steady state equations

The physical processes described in the previous subsections can all be combined together to study self-consistently the behaviour of photon and electron distributions in a region. The formalism for this was developed in a long series of papers by several groups. Important steps in the development may be found in Fabian *et al.* (1986), Lightman and Zdziarski (1987), Svensson (1987) and Done and Fabian (1989). Useful reviews may be found in Svensson (1990) and Svensson (1994).

The situation considered now is where energetic electrons (and positrons), either with a fixed Lorentz factor $\gamma_{\mathrm{max}} \gg 1$ or a power law distribution of γ with the maximum again satisfying this condition, are injected into a spherical region of radius R. The injected energy is measured in terms of an *electron compactness parameter*

$$l_i = \frac{L_i}{R} \frac{\sigma_{\mathrm{T}}}{m_e c^3}, \tag{4.77}$$

(see Equation 4.67), where L_i is the total energy injected per unit time in the form of relativistic electrons. Soft photons with luminosity L_s and compactness parameter l_s are also injected into the region. These can come from an accretion disk above which is located the region in which the processing occurs. The soft photons are inverse Compton scattered by the relativistic electrons to X-ray and γ-ray energies leading to a high energy luminosity L_h and compactness parameter l_h with

$$L_h = L_i, \quad l_h = l_i. \tag{4.78}$$

If $n(x)$ is the number density of photons of energy $x = \epsilon / (m_e c^2)$ and $N(\gamma)$ the number density of relativistic particles in the region, the steady state equations can be written as (Lightman and Zdziarski 1987)

$$\dot{n}(x) = \dot{n}_0 + \dot{n}_A + \dot{n}_C^{\mathrm{NT}} + \dot{n}_C^{\mathrm{T}} - \frac{c}{R} n(\tau_C^{\mathrm{NT}} + \tau_{\gamma\gamma}) - \dot{n}_{\mathrm{esc}}$$

$$= 0 \tag{4.79}$$

$$\dot{N}(\gamma) = \dot{N}_C + P(\gamma) + Q(\gamma) = 0. \tag{4.80}$$

In these equations $\dot{n}_0$ is the rate of production of photons of energy x per unit range in x due to soft photon injection, $\dot{n}_A$ is that due to pair annihilation, $\dot{n}_C^{\mathrm{NT}}$ that due to Compton scattering by non-thermal electrons and $\dot{n}_C^{\mathrm{T}}$ that due to Compton scattering by thermal electrons. Photons at energy x can be removed by Compton scattering against non-thermal electrons with optical depth τ_C^{NT} and through photon–photon interactions with optical depth $\tau_{\gamma\gamma}$. The removal of photons due to scattering by thermal electrons is included in $\dot{n}_C^T$.

In addition to being lost to the different interactions, photons also escape from the finite-sized region at the rate $\dot{n}_{esc}$. In the absence of scattering, unabsorbed photons would leave the system in a time $\sim R/c$. When the scattering optical depth is not negligible, the photons are repeatedly scattered and diffuse out of the system on a time scale $\tau_{esc} \sim R(1 + \tau_T)/c$ and so $\dot{n}_{esc} \sim n/\tau_{esc}$. Simple variants of this expression are used by different authors. The above escape probability formalism applies when the change in photon energy per encounter is small, i.e., when $x \ll 1$. For larger values of x, i.e., in the relativistic domain, the photon can lose a large fraction of its energy, and the process is treated as absorption. It is the escaping photons that allow the distant observer to perceive the spectrum generated in the source. Before reaching the observer, some of the escaping photons can be further processed, with important modifications made to their spectrum, as in the case of reflection from cold matter treated below.

In Equation 4.80 for the electrons, $\dot{N}_C$ is the rate of change of number density of particles due to non-thermal Compton scattering and P and Q are the rates of pair creation and particle injection respectively, all per unit volume per unit range in γ. Expressions for the various terms in the two equations have been discussed above, or may be found in Lightman and Zdziarski (1987) and the other papers cited, with different authors adopting somewhat different approximations and approaches.

The equations are solved using numerical techniques explained in Fabian *et al.* (1986), Lightman and Zdziarski (1987) and Done and Fabian (1989). For mono-energetic electrons the steady state solution is determined by the four parameters l_s, l_s/l_h, γ_{max} and the typical soft photon energy $\epsilon_s \simeq kT_i$, where it is assumed that these are black body photons from a thermal disk at temperature T_i.

The parameter space has been exhaustively explored in the various papers we have cited and the other references that may be found there. At the time of the development of the theory, observations indicated that the X-ray spectra of AGN had a simple power law form with $\alpha_x = 0.7 \pm 0.2$, at least in the observed ~ 2–$10\,\text{keV}$ range, and it was hoped that this 'universal' power law would emerge from the non-thermal pair models in a robust fashion, i.e., for a wide range of the free parameters.

Examples of spectra generated by pair processing are shown in Figure 4.4, taken from Svensson (1994). In this case the soft photon input is taken to be black body radiation with $kT_i = 5.1\,\text{eV}$, which is in the UV domain, with compactness parameter $l_{UV} \equiv l_s$. The injected relativistic electrons are assumed to be monoenergetic, with $\gamma_i = 7.5 \times 10^3$ and compactness parameter $l_i = l_h$ (see Equation 4.78) and with $l_{UV} = 2l_i$.

The soft photons are inverse Compton scattered to energies $\epsilon \lesssim 4\gamma^2 kT_i/3 \simeq 4 \times 10^5\,\text{keV}$. For electron compactness parameter l_i smaller than a few units, it follows from Equation 4.66 that the optical depth to the γ-ray photons provided by the X-rays is $\ll 1$ and pair production may be neglected. Owing to the energy loss to Compton scattering the electrons then acquire a spectrum $N(\gamma) \propto \gamma^{-2}$ and the photon

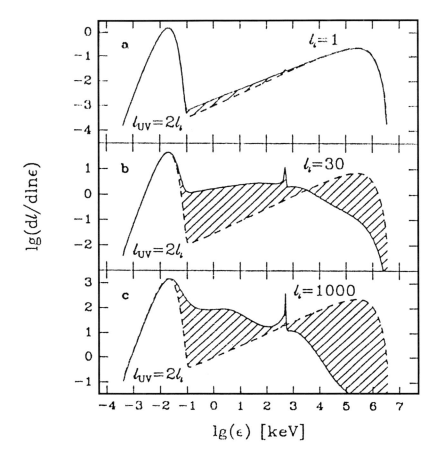

Fig. 4.4. Photon spectra produced in a region containing non-thermal relativistic pair plasma. The soft photon input has a black body spectrum with $kT_i = 5.1$ eV, and the injected electrons all have $\gamma_i = 7.5 \times 10^3$. The continuous line in each panel indicates the spectrum for the choice of parameters indicated. The broken line is the spectrum which would be produced in the absence of pairs, i.e., for $L_i \ll 1$. Reproduced from Svensson (1994).

number density $n(x) \propto x^{-1.5}$ (see Section 3.7). The energy spectrum of the photons is therefore a combination of the black body spectrum of the soft photons and a power law of index $\alpha_x = 0.5$. This is shown in the three panels in Figure 4.4 as a broken line.

When the compactness l_i increases beyond a few units, the optical depth to the γ-rays becomes significant at the highest energy and pairs are produced, the net result being an increase in the X-ray energy density at the cost of the γ-rays . The X-ray spectral index now becomes steeper, with $\alpha_x \sim 0.6$, which is shown in the middle panel of Figure 4.4. The scenario become more interesting as l_i increases to large values and the photon–photon interactions produce a large number of pairs. The pairs cool by Compton scattering of the soft radiation, and a pair cascade is obtained as described in subsection 4.6.2, with $\alpha_x \to 1$ at saturation. Photons with energy $> m_e c^2 / \tau_T^2$ are down-scattered, increasing the energy in X-rays at the cost of the higher energy photons and producing a break in the spectrum. The pairs cool to an equilibrium Compton temperature and annihilate, producing a characteristic line. It follows from Equation 4.73 that as l_i increases the scattering optical depth due to the pairs becomes significant. The break in the spectrum due to down-scattering of the photons by the pairs therefore moves to softer energy. The cool pairs then inverse-scatter the UV photons to soft X-ray energies, thus producing an excess. The spectra corresponding to $l_i = 30$ and 1000 are shown in the middle and bottom panels respectively of Figure 4.4. It should be noted here that details of the spectrum depend on the approximations made in deriving various expressions, which differ somewhat over the literature cited.

The type of spectra described above are produced when $l_s / l_i \gtrsim 1$. When the energy in the soft photons is less than the energy injected into the electrons, $l_s / l_i < 1$, the resulting spectrum can be quite different. In this case the radiation density in the X-rays produced by the scattering of the soft photons can exceed the energy density in the soft photons. Therefore it is the former which are responsible for the cooling of the relativistic electrons. The inverse Compton scattering is efficient in cooling so long as the energy of the already-once-scattered photon satisfies $\gamma_i \epsilon < m_e c^2$ (see Section 4.2). The total luminosity (or energy density) $L_{1/\gamma}$ corresponding to this subset of the once-scattered photons is obtained by integrating their spectrum, which is $\propto x^{-0.5}$ to the limit $\epsilon = m_e c^2 / \gamma_i$, i.e., $x = 1/\gamma_i$. Similarly, the total energy in *all* the once-scattered photons is obtained by integration up to $x \simeq \gamma_i^2 \epsilon_{xs}$ and is equal to L_i. The condition that $L_{1/\gamma}$ exceeds the soft photon energy input L_s is then, in terms of compactness parameters,

$$\frac{l_s}{l_i} < \gamma_i^{3/2} x_s^{1/2}. \tag{4.81}$$

The scattering of the X-ray photons to higher energy reduces the spectral index to $\alpha_x < 0.5$.

As l_i increases the pairs become important. Since there are relatively few soft photons because $l_s / l_i < 1$, there is less energy drag on the pairs, and they settle to a higher Comptonization temperature. This can lead to $y = 4kT\tau_T^2/(m_e c^2) > 1$ and the formation of a prominent Wien peak (see Equation 4.37).

It is necessary to examine the predictions of the non-thermal pair models for wide ranges of the input parameters and to compare the results with observations. In this respect we shall consider the results obtained by Done and Fabian (1989), who examined the spectra produced for soft photon energy $\epsilon_{xs} = 10^{-5}$, $\gamma_i = 10^3$ and 10^4, $l_i = 0.4$ to 400 and $l_s / l_i = 0.01, 0.1, 1, 10$. The observational constraints in the X-ray range used by Done and Fabian, which were based on the data available at that time, were $\alpha_x \simeq 0.7$ and $1 < l_s / l_x < 10$, where l_x is the compactness parameter corresponding to a 2–10 keV luminosity. In addition to this, γ-ray observations of AGN and the observed γ-ray background required a steepening in the photon spectrum beyond a few MeV, with spectral index $\alpha_\gamma \geq 1.7$ (see Section 11.7).

Done and Fabian found that the two constraints ruled out models with $l_s/l_h \gg 1$. The γ-ray constraint further requires models with $l_i = l_h > 10$, in which the γ-rays produced in the inverse scattering are depleted through photon–photon interactions. A further constraint is provided by the annihilation line. The pair yield in the models can be as high as ~ 0.1 i.e., ~ 10 per cent of the energy supplied in the form of energetic electrons can emerge in the photons with energy $\sim m_e c^2$ produced in the annihilation of electron and positron pairs (see subsection 4.6.3). It was within the capability of the observations available to have detected a sharp line with this luminosity; yet the line was not found. However, the detection of line luminosity could be made difficult by turbulent velocities in the source, which could exceed $\sim 0.3c$, making the line a broad hump on a spectrum that was poorly measured. Done and Fabian (1989) used 300 keV as an upper limit to the equivalent width of a line that would escape detection, and models which included this constraint, as well as the rest, were found to require $\gamma_i = 10^4$, $\epsilon_s = 10^{-5}$, $l_s/l_h \lesssim 1$ and $l_i \sim 50$. Also allowed were models where the first order (i.e., the once-scattered) spectrum from the energetic pairs ended at $\sim m_e c^2$. The requirement of this special set of values, for which there was no particular justification, seemed to suggest that the pair models as developed at that time were too simplistic. But as we shall see below, it soon emerged from observations that X-ray spectra were more complex than simple power laws, and that pair models could be made consistent with this new data if a fraction of the photons produced were reflected from cold matter external to the region.

4.6.5   *Compton reflection from a cold gas*

While the first measurements of the X-ray spectra of Seyfert 1 galaxies in the $\sim$ 2–10 keV range indicated a simple power law with index $\alpha_x = 0.7 \pm 0.15$, later observations, especially those from the Japanese X-ray satellite *GINGA*, have revealed a much more complex spectrum (see Section 11.1 for a discussion and references). The features include, apart from the power law with $\alpha_x \sim 0.7$, a flattening of the continuum above 10 keV, absorption edges above a mean energy of ~ 8 keV and line emission at ~ 6.4 keV with an equivalent width of ~ 150 eV. It has been possible to explain this complex shape by invoking the reflection of part of the emitted X-ray spectrum from a slab of cold matter, so that the observed spectrum is the sum of a direct and a reflected component.

The reflection of X-rays has been discussed over the years in the context of the Solar photosphere, accretion disks in X-ray binaries and so on (see George and Fabian 1991 for references). In the context of AGN, the effects of reflection from cold matter, in the form of a disk, filament or cloud, were first considered by Guilbert and Rees (1988) and their illuminating arguments were followed by several detailed calculations, some of which we will refer to below.

At energies $\lesssim 1$ keV excess emission over the extrapolation of the higher energy power law has been observed in several Seyfert 1 galaxies and quasars. The soft X-ray excess has been observed to vary on a time scale of hours and can be modelled as thermal emission from an optically thick source at a temperature of $\sim 10^5\text{--}10^6$ K. The thermal matter may be present in the form of a disk or clouds close to the central black hole. If such matter is illuminated by X-rays, photons more energetic than 1 keV ($\sim 10^7$ K), would find the matter *cold*, i.e., the photons would lose energy to the electrons in Compton scattering. X-ray photons can also be absorbed through the photoionization of heavy elements that have full K- and L-shells.

The continuum radiation For photoelectric absorption, only the innermost shells with high binding energies are relevant, because of the high energy of the X-ray photons. For absorption to occur it is necessary that the photon has at least as much energy as the binding energy of the electron. For energy ϵ large compared to this threshold, an approximate form for the K-shell absorption cross-section for atomic number Z is $\sigma_K \propto Z^5 (m_e c^2 / \epsilon)^{-3.5}$, for $\epsilon \ll m_e c^2$. When the photon energy is $\gg m_e c^2$, $\sigma_K \propto (m_e c^2 / \epsilon)^{-1}$. In obtaining the total optical depth for absorption, the contribution of the different elements in various states of ionization, weighted by element abundances, are summed over. An approximate expression for the average photoelectric absorption coefficient is given by (see Blandford 1990)

$$\alpha_{\text{ph}} = \left(\frac{Z_{\text{abun}}}{Z_{\odot,\text{abun}}} \right) \left(\frac{E}{10\,\text{keV}} \right)^{-2.5} \alpha_{\text{T}}, \tag{4.82}$$

where the first factor on the right hand side is the total abundance of the heavy elements relative to their Solar abundance and α_{T} is the Thomson scattering absorption coefficient. Absorption dominates over scattering below $\lesssim 10\,\text{keV}$, and photons of such energy are partially absorbed.

At higher energies the absorption coefficient reduces rapidly, and the photons undergo repeated scattering until they are reflected from the slab. It follows from the arguments in subsection 4.6.3 that the energy lost in Compton recoil is significant for $\epsilon \gtrsim m_e c^2 / \tau_{\text{T}}^2$, where τ_{T} is the optical depth to which the photon penetrates in the slab. Monte-Carlo calculations (George and Fabian 1991, see below) show that for the great majority of photons the optical depth reached is $\lesssim 3$, so that the energy loss first becomes significant at $\epsilon \gtrsim 50\,\text{keV}$. For incident photon energy $\gtrsim 200\,\text{keV}$ the scattering cross-section reduces significantly below the Thomson cross-section applicable at low energy (see Figure 4.2) and there is preferential scattering in the forward direction, which allows the photons to penetrate deeper into the medium. This increases the probability that the photon is absorbed or loses a considerable amount of its energy because of multiple scattering. The net effect of the processes acting at low and high energies is to produce a broad hump in the reflected spectrum between $\sim 10\,\text{keV}$ and $\sim 200\,\text{keV}$ (Lightman and White 1988). The distortion produced in an incident power law spectrum due to reflection from a slab of cold matter, including the iron fluorescence line discussed below, is shown schematically in Figure 4.5.

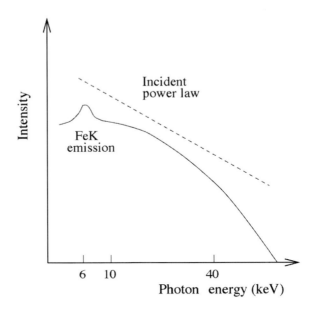

Fig. 4.5. The distortion produced in an incident power law X-ray spectrum owing to reflection from a slab of cold matter. The flux below $\sim 10\,\mathrm{keV}$ is attenuated by absorption, while the flux at energies above $\sim 50\,\mathrm{keV}$ is reduced because photons there lose energy due to Compton recoil. Iron fluorescence emission is indicated.

       Iron fluorescence   Fluorescence is the radiation emitted by an atom after photoionization. Photoionization by X-rays creates a vacancy in the inner shells, leaving the ionized atom in an excited state. The atom can reduce its energy either through radiative transitions, in which the vacated state is filled from an electron in one of the higher levels with the emission of a photon, or through the *Auger effect*. In the latter an electron makes a radiationless transition to the vacancy, the difference in energy being used to release an electron from an outer shell. The absorption edges, the energy of the fluorescent photons and the yield all depend on the ionization state of the atom, because of the screening effect of the electrons in different shells.

      The fluorescent yield of a shell is the probability that a vacancy in that shell is filled by a radiative transition rather than by the ejection of Auger electrons. For a sample of many atoms, this is equal to the number of photons emitted when vacancies in the shell are filled divided by the number of vacancies in the shell created by photoionization. For the K-shell this can be written as

$$Y_Z^K = \frac{N_K}{n_{vK}}, \tag{4.83}$$

where N_K is the number of K-shell X-ray photons emitted from the sample, n_{vK} is the number of primary vacancies and Z is the atomic number (see Bambynek *et al.* 1972

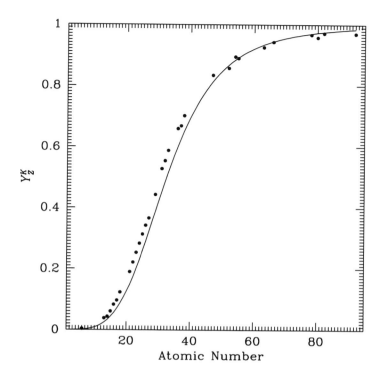

Fig. 4.6. The K-shell fluorescent yield as a function of atomic number. Data points from Bambynek *et al.* (1972) and the analytic form for Y_Z^K in Equation 4.84 are shown.

for definitions and a review). The definition of the fluorescent yield for higher shells becomes more complicated because of the existence of subshells within the shell.

A simple representation of the dependence of Y_Z^K on Z is given by

$$Y_Z^K = \frac{Z^4}{a_K + Z^4},$$ (4.84)

where $a_K = 1.12 \times 10^6$ (B52; see D90 for a discussion). Better analytical approximations which take into account relativistic effects and screening are available. A plot of the function in Equation 4.84, along with experimental data from Bambynek *et al.* is shown in Figure 4.6. In the case of the L-shell, it is found that the fluorescent yield is different for the three subshells. An approximate value for the yield for the L_3-subshell can be obtained from Equation 4.84, with a_K replaced by $a_L = 1.02 \times 10^8$ (see D90). The yields for the other subshells are similar.

The intensity of the K-shell lines of a particular element in a plasma is proportional to the product of the fluorescent yield and the abundance of the element. In Figure 4.7 we have shown this product for the interstellar abundances, relative to hydrogen, as given by Morrison and McCammon (1983). It is seen that iron is the element with

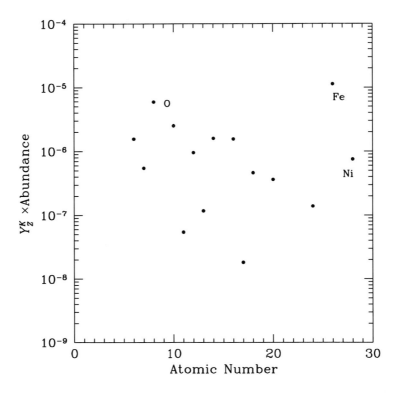

Fig. 4.7. The product of the K-shell fluorescent yield and cosmic abundance of elements.

the highest product and therefore it is iron fluorescence that is most important in the X-ray observations of quasars and AGN in the soft X-ray band.

For iron the K-shell absorption-edge energy varies from $E_K = 7.1$ keV for Fe I to 9.3 keV for Fe XXVI and the photoelectric absorption cross-section per atom decreases from 3.8×10^{-20} cm^2 for Fe I to 3.3×10^{-20} cm^2 for Fe XXVI. For Fe I the K_α fluorescence line consists of two components with energies 6.404 and 6.391 keV respectively, corresponding to transitions from the L_3- and L_2-subshells respectively. The difference in energy, as well as the line widths due to natural broadening and thermal motion, are negligible compared to the energy resolutions available in the X-ray data from space missions and it is enough simply to consider an average energy for the lines. The K_α line energy varies from 6.4 keV in Fe I to 6.9 keV in Fe XXVI. The fluorescent yield for the iron K-shell varies from 0.34 in Fe I to 0.49 in Fe XXII, then varies between 0.11 and 0.75 from Fe XXIII to Fe XXVI (see George and Fabian 1991 and Krolik and Kallman 1987 for these data and references).

When X-rays are incident on a cold slab, K-shell ionizations will be caused by photons with energies above the absorption threshold, leading to iron line emission in the manner described above. The equivalent width (see Section 6.6 for the definition) is given by

$$EW = \frac{\int d\Omega \int_{\epsilon_{\text{th}}} d\epsilon I(\epsilon) Y_Z^K (1 - e^{-\tau_{\text{th}}})}{4\pi I(\epsilon_\alpha)}. \qquad (4.85)$$

The lower limit on the integral over photon energy is the K absorption threshold. The integral over directions is confined to those photons that are emitted in directions such that they can escape to the observer. Taking $\tau_{\text{th}} \leq 1$, since most fluorescent photons from such optical depths escape being absorbed, and using Y_Z^K for iron, an approximate expression for the equivalent width is (Blandford 1990)

$$EW \simeq 300 \left(\frac{\Delta\Omega}{4\pi}\right) \left(\frac{Z}{Z_\odot}\right) \tau_{\text{T}}. \qquad (4.86)$$

For $\Delta\Omega \simeq 2\pi$, $Z = Z_\odot$ and $\tau_{\text{T}} \simeq 1$, $EW(K_\alpha) = 150\,\text{eV}$, which as we shall see in Chapter 10 is in accord with observation.

Reflection models The photons incident on a cold semi-infinite slab of matter are repeatedly scattered, until they are either turned back and re-emerge from the medium (i.e., are 'reflected'), or are absorbed when they photoionize an atom. The photoionized atom loses its energy either by emission of an Auger electron, which loses its energy to the medium, or through fluorescence. The emitted photons again diffuse through the medium. White, Lightman and Zdziarski (1988) have used the Monte-Carlo technique to derive semi-analytic approximations for Green's functions that can be used to calculate the reflected flux and spectrum. This method was applied by Lightman and White (1988) to show that there would be a broad hump in the spectrum between $\sim 10\,\text{keV}$ and $\sim 200\,\text{keV}$. George and Fabian (1991) have applied the Monte-Carlo technique to obtain the intensity, spectrum and angular dependence of the reflected radiation, including the iron fluorescent lines. These authors, however, did not include a model for the generation of the direct continuum, and simply assumed that it had a power law form. Inspired by new data from *GINGA* (see Chapter 10), Zdziarski *et al.* (1990) have examined non-thermal pair models with inclusion of reflection.

Zdziarski *et al.* have considered a compact source of non-thermal pair-processed radiation located above a semi-infinite slab of cold matter. Half the radiation produced by the source escapes directly, while the other half is intercepted by the slab and is absorbed or reflected. The total reflected *albedo*, i.e., the ratio of the number of continuum photons reflected to those incident on the slab, is small and therefore much of the incident radiation is absorbed. The absorbed energy heats the slab to some temperature T_i and the corresponding black body radiation provides the soft photon input to the region in which the Compton scattering and pair processing takes place. The parameters used for the pair model were $kT_i = 3 \times 10^{-5} m_e c^2$, $\gamma_i = 2000$, $l_s = l_i$ and $l_i = 30, 100, 300$. The steady state equations in subsection 4.6.4 were solved using the formalism in Lightman and Zdziarski (1987), and the reflected continuum was

Table 4.1. *Parameters for the models of non-thermal pair cascades with Compton reflection shown in Figure 4.8*

Panel	Compactness parameter l_i	Albedo	Equivalent width iron K_α, eV
top	30	0.17	111
middle	100	0.15	108
bottom	300	0.10	100

evaluated using the Green's function formalism mentioned above. The fluorescent line contribution was evaluated using the Monte-Carlo technique in George and Fabian (1991). The absorption was evaluated using the prescription of Morrison and McCammon (1983) and iron K_α and K_β and nickel K_α lines were used in evaluating the fluorescence emission.

The results of the calculation are shown in Figure 4.8. Because of the high l_i, pair cascades are produced and the direct photon spectrum becomes soft, $\alpha_x \simeq 0.8$–1. But because of the hump in the reflected component, the composite spectrum is harder, with $\alpha_x \sim 0.7$. Above a few tens of keV, there is a break in the spectrum because of the down-scattering of photons by cool pairs, as discussed in subsection 4.6.3. Since nearly all high energy γ-ray photons are lost to pair production, the spectrum steepens sharply above $\sim 2\,\mathrm{MeV}$, with $\alpha_\gamma \simeq 2$. All the features described here are anticipated from the theory developed above. What is interesting is that the combination of direct and reflected components produces generic features in the continuum, which, as we will see in Section 11.1, are in accord with data from *GINGA*.

Fluorescence in the cold matter following photoelectric absorption produces iron and nickel lines at ~ 6–$8\,\mathrm{keV}$ which are prominently seen in Figure 4.8. The lines are predicted to have equivalent widths in the range ~ 100–$110\,\mathrm{eV}$ and to have low energy tails because of (1) the down-scattering of the fluorescent photons during their diffusion through the cold slab, (2) rotation of the cold matter and (3) gravitational redshift. A prominent pair annihilation line is seen at $\sim 500\,\mathrm{keV}$ and, to avoid conflict with observation, this will have to be broadened due to motions in the plasma so that it appears as a broad bump, as mentioned in subsection 4.6.4. The soft photons can be boosted to higher energy as a result of repeated Compton scatterings by pairs in the plasma cooled to the Compton temperature. This produces a soft excess as the power law tail of the black body distribution. We have listed in Table 4.1 some parameters for the spectra in the three panels in Figure 4.8.

It has been assumed in these models that the radiation incident on the slab is half that emitted by the source. This fraction can be increased either if the slab subtends a solid angle $> 2\pi$ by being concave towards the source, or if more flux is emitted towards the slab than away from it. The latter happens because of anisotropic Compton scattering (Ghisellini *et al.* 1991). An increase in the reflected component due to these effects hardens the spectrum, as may be seen in the bottom panel of Figure 4.8.

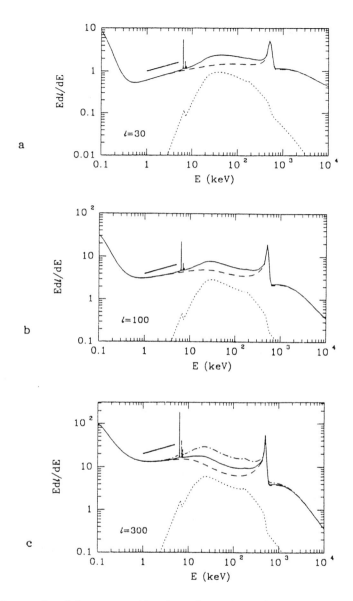

Fig. 4.8. Spectra produced from a combination of non-thermal pair cascades and Compton reflection from a slab of cold matter. The numbers on the y-axis are proportional to $EL(E)$. The parameters for the models are given in the text. The models differ only in the values of l_i, which are indicated in the panels. The dotted lines are the reflected part, the broken lines are the direct part and the continuous lines are the net spectra produced from combining the two. The short solid lines indicate a power law spectrum with $\alpha_x = 0.7$. The fluorescence lines are shown only on the net spectrum for clarity. The dotted and broken line in the bottom panel shows the result of increasing the reflected component by a factor of 3. Reproduced from Zdziarski *et al.* (1990).

4.7   Thermal pair models

The electron–positron pairs considered in Section 4.6 were produced from photon–photon interactions involving γ-rays , which were themselves the result of the inverse Compton scattering of soft photons by highly relativistic electrons. Pairs can also be produced when the temperature of a plasma becomes relativistic, i.e., $\theta \equiv kT/(m_e c^2) \gtrsim 1$, in encounters between particles such as $e^- e^- \rightarrow e^- e^- e^+ e^-$. When the rate of pair production is equal to the rate of pair annihilation, equilibrium can be maintained, but there is a maximum temperature beyond which this is not possible.

For a relativistic plasma which is dominated by pairs, simple expressions for the pair production rate $\dot{n}_{\mathrm{pp}}$ and pair annihilation rate $\dot{n}_A$ are

$$\dot{n}_{\mathrm{pp}} = n_e^2 c\sigma_T \alpha_f^2 \frac{6}{\pi^2}(\ln\theta)^3, \tag{4.87}$$

$$\dot{n}_A = n_e^2 c\sigma_T \frac{3}{16}\theta^{-2}\ln\theta, \tag{4.88}$$

where α_f is the fine structure constant (see Svensson 1990 and Fabian 1994 for reviews and references). The pair production rate increases monotonically with temperature, while the annihilation rate has a maximum at $kT = 1.65 m_e c^2$. For $\theta > 24$, the pair production rate exceeds the annihilation rate and above $\theta = 24$ it is not possible to have a thermal pair plasma in equilibrium, unless the pairs produced can escape from the region.

4.7.1   *Maximum compactness*

When the compactness parameter of a source is large, photon–photon pair production, as well as the scattering of photons by electrons, becomes important. When the Compton scattering time scale becomes shorter than the bremsstrahlung time scale and the Comptonization parameter $y \gg 1$, Wien equilibrium is reached. For a plasma consisting only of the pairs and radiation, the ratio of the photon density to the electron density is given by (Svensson 1994)

$$\frac{n_\gamma}{n_e} = \frac{8}{\pi}\theta^{3/2}e^{1/\theta}g^{-1}, \tag{4.89}$$

where a numerically fitted expression for g, accurate to 0.06 per cent, is

$$g = 1 + 3.7762\theta + 5.1054\theta^2 + \frac{8}{\pi}\theta^3. \tag{4.90}$$

For relativistic temperatures $\theta \gg 1$, $g \simeq 8\theta^3/3$ and $n_\gamma/n_e \rightarrow 1$. For non-relativistic temperatures $\theta \ll 1$, $g \simeq 1$.

In the steady state, the luminosity of a spherical source of radius R is given by

$$L = \frac{4\pi R^3}{3}3kT\dot{n}_\gamma, \tag{4.91}$$

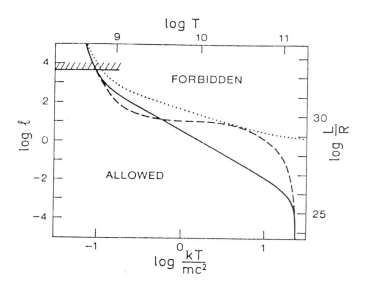

Fig. 4.9. Maximum compactness as a function of temperature. Thermal equilibrium can occur only in the regions below the curves. The dominant radiation mechanism is Comptonized bremsstrahlung for the broken curve, and Comptonized soft photons with $\alpha_x = 0.7$ for the other two curves. The dominant pair destruction mechanism is pair escape for the dotted curve and pair annihilation for the other two curves. Reproduced from Svensson (1986).

where $3kT$ is the average energy of photons with the Wien distribution (see subsection 4.3.3) and $\dot{n}_\gamma$ is the rate at which photons are produced, which in the steady state is equal to the rate at which photons diffuse out of the system. Therefore, from the discussion in subsection 4.6.4, $\dot{n}_\gamma \simeq n_\gamma / \tau_{\text{esc}}$. Following Svensson (1990),

$$\tau_{\text{esc}} = \frac{R}{3c}\tau_T g_\tau, \tag{4.92}$$

where $\tau_T = 2n_e \sigma_T R$ and g_τ is the Klein–Nishina correction to the Thomson scattering cross-section, given by

$$g_\tau = 1 + 5\theta + 0.4\theta^2, \quad \theta \leq 1, \tag{4.93}$$

and

$$g_\tau = \frac{3}{16}\theta^{-2}\left[\ln(1.12\theta) + 0.75\right]\left(1 + \frac{0.1}{\theta}\right)^{-1}, \quad \theta > 1. \tag{4.94}$$

From Equations 4.91 and 4.92 and the definition of the compactness parameter l in Equation 4.67, it follows that (Svensson 1990)

$$l = 4(2\pi)^{1/2}\theta^{5/2}e^{1/\theta}(g_\tau g/3)^{-1}. \tag{4.95}$$

This represents the maximum compactness for a temperature T for which thermal pair equilibrium is possible. For different radiation processes, the maximum compactness takes on a different form. Some of these are shown in Figure 4.9. For high values of the compactness parameter, as are found in the case of AGN (see subsection 4.6.1), the curves in Figure 4.9 imply that the temperature $\theta \lesssim 0.3$. This is consistent with the observational requirement that the spectra steepen around a few hundred keV. Thermal plasma, together with reflection from a cold slab to account for the line emission and other features as described in subsection 4.6.5, appear to be able to explain the observed high energy spectra.

4.8   Concluding remarks

In this chapter, and the previous one, we have described the more important aspects of radiative processes and relativistic beaming used in the interpretation of quasar and AGN spectra over the vast frequency range which is now accessible to observation. In the next chapter we shall consider the 'standard model' of the central engine which is the ultimate origin of the tremendous power produced by these sources. We will then go on to consider observations in different wavelength regions and the theoretical models used in their interpretation.

5 The standard model

5.1 Introduction

The early ideas of Hoyle and Fowler (1963) concerning gravitational collapse to a compact object that would serve as an energy reservoir for a quasar found a modified expression in the black hole accretion disk paradigm, a few years later. This paradigm had been invoked and worked well in the understanding of binary X-ray sources in the Galaxy. In the binary star context the compact member is taken to be either a neutron star or a black hole with mass of stellar order. For quasars and AGN, the compact masses would have to be several orders of magnitude higher, as already pointed out by Hoyle and Fowler (1963). The scenario here had therefore to explain how such objects form in the first place, how they generate an accretion disk and jets, and how and with what efficiency is the gravitational energy converted to the observed radiant energy.

In this brief review of the current thinking on the subject we shall follow the excellent account given by Rees (1984) whose basic tenets have remained more or less the same since then.

5.2 The formation of a massive black hole

As first pointed out by Hoyle and Fowler (1963), the energy source of a quasar or AGN is gravitational and could arise from a highly collapsed object or a massive black hole. This much is broadly agreed by most workers in the field. The question is, in the first place how does a collapsed massive object come about?

Figure 5.1, taken from Rees's article mentioned above, lays out a variety of scenarios all having as initial conditions known astrophysical systems such as gas clouds, dense star clusters etc. It is not known in quantitative terms how the sequences of development would follow. The general expectation is that given the large masses of these systems, they will evolve in such a way that eventually gravity will begin to dominate.

In Newtonian gravitation, we know that the gravitational potential energy of a mass M having radius R is of order

$$\Omega = -\frac{GM^2}{R}. \tag{5.1}$$

Notice the sign in the above equation signifying that the energy is negative and goes on decreasing as R decreases. The gravitational force itself acts in such a way as to

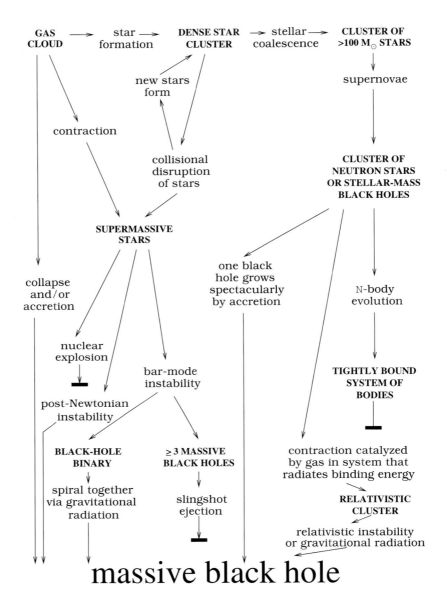

Fig. 5.1. Scenarios leading to the formation of a collapsed massive object. After Rees (1984).

bring about a decrease of R. Thus the system is unstable, in contrast to a situation where an attractive force of positive energy is operating; e.g., in the case of a stretched elastic spring the two ends of the spring are attracted towards each other, but the force of attraction disappears as the spring shrinks to its natural length with its potential energy attaining a *finite minimum* value. Not so in the gravitational case, where the force of attraction *grows* as the system evolves towards lower energies; thus a *gravitational collapse* ensues. The Newtonian gravitational collapse is relatively simple to understand. The collapsing object eventually shrinks to a point and thus to a state of infinite energy within a finite time. In general relativity, however, the scenario is more complex.

The relativistic equations are complicated, but can be handled to some extent in the case of the spherical collapse of a dust ball. The classic problem was first solved by B. Datt (1938) and applied in the context of neutron stars by J.R. Oppenheimer and H. Snyder (1939). For detailed solutions we refer the reader to standard texts in relativity. The special solutions referred to above, however, indicate the general scenario, namely that the collapsing object is enveloped within an *event horizon* that prevents the transmission of any signals from inside to outside. The general expectation is (though no rigorous proof exists) that the only items of information about such an object that are left for the outside observer are its mass, angular momentum and electric charge. This is when the object is said to become a *black hole*.

We will hop across the unworkable details of this scenario and assume that the end product, a massive black hole, is somehow formed. Since in general we do not expect any charge separation to survive, the black hole will have no electric charge but will have mass and angular momentum. A black hole of this kind is called a *Kerr* black hole, while a black hole with no angular momentum is called a *Schwarzschild* black hole. These names are given to recognize the pioneering work done by Roy Kerr (for the axially symmetric solution) and by Karl Schwarzschild (for the spherically symmetric solution) in general relativity.

5.3   The black hole environment

Before considering the general relativistic equations, let us briefly examine the Newtonian scenario. Consider a particle moving in the gravitational field of a spherically symmetric mass M, its radial position with respect to the centre of M at time t being given by $\mathbf{r}(t)$. Then the energy equation is

$$\frac{\dot{\mathbf{r}}^2}{2} - \frac{GM}{r} = E, \quad r = |\mathbf{r}|,$$ (5.2)

E being the total energy of the particle. If the angular momentum of the particle per unit mass is h then the Newtonian laws of motion give

$$\frac{\dot{r}^2}{2} + V(r) = E,$$ (5.3)

where

$$V(r) = \frac{h^2}{2r^2} - \frac{GM}{r}$$ (5.4)

is the *effective potential* for the motion. When $h = 0$ the effective potential reduces to the usual Newtonian potential, while for $h \neq 0$ it allows the repulsive centrifugal force to be taken into account. A particle having an orbit with $E < 0$ cannot escape to infinity, since as $r \to \infty$ $V(r) \to 0$ and $\dot{r}^2$ would become negative. Orbits with $E < 0$ are therefore bound, with the particle moving in an ellipse having M at one of the foci. When $E \geq 0$, the particle can escape to infinity, $E = 0$ being the marginally free case. The notation used here is commonly found in most standard texts in relativity. Effective potentials are discussed in detail in Misner, Thorne and Wheeler (MTW73). We will freely use results from this and other texts.

In general relativity we use the Schwarzschild line element to describe space–time in the vicinity of the mass M (assumed here to be spherically symmetric):

$$ds^2 = \left(1 - \frac{2GM}{c^2 r}\right) c^2 dt^2 - \left(1 - \frac{2GM}{c^2 r}\right)^{-1} dr^2 - r^2 \left(d\theta^2 + \sin^2 \theta d\phi^2\right). \qquad (5.5)$$

When $GM/(c^2 r) \ll 1$ the gravitational field is weak and the Newtonian approximation applies. When $GM/(c^2 r) \simeq 1$, the gravitational field is strong and it is necessary to use general relativity. We will consider this case now, assuming that the typical 'test' particle has mass small enough that it does not disturb the space–time as described by Equation 5.5.

The equation of motion of a test particle in the Schwarzschild space–time has the following first integrals:

$$\theta = \text{constant} = \pi/2, \qquad (5.6)$$

which states that the particle moves in a plane (this may be taken to be $\theta = \pi/2$ without any loss of generality);

$$r^2 \frac{d\phi}{ds} = \text{constant} = h, \qquad (5.7)$$

which expresses the conservation of angular momentum, hc being the angular momentum of the test particle per unit mass, and

$$\frac{dt}{ds} = \frac{E}{c} \left(1 - \frac{2GM}{c^2 r}\right)^{-1}, \qquad (5.8)$$

which is the statement of the conservation of energy, with E the dimensionless energy per unit mass. These first integrals can be used in Equation 5.5 to give

$$\left(\frac{dr}{ds}\right)^2 + V^2(r) = E^2, \qquad (5.9)$$

where $V(r)$ is the *relativistic effective potential*, with

$$V^2(r) = \left(1 - \frac{2GM}{c^2 r}\right) \left(1 + \frac{h^2}{r^2}\right). \qquad (5.10)$$

The effective potential is plotted in Figure 5.2, for various values of the parameter $H \equiv c^2 h/(GM)$, along with the Newtonian expression. Normally the physical radius of an object of mass M would be much larger than $2GM/c^2$ and the details noted in Figure 5.2 would not matter, but for strong-field conditions near a black hole they do.

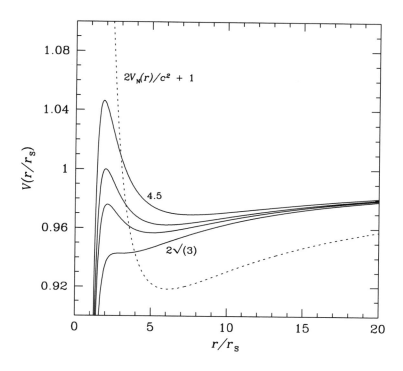

Fig. 5.2. The general relativistic effective potential as a function of the radial distance from a black hole. Potentials for $H = 4.5$ and $H = 2 \times 3^{1/2}$ are labelled. The other two curves are for $H = 4$ and $H = 3.8$ respectively. The dotted line is a Newtonian potential well for an angular momentum per unit mass of $12.25GM/c$, where M is the mass of the black hole. The function actually plotted is indicated by the label.

For motion to be possible, the energy constant E has to exceed $V(r)$. $V(r)$ as a function of r has a minimum as well as a maximum when the angular momentum per unit mass $h \geq 2\sqrt{3}GM/c^2$, i.e., $H \geq \sqrt{12}$. It is therefore possible to have stable as well as unstable circular orbits, the former being at the minimum in the potential, with $E^2 = V^2(r_{min})$, and the latter at the maximum with $E^2 = V^2(r_{max})$, where r_{min} and r_{max} are the radii at which the minimum and maximum in the potential occur. For $E^2 \geq V^2(r_{min})$ bound orbits are possible, the radius r remaining between turning points at which $E^2 = V^2(r)$. An important difference between Newtonian and general relativistic bound orbits is that the former are closed while the latter are not, which leads to a precession of the orbit in the latter case. Observation of the precession of the perihelion of Mercury was crucial to making Einstein's theory acceptable.

For $h < 2\sqrt{3}GM/c^2$ there is no minimum in $V(r)$ and a particle moving inwards always falls into the black hole. Also, for *any* h when $E^2 > V^2(r_{max})$ there are no turning points, and again a particle moving inwards always falls into the black hole. This is unlike the Newtonian case, where $V(r) \to \infty$ as $r \to 0$ for any value of $h \neq 0$

and a particle with non-zero angular momentum always manages to avoid falling into the centre.

Stable circular orbits occur at

$$r_{\rm sco} = \frac{GM}{2c^2}\left[H^2 + (H^4 - 12H^2)^{1/2}\right], \tag{5.11}$$

provided, of course, that h exceeds the threshold mentioned above, i.e., $H^2 \geq 12$. On a circular orbit the specific angular momentum is

$$l = hc = \sqrt{\frac{GMr^2}{r - 1.5r_{\rm S}}}, \tag{5.12}$$

where $r_{\rm S} = 2GM/c^2$ is the Schwarzschild radius. Since $dr/ds = 0$ for a circular orbit, the energy is given by $E_{\rm sco} = V(r)$ with the effective potential $V(r)$ as in Equation 5.10 and h as in Equation 5.12:

$$E_{\rm sco} = \mu c^2 \frac{r - r_{\rm S}}{\sqrt{r(r - 1.5r_{\rm S})}}. \tag{5.13}$$

The difference between the rest mass energy and the binding energy is extracted from the particle, and the efficiency for the process is

$$\epsilon = (\mu c^2 - E_{\rm sco})/\mu c^2. \tag{5.14}$$

The innermost stable circular orbit has $H^2 = 12$, and is situated at $r_{\rm sco,min} = 6GM/c^2$. On this orbit the energy is $\sqrt{8}\mu c^2/3$ and the efficiency with which energy is extracted up to this point is given by

$$\epsilon_{\rm max} = \left(1 - \frac{\sqrt{8}}{3}\right)\frac{\mu c^2}{\mu c^2} = 0.057. \tag{5.15}$$

One can assume that the gas that spirals through a thin accretion disk towards the central black hole, releasing energy in the process, does so through a series of approximately circular orbits. The energy release per unit mass when the gas reaches the last circular orbit is $\sqrt{8}/3$. As the energy decreases further, since there are no more stable orbits the gas quickly falls into the hole, on a time scale short compared to the radiative time scale, so that no further energy is available to the outside world through the infall. The maximum efficiency for energy extraction for a Schwarzschild black hole is therefore 0.057 as in Equation 5.15. The circular orbit with $r = 2r_{\rm S}$ has $l = 2cr_{\rm S}$, i.e., $h = 2r_{\rm S}$, $H = 4$ and zero binding energy.

The case of the rotating black hole is more involved in its details but similar in principle. Such a black hole is described by the Kerr line element, which is given in *Boyer–Lindquist coordinates* by

$$ds^2 = \frac{\Delta}{\rho^2}(cdt - a\sin^2\theta d\phi)^2 - \frac{\sin^2\theta}{\rho^2}[(r^2 + a^2)d\phi - acdt]^2 - \frac{\rho^2}{\Delta}dr^2 - \rho^2 d\theta^2. \tag{5.16}$$

Here ac is the angular momentum per unit mass of the black hole about the polar axis and

$$\Delta = r^2 + a^2 - 2GMr, \quad \rho^2 = r^2 + a^2 \cos^2 \theta. \tag{5.17}$$

In the case of the rotating black hole the effective potential in the equatorial plane is defined as the minimum energy per unit mass which is required for possible motion at that point, and is given by

$$
\begin{aligned}
V(r) &\equiv E_{\min}(r) \\
&= \frac{(r^2 - 2mr + a^2)^{1/2}\{r^2h^2 + [r(r^2 + a^2) + 2a^2m]r\}^{1/2} + 2ahm}{[r(r^2 + a^2) + 2a^2m]},
\end{aligned} \tag{5.18}
$$

where $m = GM/c^2$. The radius of the *event horizon* of the Kerr black hole, inside which particles are permanently trapped and cannot escape to the outside, is

$$r_+ = m + \sqrt{m^2 - a^2}. \tag{5.19}$$

For zero angular momentum this reduces to the Schwarzschild radius of a spherically symmetric black hole. The event horizon disappears when $a = m$ i.e., the angular momentum per unit mass is mc. Since the event horizon is required to exist by the *cosmic censorship hypothesis*, one must always have $a < m$, and the case with $a \to m$ is called the *extreme Kerr* or *maximally rotating* black hole.

We now look at a situation similar to that described by Figure 5.2, with the difference that there we had only the case $a = 0$ while now we consider $a \neq 0$. The innermost stable circular orbit for the rotating black hole occurs at a radius $r_{\text{sco,min}}$ given by

$$r_{\text{sco,min}} = m \left[3 + B \mp \sqrt{(3 - A)(3 + A + 2B)} \right], \tag{5.20}$$

where

$$A = 1 + (1 - x^2)^{1/3} \left[(1 + x)^{1/3} + (1 - x)^{1/3} \right], \quad B = (3x^2 + A^2)^{1/2}, \tag{5.21}$$

with $x = a/m$ (for details see Bardeen 1973). The minus and plus signs denote the two cases: minus for the case where the particle is orbiting in the same angular sense as the rotation of the black hole and plus for the counter-orbiting case. For the extreme Kerr case $a \to m$ we get $r_{\text{sco,min}} = m$ in the co-rotating case and $r_{\text{sco,min}} = 9m$ in the counter-rotating case.

The last stable circular orbit in the equatorial plane corresponds to the maximum efficiency of energy extraction:

$$\epsilon_{\max} = 1 - \frac{r_{\text{sco,min}} - 2m \pm a(m/r_{\text{sco,min}})^{1/2}}{\sqrt{r_{\text{sco,min}} \left[r_{\text{sco,min}} - 3m \pm 2a(m/r_{\text{sco,min}})^{1/2} \right]}}. \tag{5.22}$$

By taking the limit of this expression as $a \to m$ we get a maximum efficiency of about 42 per cent for the co-rotating case. However, one should bear in mind that these are not situations common in astrophysics. Indeed if one goes by the laws of black hole physics, which are akin to the laws of thermodynamics, then attainment of the extreme Kerr state is as difficult as attainment of the absolute zero of temperature. These calculations are generally given to justify the expectation that the black hole may allow energy extraction up to around ten per cent of its rest mass energy. This may be compared to the nuclear fusion case where a fraction ~ 0.007 of the rest mass energy of hydrogen is converted to radiation.

The Kerr black hole has a critical radius, called the *static limit*, inside which the light cones always point in the ϕ-direction, because of the dragging of inertial frames by the spinning hole. The result is that particles which move on time-like geodesics with fixed r, θ are necessarily in orbit with respect to non-rotating observers at infinity. The static limit is given by

$$r_{\mathrm{E}} = m + \sqrt{m^2 - a^2 \cos^2 \theta}. \tag{5.23}$$

For $\theta = 0$ the static limit has the same radius as the event horizon, while for all other angles it is outside the event horizon. The region between the event horizon and the static limit is called the *ergosphere*.

Energy extraction   The ergosphere of a rotating black hole has the property that particles moving in it with orbits crossing the event horizon can have negative total energy, which in the Newtonian approximation includes the gravitational, kinetic and rest-mass energy. These orbits are contained within the static limit. When a particle on such an orbit crosses the event horizon, the mass of the hole decreases. This can be used in principle to extract energy from the black hole via the *Penrose process* (Penrose 1969), in which a particle with positive energy enters the ergosphere and splits into two particles, one of which has negative energy and is in an orbit that takes it into the event horizon. The other particle acquires more positive energy than the original particle had, and leaves the ergosphere. The net result is that energy is extracted from the black hole, which then has less angular momentum than it did before the event. The Kerr black hole has two kinds of energy: that due to the spin of the hole, and that due to an irreducible mass. The latter is equal to $m/\sqrt{2}$, so that the fraction which can be extracted is $(\sqrt{2} - 1)/\sqrt{2}$, i.e., 29 per cent of the rest mass of the black hole. Note that this is the efficiency with which the mass–energy of the rotating black hole is extracted, while the efficiency in Equation 5.22 refers to the extraction of energy from a particle as it falls from infinity into the last stable orbit, while co-rotating with the black hole.

Though the Penrose process is very efficient and could provide immense sources of energy, it has not been possible to imagine realistic scenarios in which the energy can actually be extracted. Some simplified processes involve the scattering of photons by charged particles inside the ergosphere. In the absence of an electromagnetic field the Penrose process requires that for a fragment to attain negative energy it should have a speed exceeding $c/2$ relative to the other fragment; this translates to unacceptable requirements and low efficiency for the energy extraction. Dadhich and his co-workers (see e.g. Wagh, Dhurandhar and Dadhich 1985; Wagh and Dadhich 1989) have considered the Penrose process in the presence of a magnetic field associated with the accretion disk. The twisting magnetic field lines give rise to a quadrupole electric field. The threshold energy required to get a particle into a negative energy orbit can now come from electromagnetic interaction, so that an exorbitant amount of kinetic energy is not required. It has been demonstrated that the process can have very high efficiency when the accretion of discrete particles is considered, but details of the complex magnetohydrodynamic processes have not been worked out.

Another process for extracting black hole rotational energy using magnetic fields is the Blandford–Znajek mechanism (Blandford and Znajek 1977), which depends on the fact that the event horizon of a black hole behaves like a spinning conducting surface, though not a perfect one, with surface resistivity 377 ohms. A spinning hole embedded in a magnetic field acquires a quadrupole distribution of electric charge and a corresponding poloidal electric field. In analogy with a unipolar inductor, power can be extracted by having a current flow between the spinning hole's equator and poles. When the angular velocity of the field lines at infinity is zero, the maximum power extracted is $\sim B^2 a^2 c$, where B is the uniform magnetic field in which the hole is embedded. The efficiency of energy extraction in slowing down a hole which starts off spinning maximally is 9.2 per cent. Descriptions of the Blandford-Znajek mechanism can be found in Blandford (1990), Rees (1984) and Wiita (1991). We will see in Section 5.6 that the Blandford–Znajek mechanism can operate in *ion supported tori*, which act as geometrically thick accretion disks.

Lens–Thirring precession This is an interesting effect due to the coupling of the angular momentum **L** of an orbit to the spin angular momentum **S** of a rotating black hole. In the post-Newtonian approximation, the torque **G** exerted on a ring of matter in orbit around a slowly rotating black hole is given by

$$\mathbf{G} = \frac{2\mathbf{S} \times \mathbf{L}}{r^3}. \tag{5.24}$$

When the orbit is *not* in the equatorial plane, it precesses with the frequency

$$\Omega_P = \frac{2\mathbf{S}}{r^3}. \tag{5.25}$$

Close to the black hole, when the precession period $2\pi/\Omega_P$ is less than the inflow time scale in a disk, the flow will develop axial symmetry relative to the spin axis of the hole, irrespective of the original angular momentum of the infalling material. If the axis of a radio jet is coupled to the disk axis, then its precession due to the Lens–Thirring effect would affect the morphology of the large scale radio structure.

5.4 Characteristic quantities

While there is great variety in the properties exhibited by quasars and AGN, there are several characteristic quantities that have typical values which must be reproduced by all models. We will consider some of these before proceeding to the theory of accretion disks.

An important characteristic is the Eddington luminosity L_{Edd}, which is associated with spherical accretion. Photons which are emitted due to the energy released during accretion exert an outward pressure on the electrons in the matter due to Compton scattering, while gravity exerts an inwards force on the protons.[1] As the luminosity of an object increases, the radiation pressure on the electrons can increase so much that the net force on electron–proton pairs acts outwards, so that there is no further accretion. This reduces the luminosity until the forces again become equal; the luminosity L_{Edd} at which this happens is the maximum achievable luminosity. For spherically symmetric accretion with Thomson scattering the Eddington luminosity is given by

$$L_{Edd} = \frac{4\pi G M m_p c}{\sigma_T} \simeq 1.3\times10^{46} \left(\frac{M}{10^8 M_\odot}\right) \text{ erg sec}^{-1}, \tag{5.26}$$

where M is the mass of the accreting body. There are several useful quantities that are associated with this luminosity. The Eddington accretion rate, at which emission at the Eddington luminosity can be sustained, given unit efficiency for conversion of mass into radiation, is

$$\dot{M}_{Edd} = \frac{L_{Edd}}{c^2} \simeq 0.23 \left(\frac{M}{10^8 M_\odot}\right) M_\odot \text{ yr}^{-1}. \tag{5.27}$$

Using Equation 5.26 and again assuming unit efficiency, we get

$$\dot{M}_{Edd} = \left(\frac{4\pi G m_p}{c\sigma_T}\right) M = \frac{M}{t_{Edd}}, \tag{5.28}$$

where the Eddington time t_{Edd} is defined as the time taken for a body to radiate its entire rest mass at the Eddington rate and is given by

$$t_{Edd} = \frac{M}{\dot{M}_{Edd}} \simeq 4\times10^8 \text{ yr.} \tag{5.29}$$

For a black hole radiating at the Eddington rate, the rate of increase of mass is

$$\frac{dM}{dt} = \dot{M}_{Edd} = \frac{M}{t_{Edd}}, \tag{5.30}$$

which can be integrated to give

$$M = M_0 \exp\left(\frac{t}{t_{Edd}}\right), \tag{5.31}$$

where M_0 is the mass of the black hole at the onset of accretion at $t = 0$.

The Eddington temperature T_{Edd} is the characteristic black body temperature required for radiation at the rate L_{Edd} from a body with the Schwarzschild radius:

$$T_{Edd} = \left(\frac{L_{Edd}}{4\pi\sigma r_S^2}\right)^{-1/4} \simeq 6.6\times10^5 \left(\frac{M}{10^8 M_\odot}\right)^{-1/4} \text{ K.} \tag{5.32}$$

---

[1] The force of gravity on the electrons can be neglected owing to their small mass relative to the protons, while the scattering cross-section for the proton is lower by a factor $(m_e/m_p)^2$ than the cross-section for electrons.

The Eddington magnetic field strength B_{Edd} has an energy density equal to the radiant energy density at the Schwarzschild radius of a body emitting at the Eddington limit:

$$B_{Edd} = \left(\frac{2L_{Edd}}{r_S^2 c} \right)^{-1/2} \simeq 6 \times 10^4 \left(\frac{M}{10^8 M_\odot} \right)^{-1/2} \text{G.} \tag{5.33}$$

The field strengths induced by accretion flows can be expected to be of this order.

The Schwarzschild radius of a body of mass M is

$$r_S = \frac{2GM}{c^2} \simeq 3 \times 10^{13} \left(\frac{M}{10^8 M_\odot} \right) \text{cm.} \tag{5.34}$$

The characteristic minimum time scale for variability is the light travel time across the Schwarzschild radius:

$$t_S = \frac{r_S}{c} \simeq 10^3 \left(\frac{M}{10^8 M_\odot} \right) \text{sec.} \tag{5.35}$$

We end this section by considering an useful observational constraint on the efficiency of spherical accretion, obtained by A. Fabian (1979). Consider a source with luminosity L, which is variable on a time scale Δt. For an optically thin source, the requirement that different parts of the varying source are causally connected translates to the constraint $R \lesssim c\Delta t$. But if the source has significant optical depth due to electron scattering, $\tau_{sc} \geq 1$, then the effective diffusion velocity of radiation in the source is $c/(1 + \tau_{sc})$ (see Equation 4.33). The restriction on the source size is therefore given by

$$R \lesssim \frac{c\Delta t}{1 + \tau_{sc}}. \tag{5.36}$$

The mass involved in the emission from the source is

$$M \lesssim \tfrac{4}{3}\pi R^3 n m_p, \tag{5.37}$$

where n is the number density of particles and m_p the proton mass. The luminosity is related to M by

$$L = \eta M c^2 / \Delta t, \tag{5.38}$$

Using these relations and $\tau_{sc} = n\sigma_T R$, we get the following lower limit on the efficiency factor:

$$\eta \geq \frac{eL\sigma_T}{4\pi c^4 m_p \Delta t} \geq 0.12 \left(\frac{L}{10^{42} \text{ erg sec}^{-1}} \right) \left(\frac{\Delta t}{1 \text{ sec}} \right). \tag{5.39}$$

Some very rapidly varying strong sources violate this constraint because for them the efficiency factor η becomes greater than unity. Relativistic beaming and rotational anisotropy have to be invoked to get round these problems. We will return to this issue in the final chapter.

5.5   The accretion disk

We will consider in this section some essential features of accretion disks, which are central to the standard scenario for energy generation in a quasar or an AGN. Considerable literature exists on this, both at the professional and pedagogic level, and our treatment will be very sketchy, the main aim being to summarize the results that we will use elsewhere in the book. Details may be found in many excellent papers, review articles (e.g. Shakura and Sunyaev 1973, Novikov and Thorne 1973, Pringle 1981, Rees 1984, Blandford 1990) and books (e.g. ST83, FKR92).

The idea of accretion onto stars was first discussed by Bondi, Lyttleton and Hoyle in the 1950s (for a good reference see H. Bondi 1952). The idea involves spherical accretion and gas dynamics in a gravitational field. The spherical case shows that the steady accretion rate $\dot{M}$ onto a star of mass M is determined by the ambient conditions in the interstellar medium at large distances and the flow condition (whether subsonic or supersonic) at the surface of the star. The resulting value of $\dot{M}$ is, however, too small to be of any observational significance.

The situation improves of course for close binary systems, where the tidal force of a compact component (a neutron star or black hole) generates an appreciable flow of material from the extended companion. In the Roche model, which applies to a pair of stars that go around each in other in circular orbits with the spin period synchronized with the orbital period, the flow begins when the extended star fills a critical equipotential surface called the Roche lobe (see ST83, FKR92). The flow occurs through the *Lagrangian point* L_1, at which the critical surfaces around the two stars touch each other and gravity is balanced by the centrifugal force, leaving the pressure force to squirt the matter towards the compact object.

The matter at L_1 has considerable specific angular momentum, and forms a ring around the compact object.[1] Due to viscous dissipation in the ring, the matter in the ring gets heated and radiates energy and sinks further into the black hole to compensate for the energy loss with an increase in the gravitational binding energy. The decrease in the angular momentum of the sinking matter is compensated by expansion of the outer parts. The time scale for energy loss is far shorter than the time scale for loss of angular momentum and the gas may always be assumed to be in the lowest energy state permitted for the specific angular momentum, i.e., it is at the bottom of the effective potential wells in Figure 5.2 and the orbits may be assumed to be circular. The net result is the formation of an accretion disk, with the inner edge at the radius $r_{\mathrm{sco,min}}$ corresponding to the last stable orbit when the compact object is a black hole, and at the radius of the star for other kinds of object. At radii less than $r_{\mathrm{sco,min}}$ the infall time scale is less than the time scale of energy loss, so that the matter may be assumed to fall into the black hole without further emission.

For a massive black hole at the nucleus of a galaxy, the scenario is not so clear.

---

[1]   When the compact object is a black hole, the ring occurs when the specific angular momentum exceeds the minimum value $2\sqrt{3}GM/c^2$ required for a stable orbit (see Section 5.3).

If the black hole is spherical we might have spherical accretion. For a spinning black hole, angular momentum considerations lead to the formation of a disk around the black hole. The disk may be thin or thick, depending on whether its thickness H is small compared to or comparable to its radius R.

5.5.1 *Thin accretion disks*

The theory of thin disks is reasonably well worked out, and excellent reviews may be found in the references cited at the beginning of this section. It is assumed that the infalling gas is in a nearly circular orbit at each radius r. The circular velocity at r is $v_\phi = r\Omega(r)$, where $\Omega(r)$ is the angular velocity. We will see below that $\Omega(r)$ is close to the Keplerian value:

$$\Omega(r) \simeq \Omega_K(r) = \left(\frac{GM}{r^3}\right)^{1/2}.$$
(5.40)

The fluid rotates differentially, and therefore viscosity and any other dissipative processes that may be active cause energy to be dissipated into heat and then radiated away. As gas at radius r loses energy in this manner, it sinks deeper into the gravitational potential well, and so has a small radial drift velocity v_r.

The equations that describe the time evolution of the surface density of the gas, $\Sigma \equiv \Sigma(r, t)$, can be obtained from mass and angular momentum conservation and are given by

$$r\frac{\partial\Sigma}{\partial t} + \frac{\partial}{\partial r}(rv_r\Sigma) = 0,$$
(5.41)

and

$$r\frac{\partial}{\partial t}(r^2\Omega\Sigma) + \frac{\partial}{\partial r}(rv_r r^2\Omega\Sigma) = \frac{1}{2\pi}\frac{\partial}{\partial r}G,$$
(5.42)

where v_r is the radial infall velocity of the gas. $G \equiv G(r, t)$ is the torque acting on a thin annulus at r due to the viscous drag of the neighbouring outer annulus. The torque is given by

$$G = 2\pi r\left(v\Sigma r\frac{d\Omega}{dr}\right)r,$$
(5.43)

where v is the kinematic viscosity. Using this expression for $G(r, t)$ in Equation 5.42 and eliminating v_r using Equation 5.41 gives the single equation

$$\frac{\partial\Sigma}{\partial t} = \frac{3}{r}\frac{\partial}{\partial r}\left[r^{1/2}\frac{\partial}{\partial r}(vr^{1/2}\Sigma)\right].$$
(5.44)

The kinematic viscosity v is in general a function of Σ, r and t, and Equation 5.44 is a non-linear diffusion equation for Σ. If v does not depend on Σ, the equation is linear in Σ. The effect of viscosity is clearly seen by taking v to be constant and then solving Equation 5.44, assuming that initially the matter is distributed in the form of a thin ring. Examination of the solution (Pringle 1981) shows that the ring spreads out due to the action of the viscosity, with most of the mass losing energy and angular momentum and moving inwards. A small amount of mass moves to larger radii so that the total angular momentum is conserved. The specific angular momentum $r^2\Omega(r)$ is $\propto r^{1/2}$, and becomes very large at large radius, so that as Pringle (1981) puts it, '...eventually all of the matter initially in the ring ends up at the origin and all of the angular momentum is carried to infinite radius by none of the mass!'.

In applying the theory of accretion disks to stellar systems such as X-ray binaries or to energy generation in quasars and AGN, one is concerned with *steady disks* and not their time evolution. It is assumed that in the steady state the matter in the disk moves along almost circular orbits, with only a slow drift towards the centre. There is essentially no motion perpendicular to the disk, and the dependence of the pressure p and density ρ in this z-direction can be obtained by using the equation of hydrostatic equilibrium:

$$\frac{1}{\rho}\frac{\partial p}{\partial z} = \frac{\partial}{\partial r}\left[\frac{GM}{(r^2+z^2)^{1/2}}\right]. \tag{5.45}$$

For thin disks, $z \ll r$ and

$$\frac{1}{\rho}\frac{\partial p}{\partial z} = -\frac{GMz}{r^3}. \tag{5.46}$$

The pressure p is in general given by the sum of the gas pressure p_g and the radiation pressure p_r:

$$p = p_g + p_r = \frac{\mathscr{R}\rho T_c}{\mu} + \frac{aT_c^4}{3}, \tag{5.47}$$

where $\mathscr{R}$ is the gas constant, T_c the temperature on the $z = 0$ surface, a is Stefan's constant and μ is the molecular weight. From Equation 5.46 we get an approximate relation for the thickness H at radius r:

$$\frac{H}{r} \sim \frac{c_s}{v_\phi} \sim \frac{1}{\mathscr{M}}, \tag{5.48}$$

where $c_s \sim (p/\rho)^{1/2}$ is the sound speed and $\mathscr{M} = v_\phi/c_s$ is the Mach number. For a thin disk $H \ll r$ and therefore $\mathscr{M} \gg 1$, i.e., the local circular velocity is highly supersonic.

For a steady flow, the radial component of the Euler equation equals zero:

$$v_r\frac{\partial v_r}{\partial r} - \frac{v_\phi^2}{r} + \frac{1}{\rho}\frac{\partial p}{\partial r} + \frac{GM}{r^2} = 0. \tag{5.49}$$

Assuming that the radial velocity v_r is subsonic and v_ϕ is supersonic, it follows from the Euler equation that

$$v_\phi = \left(\frac{GM}{r}\right)^{1/2}\left[1 + \mathscr{O}\left(\frac{1}{\mathscr{M}^2}\right)\right], \tag{5.50}$$

i.e., the circular velocity is close to Keplerian since $\mathscr{M} \gg 1$.

In the steady state Equation 5.41 can be integrated to give

$$\dot{M} = -v_r \Sigma 2\pi R. \tag{5.51}$$

This describes the steady flow of matter into the central object, with $\dot{M} > 0$ since $v_r < 0$. Integration of Equation 5.44 gives

$$-\frac{d\Omega}{dr} v\Sigma = -v_r \Omega\Sigma - \frac{C}{2\pi R^3}, \tag{5.52}$$

where C is the constant of integration and represents the steady flow of angular momentum into the disk.

For the case of an accretion disk around an object of mass M and radius r_*, the circular velocity remains close to Keplerian until a narrow boundary layer of size $L \ll r_*$ is reached; it then quickly acquires the angular velocity of the surface of the star, Ω_*. The gradient of the angular velocity $d\Omega/dr$ therefore vanishes at some point close to the stellar surface, and one can assume that Ω has the Keplerian value $(GM/R_*^3)^{1/2}$ at that point. The integration constant is then given by $C = \dot{M}(GMr_*)^{1/2}$ and Equation 5.52 becomes

$$v\Sigma = \frac{\dot{M}}{3\pi}\left[1 - \left(\frac{r_*}{r}\right)^{1/2}\right]. \tag{5.53}$$

For $r \gg r_*$ the radial velocity is $v_r \simeq -3v/(2r)$, and the infall time scale is

$$t_{in} \sim \frac{r}{v_r} \sim \frac{r^2}{v} \sim t_{vis}, \tag{5.54}$$

where t_{vis} is the viscous dissipation time scale on which a ring of matter at radius r would spread due to viscosity.

From the standard prescription, the dissipation $D(r)$ per unit area of the disk per unit time due to the viscosity is given by

$$D(r) = \frac{1}{2}v\Sigma\left(r\frac{d\Omega}{dr}\right)^2 = \frac{3GM\dot{M}}{4\pi r^3}\left[1 - \left(\frac{r_*}{r}\right)^{1/2}\right]. \tag{5.55}$$

The rate of dissipation due to viscous stresses is three times the rate at which gravitional energy is released locally. This local imbalance is compensated at the inner region of the disk by the release of binding energy at a rate that exceeds the dissipation (see FKR92, p. 74). The rate of dissipation is independent of the viscosity, because it has been assumed in the derivation that the viscosity adjusts itself to provide a steady inward flow of matter at the rate $\dot{M}$. The viscosity, however, appears in expressions for various local disk properties like surface density and opacity. The viscosity is thought to be due to turbulence or magnetic stresses. The details are highly uncertain and therefore a simple prescription due to Shakura and Sunyaev (1973) is generally used (see Pringle 1981 for a discussion): for dissipation due to turbulence, the viscosity is

$$v = \alpha c_s H, \tag{5.56}$$

where c_s is the sound speed and the constant $\alpha \leq 1$.

The total energy dissipated in the disk per unit time, which in the steady state is the total disk luminosity, is given by

$$L_{\text{disk}} = 2\pi \int_{r_*}^{\infty} rD(r)\,dr = \frac{GM\dot{M}}{2r_*}.$$ (5.57)

This is half the total gravitational energy release $GM\dot{M}/r_*$. The other half is accounted for by the kinetic energy of the material just outside the boundary layer. This energy is liberated from the boundary layer and therefore emission from the inner edge of the accretion disk is as important as emission from the disk itself.

The heat energy released due to dissipation is transported outwards, principally in the z-direction because the disk is thin. The net spectrum of the radiation emitted by the disk is obtained by integrating the local spectrum over the disk surface. If the disk is optically thick, the energy emitted per unit area of *each* disk surface per unit time is given by $\sigma T_s^4(r)$, where σ is the Stefan–Boltzmann constant. In the steady state this is equal to the energy dissipation rate $D(r)/2$, where the factor 2 comes in because we are considering the radiation from each surface of the disk. The radial dependence of the temperature is therefore given by

$$T_s(r) = \left\{ \frac{3GM\dot{M}}{8\pi r^3 \sigma} \left[1 - \left(\frac{r_*}{r}\right)^{1/2} \right] \right\}^{1/4}.$$ (5.58)

For $r \gg r_*$,

$$T_s(r) \simeq T_* \left(\frac{r}{r_*}\right)^{-3/4},$$ (5.59)

where $T_* = [3GM\dot{M}/(8\pi r_*^3 \sigma)]^{1/4}$. Normalized to parameters pertinent to quasars and AGN, this temperature at the inner edge of the disk is given by

$$T_* = 2\times 10^5 \left(\frac{M}{10^8 M_\odot}\right)^{1/4} \left(\frac{\dot{M}}{M_\odot \, \text{yr}^{-1}}\right)^{1/4} \left(\frac{r_*}{10^{14}\,\text{cm}}\right)^{-3/4} \text{K}.$$ (5.60)

The characteristic temperature at the inner edge of the disk is therefore $\sim 10^5$ K.

At each r the frequency dependence of the spectrum is given by the Planck form

$$B_\nu(T_s) \propto \frac{\nu^3}{\exp(h\nu/kT_s) - 1},$$ (5.61)

and the net integrated spectrum is

$$I(\nu) \propto \int_{r_*}^{r_{\text{out}}} B_\nu(T_s(r))r\,dr,$$ (5.62)

where r_{out} is the outer disk radius. The functional form obtained from Equation 5.62 is shown in Figure 5.3.

The approximate form of the spectrum in various frequency domains can be determined as follows. For $\nu \ll kT_s(r_{\text{out}})/h$, the Rayleigh–Jeans approximation applies even for the coolest part of the disk and we can take $B_\nu(T_s(r)) \propto \nu^2$ and therefore $I(\nu) \propto \nu^2$.

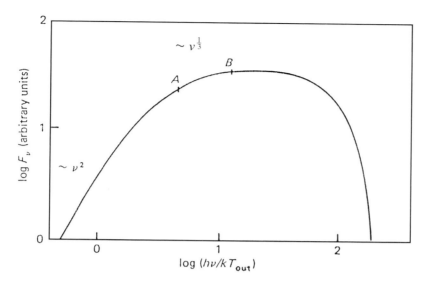

Fig. 5.3. The geometrically thin, optically thick, accretion disk spectrum as a function of frequency. The units are arbitrary. The approximate forms discussed in the text are indicated. Reproduced from FKR92.

For $v \gg kT_*/h$, the Wien approximation $B_v(T_s(r)) \propto v^3 \exp(-hv/kT)$ is applicable. The integral in Equation 5.62 is dominated by the hottest part and the result is that the spectrum drops exponentially. For $kT_s(r_{out}) \ll v \ll kT_*/h$, the radiation comes predominantly from the cooler regions at large r and we can assume $r \gg r_*$, so that the approximation in Equation 5.59 applies. Introducing the variable

$$x = \frac{hv}{kT_s(r)} = \frac{hv}{kT_*}\left(\frac{r}{r_*}\right)^{3/4}, \tag{5.63}$$

the intensity can be expressed as

$$I(v) \propto v^{1/3} \int_0^\infty \frac{x^{5/3}\, dx}{e^x - 1}, \tag{5.64}$$

where the lower limit has been set to zero since $x_* = hv/kT_* \ll 1$ and the upper limit to ∞ since $x_{out} = hv/kT_s(r_{out}) \gg 1$. In this frequency region therefore $I(v) \propto v^{1/3}$, which is usually taken to be the characteristic accretion disk spectrum. These approximate forms are indicated in Figure 5.3.

We will see in subsequent chapters that the typical broad band quasar or AGN spectrum has a shape that is quite different from the accretion disk spectrum described here. The overall spectrum is built out of contributions from a number of different processes that come into play in different parts of the spectrum. But, as we shall see in Section 8.5, it can be argued that the *big blue bump* in the ultraviolet part of the spectrum is emitted by the accretion disk.

Disk models can be given by assuming the α-prescription of viscosity and an opacity law. The models provide various disk quantities, such as the surface density, disk thickness, temperature on the equatorial plane, optical depth etc., in terms of α, the accretion rate, the central mass and the radial distance r. Thin-disk models scaled to characteristic AGN values can be found in FKR92. Disk structures using general relativistic equations have been considered, amongst others, by Novikov and Thorne (1973).

5.6   Thick disks

In the outer region of a thin disk, the Kramers opacity, which is in general due to free–free and photoelectric absorption processes, dominates. In the inner parts of the disk, electron scattering becomes more important. In the outer Kramers region, the radiation pressure p_{r} is much less than the gas pressure p_{g}. At smaller radii the importance of p_{r} increases, especially when electron scattering becomes the dominant source of opacity, and at small enough radii $p_{\mathrm{r}} \gtrsim p_{\mathrm{g}}$. In such a region the disk thickness is given by (FKR92)

$$H \simeq \frac{3r_*}{4} \left(\frac{\dot{M}}{\dot{M}_{\mathrm{Edd}}} \right) \left[1 - \left(\frac{r_*}{r} \right)^{1/2} \right], \tag{5.65}$$

where $\dot{M}_{\mathrm{Edd}}$ is the critical Eddington accretion rate in Equation 5.27, obtained by assuming unit efficiency for the conversion of accreted matter to energy, and r_* is the 'radius' of the central object. For accretion rates $\dot{M} \gtrsim \dot{M}_{\mathrm{Edd}}$, the thin-disk approximation $H \ll r$ breaks down for small radii. The matter is now distributed in the shape of a *torus*, which can be imagined to be a puffed-up accretion disk. The surface of the torus defines a funnel that provides a natural collimator for ejecta from the central engine, which makes the thick disk a very attractive idea.

To understand how the shapes of thick disks can be investigated, we will consider an axially symmetric rotating mass of fluid with no viscosity. In cylindrical coordinates R, ϕ, z, the equation for hydrostatic equilibrium is

$$\frac{1}{\rho} \nabla p = -\nabla \Phi + \Omega^2 \mathbf{R} = \mathbf{g}_{\mathrm{eff}}, \tag{5.66}$$

where Φ is the gravitational potential, the self-gravity of the disk is assumed to be negligible, the effective gravity $\mathbf{g}_{\mathrm{eff}}$ is the vector sum of the gravitational and centrifugal acceleration, and the angular velocity Ω is in general a function of R and z. The velocity field of the rotating mass is given by

$$v_{\mathrm{r}} = 0, \quad v_{\phi} = R\Omega, \quad v_{\mathrm{z}} = 0. \tag{5.67}$$

We shall consider *barytropic* systems in which the pressure and density are related by

$$p = p(\rho). \tag{5.68}$$

Taking the curl of Equation 5.66 and using standard vector identities and Equation 5.67, we get

$$\nabla \frac{1}{\rho} \times \nabla p = 2\Omega \nabla \Omega \times \mathbf{R} = 2\frac{\partial \Omega}{\partial z}\mathbf{v}. \tag{5.69}$$

For barytropic systems, the surfaces of constant p and ρ coincide. On such surfaces, it follows from Equation 5.69 that Ω is independent of z, i.e.,

$$\Omega = \Omega(R), \tag{5.70}$$

and it is possible to express the centrifugal term as the gradient of a potential,

$$\Omega^2(R)\mathbf{R} = \nabla \Phi_{\rm rot}, \quad \Phi_{\rm rot} = \int^R \Omega^2(R')\,R'\,dR'. \tag{5.71}$$

From Equation 5.66 the effective gravity is then given by

$$\mathbf{g}_{\rm eff} = -\nabla \Phi_{\rm eff}, \quad \Phi_{\rm eff} = \Phi - \Phi_{\rm rot}. \tag{5.72}$$

On surfaces of constant pressure (density) $\mathbf{g}_{\rm eff}$ vanishes and

$$\Phi_{\rm eff} = \Phi(R, z) - \Phi_{\rm rot}(R) = {\rm constant}, \tag{5.73}$$

i.e., the surfaces of constant density, pressure and effective potential all coincide. This is known as *von Zeipel's theorem*. Given the distribution $\Omega(R)$ of angular velocity as a function of R, Equation 5.73 can be used to get the cross-section $z_s(r)$ of the equipotential surfaces given the value of $\Phi_{\rm eff}$ on them. A detailed discussion on the nature of these surfaces with examples can be found in FKR92.

The Newtonian tori described above cannot really represent thick accretion disks as they are not affected by viscosity and have no radial motion of the fluid, so that there is no accretion onto the central mass.

The theory of relativistic configurations too can be developed along the lines followed in the Newtonian case (see Blandford 1990 for a review and references). The equation for hydrostatic equilibrium in the relativistic case is

$$\frac{-\nabla p}{w} + \nabla e + \frac{\Omega \nabla l}{1 - \Omega l} = 0, \tag{5.74}$$

where w is the enthalpy per unit volume. The specific energy e and specific *fluid angular momentum* are defined in terms of the covariant components of the fluid four-velocity $\mathbf{u}$ by

$$e = -u_0, \quad l = -\frac{u_\phi}{u_0}. \tag{5.75}$$

The angular velocity is given in terms of contravariant components by $\Omega = u^\phi/u^0$. In the relativistic case a barytropic torus is defined by $p = p(w)$. In the Newtonian case, we saw that a consequence of the barytropic assumption is that the angular momentum is a function of R only, i.e., it is constant on a cylindrical surface. In the relativistic case the corresponding result is

$$\Omega = \Omega(l), \tag{5.76}$$

i.e., the specific fluid angular momentum is constant on surfaces of constant Ω. Given a function $\Omega(l)$, surfaces of constant p, i.e., constant w, can be obtained. The surface of the torus is given by such an isobar. Accretion disks that are thick and have low accretion rates, so that the matter can be assumed to be moving in almost circular orbits, could be represented by such tori.

We have seen in Section 5.3 that close to the event horizon the behaviour of the general relativistic effective potential is quite different from the Newtonian kind. The form of the potential is such that there are no stable orbits with angular momentum less than a critical value. Particles with subcritical angular momenta which move towards a black hole end up plunging into the event horizon. The effect of this on the accretion torus is the formation of a cusp, close to the event horizon, in the equipotential surfaces corresponding to a given angular momentum distribution. For the simple case of constant angular momentum, the cusp lies between the marginally bound and marginally stable orbits. Matter from a thick accretion disk will expand until the matter closest to the hole comes close to the cusp. At the cusp, matter can spill into the hole, like the flow of material through the Lagrangian point in the critical Roche potential in a binary system. Matter that passes through the cusp flows into the hole without emitting much energy. The efficiency for converting accreting mass into energy is therefore given by the binding energy of the matter at the cusp. For the special case of a torus with constant angular momentum distribution, $l = $ constant, the binding energy is given by r_S/R_0, where R_0 is the outer radius of the torus (see Rees 1984 for a discussion), i.e., the accretion efficiency is very low. Thick accretions disks in general have low accretion efficiency.

The shapes of the radiation-supported tori discussed above naturally provide a funnel that could collimate radiation produced in the inner region of the disk. This is shown in Figure 5.4. The luminosities of such tori can exceed the Eddington limit by large factors, and can in principle accelerate matter in the funnels, producing relativistic beams of particles. However, detailed modelling of thick disks with constant specific angular momentum shows that an optically thin plasma consisting of electrons and positrons can be accelerated only to a bulk Lorentz factor $\gamma \lesssim 3$. If even a small fraction of the plasma is protons, the maximum bulk velocity reached drops substantially and $\gamma \sim 1.1$ (Sikora and Wilson 1981). These Lorentz factors are smaller than required by the observed superluminal motion and related phenomena (see Section 3.8). The jet velocity can also be reduced through the Compton drag of the disk radiation.

A great difficulty with radiation tori is that they are subject to global, dynamical instability. This was shown by Papaloizou and Pringle (1984), who considered a torus of uniform entropy and specific angular momentum, and showed that the tori are unstable to low order non-axisymmetric modes. These instabilities are global and their existence cannot be deduced from a local analysis or from consideration of axisymmetric modes. The unstable modes grow on a dynamical time scale $t_d \sim \Omega^{-1}$. The disk can be destroyed as a consequence of this instability or, alternatively, the net result might be to produce an effective viscosity that can drive a steady flow onto the hole. Further discussions on the instability can be found in Blandford (1990), Wiita (1991) and FKR92.

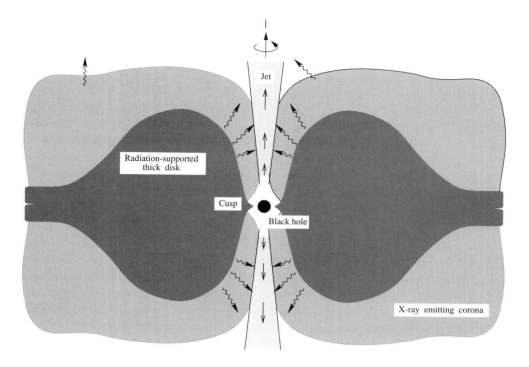

Fig. 5.4. Schematic diagram of a radiation-supported thick accretion disk. After Wiita (1991).

Another kind of thick accretion disk in which the pressure support is provided by hot ions has been proposed by Rees *et al.* (1982). When the inflow time scale of accreting matter is much shorter than the cooling time scale, the Compton coupling between ions and electrons can become ineffective. In this case, even though the electrons cool the ions retain their energy and temperature at the virial value $T_i \sim GMm_p/(kr)$. The ineffective coupling requires low matter density, i.e., low accretion rates. The temperature of the ions causes the disk to swell up and form an *ion-supported torus*. The low accretion rate, and the inefficient conversion of accreted mass to luminosity in thick accretion disks, means that the directly radiated luminosity cannot be high. But the magnetic field anchored in the torus can extract rotational energy from a spinning black hole, which gets collimated through the funnel into twin relativistic beams of charged particles. Ion-supported tori could therefore power radio galaxies, which have twin jets and relatively weak nuclei. These tori could, however, be susceptible to the same global instabilities as radiation-supported tori. If radio galaxies are powered with ion-supported tori, and quasars, because of their high accretion rates, are powered with radiation-supported tori, the two would be intrinsically different kinds of object. This would go against the unification scheme discussed in Section 12.4.

6   Surveys

6.1   Introduction

The 1993 edition of the catalogue of quasi-stellar objects by Hewitt and Burbidge contains 7315 quasars and BL Lacs, which is an order of magnitude more than the number included in their first catalogue, published in 1977. Many more quasars have appeared in the literature since the publication of the 1993 catalogue, and the number continues to increase as observational and data analysis techniques become more efficient. Another catalogue due to Hewitt and Burbidge (1991) contains 935 galaxies that have nuclei with properties similar to those of quasars. The large numbers of quasars and galaxies with active nuclei (AGN) were discovered because of their peculiar colours, morphology or strong emission in the radio, X-ray or infrared bands, or in one of the surveys specially designed to identify objects that have properties different from Galactic stars and 'normal' galaxies. A small number of quasars have also been found through their variability or lack of proper motion and these provide samples independent of the biases inherent in the other techniques.

We shall consider in this chapter the various techniques used in the identification process, but first we provide brief definitions of some of the types of object that appear in the book. The definitions do not provide a water-tight division into physically distinct classes of object, and there are many marginal cases that could belong to more than one class. We shall elaborate on the definition and properties of the different kinds of object in other chapters.

Quasars   We have used the terms 'quasar' and 'quasi-stellar objects' (QSO) without discrimination as per recent practice. Historically quasars were star-like objects identified with strong radio sources, while QSOs included objects which had the optical properties of quasars but which did not necessarily have strong radio emission. For our purposes a quasar can be defined as a star-like object that has strong emission lines with high redshift. Hewitt and Burbidge take the limiting redshift to be 0.1. The exact value is not important, and in fact if this criterion is omitted from the definition of a quasar, only 28 more objects would be added to the catalogue.

High dynamic range optical observations usually reveal a faint fuzz around objects classified as quasars and having redshifts $\lesssim 1$. However, the classification depends only on the appearance of the objects on the discovery plates, and not on the subsequent detection or non-detection of fuzz. In order to avoid confusing low luminosity quasars with objects such as Seyfert galaxies, which are the most luminous representatives of their class, a limiting absolute magnitude is often introduced in surveys and only objects brighter than this limit are accepted as quasars.

Seyfert galaxies   These form a class of galaxies with bright star-like nuclei and peculiar morphology. Seyfert (1943) found that these galaxies had strong broad emission lines. Seyfert galaxies are found in the lists of Markarian galaxies, which are objects with a blue excess and strong ultraviolet continuum discovered in low dispersion objective-prism surveys (see Hewitt and Burbidge 1991 for references). Seyfert galaxies are also found in redshift surveys and as optical counterparts of strong X-ray and infrared sources.

A basic subdivision of the class of Seyfert galaxies can be made on the basis of their emission line properties. A *Seyfert type 1* galaxy has a bright star-like nucleus that emits strong continuum emission from the far infrared to the X-ray band. It has strong emission lines, some of which are broad, with a full width at half maximum (FWHM) greater than a few thousand kilometres per second, while other lines are relatively narrow, with widths greater than a few hundred kilometres per second. A *Seyfert type 2* galaxy has a weak continuum and only strong narrow emission lines. Seyfert galaxies with properties intermediate between those of types 1 and 2 are designated as types 1.2, 1.5 and so on. Seyfert galaxies appear to have the morphology of spiral galaxies.

Radio galaxies   These are galaxies identified with sources in radio catalogues. Many of these galaxies appear to be elliptical, and there has been much speculation about why strong radio emission seems to arise only in ellipticals. In fact high dynamic range observations show that many radio galaxies have a complex and disturbed morphology (see Section 9.3) and are far from being smooth ellipticals. An optical nucleus emitting strong continuum radiation and lines can be detected in some radio galaxies. *Broad line radio galaxies* (BLRG) have a continuum and emission lines resembling those from Seyfert 1 galaxies, while *narrow line radio galaxies* (NLRG) have spectra like the Seyfert 2 galaxies.

BL Lac objects   These have strong nuclear continuum radiation that shows high polarization and rapid variability. Emission lines are, however, absent or weak. BL Lacs are found mainly in the optical identifications of radio or X-ray sources, and all known examples are strong radio sources. BL Lacs are sometimes found to be the nuclei of elliptical galaxies (see Section 9.9 for a detailed description).

Blazars   This class of objects consists of BL Lacs and quasars that show rapid variability and are known as *optically violent variable* (OVV) quasars. The continuum radiation from this class is dominated by non-thermal emission and is believed to be relativistically beamed owing to the bulk motion of the emitting region (see Section 9.9).

Table 6.1. *Local space densities (Osterbrock O89)*

Type of object	Number density Mpc^{-3}
field galaxies	10^{-1}
luminous spirals	10^{-2}
Seyfert galaxies	10^{-4}
radio galaxies	10^{-6}
quasars	10^{-7}
radio quasars	10^{-9}

LINERs   These are the *low ionisation emission line regions* that are found in a substantial fraction of galaxies when sensitive observations are made. They have lines with low excitation, and when this class was first identified, it was thought that the lines could only be produced by shock heating. But it is also possible to treat LINERs as low luminosity Seyfert 2 nuclei with smaller ionization parameters and a steeper photoionizing continuum than the other kinds of AGN.

LINERs occur in $\lesssim$ 80 per cent of spiral galaxies of the type Sa or Sb, and in $\lesssim$ 20 per cent of Sc galaxies. Barring a small percentage of galaxies with active nuclei, the rest are *H II region galaxies*, which are emission line objects in which the photoionizing continuum is provided by hot stars. Extreme examples of the latter are *starburst galaxies*, in which the rate of star formation is much higher than the average star formation rate over the life of the galaxy. The star bursts can occur in the nuclear region, in rings around the nucleus or spread throughout the disk.

A detailed description of the different kinds of emission line galaxies and the spectroscopic diagnostics used to distinguish between the classes may be found in Lawrence (1987), Osterbrock (O89), Netzer (1990) and Woltjer (1990). The approximate space densities, in our local neighbourhood, of the different kinds of object defined above is given in Table 6.1, which is based on data in Osterbrock (O89). These numbers are just indicative, and the discussion in this chapter and the next describes some of the effort that goes into counting objects and the difficulties and pitfalls met with in the process.

The long list of quasars, and other types of object, discovered by different techniques is useful in assessing the range of various parameters which characterize them, in the selection of smaller samples for further observations and in the theoretical analysis of data where the use of a large number of objects mitigates the inhomogeneity and statistical incompleteness of the sample. However, the determination of the surface density as a function of magnitude, or the distribution in space and possible evolution as a function of redshift, requires the use of carefully selected *complete* samples of quasars, with a fair assessment of the degree to which unavoidable observational biases affect the sample. In this chapter we will present a review of the principal methods employed in finding quasars and will discuss their surface density. In the following chapter we will consider the methods used in the determination of the space density and the possible patterns of evolution. We shall devote most of the attention to quasars.

6.2   Optical surveys

As described in Chapter 1, the earliest quasars were discovered as star-like optical counterparts of strong radio sources (3C 48, 3C 196, 3C 273, 3C 286) from the third Cambridge catalogue. Radio quasars have excess ultraviolet (UVX) radiation relative to most stars. It was found by Ryle and Sandage that when a UVX object was found close to a known radio position, in nearly every case it could be identified as a quasar associated with the radio source. During these observations a number of UVX objects were found that had no radio counterparts, and these were shown to have properties similar to those of radio quasars except that they had no discernible radio emission. The class of *radio-quiet quasars* was thus established and it became apparent that radio-loud quasars constituted only a small subset of the entire quasar population. Searching for objects with UVX has proved to be very fruitful in the discovery of new quasars, but the method fails for redshift $z > 2.2$ because then the effect of emission lines reduces the excess ultraviolet radiation. At high redshift most quasars are discovered using the techniques of slitless spectroscopy, multicolour surveys and more recently, their extreme red colour. Even though discovery methods using selection in other bands such as the X-ray band are often used, most quasars are still found using optical selection of one kind or another.

6.2.1   *Ultraviolet excess surveys*

In the optical region, the continuum spectrum of quasars can crudely be approximated by a power law of the form

$$F(v) \propto v^{-\alpha}, \qquad 0.5 \lesssim \alpha \lesssim 1. \tag{6.1}$$

The $U - B$ colour for such a continuum is in the range $-0.8 \lesssim U - B \lesssim -0.7$. Quasars by and large have lesser values of $U - B$ than most Galactic stars, and they are said to be *bluer* or to have an *ultraviolet excess* (UVX); this property can be used to distinguish them from stars in surveys. The colour distribution for different kinds of object may be seen in Figure 6.1 (after Sandage 1965), which shows the distribution of $U - B$ and $B - V$ for quasars from Hewitt and Burbidge (1993), white dwarfs and stars of various types. Most quasars have $U - B \lesssim -0.4$, and choosing star-like objects with colours in this range produces a reasonably complete list of quasar candidates, contaminated principally by white dwarfs.

To find quasars one begins by identifying all star-like objects (i.e., those that do not have discernible extended structure) from a photographic plate or other imaging detector. The colours of the chosen objects are then obtained, and those that are bluer than some fixed $U - B$ limit are identified. These are called *quasar candidates* since not all of the UVX objects are expected to be quasars. The main contaminants in the population are, as mentioned above, white dwarfs and other hot stars, whose number, however, drops rapidly at faint magnitudes, the fraction reducing to ~ 15 per cent at $B \sim 18$ (Bracessi *et al.* 1980). The presence of contaminants means that all objects

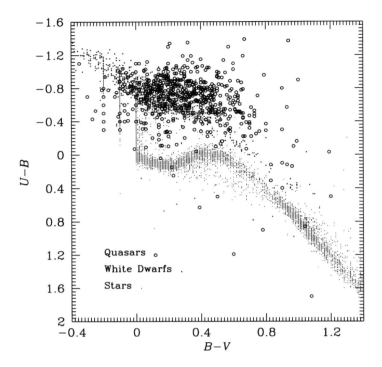

Fig. 6.1. A plot of $U - B$ against $B - V$ for quasars, white dwarfs and stars of various types.

chosen in a UVX survey have to be examined spectroscopically for the presence of broad emission lines with high redshift before they can be accepted as quasars. This is a long and tedious process with a great deal of large-telescope time involved, because of the faintness of the candidates. The usefulness of any technique for finding quasars depends on what fraction of the candidates are rejected after spectroscopic observations.

The efficacy of the UVX technique is apparent from Figure 6.2, which shows a plot of $U - B$ against redshift for all quasars from the Hewitt and Burbidge catalogue with known magnitudes. For $z \lesssim 2.2$ essentially all known quasars would be found with the criterion $U - B < -0.4$, while for higher redshift many quasars appear redder than this value, and would be missed in a UVX survey. The reason for the reddening is the contribution to the detected flux from strong emission lines as they are redshifted in and out of the relevant filter band-passes. The effect is most pronounced for the Ly α line. At $z \gtrsim 2$ the observed wavelength of this line is $\gtrsim 3648 \, \text{Å}$ and it moves out of the U band into the B band. The observed equivalent width of several hundred ångstrom of the Ly α line is then large enough to make a substantial contribution to the flux in the B band. This increases the value of $U - B$, and it can move the quasar out of the UVX selection region. The effect is clear in Figure 6.6, where a number of quasars

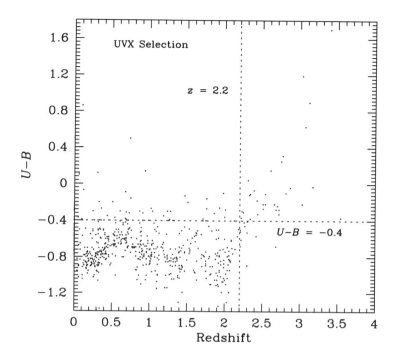

Fig. 6.2. $U - B$ colour as a function of redshift for quasars from Hewitt and Burbidge (1993).

are seen to be redder than the usual UVX limit of $U - B = -0.4$ for $z \gtrsim 2.2$. Simply making the limit redder does not help, as the number of non-quasars amongst the candidates increases sharply, making the technique inefficient. Therefore at these high redshifts the UVX technique cannot be effectively used to find quasars.

At lower redshift the effect of various emission lines as they march in and out of the U and B band-passes is apparent in the wave-like form of the distribution in Figure 6.2. For a given selection limit, a quasar close to the limit can be lost because measurement errors make it too red to be included. The chance that a quasar is scattered out of the selection region owing to measurement errors increases as the mean value of $U - B$ becomes increasingly red (i.e. increasingly positive). The fraction of missed quasars, i.e., the degree of incompleteness of the sample, is therefore a function of redshift. Completeness levels can be estimated from the mean value and dispersion of $U - B$ as a function of z, and also the distribution of errors in the observed colours. The incompleteness that arises from the selection biases inherent to the UVX technique can also be studied by comparison with samples selected without reference to the $U - B$ colour.

The speed with which samples of quasars can be collected has increased enormously with the use of automatic plate measuring (APM) machines that select star-like images with the appropriate colours, and fibre optics systems that obtain in a single observation the spectrum of a large number of the selected objects. Owing to these developments, new surveys containing hundreds of quasars with well-defined properties and completeness estimates have become available. We will consider some of these surveys in this and the following sections.

The Palomar Bright Quasar Survey (BQS)   This was a survey for star-like objects with ultraviolet excess (Schmidt and Green 1983). It was based on 266 U and B plates obtained using the Palomar 46 cm Schmidt telescope. The survey covered 10714 deg^2 of the sky with Galactic latitude $|b| > 30$ deg. The selection of quasar candidates was made using the UVX criterion $U - B < -0.44$. The spectra of the selected objects were obtained and an object was accepted as a quasar if it satisfied the following criteria. (1) The object has a dominant star-like appearance on B prints of the *Schmidt Sky Atlas*. (2) The object has broad emission lines with substantial redshift. (3) The object has blue absolute magnitude $M_B < -23$ ($H_0 = 50$ km sec^{-1} Mpc^{-1}). This last criterion was introduced to distinguish quasars from Seyfert 1 nuclei. The BQS contains 92 quasars that satisfy all three criteria. The number of objects that satisfy the first two criteria but have $M_B > -23$ is 22. These are classified as Seyfert 1 nuclei. The limiting magnitude of the survey is $B = 16.16$. The BQS quasars have been observed at many different wavelengths in order to establish correlations between emission at different wavelengths. However, later surveys have shown that the BQS could be severely incomplete (see subsection 6.7.2).

The survey of Boyle, Shanks and Peterson   This survey (Boyle, Shanks and Peterson 1988) is based on the U and J plates of seven 5×5 deg^2 fields at high Galactic latitude, obtained with the United Kingdom Schmidt telescope. The COSMOS machine in Edinburgh was used to obtain U and J magnitudes of $\sim 30\,000$ star-like images on each plate. The magnitudes were converted to U and B and these were used to identify a total of ~ 1400 UVX objects, using a limiting $U - B$ colour that varied from field to field in the range $(-0.5, -0.3)$. The relatively red limiting colours used ensured greater completeness of the survey, but also increased the number of candidate quasars. However, the spectra of a number of objects could be obtained simultaneously with multi-object spectroscopy using a fibre-optic system. From the spectra 420 objects were confirmed as quasars, with a limiting magnitude of $B = 20.9$. The incompleteness of the survey is estimated to be ~ 5 per cent for $z < 2.2$, rising to ~ 10 per cent in the range $0.6 < z < 0.9$.

6.2.2   *Multicolour surveys*

In these surveys more than one colour is used to discriminate between quasars and non-quasar objects with a star-like appearance. Quasars and other star-like objects, say white dwarfs, have continua with intrinsically different shapes. So even though they may have the same colour in some wavelength region, the colours in other regions are different. Using more than one colour as a discriminant between objects of stellar appearance therefore increases the probability that a candidate will turn out to be a quasar. Moreover, multicolour surveys allow red quasars to be detected, unlike the UVX surveys, and are therefore suited to finding high redshift quasars that have red colours.

At $z \lesssim 3$ the range of quasar colours is determined by the shape of the continuum and the contribution of important emission lines to the different filter band-passes. When the redshifts are such that wavelength regions covered by different optical band-passes correspond to emission wavelengths $< 1200\,\text{Å}$, absorption due to any neutral hydrogen along the line of sight strongly affects the observed colours. At higher redshifts the density of the absorption lines increases, and when the region of the quasar rest frame spectrum shortwards of $912\,\text{Å}$ enters the observed band, the effects of continuum absorption due to the intervening optically thick clouds further redden the spectrum. This moves quasars away from the locus of Galactic stars in a multicolour space, making it easier to identify quasar candidates. Multicolour surveys have been carried out by a number of groups over the years, including Koo and Kron (1982), Koo, Kron and Cudworth (1986), Marano, Zamorani and Zitelli (1988), Cristiani *et al.* (1989), Goldschmidt *et al.* (1992) (other references may be found in Table 6.3). Descriptions of some recent surveys are given in Crampton (C91). We will now describe a few multicolour surveys that have produced large samples of high redshift quasars.

The Wide-Field Multicolour Survey The aim of this survey (Warren, Hewitt and Osmer 1994) was to find high redshift quasars with $z \geq 2.2$ using a multicolour approach. It covers two fields with a total effective area of $43\,\text{deg}^2$. Photographic plates of the fields were obtained in the u, b_j, v, or, r, i bands.[1] The plates were scanned with an automatic plate-measuring machine and only star-like objects with magnitudes in the range $16 \leq m_{or} \lesssim 20$ were retained for further consideration.

The six broad band magnitudes of each object may be considered as coordinates in a six-dimensional space. Most stars lie in an elongated clump in this space (see Figure 6.3), while points away from the main locus include quasars, some rare types of star and unresolved galaxies, all of which have spectra different from the main stellar population. Warren *et al.* obtained the Euclidean distance of each star-like object in this space from its tenth nearest neighbour and used it to quantify the degree of isolation of various points from the locus of common stars. A point was taken to be a candidate quasar when this distance exceeded a threshold value.

The six-dimensional approach to locating candidate quasars is more effective than a technique that uses two-colour diagrams: the latter are projections of the higher dimensional space, and depending on which axes are chosen for the projection, some quasars may well be hidden amongst stars. This effect is clearly seen in Figure 6.3, with high redshift quasars found by the multidimensional technique lying within the locus

---

[1] Transformations from these bands to the Johnson–Cousins system $U, B, V \ldots$ are given in Warren, Hewitt, Irwin and Osmer (1991).

UB$_J$V colour−colour plot

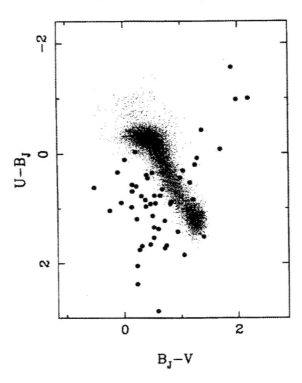

Fig. 6.3. Colour–colour diagram showing the locus of stars (small dots) and identified quasars (filled circles) from the survey of Warren, Hewett and Osmer (1991). The diagram represents a two-dimensional projection of the five-dimensional $UBVRI$ space of colours used in separating quasars from other star-like objects. Figure kindly supplied by Paul Hewett.

of stars in the two-colour diagram shown. Though the quasars are mixed up with the stars in the two-dimensional projection, they appear as outliers when the full space is considered.

The colour selected candidates were subject to further selection, bright stars being eliminated on the basis of their objective prism spectra. Star-like objects bluer than a specified value of $u - b_j$ were also eliminated, since these were very likely to be low redshift quasars that were not to be included as part of the survey. Intermediate resolution spectra were obtained for the remaining candidates, which confirmed the identity of a large number of high redshift quasars. Together with quasars which were already known and which appeared in the survey region, the net sample consists of 85 quasars with $z \geq 2.2$. The effective area of the survey was reduced to 43 deg^2 to exclude 18 outlying objects that remained unclassified.

In the survey of Warren *et al.* the completeness is carefully assessed by making contour plots of the probability p, as a function of redshift and absolute magnitude, that any quasar in the survey would be found, given its spectral energy distribution and the conditions under which it was observed. The contribution of each quasar to the surface density is then weighted by $1/p$, to allow for quasars which are similar to the ones found but that have been missed in the survey.

The APM Survey In this survey due to Irwin, McMahon and Hazard (1991) use is made of the fact that for quasar redshift > 3.9 the region of the spectrum absorbed by neutral hydrogen is redshifted to the B band. A majority of quasars with such redshifts have $B - R$, $R - I$ colours that are extremely red and different from the colours of normal stars. Irwin *et al.* note that for $z > 4.2$ the quasars lie so far from the stellar positions in the colour–colour diagram that selection in the single colour $B - R$ is sufficient to identify them with high probability. This technique, of selecting quasars with red excess ($B_J - R > 3, B_J = B + 0.1$), is analogous to the UVX selection used for $z < 2.2$: the latter relies on the excess blue and UV colour of quasars relative to stars, while the former relies on their excess red colour when at high redshift. The APM survey covers $\sim 2000 \, \mathrm{deg}^2$, with about half the area being surveyed in two colours, and the rest in the single colour $B_J - R$. A substantial number of high redshift quasars that would have been missed in the UVX selection have been discovered in the various multicolour surveys.

POSS II quasars Kennefick *et al.* (1995) have carried out a multicolour survey with the aim of determining the space density of bright quasars ($M_B < -27$) at $z > 4$. The survey used catalogues produced from scanning of the *Second Palomar Observatory Sky Survey* (POSS II) plates. Gunn–Thuan g, r, i magnitudes are available for objects in these catalogues. The multicolour technique for candidate selection uses the fact that for $4.0 < z < 4.8$, the Lyα line lies in the r band and the drop in the continuum bluewards of the line due to the Lyα forest results in large $g - r$ colours. High redshift quasars therefore lie away from the locus of Galactic stars in $g - r, r - i$ colour–colour diagrams and can be picked up as in the other multicolour surveys described above. Kennefick *et al.* have covered an area of $681 \, \mathrm{deg}^2$, obtained spectra for about half the selected candidates and found 10 quasars with $z > 4$; further work is in progress at the time of writing.

6.2.3 *Slitless spectroscopic surveys*

These surveys depend on the detection of strong emission lines in quasar spectra for the primary identification. Objective prism surveys use a thin prism over the objective of a Schmidt telescope to disperse images at the focal plane into low resolution spectra (with $1000 - 2000 \, \text{Å} \, \mathrm{mm}^{-1}$ dispersion). The images are photographed with a fine-grained emulsion, and the plates are searched visually using a binocular microscope. Quasars are identified using one or more emission lines or an unusually blue continuum or even broad absorption features. The line normally detected is Lyα (1216 Å), and for the emulsion normally used (Kodak III a-J), the quasars found are in the range $1.8 \lesssim z \lesssim 3.3$. Here the lower limit is set by transmission through the atmosphere, and the upper by the response of the emulsion. The Lyα line can be detected to a redshift as high as 4.7 by using a red-sensitive emulsion (Kodak III a-F), but sky background increases with wavelength, making it more difficult to detect faint objects so that one has to make do with a brighter limiting magnitude. Instead of the objective prism, it is possible to use either a grating–prism combination (grism) or a grating–lens combination (grens) at the focal plane of a 4 m class telescope. Surveys with such elements produce linear dispersion, and extend to fainter magnitudes than objective-prism surveys.

The slitless spectroscopic technique has been highly successful in finding quasars with $z \gtrsim 2$ and it nicely complements the UVX technique. Though prolific in its output of high redshift quasars, slitless spectroscopy is beset with many selection effects, which leads to incompleteness. The possibility of detecting emission lines depends on the seeing conditions, the emulsion response and the dispersion that is wavelength dependent. Objects with strong emission lines are easier to detect, and this can introduce a redshift-related bias as the strong lines are redshifted in and out of the observing band-pass. It is also difficult to define a unique limiting magnitude for a slitless spectroscopic survey, since this depends on line strength. The problem of incompleteness can be mitigated by defining precisely the selection criteria in advance and using automated searches for the features on which the selection of candidates is based. We shall now consider a few important surveys based on slitless spectroscopy.

The Palomar Transit Grism Survey   This survey (PTGS) was designed to produce a large sample of quasars with $z \geq 2.7$. The observations were obtained with the 5 m Hale telescope with four 800×800 CCDs (charge coupled devices) being used as the detectors. A grism was used to produce low dispersion (~ 77 Å arcsec^{-2}) spectra in the range ~ 4400–7500 Å from which emission line objects were identified. A summary of the observational techniques used in the survey and data on the emission line objects found is given by Schneider, Schmidt and Gunn (1994).

A novel feature of the survey was that data were taken in the transit mode, with the telescope pointing in a required direction and the tracking halted.[1] Due to the absence of tracking, the sky drifts across the CCD in the east–west direction, and a given object is in turn exposed on different portions of it. The CCD is read out continuously at a rate such that a whole frame is read in the time it takes for a given portion of the sky to drift across it, which depends on the declination of the direction in which the telescope is pointing. The four-CCD combination is oriented in such a way that each object is exposed on two of the CCDs. The result of exposure over a period of time is the observation of a strip of the sky that is ~ 8.5 arcmin wide, with the length of the

---

[1] Tracking is the motion of the telescope required to keep the telescope pointed at a given object by compensating for the rotation of the Earth.

strip covered depending upon the duration of the observation. For a given part of the sky covered the exposure is equal to the time taken for a drift across the two CCDs and is $\propto 1/\cos\delta$, where δ is the declination.

The PTGS consists of six strips of the sky obtained in this transit mode, covering a total area of $61.47\,\mathrm{deg}^2$ and having exposures in the range 35–64 sec. Each field was observed twice: once in good photometric conditions and stable seeing with a grism to obtain low resolution spectra, and once to obtain direct images. The latter provided positions of the zero-order images in the spectra. Each pixel of every grism spectrum was searched for an emission line centred on that pixel by fitting a Gaussian line profile also centred on it, after subtracting out a continuum estimated from neighbouring pixels. An emission line was assumed to have been found only if the fitted feature had a high enough signal-to-noise ratio and equivalent width. Slit spectra were obtained of candidate emission line objects to confirm the presence of lines and measure accurately wavelengths and redshift.

The survey found 928 emission line objects, of which ~ 620 are emission line galaxies with $z < 0.45$, while 90 high redshift quasars with redshifts in the range 2.75–4.75 were detected via their Lyα + N V emission lines. The survey was not optimized for the detection of quasars with $z < 2.1$, which in emission line surveys are detected through the C III] and Mg II lines. Nevertheless the survey includes about 100 low redshift quasars with strong lines.

The Large Bright Quasar Survey This survey (LBQS) was initiated in 1986 with the objective of building up a large sample of bright quasars over a wide redshift range. The completed survey has 1055 quasars with magnitude $16 \le B_J \lesssim 18.5$ and redshift in the range $0.2 \le z \le 3.4$ and covers an effective area of $453.8\,\mathrm{deg}^2$. It is presently the largest single survey of quasars with well defined selection criteria, with broad band magnitudes accurate to ± 0.15, coordinates known within 0.1 arcsec and spectra signal-to-noise ratio, in the continuum at 4500 Å, of ~ 10. A summary of the various techniques adopted in the survey and data on all the quasars in it has been provided by Hewett, Foltz and Chaffee (1995).

The LBQS is based on direct and objective-prism plates taken in 18 fields with the United Kingdom Schmidt Telescope. The primary selection of quasar candidates was made from the overall shape of the objective-prism spectra. For each star-like object in the survey a *median wavelength* is defined as the wavelength that bisects the objective-prism spectrum into two parts of equal total intensity. Where this half-power point is situated depends on the shape of the spectrum.[1] A plot of the magnitudes measured from the objective-prism plates against the median wavelength (half-power point), for star-like objects from one of the survey fields, is shown in Figure 6.4. The boundary, which is indicated by a broken line, approximately follows the rather well-defined edge of the distribution of F-type stars, and extends to smaller values of the

---

[1] This can most easily be seen for a power law spectrum, where the median wavelength is straightforwardly related to the spectral index.

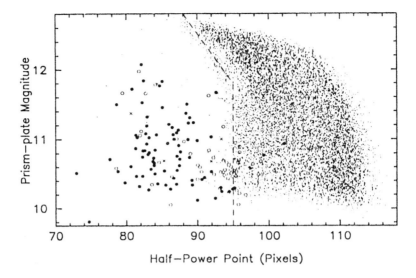

Fig. 6.4. Half-power point against objective-prism plate magnitudes. The broken line indicates the well defined edge of the distribution of F-type stars. See text for details. Reproduced from Hewett *et al.* (1995).

half-power point for bright magnitudes. This allows for the effects of saturation on the spectra of the brightest objects (see Hewett, Foltz and Chaffee 1995 for a detailed discussion). Objects that are to the left of the boundary, i.e., are on the bluer side of it, are taken to be candidate quasars. A number of quasar candidates on the left of the boundary are Galactic stars such as DA white dwarfs, A stars and blue horizontal-branch stars. These can be identified by matching the candidate spectra with template objective-prism spectra of the corresponding objects obtained from high signal-to-noise data. Identifiable stars are eliminated from the list of candidates for final spectroscopic identification.

The procedure of selecting quasar candidates from bluewards of the boundary line does not work for a majority of the quasars with $z \gtrsim 2.4$, as these have half-power points to the right of the boundary. Candidates from the right are therefore selected on the basis of unusual features in their spectra, such as emission lines, absorption lines or breaks in the continuum spectrum. These objects with unusual features are identified by comparison of the catalogue of feature detections amongst objects within half-magnitude bands. The final identifications of quasars were made on the basis of intermediate resolution (6–10 Å) spectroscopy performed on candidates selected on the basis of either their blue half-power points or the presence of unusual features in the objective-prism spectra.

There are very few quasars in the sample with $z \gtrsim 3.4$, at which the Ly α emission line moves out of the B_J filter band-pass. This is taken to be the effective redshift of the survey because of the difficulty in identifying higher redshift quasars with the techniques adopted. In view of the effectiveness of the survey in finding previously known quasars in the survey region, from the surface density as a function of the magnitude of the identified quasars (see Section 6.7), and comparison with the results of a *ROSAT* X-ray study of the survey region, Hewett *et al.* estimate that the LBQS contains ~ 90 per cent of the quasars which are situated in the area of the sky covered by the survey and which have $16 \leq B_J \lesssim 18.5$ and $0.2 < z \leq 3.4$.

6.2.4   *Proper motion and variability surveys*

Because quasars are very far away, they show no proper motion even over many years. This absence of motion can be used to separate them from stars in our Galaxy. Kron and Chiu (1981) surveyed an area of $0.1 \deg^2$ in the field Special Area 57 (SA 57) to $B = 21$ with an interval of 25 years. They found that the survey missed two of the seven known quasars in the field, while it included two subdwarfs. Continuing this work, Majewski *et al.* (1991) carried out a proper motion survey of a $0.3 \deg^2$ field at the north Galactic pole (NGP) to $B \leq 21$. For all 1185 objects in the field brighter than the limiting magnitude they have measured the proper motion with a 1σ error of ~ 1 milliarcsec yr^{-1}. Of the 46 known quasars or active galaxies in the field, all except one were found to have proper motion $< 3\sigma$. The remaining object had a proper motion of 3.4σ, and one such object is indeed expected for a Gaussian distribution of errors. The astrometric precision reached in the survey is therefore sufficient to produce a complete sample of quasars through the lack of measured proper motion. However, the candidate sample of stationary objects also contains a number of stars, which have to be eliminated through spectroscopy, and the efficiency of the technique is therefore only ~ 20 per cent.

Variability is a characteristic of quasars (see Chapter 8), and this may be used to find them using photometric data on a given field at different epochs. Majewski *et al.* have detected variability in 30 of the previously known 31 quasars in the NGP sample, and conclude that all quasars are variable over the 16-year duration they use. This conclusion, however, depends on the criterion chosen to separate true variability from the change in flux due to variable noise on the plates used, since these are taken over a number of years. The large percentage of quasars detected as variable means that variability too can be used to find them, but the procedure is again inefficient because of the contamination by variable stars. Majewski *et al.* have argued that a search for stationary objects that show variability can be a powerful technique for efficiently finding quasars.

Variability and proper motion surveys have the advantage of being unbiased in colour or emission lines. However, Hook *et al.* (1994) have found, by studying the variability properties of ~ 300 quasars at the south Galactic pole (SGP) that the most luminous quasars vary the least. Variability surveys would therefore be incomplete for highly luminous quasars, i.e., at high redshift.

6.3   Radio surveys

We mentioned at the beginning of Section 6.2 that the first quasars were discovered as star-like optical counterparts of strong radio sources. It soon became clear that such objects constitute only a small fraction of all quasars. Sensitive observations with the Very Large Array (VLA) of samples of optically selected quasars have shown that at a level of $\sim$ 10 mJy only $\sim$ 10 per cent of quasars are radio loud, and the detected fraction increases rather slowly with decreasing flux (see Section 9.8). In spite of this, radio selection does provide a useful way to discover quasars, as we shall now see.

To produce a sample of quasars with the primary source selection in a band other than the optical, one proceeds as follows. First, a catalogue of sources that is complete to some flux level in the band is made, with source positions determined as accurately as possible. The positional error boxes are then searched for optical counterparts, on survey plates such as those of the Palomar Observatory Sky Survey or the SERC surveys, with star-like appearance, peculiar colours and so on. The candidates selected in this manner are examined spectroscopically until a confirmed quasar is found. The process of candidate selection can be made simpler by using digitized surveys, but spectroscopic confirmation remains a formidable problem when each error box contains several faint candidates. In radio astronomy, however, it is straightforward to determine positions to subarcsecond accuracy, especially in the case of compact radio sources. The radio error boxes then usually contain at most one object, and it is possible to make an optical identification on the basis of positional coincidence alone, with spectroscopy providing the redshift. Quasars discovered in this manner are free from the biases involved in optical selection. This is especially useful in the discovery of high redshift quasars which would be missed in the UVX surveys. The first quasars to be discovered with redshift $z > 3$ were found as optical counterparts of radio sources with accurately known positions.

Radio spectra are quite often found to be of the power law type, with the flux $F(v) \propto v^{-\alpha}$. Extended radio structures, such as the lobes associated with radio galaxies, have steep radio spectra (this is conventionally taken to mean $\alpha > 0.5$). Compact radio sources, however, have flat, inverted or complex radio spectra owing to synchrotron self-absorption, which sets in below a limiting frequency related to the observed angular size of the source (see subsection 3.5.1). Flat spectrum sources therefore dominate surveys made at high radio frequency ($\sim$ GHz), while steep spectrum sources dominate low frequency ($\sim 10^2$ MHz) surveys. Although a certain fraction of quasars have extended radio structures, quite often the radio emission in them is confined mainly to a compact ($\ll$ 1 arcsec), self-absorbed, nuclear region. The discovery rate of quasars can therefore be increased by looking for optical counterparts of flat spectrum sources in high frequency surveys. An extensive search of this kind has been carried out by Savage *et al.* (1988), who investigated flat spectrum radio sources from the Parkes 2.7 GHz survey; the latter is complete to a flux limit of 0.5 Jy, covers $\sim$ 15 000 deg^2 and comprises 403 sources. Optical identifications are made from the SERC/UKST III a-J sky survey, and owing to the accurate radio positions, positional coincidence alone is sufficient to fix the optical counterparts to the ~ 22.5 magnitude limit of the survey. The ~ 286 quasars discovered include ~ 3 per cent with redshift > 3. The optical counterparts of empty fields, i.e., radio sources that remain unidentified at the limit of the optical survey used, have to be found in deep CCD exposures. McMahon (1991) has pointed out that the discovery of high redshift radio-loud quasars is helped by examining spectroscopically the reddest quasar candidates in a radio survey.

Radio quasars discovered through positional coincidence are free from the biases present in optical surveys. However, there may be biases inherent in the radio surveys themselves. Optical identifications have largely been limited to flat spectrum sources because the chance of discovering a quasar is the highest here. But this discriminates against quasars with large scale structures. The proportion of the two kinds of quasar may change with environment, redshift and other factors, and large unbiased samples are necessary to study the possibilities. It is also not clear that the non-radio properties of radio-loud quasars are representative of the entire quasar population. Such properties may be affected by the processes that produce the strong radio emission in the first place. A detailed comparison of large samples of radio and optically selected quasars is therefore necessary.

Some of these problems are addressed by the FIRST[1] survey, which is under way at the time of writing (Becker, White and Helfand 1995). The survey is designed to observe the radio sky at 20 cm with the VLA in its B configuration, with an angular resolution better than 5 arcsec. It will cover a total of 10 000 deg^2 of the northern Galactic cap to a flux density limit of 1 mJy. When complete, the survey will provide source position to subarcsecond accuracy, flux density at 20 cm and morphological information on $> 10^6$ sources.

A subset of the FIRST survey, which covers 306 deg^2 of the sky, in a narrow strip through the north Galactic pole, and contains $\sim 27 000$ sources brighter than the survey flux limit of 1 mJy has been searched for quasar candidates by Gregg et al. (1996). This was done by matching source positions with those from the digitized APM version of the Palomar Observatory Sky Survey (POSS I). Objects classified as stellar on either the blue or red plates of POSS I, located within 2 arcsec of a FIRST source and brighter than magnitude 17.5 on the red plate, were chosen as candidate quasars. A total of 219 quasar candidates (0.8 per cent of all the sources in the subset) were identified in this manner, of which 25 were found to be known quasars or galaxies. Spectroscopy on 151 of the remaining candidates has produced 69 quasars and Seyfert 1 galaxies, three BL Lacs, 32 emission line galaxies, 41 galaxies with only absorption lines and 31 stars.

Defining $L(21 \text{ cm}) = 10^{32.5} \text{ erg sec}^{-1} \text{ Hz}^{-1}$ ($H_0 = 50 \text{ km sec}^{-1} \text{ Mpc}^{-1}, q_0 = 0.5$) as the dividing line between radio-loud and radio-quiet quasars (see Section 9.8), Gregg et al. find that the fraction of radio-loud objects increases with redshift, from a low of 15 per cent for $z < 0.5$ to ~ 75 per cent for $z > 1.5$. Eleven quasars have extended emission on a scale > 10 arcsec.

It is clear that the FIRST survey forms the basis for an efficient survey of quasars

---

[1] Faint images of the radio sky at twenty centimetres.

unbiased by colour selection and covering a substantial fraction of the total quasar population. Quasars from this survey can be used to probe the bivariate radio and optical luminosity functions of quasars to faint levels and to study the distinction between radio-loud and radio-quiet quasars using large samples. The survey will also provide large samples for the study of quasar absorption lines, bivariate radio and X-ray luminosity functions for different kinds of object, through cross-correlation of the *ROSAT* All-Sky Survey, and unified models of AGN and quasars using large homogeneous samples of the different kinds of object.

Another important survey for finding faint radio selected quasars and AGN is the NRAO–VLA Sky Survey (NVSS) which covers ~ 10.3 steradian of the sky with declination $\delta \geq -40$ deg. The survey was made during 1993–6, by J.J. Condon and his collaborators, at 1.4 GHz with the VLA in the compact D and DnC configurations, which provide high sensitivity but low angular resolution. The beam width used was ~ 45 arcsec, which is larger than the median size of faint extragalactic sources. This helps in determining source fluxes accurately and ensuring flux-limited completeness. The survey is complete to a flux level of ~ 2 mJy. Total intensity and linear polarization images are available, and ~ 2 million sources are catalogued. The optical counterparts of these sources will include quasars, radio galaxies, AGN with low radio luminosity and normal galaxies.

6.4   X-ray surveys

It is known from observations with the imaging proportional counter (IPC) and high resolution imager (HRI) aboard the *EINSTEIN* X-ray observatory (see Section 10.2 for a description) that strong X-ray emission with luminosity $\sim 10^{44}$–10^{48} erg sec^{-1} in the energy range ~ 0.5–4.5 keV is a property common to most quasars. X-ray surveys should therefore provide another route to their discovery. There have been all-sky surveys by *HEAO 1* A2 (Piccinotti *et al.* 1982), deep IPC X-ray surveys of fields in the constellations Draco and Eridanus (Giacconi *et al.* 1979) and Pavo (Griffiths *et al.* 1983), and more recently with the position sensitive proportional counter (PSPS) on *ROSAT* (Georgantopoulos *et al.* 1993, Hasinger *et al.* 1993, Branduardi-Raymont *et al.* 1994). These deep surveys, however, cover only small areas of the sky and do not provide large samples of quasars. Their main aim has been to estimate the surface density of quasars for faint limiting flux and the quasar contribution to the X-ray background.

An exposure with the IPC detector provides an X-ray image ~ 1 deg^2 in angular size with a resolution of ~ 1 arcmin. Besides the X-ray source for which the observation was made, the field often contains other sources, some of which could be quasars. The images in the IPC data bank are therefore a rich source of X-ray selected quasars.

The *EINSTEIN* observatory Extended Medium Sensitivity Survey (EMSS) (see Gioia *et al.* 1990, Maccacaro *et al.* 1991) is a flux-limited complete sample of 835 such serendipitous[1] sources discovered in IPC fields centred on targets at high galactic latitude ($|b| \geq 20$ deg). The sources were detected at a level $\geq 4\sigma$ above the background in fields covering 780 deg². The limiting sensitivity of the exposures ranges over $\sim 5 \times 10^{-14}$ to $\sim 3 \times 10^{-12}$ erg cm⁻² sec⁻¹ in the 0.3–3.5 keV X-ray band. The X-ray sources have positional accuracy of 35–70 arcsec, and there are typically 1–8 optical candidates visible in the error circle on the Palomar Sky Survey at high Galactic latitude. A variety of optical and radio observations has been used to determine the optical counterparts (Stocke *et al.* 1991), and secure identifications are available in 96 per cent of the X-ray sources in the survey. Of the EMSS sources, 51 per cent are found to be quasars and Seyfert galaxies, while the rest are BL Lacerta objects, clusters of galaxies, normal galaxies, cooling flow galaxies and stars. About four per cent of the sources remain unidentified. The EMSS has thus provided a large complete sample of X-ray selected quasars, which can be used to obtain the X-ray luminosity function and to study the possible cosmological evolution of the X-ray properties (see Chapter 10). Many more quasars of this type are expected from the optical identification of X-ray sources in various *ROSAT* surveys. We shall further consider X-ray surveys in Section 10.4.

6.5 Completeness of samples

The aim of a quasar surveyor is to employ one of the techniques outlined in the previous section, or perhaps a new one, to find quasars in a limited area of the sky (all-sky surveys for quasars are not feasible at the present time). In order not to bias the results of a survey it is necessary to define the selection criteria carefully, and to make sure that *all* quasars that satisfy these rules are found. These criteria always include a flux (magnitude) limit such that all quasars in the survey area brighter than this limit are identified. On the one hand, if this limit is taken to be too bright, there is loss of data in that quasars fainter than the putative flux limit, but which are present on the plate and could have provided a larger complete sample, are missed. On the other hand, if the flux limit is chosen to be too faint, not all quasars brighter than the limit in the survey area will be identified, and the resulting incompleteness would affect deductions made from the observation.

In addition to the limiting flux, there are always some other criteria that go into the choice of candidate quasars. In the case of UVX surveys, the selection criterion is simply defined as a limiting colour, say $U - B = -0.4$, and all star-like images bluer than this are made candidates. In multicolour surveys there is an element of subjectivity because the observer has to choose sources that are located in a manner different from normal stars in the multicolour space, and there is no unique way in which the difference can be quantified. In this case one has to choose between having a well-defined criterion, thereby restricting the kind of quasar that is selected, and having

---

[1] Serendipity: the faculty of making happy chance finds.

a wider sample that is incomplete. The slitless spectroscopic surveys are of course the most prone to incompleteness, owing to the reasons cited in subsection 6.2.3.

Some of the reasons for the loss of quasars from a sample, and the degree of incompleteness introduced due to them, are as follows.

Reddening   This affects the UVX surveys, and occurs due to the contribution of emission lines to the B filter band-pass, which reddens the $U - B$ colour, scattering some quasars out of the selection region. The correction to be applied to account for this effect is described in Section 6.6.

Photometric error   Consider some fixed apparent magnitude, say B. Owing to photometric errors, some quasars brighter than B will cross over to the fainter side, while some quasars fainter than B will be counted as being brighter. As will be seen below, the surface density of quasars is a steeply increasing function of magnitude. Therefore the number of quasars that make the transition to the brighter side of B is higher than the number that go the other way. This leads to an apparent excess in the number of quasars brighter than B, i.e., there is *over-completeness* rather than incompleteness. A correction to the surface density due to this effect has been provided by Eddington (1940) and Schmidt and Green (1983). For a Gaussian distribution of errors with dispersion σ in the measured magnitude, it can be shown that the correction is

$$\Delta N(B) = -\frac{\sigma^2}{2}\frac{d^2 N}{dB^2},\tag{6.2}$$

where $N(B)$ is the surface density. For a relation of the form $\log N(B) = aB + \text{constant}$, we get

$$\frac{\Delta N(B)}{N(B)} = -\frac{\sigma^2}{2}(a\ln 10)^2.\tag{6.3}$$

For $a \simeq 1$ (see below) and $\sigma = 0.1$ magnitude, $\Delta N(B)/N \sim 2.5$ per cent. Because of this effect the observed source counts are steeper than is actually the case, towards the limiting faintness of the survey.

Variability   The variation in time in the flux from a quasar can affect a survey in two ways. First, in the case of a UVX survey the plates used to obtain colours could be taken as much as several years apart. For quasars that are variable over a shorter time scale than the interval between observations, this can make the measured colour too red for some of the quasars to be included in the sample. The influence of such reddening on their UVX survey has been estimated by Boyle *et al.* (1988) to cause an incompleteness of $\lesssim 2$ per cent. The second effect of variability is to move quasars in and out of a given magnitude bin as their flux becomes brighter or fainter. When the number of quasars increases steeply with magnitude, their variability, like photometric errors, leads to an overestimation of their surface density. Estimating from available data that 30 per cent of quasars are variable at the level of 0.15 magnitude, 20 per cent at 0.25 magnitude, another 20 per cent at 0.35 magnitude and that the variation in the rest can be neglected, Hartwick and Schade (1990) find using Equation 6.2 that variability leads to ~ 10 per cent over-completeness.

Confusion with Seyfert I nuclei Quasars that are nearby are known to exhibit extended structure, and therefore low luminosity objects ($M_B \gtrsim -23$) can be confused with the nuclei of Seyfert I galaxies and so be omitted from a survey which selects star-like images. In the case of faint quasars, the extended structure, which may very well be a host galaxy with normal colours, can dominate the emission. This can make the galaxy–quasar combination redder than the quasar by itself. Since the extended structure is very faint, it is the total colour that is observed and attributed to the quasar. If the total colour is redder than the limit used to identify quasar candidates, this leads to the exclusion of the quasar from the survey. To avoid these problems, following Schmidt and Green (1983) only objects brighter than $M_B = -23$ are usually retained as quasars in a sample (see subsection 6.7.2 for biases introduced by this procedure).

Equivalent width distribution It is possible for quasars with low emission line equivalent widths to be missed in slitless spectroscopic surveys. If this effect is significant, quasars selected spectroscopically should have larger mean equivalent width than those selected by other means. It has been reported by Peterson (1988) that radio selected quasars and those found by their variability have smaller Lyα equivalent widths than quasars from spectroscopic surveys. But this difference may not be real, and could arise due to the combining of measurements that de-blend the Lyα line from the neighbouring N V (1260 Å) line with measurements where the lines are blended (Hartwick and Schade 1990).

A detailed study of redshift-dependent selection biases that influence the discovery of quasars and the evolution laws that are derived from them has been made by Wampler and Ponze (1985).

6.6   Corrections to measured magnitudes

The properties and distribution of quasars in the sky are generally expressed as a function of their continuum magnitude or luminosity. Before a measured standard magnitude can be used in this manner it needs to be corrected for the effects of Galactic extinction and of absorption due to intervening galaxies, which reduce the observed flux, and for the effects of emission lines, which distort the continuum colours by contributing to the observed flux in wavelength regions that depend on the quasar redshift. The measured continuum has also to be corrected for absorption due to the Lyα forest and due to Lyman limit systems. The continuum absorption shortward of the Lyman limit causes discernible effects when the redshift is high enough for the region to appear in the band-pass.

A simple prescription for the Galactic extinction correction is provided by Schmidt

(1968):

$$\Delta V = -0.18 \operatorname{cosec} |b|,$$
$$\Delta(B - V) = -0.06 \operatorname{cosec} |b|, \qquad (6.4)$$
$$\Delta(U - B) = 0.72\Delta(B - V),$$

where b is the Galactic latitude of the quasar. The variation of extinction with Galactic coordinates can be taken into account by using the Burstein and Heiles (1982) reddening maps constructed from neutral hydrogen radio emission observations or the machine-readable data from the Bell Laboratories H I survey by Stark *et al.* (1992) for declinations > -40 deg. The extinction correction is $\lesssim 0.1$ magnitude for Galactic latitude $|b| \gtrsim 30$ deg, and can often be neglected relative to the photometric uncertainties that arise owing to the use of an approximate power law shape for the continuum spectrum.

Clouds of neutral hydrogen in the line of sight to the quasar lead to absorption due to the Lyman lines shortwards of $1216\,\text{Å}$ and Lyman continuum absorption below $912\,\text{Å}$. If the spectral region affected by absorption is redshifted into one of the filter band-passes, the observed magnitude and colour can be seriously affected. A quantitative estimation requires a model of the distribution of the absorbers as a function of redshift.

Emission lines contribute to the measured magnitudes, since they are redshifted through the filter band-passes, and their flux has to be subtracted out before the continuum flux is known. This correction can be made in the following simple manner. The detected flux due to a particular line from a quasar at redshift z in a given filter band-pass is

$$(EW)_e(1 + z)F(\lambda_e(1 + z))R(\lambda_e(1 + z)), \qquad (6.5)$$

where λ_e and $(EW)_e$ are the rest frame wavelength and equivalent width[1] in wavelength units respectively, $F(\lambda)$ is the continuum flux and $R(\lambda)$ is the response function of the filter and detector combination being used. Assuming that the emission lines are sufficiently narrow to be wholly included in the filter band-pass, the total contribution Δf_e due to emission lines can be obtained by summing the above expression over all the lines that have $(1 + z)\lambda_e$ in the band-pass.

The continuum flux detected in the filter is

$$f = \int_{\text{filter}} F(\lambda)R(\lambda)\,d\lambda. \qquad (6.6)$$

The effect of the emission lines is to be subtracted from the total flux detected. When the emission line flux is only a small fraction of the continuum flux, the subtraction is achieved by *adding* to the measured magnitude the term

---

[1] The equivalent width of a line is the width (EW) of the continuum, at the position of the line, which would be required to produce the total flux $F(\lambda)$ in a line with central wavelength λ: $F_{\text{line}}(\lambda) = (EW) \times F(\lambda)$. The observed equivalent width $(EW)_o$ and the rest frame equivalent width $(EW)_e$ are related through $(EW)_o = (1 + z)(EW)_e$.

Table 6.2. *Important emission line equivalent widths*

Emission line	Wavelength (Å)	Equivalent width Å
Ly α + N V	1216 + 1240	52
C IV	1549	37
Al III + C III]	1858 + 1909	22
Mg II	2798	50
H β	4861	58

$$\Delta m \simeq \frac{\Delta f_e}{f} \simeq \frac{1}{f} \sum_{\lambda_e} (EW)_e (1+z) F(\lambda_e(1+z)) R(\lambda_e(1+z)). \tag{6.7}$$

It is usually sufficient to assume that the continuum flux is constant over the relatively narrow band-pass of the standard filters used. In this case the correction term reduces to

$$\Delta m \simeq \sum_{\lambda_e} ((EW)_e (1+z) R(\lambda_e(1+z)); \tag{6.8}$$

the area under the response function is assumed to be normalized to unity.

The equivalent widths of lines that make significant contributions can be found from the observations of a number of quasars. The widths can also be found from measurements on a composite spectrum made by averaging the spectra of many quasars; this has the advantage of preserving features such as the blue bump and the Ly α forest. We use the high signal-to-noise ($S/N \sim 400$) composite spectrum produced by Francis *et al.* (1991) which extends from the rest frame ultraviolet to the optical region (800–6000 Å). The composite was built with the spectra of 718 individual quasars from the Large Bright Quasar Survey (LBQS, see description above) and is shown in Figure 6.5. The equivalent widths of the most significant lines obtained from the composite are given in Table 6.2. The equivalent widths of many other lines have also been derived from the composite spectrum, but these are too low to make a difference to the derived correction. The emission line corrections for different standard filters as a function of redshift are shown in Figure 6.6. It is sometimes necessary to correct the measured magnitude for vignetting, i.e., shadowing of the image caused by components of the optical system of the telescope. The details may be found in Marshall, Tananbaum *et al.* (1983). The magnitude in a given band-pass including all the corrections is given by

$$m_C = m + \Delta m_{ex} + \Delta m_{em} + \Delta m_{vig}. \tag{6.9}$$

6.7   The surface density of quasars

A survey produces a list of quasars with known magnitude (flux) and redshift that have all been selected in the same manner. Ideally there is a limiting magnitude for the survey and all quasars brighter than this limit in the survey area are iden-

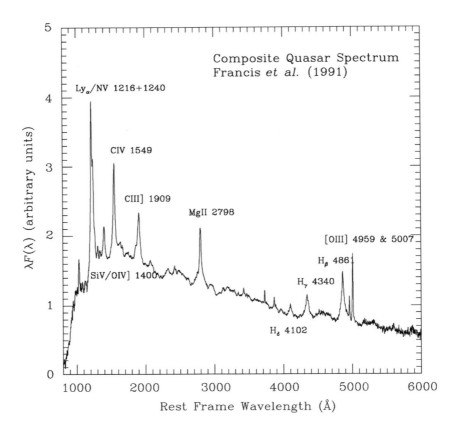

Fig. 6.5. The composite quasar spectrum of Francis *et al.* (1991). The rest frame wavelengths of important emission lines are shown. Data kindly supplied by Paul Francis.

tified. The surface density of quasars, i.e., the number of quasars per unit area of the sky, as a function of magnitude is then determined by simple counting. The ultimate aim of a survey is to obtain the space density of quasars as a function of their luminosity, redshift and any other property (say radio spectral index) that may be relevant. This could be done by choosing small bins defined by the parameters, counting the number of quasars in each bin and converting the number to a density. However, the total number of quasars is usually too small to populate the bins significantly even when one considers only two-dimensional bins defined by luminosity and redshift. The space density has therefore to be determined less directly (see Chapter 7) and in such a case the surface density provides important constraints on the parameters that define the space density and its possible evolution with redshift. Determination of the surface density is necessary for the investigation of the spatial correlation of quasars with other quasars or with galaxies or clusters of galaxies. It also helps in the estimation of the

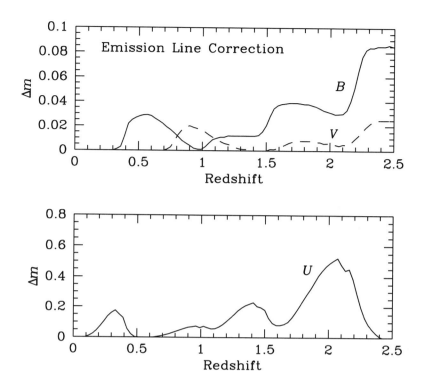

Fig. 6.6. Emission line corrections for different filters.

quasar contribution to the X-ray and other background radiations. We will now summarize the present observational status of our knowledge of the surface density.

6.7.1 The surface density for $z < 2.2$

We have seen in Section 6.2 that the UVX technique is highly effective in finding quasars for $z < 2.2$, and that most known quasars in this redshift range have been discovered using it. The selection effects besetting UVX surveys and the consequent incompleteness are well understood, and it is possible to obtain a reliable estimate of the surface density of quasars found with this technique. The situation at higher redshift is less certain and therefore it helps to consider the surface density separately for redshift $z < 2.2$ and > 2.2. The boundary has no fundamental significance and simply represents the approximate redshift at which emphasis shifts from the UVX method to the slitless spectroscopic, multicolour and other surveys.

The current knowledge for $z < 2.2$ is based mainly on two UVX surveys which we have described in subsection 6.2.1: the Palomar Bright Quasar Survey (BQS) (Schmidt and Green 1983) and that due to Boyle, Shanks and Peterson (1988). The

BQS contains 92 quasars to an average limiting magnitude $B = 16.16$, while the Boyle, Shanks and Peterson sample contains 420 quasars brighter than $B = 20.9$ (1988). About 57 quasars with $z < 2.2$ have been found by Boyle et al. (1991) in a multicolour survey covering 0.85 deg^2 of the sky and complete to $B < 21$. A sample of ~ 40 quasars with $B < 22$ and $z < 2.2$ has been obtained by Zitelli et al. (1992) as the result of spectroscopic observation of a sample selected by Marano et al. (1988) using multicolour search, slitless spectroscopy (the grism method, see subsection 6.2.3) and variability. Both these complete surveys contain some quasars with $z > 2.2$.

Besides the surveys mentioned here, a number of UVX and other major surveys are listed in Table 6.3, in which are indicated their selection technique, the area of the sky covered, their limiting magnitude and redshift range. Hartwick and Schade (1990) have made a detailed statistical comparison of the magnitude and redshift distribution of different surveys available to them, in an attempt to detect any incompleteness. The comparison is vitiated by the fact that the overlap in the magnitude–redshift space between various surveys is often not considerable. It was however found that, barring exceptions, the surveys agreed with each other in the distribution of surface density as a function of magnitude. Hartwick and Schade have combined the different samples to produce the surface density of quasars $N(B)$ in the range $13 < B < 22.5$. The surface density for $B > 21$ has been estimated using candidate quasars from the surveys of Koo and Kron (1988) and Marano et al. (1988). The spectroscopic confirmation of the objects in these samples was not complete at the time of the estimate. This was allowed for by scaling the sky area of the survey by the fraction of objects in the survey with spectroscopic data, but the numbers should be treated with caution.

We have listed in Table 6.4 the values of the integral surface density $N(< B)$ for $z < 2.2$ obtained by Hartwick and Schade, and show these graphically in Figure 6.7. The numbers agree with those obtained in the Boyle, Shanks and Peterson survey alone. For blue magnitude brighter than ~ 19, a straight-line fit to the logarithmic surface density per square degree gives

$$\log N(< B) = 0.88B - 16.11, \qquad B < 18.75, \tag{6.10}$$

the 1σ error estimates for the slope and intercept being 0.02 and 0.51 respectively. The function $N(< B)$ flattens significantly at $B \sim 19$, and is given by

$$\log N(< B) = 0.31B - 5, \qquad B > 19.5, \tag{6.11}$$

with 1σ errors of 0.05 and 1.14 in slope and intercept. We have obtained the fit in Equation 6.11 by finding the slope from the data and matching the counts given by Equation 6.10 at $B = 19.5$. The surface densities for $z < 2.2$ obtained by Boyle et al. (1990), 67.6 ± 8.8 per square degree for $B < 22$, and Zitelli et al. (1992), 86.3 ± 14.5 per square degree for $B < 22$, agree well with the value given by Hartwick and Schade; the latter is higher, but it lies in the region where the estimate of these authors was based on uncertain data. In the early 1980s, it was believed on the basis of rather

Table 6.3. *Major optical surveys for quasars*

Ref.[a]	Magnitude limit	Redshift range	Area deg^2	Number	Selection[b]	Survey name
1	$B < 18.2$	$z < 2.2$	35.5	22	UVX	AB
2	$B < 16.2$	$z < 2.2$	10714	114	UVX	BQS
3	$B < 19.5$	$z < 2.2$	1.72	32	UVX	BF
4	$B < 17.7$	$z < 2.2$	109	32	UVX	MBQS
5	$B < 20.9$	$z < 2.2$	11.2	351	UVX	
6	$B < 21$	$z < 2.2$	11.9	420	UVX	
7	$B < 16.5$	$0.3 < z < 2.2$	330	8	UVX	
8	$B < 19.8$	$z < 2.2$	10	99	MC	
9	$B < 22$	$0.6 < z < 2.9$	0.85	66	MC	
10	$B < 22.1$	$z < 3$	0.35	54	MC,GV	
11	$R < 19$	$4 < z < 5$	2000	30	MC,BRX	APM High z
12	$R < 20$	$2.2 < z < 4.5$	43	86	MC	
13	$r < 19.6$	$4 < z < 4.8$	681	10	MC	POSS II
14	$B < 20.5$	$0.3 < z < 3.4$	9.4	268	GS	CFHT
15	$R < 20.5$	$2.0 < z < 4.7$	61.47	141	GM	
16	$B_J < 18.5$	$0.2 < z < 3.4$	453.3	1055	OP	LBQS

[a]References: 1, Marshall *et al.* (1983); 2, Schmidt and Green (1983); 3, Marshall *et al.* (1984); 4, Mitchell *et al.* (1984); 5, Boyle *et al.* (1987); 6, Boyle *et al.* (1988); 7, Goldschmidt *et al.* (1992); 8, Cristiani *et al.* (1989); 9, Boyle *et al.* (1991); 10, Zitelli *et al.* (1992); 11, Irwin *et al.* (1991); 12, Warren *et al.* (1994); 13, Kennefick *et al.* (1995); 14, Crampton *et al.* (1989); 15, Schneider *et al.* (1994); 16, Hewett *et al.* (1995).

[b]Selection: UVX, ultraviolet excess; MC, multicolour; GS, grens; V, variability; BRX, red excess; GM, grism; OP, objective prism; GV, grism and variability.

scanty data that the steep slope continued to faint magnitudes. Given the mean ratio of quasar X-ray luminosity to optical luminosity, this would have implied that the quasar contribution to the X-ray background exceeded the total observed value. For a Euclidean, non-evolving universe with a uniform space distribution of quasars,

$$\frac{dN(< B)}{d \log B} = 0.6. \tag{6.12}$$

The actual steeper slope obtained for the brighter quasars straight away implies that one or more of the assumptions that go into this simple model must be wrong. The flattening in the count at $B \sim 19$ likewise has an important bearing on the form of the evolution of the space density with redshift. In particular, the break rules out the simple *density evolution* model, which requires a constant slope. We will discuss these matters in detail in Chapter 7.

Table 6.4. *The integral surface density of quasars*

B	N(< B) deg⁻²	
	$z < 2.2$	$2.2 < z < 3.3$
13.0	0.00010 ± 0.00001	
13.5	0.00020 ± 0.0002	
14.0	0.00028 ± 0.0006	
14.5	0.00058 ± 0.0003	
15.0	0.0017 ± 0.0004	
15.5	0.0041 ± 0.0007	
16.0	0.010 ± 0.001	
16.5	0.020 ± 0.003	
17.0	0.058 ± 0.01	
17.5	0.19 ± 0.03	0.016 ± 0.01
18.0	0.58 ± 0.05	0.04 ± 0.02
18.5	1.3 ± 0.09	0.20 ± 0.04
19.0	4.3 ± 0.3	0.76 ± 0.2
19.5	8.5 ± 0.6	1.3 ± 0.3
20.0	16.2 ± 0.7	2.4 ± 0.5
20.5	25 ± 1.7	4.2 ± 1.5
21.0	33 ± 2.3	7.6 ± 2.4
21.5	53 ± 12.4	13 ± 7
22.0	74 ± 17	20 ± 11
22.5	129 ± 27	32 ± 18

6.7.2   *The surface density of bright low-redshift quasars*

The estimate of the surface density of quasars brighter than $B \sim 16$ is based mainly on the BQS of Schmidt and Green (1983). However, later work has shown that the BQS may be substantially incomplete. Goldschmidt *et al.* (1992), using the UVX Edinburgh survey, have found five quasars brighter than $B = 16.07$ in an area of 330 deg² contained within the boundaries of the BQS. In contrast, there is only one BQS quasar in the area. The surface density of quasars, with $B < 16.5$ and $0.3 < z < 2.2$, in the Edinburgh survey is 0.024 deg⁻², which is consistent with the best-fit line defined by Equation 6.10, but considerably higher than the prediction of the BQS for this magnitude limit.

Higher surface densities at bright magnitudes have also been found by Köhler *et al.* (1997), in a flux-limited sample drawn from the Hamburg/ESO survey, covering an area of 611 deg². The candidates in this survey were selected from digitized objective prism plates taken with the ESO Schmidt telescope. The selection of candidates was based on ultraviolet excess as measured from the objective prism spectra. The candidates were either confirmed on the basis of follow-up spectroscopy, or were already known to be quasars or Seyfert 1 galaxies from other surveys. An important aspect of this survey is that objects with point-like as well as extended morphology have been included, so long as they had quasar or Seyfert-1-like spectra. The complete sample consists of 48 quasars and Seyfert 1 galaxies with $z > 0.07$. The lower limit on redshift was imposed

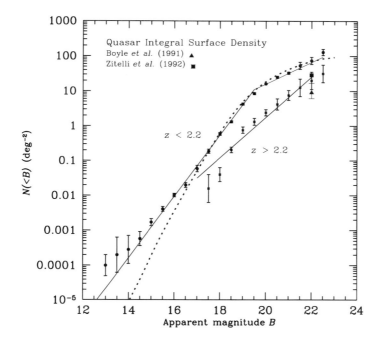

Fig. 6.7. The integral surface density of quasars. The solid circles indicate the densities listed in Table 6.4. Some points not in the table are separately indicated. The solid lines are best fits as described in the text. The dotted line is the integral source count obtained from the luminosity function of Boyle *et al.* (1991), as described in Section 7.7.

to avoid possible incompleteness owing to host galaxy contamination. For the 33 plates used in the survey, the value of the limiting blue magnitude for completeness ranged between 17.66 and 16.44.

Köhler *et al.* (1997) find that at all magnitudes covered by their survey, there is a significant excess of surface density over the corresponding values for the BQS, with only the $z > 0.07$ quasars retained in the latter sample. At $B = 16$, their surface density is higher by a factor 3.6 than the BQS value. The surface density at $B = 16.5$ is consistent with the value of Goldschmidt *et al.* (1995) and at fainter magnitudes it agrees well with the densities in other surveys such as the LBQS.

The reasons for the discrepancy between the BQS and the later surveys, with overlapping redshift and absolute magnitude ranges, are not fully understood. The BQS magnitudes have relatively large errors, which result in incorrectly large $U - B$ colours, scattering the quasars out of the colour selection region. The extent of correction due to this effect remains to be established. Another point of difference is that BQS objects were selected on the basis of their *dominant-star like appearance*, and objects with $M_B > -23$ were omitted from the final sample, while the Hamburg/ESO sample contains all objects which have $z > 0.07$, an ultraviolet excess and quasar or Seyfert-1-

like spectra. Köhler *et al.* (1995) have obtained surface densities for a subset of objects with $z > 0.2$, which excludes most border-line cases between quasars and Seyfert galaxies. The surface densities obtained for the subset case are only slightly less than for the whole sample, showing that the differences in sample selection criteria cannot explain the discrepancy between this survey and the BQS.

6.7.3   *The surface density for $z > 2.2$*

It has been difficult to determine accurately the surface density at redshifts $z > 2.2$ because of the many selection effects involved in the emission line surveys, which were used in the past in the discovery of high redshift quasars. As explained in subsection 6.2.3, because of the selection effects it was hard to estimate the nominal limiting magnitude of the survey, as well as the degree of completeness. This situation has changed for the better with the availability of large samples of high redshift quasars from emission-line-based surveys such as the Large Bright Quasar Survey (LBQS) and the Palomar Transit Grism Survey (PTGS), which we have described in subsection 6.2.3. In these surveys quasars are selected in machine searches based on quantitatively defined selection procedures, minimizing the subjective element. The selection effects are therefore better understood and the degree of completeness can be estimated more reliably than before. The magnitude limits now, however, refer to emission line fluxes, and it is necessary to relate these carefully to continuum fluxes so that comparison with other surveys may be made. Large samples of high redshift quasars have also become available through multicolour surveys, such as those due to Warren, Hewett and Osmer (1994), in which again the level of completeness is carefully assessed (see subsection 6.2.2). The procedure developed by Irwin, McMahon and Hazard (1991) to find high redshift quasars through their excess red colour is particularly simple, and has so far produced 30 bright quasars with $z > 4$. The colour-based surveys provide a valuable comparison with the emission line surveys.

Hartwick and Schade (1990) have enumerated the surface density of quasars using the surveys available to them, and these are shown in Figure 6.7 and listed in Table 6.4. We have also listed the surface density for $2.2 < z < 2.9$ as found by Boyle *et al.* (1991) and Zitelli *et al.* (1992). Giving equal weight to all the points available to them, Hartwick and Schade find for $z > 2.2$ the best-fit line

$$\log N(< B) = 0.58B - 8.36, \qquad B < 18.75. \tag{6.13}$$

For a given apparent magnitude, the surface density for $z > 2.2$ is a factor ~ 4 less than at $z < 2.2$. This is a reflection of the flattening in the evolution of the space density as a function of redshift beyond $z \sim 2$, to be discussed in Chapter 7. The estimates of the surface density for $z > 2.2$ provided by Hartwick and Schade should be viewed as at best tentative and subject to revision in the future.

Warren, Hewett and Osmer (1994) have discussed in some detail the surface density of quasars obtained from the LBQS. They examine the surface density separately for the redshift ranges $0.2 \leq z \leq 2.2$ and $2.2 \leq z \leq 3.4$. They find overall good agreement with the surface densities derived by Hartwick and Schade, the only formally significant disagreement occurring at faint magnitudes in the low redshift part, where the LBQS surface densities are found to be lower by ~ 60 per cent. Warren *et al.* believe that this could be due to statistical fluctuation and is not indicative of a serious incompleteness in their sample. They find a distinct change in slope at $B_J \simeq 17.8$, for the low redshift part, the slope being flatter at the fainter end.

7 Luminosity functions

7.1 Introduction

The luminosity function of a population of discrete sources describes the distribution of the objects in space as a function of their luminosity. Apart from luminosity, such a function may depend on many other properties, the environment and the evolutionary state of the universe. Surveys for a particular kind of object provide the surface density of the objects in the sky as a function of their magnitude, redshift and perhaps some other attributes. Using redshift as the distance indicator, the surface density can be deprojected to provide their number per unit volume of space, which is the more fundamental quantity. Owing to the limited data available, the deprojection involves a number of techniques, assumptions and models, some of which we will describe below.

The space density of quasars is a fundamental quantity because it could help link the quasar phenomenon to other objects, such normal galaxies, in the universe. On the one hand, the properties of quasars are very similar to those of the active galactic nuclei (AGN) of Seyfert galaxies and radio galaxies. On the other hand, nebulosities which resemble galaxies have been discovered around some low redshift quasars (see Chapter 8). Given these two facts, it is generally believed that quasars represent the extreme end of the active galaxies population, in which the luminosity of the active nucleus overwhelms the luminosity of the rest of the galaxy. Active galaxies similarly are considered to be an extreme subset of all galaxies. From the space density of the different kinds of active objects it will be possible to determine whether this is indeed the correct picture.

At low redshift, the number density of quasars is only a rather small fraction of the number density of bright galaxies. The data from quasar surveys seem to indicate that at a given luminosity, the density of quasars increases towards higher redshift, at least for $z \lesssim 2$. This means either that the density of quasars was higher in the past, or that they were brighter (any combination of these two effects is also possible). Also, given the small ratio between the number densities of quasars and galaxies, the quasar phase, if it is part of the evolution of a galaxy, lasts for a comparatively short time scale. At the present epoch many galaxies may therefore contain remnants of once-active quasars, even though they may not show overt signs of quasar-like activity. The number of galaxies with such remnants depends upon the pattern of evolution and it is even possible that *every* galaxy contains a quiescent quasar or AGN in its nucleus.

The dormant quasars could be triggered into activity, given the right circumstances such as availability of fuel, either in only a fraction of the galaxies, or in all galaxies but only for a small fraction of the galaxy lifetime. As seen in the previous chapter, there are indications in the data that the number density of quasars stops increasing beyond $z \lesssim 2$ and that it may even be decreasing at higher redshifts. If this is true, we may already have seen the epoch of quasar formation.

A quasar could form at the centre of a galaxy after the galaxy is wholly formed, or in the earliest stages of the galaxy's development. In either case it could have a serious effect on the physical processes, such as star formation, taking place inside the galaxy, especially close to the centre. According to these ideas the phenomenon of quasar formation can provide constraints on the formation of galaxies and clusters. It would of course be an embarrassment to the standard picture of a quasar as the active nucleus of a galaxy, if quasars were found in abundance at such high redshifts that no galaxies or their forebears could have formed at that epoch. In order to distinguish between the possible scenarios for the origin and evolution of the quasar population, and to assess their effect on the environment within galaxies as well as in the intergalactic regions, we need to estimate carefully their distribution in space as a function of their own properties, of their environment and the epoch.

7.2   Definition of the luminosity function

We have seen in Section 2.8 that the comoving (i.e., coordinate) volume of a region of space and the corresponding proper volume are related by (proper volume) $\propto$ (comoving volume)$\times(1 + z)^{-3}$. While proper volumes decrease $\propto (1 + z)^{-3}$, coordinate volumes remain constant. This means that the comoving density of a non-evolving population of sources is a constant too and it is convenient to use it in studying the distribution of sources as a function of redshift. The luminosity function is defined as the number of quasars per unit comoving volume of the universe. Any change in the comoving density as a function of redshift then implies a *real* change, as opposed to a change that follows from the expansion of the universe.

The luminosity function in general depends on the luminosity in various bands of the spectrum and on other properties such as the continuum spectral index, the redshift and the environment. The data available from quasar surveys are usually so sparse that one has to consider the dependence on just a few parameters, say the optical and X-ray luminosity and redshift, and to average over all the others. For radio quasars it has been possible to distinguish between the evolution of steep and flat radio spectrum objects. But in spite of the relatively large samples that have recently become available, our knowledge of the luminosity function still remains rather sketchy.

We will denote by $\Phi(L, \ldots, z)$ the luminosity function or comoving space density of quasars, where the ellipses indicate all the variables, other than the optical luminosity and redshift, on which the function depends. We will drop the ellipses in the following for convenience, and introduce the other variables explicitly where necessary. The number of quasars in an infinitesimal comoving volume element $dV(z)$ is given by

$$dN(L, z) = \Phi(L, z) \, dL \, dV(z). \tag{7.1}$$

It is often convenient in the optical case to express Φ as a function of absolute magnitude. In that case

$$\Phi(M, z) \, dM \, dV(z) = -\Phi(L, z) \, dL \, dV(z), \tag{7.2}$$

with

$$M = -2.5 \log L + \text{constant}. \tag{7.3}$$

The negative sign in Equation 7.2 appears because M becomes more negative as the luminosity increases. The form of Φ as a function of M is different from its form as a function of L, and

$$\Phi(M, z) = 0.921\Phi(L, z)L, \tag{7.4}$$

but it is convenient to use the same symbol in both cases. Φ is sometimes expressed as a function of $\log L$ per unit logarithmic luminosity. In that case

$$\Phi(\log L, z)d(\log L) = \Phi(L, z) \, dL, \tag{7.5}$$

i.e.,

$$\Phi(\log L, z) = (\ln 10)\Phi(L, z)L. \tag{7.6}$$

7.3   Evolution of the luminosity function

The luminosity function can always be expressed as

$$\Phi(L, z) = \Phi(L)\rho(L, z), \tag{7.7}$$

with

$$\Phi(L) \equiv \Phi(L, 0), \quad \rho(L, 0) = 1. \tag{7.8}$$

$\Phi(L)$ is the *local luminosity function*, while $\rho(L, z)$ is the evolution function which describes the change in the comoving number density of quasars relative to the local density. The evolution function in general depends on the luminosity as well as the redshift, so that the comoving densities of quasars with different luminosity can evolve at different rates. The evolution function can also depend on other parameters. For example, radio quasars with a flat continuum spectrum, which have most of their emission confined to a compact region, could be evolving at a different rate from steep spectrum radio quasars, which have most of their emission in extended radio structures far from the nucleus. We should here emphasize the obvious fact that while a typical quasar, like any astronomical object, is expected to evolve, the evolution of a large population can have cosmological implications only if the origin of such a population is shown to be linked with the large scale structure of the universe. When fitting model luminosity functions to the data, one tries to use as simple an evolution function as possible. We will now discuss two particularly simple forms of the luminosity function, which were the first to be used and are at the basis of many other more complicated forms in vogue now.

7.3.1 *Pure density evolution*

In this case the evolution function is taken to be independent of L, so that

$$\Phi(L, z) = \Phi(L)\rho(z). \tag{7.9}$$

The variables L and z are therefore separable, and the distribution of L is the same at all epochs. The total comoving number of quasars with $L_{min} < L < L_{max}$ in a comoving volume element $dV(z)$ at a given epoch is

$$dN(L_{min}, L_{max}; z) = \rho(z)\, dV(z) \int_{L_{min}}^{L_{max}} \Phi(L)\, dL, \tag{7.10}$$

justifying the name *density evolution*. Considering Φ to be a function of absolute magnitude, we have

$$\log \Phi(M, z) = \log \Phi(M) + \log \rho(z), \tag{7.11}$$

so that in the $(\log \Phi)M$-plane the luminosity function at any epoch is obtained simply by sliding the local luminosity function along the $\log \Phi$-axis by a constant amount which depends on the redshift.

Density evolution in the form $\rho(z) \sim (1 + z)^n$, $n \simeq 6$, was used by Schmidt (1970) to describe the evolution of radio quasars from the 3CR sample, and by Braccesi *et al.* (1980) to fit an assumed power law luminosity function to the observed surface density of optically selected quasars. Density evolution, however, fails to reproduce the observed flattening in the quasar counts at $m_B \simeq 19.5$ (see Figure 6.7), and is now mostly discarded in favour of luminosity evolution.

7.3.2 *Pure luminosity evolution*

In this pattern of evolution it is assumed that the comoving density of quasars remains constant, while there is a change in their luminosity with redshift. In models in which the lifetime of a quasar is short compared to the cosmological time scale $1/H_0$, the change in luminosity is to be understood in a statistical sense to reflect the change in the mean luminosity of an ensemble of quasars. The luminosity function satisfies

$$\Phi(L(z), z)\, dL(z) = \Phi(L_0)\, dL_0, \tag{7.12}$$

where

$$L(z) = L_0 \psi(L_0, z), \quad \Phi(L_0) \equiv \Phi(L_0, 0). \tag{7.13}$$

$L_0 = L(0)$ is the luminosity at the present epoch and ψ is the function that describes how the luminosity of a source evolves with redshift. In terms of the absolute magnitude,

$$\Phi(M, z) = \Phi(M_0, 0), \quad M = M_0 - 2.5 \log \psi(M_0, z). \tag{7.14}$$

If the local luminosity function is defined between the limits $L_{\min,0}$ and $L_{\max,0}$ then, owing to luminosity evolution, at redshift z the limits map into $L_{\min}(z)$ and $L_{\max}(z)$ respectively. The total number of quasars in a comoving volume element $dV(z)$ is then given by

$$dN(L_{\min}(z), L_{\max}(z); z) = dV(z) \int_{L_{\min}(z)}^{L_{\max}(z)} \Phi(L, z) \, dL$$

$$= dV(z) \int_{L_{\min,0}}^{L_{\max,0}} \Phi(L_0) \, dL_0, \tag{7.15}$$

i.e., the total number of quasars remains constant. It is this conserved comoving density that makes luminosity evolution fundamentally different from density evolution.

When the evolution function depends only on z we have *pure luminosity evolution*. In this case, in the $(\log \Phi)M$-plane the luminosity function at any epoch is obtained simply by sliding the local luminosity function along the M-axis by a constant amount that depends on the redshift. We will see in Section 7.7 that a luminosity function which has a power law form with different slopes at faint and bright regions and which incorporates pure luminosity evolution best describes quasar data for $z < 3$.

7.3.3   *The equivalence of pure density and pure luminosity evolution*

We will now discuss the circumstances in which pure density evolution may be considered to be equivalent to pure luminosity evolution. Using absolute magnitude as a variable, the two patterns of evolution can be expressed as

$$\Phi(M, z) = \Phi(M)\rho(z), \quad \Phi(M, z) = \Phi(M + 2.5 \log \psi(z)) \tag{7.16}$$

respectively, where as usual the Φ on the right hand side of each equation, written as the function of a single variable, represents the local luminosity function. If the two views are to be equivalent, we must have

$$\Phi(M)\rho(z) = \Phi(M + 2.5 \log \psi(z)), \tag{7.17}$$

or

$$\log \rho(z) = \log \Phi(M + 2.5 \log \psi(z)) - \log \Phi(M), \tag{7.18}$$

for all values of M and z. In this equation the right hand side has the absolute magnitude as a variable, while the left hand side does not. The two sides will be consistent only if the right hand side has no net dependence on M. This is the case when $\log \Phi$ is linear in M:

$$\log \Phi(M) = \beta M + \delta, \tag{7.19}$$

where β and δ are constants, i.e., when

$$\log \Phi(L) \propto L^{-\beta}. \tag{7.20}$$

Pure density and pure luminosity evolution are therefore equivalent if the luminosity function is of power law form.

The equivalence of the two forms of evolution in the special case of a power law seems to contradict the property that in the case of luminosity evolution the number of sources is conserved, while in density evolution it is not. There is no contradiction, however, because it is not possible for a pure power law to span all luminosities from the arbitrarily small to the arbitrarily large, since then the number of sources would diverge at one of the ends. This divergence can be avoided by introducing a cutoff either at low or at high luminosity, depending on the power law index, or by having a break so that the slope at the low luminosity end is flat while at the high luminosity end it is steep. Any such operation introduces a departure from a structureless power law, and destroys the equivalence, because any structure would mean that translations along the M-axis are not equivalent to translations along the $\log \Phi$-axis. The equivalence can only be local, and can apply only to small sections of the luminosity function that can be approximated by a power law and are far from the points of departure from a structureless form.

7.4   Surface density and luminosity function

Given the luminosity function, it is a simple matter to obtain the surface density as a function of observed flux or apparent magnitude. Consider quasars with luminosity in the interval $(L(v), L(v) + dL(v))$ at redshift z, where the emitted frequency v and the luminosity are in the reference frame of the quasar. The observed frequency is $v_0 = v/(1 + z)$ and the observed flux $F(v_0)$ is related to $L(v)$ through Equation 2.61. Assuming for simplicity that the luminosity has a power law form, $L(v) \propto v^{-\alpha}$, we have $L(v) = (1 + z)^{-\alpha}L(v_0)$, and both the flux as well as the luminosity can be expressed at the observing frequency v_0. In the following we will drop the argument v_0 from $L(v_0)$ and $F(v_0)$ for convenience.

Given the luminosity and redshift distribution, we want to obtain the surface density of quasars as a function of the flux F. If a quasar with luminosity L is to have flux greater than some value F, it must have redshift $z < z(F, L)$, the limiting redshift being obtained as a solution to Equation 2.14. The number of quasars with flux $> F$ per unit area of the sky (steradian) is therefore given by

$$N(> F) = \frac{1}{4\pi} \int_{L_{\min}}^{L_{\max}} dL \int_0^{z(F,L)} dz \, \frac{dV(z)}{dz} \Phi(L, z), \tag{7.21}$$

where $dV(z)$ is the comoving volume element given in Equation 2.74. The surface density $N(F) \, dF$ of objects with flux in the range $(F, F + dF)$ can be obtained by differentiating this equation with respect to F:

$$N(F) = \left| \frac{dN(>F)}{dF} \right| \tag{7.22}$$

Note that we use the same symbol N to indicate the differential as well as the integral count; we shall distinguish between the two functions by always including the argument of the function. We will now consider the special form of Equation 7.21 applicable to a Euclidean universe populated by objects with a single luminosity L_0 and constant comoving density:

$$\Phi = \Phi_0 \delta(L - L_0), \quad \Phi_0 = \text{constant}, \tag{7.23}$$

where δ is the Dirac delta function. The Euclidean results can be obtained by restricting Equations 2.61 and 7.21 to $z \ll 1$, in which case

$$F = \frac{L}{4\pi(c/H_0)^2 z^2}, \quad dV(z) = 4\pi(c/H_0)^3 z^2 dz, \tag{7.24}$$

and

$$N(>F) \propto F^{-3/2}, \quad \log N(>F) = -1.5 \log F + \text{constant}. \tag{7.25}$$

The corresponding relations for $N(F)$ are

$$N(F) \propto F^{-5/2}, \quad \log N(F) = -2.5 \log F + \text{constant}. \tag{7.26}$$

In the non-Euclidean geometry of the expanding universe and when there is evolution, there is departure from the simple 3/2 law. Even then it is usual to express the surface density as a power law, with the index possibly changing over the range of flux considered:

$$N(>F) \propto F^{-\beta}, \quad N(F) \propto F^{-(\beta+1)}. \tag{7.27}$$

In the optical case it is usual to express the surface density as a function of magnitude m. Using $m = -2.5 \log F + \text{constant}$, Equation 7.27 becomes

$$\log N(<m) = \log N(>F) = 0.4\beta m + \text{constant}, \tag{7.28}$$

from which it follows that

$$\frac{N(<m+1)}{N(<m)} = 10^{0.4\beta}. \tag{7.29}$$

In the non-evolving Euclidean case with $\beta = 1.5$ the slope of the linear relation between $\log N(<m)$ and m is 0.6, i.e., there is an increase by a factor four in the surface density per unit increase in magnitude. The observed surface density of quasars given in Equations 6.10 and 6.11 corresponds in this notation to $\beta = 2.20$ for blue magnitude $B < 18.75$ and $\beta = 0.78$ for $B > 18.75$. In the steep part there is an increase by a factor 7.6 in the surface density of quasars per unit increase in magnitude.

It is possible to find the parameters of a model luminosity function by predicting the surface density as a function of flux using Equation 7.21, and fitting this to the observed surface density. Such a procedure was adopted by radio astronomers to infer the luminosity function from radio source counts for most of which no optical identifications, and hence no redshift, was available (see e.g. Longair 1979). For a sample of quasars the redshift of each object is always known (barring a few cases when only an estimate may be available, especially when the identification is based on a single spectral line), and more direct methods for determining the space density can be used. It is clear that a direct determination of the luminosity function will always be superior to obtaining it from a projection of the space density on the plane of the sky.

Samples of quasars that are based on source surveys in a band other than the optical have two flux limits. First, there is the limit of the survey in which the objects are first discovered, say in the radio band. Then there is the magnitude limit up to which optical identification of the quasar content of the survey is complete. The presence of the two limits makes it necessary to adopt a procedure such as the accessible volume method discussed below (Rees and Schmidt 1971).

7.5   The V/V_m method

The V/V_m test was first used by Maarten Schmidt (1968) to study the space distribution of a complete sample of radio quasars from the 3CR catalogue. A similar test was proposed earlier by Kafka (1967), and later Lynden-Bell (1971) proposed the related C-method.

7.5.1   *Euclidean geometry*

To understand how the method works, we will first limit ourselves to Euclidean geometry. Let us consider a sample of optically selected quasars that are uniformly distributed in space, and let F_m be the limiting flux of the survey. To every quasar in the sample there correspond two volumes,

$$V(r) = \frac{4\pi}{3} r^3, \qquad V_m = \frac{4\pi}{3} r_m^3, \tag{7.30}$$

where r is the radial distance to the quasar and

$$r_m = \left(\frac{L}{4\pi F_m} \right)^{1/2}, \tag{7.31}$$

is the limiting distance at which the flux of a quasar with luminosity L reduces to the limiting value F_m. For $r > r_m$ the quasar would no longer be part of the sample, as its flux would reduce to a level below the flux limit. $V(r)$ is the volume enclosed at the distance of the quasar, while V_m is the maximum volume accessible to it under the condition that it is a part of the complete sample. Now a quasar from the uniformly distributed population is equally likely to be anywhere in the accessible volume, and therefore the values of V/V_m for the sample are uniformly distributed in the interval

[0, 1]. This is true for all luminosities, and therefore for the whole sample the mean value $\langle V/V_m \rangle = 0.5$.

More formally, for a large sample drawn from a uniformly distributed population

$$\langle V/V_m \rangle = \frac{\int_{L_{min}}^{L_{max}} dL\Phi(L) \int_0^{r_m(L)} (V/V_m)4\pi r^2 \rho_0 dr}{\int_{L_{min}}^{L_{max}} dL\Phi(L) \int_0^{r_m(L)} 4\pi r^2 \rho_0 dr}, \tag{7.32}$$

$r_m(L)$ is the limiting distance for objects with luminosity L, and ρ_0 is the uniform density. Using Equations 7.30 and 7.31 it follows that in the Euclidean case

$$\langle V/V_m \rangle = 1/2. \tag{7.33}$$

It can similarly be shown that $\langle (V/V_m)^2 \rangle = 1/3$, so that the standard deviation is

$$\sigma\left(V/V_m\right) = 1/\sqrt{12}. \tag{7.34}$$

The σ in this equation gives the spread of the sample values of V/V_m about $\langle V/V_m \rangle$. If samples of the size N are repeatedly drawn from the parent population, then it can be shown using the theory of statistics (see [HC89], p173) that the mean value of $\langle V/V_m \rangle$ over the repeated trials is again $1/2$, while the variance in the mean values is given by

$$\sigma\left(\langle V/V_m \rangle\right) = 1/\sqrt{12N}. \tag{7.35}$$

If a survey for a uniformly distributed population misses faint objects, it will in effect be missing objects which are at the more distant parts of the accessible volume, making $\langle V/V_m \rangle < 1/2$. This can be used to test the completeness of a survey, when the parent population is a priori known to be uniformly distributed. This is the case when one is counting galaxies to distances $\ll c/H_0$ (see e.g. Huchra and Burg 1992). When the parent population has a density which effectively increases outwards, there is a piling up of objects towards the far end of the accessible volume, and $\langle V/V_m \rangle > 1/2$.

7.5.2 V/V_m *in an expanding universe*

Consider a quasar with luminosity L and redshift z, which is part of a complete sample with limiting flux F_m. As in the Euclidean case, we can define two volumes: $V(z)$, which is the comoving volume enclosed at the redshift z, and $V_m \equiv V(z_m)$, the accessible comoving volume up to the redshift z_m at which the flux reduces to the limit of the survey. For a sample of quasars drawn from a parent population with luminosity function $\Phi(L, z)$, the mean value of V/V_m is

$$\langle V/V_m \rangle = \frac{\int_{L_{min}}^{L_{max}} dL \int_0^{z_m(L)} (V/V_m)\Phi(L, z)\, dV(z)}{\int_{L_{min}}^{L_{max}} dL \int_0^{z_m(L)} \Phi(L, z)\, dV(z)}. \tag{7.36}$$

When the luminosity function is independent of z this reduces to

$$\langle V/V_m \rangle = \frac{\int_{L_{min}}^{L_{max}} dL\Phi(L)V_m(L) \int_0^1 (V/V_m)d(V/V_m)}{\int_{L_{min}}^{L_{max}} dL\Phi(L)V_m(L) \int_0^1 d(V/V_m)} = \frac{1}{2}. \tag{7.37}$$

When Φ is independent of z, the number of quasars per unit comoving volume is the same everywhere in the accessible volume. V/V_m is therefore uniformly distributed in the interval $[0, 1]$ and again $\langle V/V_m \rangle = 1/2$. The dispersion in V/V_m about $\langle V/V_m \rangle$ and the dispersion in $\langle V/V_m \rangle$ for repeated trials are as in Equations 7.34 and 7.35 respectively.

For a complete sample of optically selected quasars there is a well-defined limiting magnitude, and the data on each quasar consist of the measured apparent magnitude (flux) and redshift. Assuming a continuum spectral index α, Equation 2.61 can be used to obtain the limiting redshift z_m for each quasar. The comoving volumes $V(z)$ and V_m are then obtained by integrating the comoving volume element in Equation 2.74, and

$$\langle V/V_m \rangle = N^{-1} \sum_i (V/V_m), \tag{7.38}$$

where the summation extends over all N quasars in the sample. A value of $\langle V/V_m \rangle = 1/2$ with specified confidence level is then consistent with no evolution, while $\langle V/V_m \rangle > 1/2$ indicates that the comoving density of quasars effectively increases towards higher redshift, independently of the exact pattern of evolution. The V/V_m test therefore allows the inspection of a sample for possible evolution even when the sample size is too small for the luminosity function to be determined. It is of course necessary for the sample to be complete, as otherwise the values of V/V_m will be biased.

7.5.3 *Multiple limiting fluxes*

The test can be easily generalized to the situation when there is a limiting flux at more than one frequency defining the sample, as in the case of a complete sample of radio selected quasars. Here there is a limiting radio flux as well as a limiting optical magnitude, and the flux in each band has to be greater than the corresponding limit for a quasar to belong to the sample. There will therefore be two limiting redshifts, $z_{m,opt}$ and $z_{m,radio}$, each defining an accessible volume. In such a case we simply take z_m to be the minimum of the two redshifts:

$$z_m = \min(z_{m,opt}, \, z_{m,radio}), \tag{7.39}$$

for each quasar, and use the corresponding accessible volume to evaluate V/V_m. It is clear that for a non-evolving population this procedure will lead to $\langle V/V_m \rangle = 1/2$ and that it can be generalized to any number of limiting fluxes.

7.5.4 *Weighted volumes*

When evolution is expected to be present, it is convenient to define a volume V' weighted by the luminosity function as follows (Schmidt 1968):

$$V'(L) = \int_0^z \Phi(L, z') \, dV(z'). \tag{7.40}$$

The weighted maximum volume V'_m is defined as before by extending the integration to z_m. The number of objects with V'/V'_m in a small interval $d\left(V'/V'_m\right)$ is equal to the number of objects with V/V_m in the small interval $d\left(V/V_m\right)$, with the two sets of volumes related through Equation 7.40. Therefore if $P\left(V'/V'_m\right)$ and $P\left(V/V_m\right)$ are the probability distributions of the two variables, we have

$$P(V'/V'_m)d(V'/V'_m) = P(V/V_m)d(V/V_m) \qquad (7.41)$$

But $P(V/V_m) \propto \Phi(L,z)$ and

$$P(V'/V'_m) = P(V/V_m)\frac{dV}{dV'} \propto \frac{\Phi(L,z)}{\Phi(L,z)} = \text{constant.} \qquad (7.42)$$

V'/V'_m therefore has a uniform distribution and $\langle V'/V'_m \rangle = 1/2$.

When $\langle V/V_m \rangle > 1/2$ for a sample, one can try to find an evolution function that leads to a uniform distribution of V'/V'_m. Such a function is a model for the evolution but it is not unique, and many different evolutionary forms may be consistent with the data. The virtue of this method is that it provides some quantification of the evolution even when the sample size is small. In his pioneering analysis Schmidt (1968, 1970) used the V/V_m test to infer that radio quasars from the 3CR sample were subject to pure density evolution. Though this pattern of evolution is not consistent with the present data, Schmidt's methods provided the foundation for later work.

7.6   Determination of the luminosity function

We will consider in this section several methods to determine the luminosity function. The treatment will be restricted to optical luminosity functions, but will apply to other bands as well, examples of which we will deal with in later chapters. Bivariate luminosity functions will be introduced in Section 7.11. Consider quasars that are observed to be in small intervals $(m, m + dm)$ and $(z, z + dz)$ of apparent magnitude and redshift. These quasars must be situated in a small comoving volume $dV(z)$, and in the absolute magnitude interval $(M, M + dM)$, where m and M are related through Equation 2.63. If $n(m,z)$ is the number of quasars observed per unit apparent magnitude and redshift interval, we have

$$n(m,z)\, dmdz = \Phi(M,z)\, dMdV(z), \qquad (7.43)$$

from which Φ may be directly inferred (the actual numbers obtained will depend on the assumed continuum spectral index and on the values of H_0 and q_0). There are, however, several difficulties in using this simple approach. A given survey covers only a limited region of the mz-plane, the extent of which depends upon the technique used in finding quasars and the limiting magnitude reached. Only a portion of the Mz-plane is therefore directly accessible to the survey. The coverage can be extended by combining different surveys, but this is difficult because of the different selection effects involved. The corrections to be made depend upon details of the continuum and line spectra, which can vary from one quasar to another, and on knowledge of the exact procedures employed in the surveys; these procedures are not always available in a quantitative form. Another difficulty in applying the direct method is that even the largest samples available do not have enough quasars to provide statistically meaningful numbers in small redshift intervals. It is therefore necessary to average over finite redshift shells. In this case the existence of a magnitude limit requires a careful treatment using the maximum accessible volumes.

Knowledge about the luminosity function gained in the above manner is by itself insufficient to provide an analytical form and therefore not much can be said about regions of the Mz-plane that remain unsampled. An alternative is to assume a functional form for Φ and to determine the parameters in it by fitting it to the observed distribution of magnitude and redshift. The disadvantage here is that results depend on the assumed functional form, which may not have any physical basis since it is chosen for its simplicity and mathematical convenience.

7.6.1 *Luminosity function from accessible volumes*

This method was first used by Schmidt (1968). Consider a sample of optically selected quasars that is complete to a certain limiting magnitude m_{lim}. We will first assume that there is no evolution, so that the luminosity function is independent of redshift. A quasar in the absolute magnitude interval dM around M will have apparent magnitude brighter than m_{lim}, and hence will be a part of the sample, when $z < z_{\text{m}}$, where z_{m} is the limiting redshift as in Section 7.5. Therefore the comoving volume of space that has been surveyed for quasars with absolute magnitude M is $V_{\text{m}} = V(z_{\text{m}})$. If the number of quasars in the sample in the absolute magnitude interval dM is $N(M)\,dM$, the comoving number density of quasars at M is given by

$$\Phi(M) = \frac{N(M)}{V_{\text{m}}}. \tag{7.44}$$

It is assumed in this equation that the limiting redshift z_{m} and hence V_{m} remain constant over the interval dM. In practice half- or one-magnitude-wide bins are generally used so that the number per bin is sufficient to provide good statistics. However, z_{m} and therefore V_{m} can change significantly over such an interval dM. In such a case one evaluates the limiting volume $V_{\text{m},j}$ individually for the jth member in a sample of n quasars in the interval $(M, M + dM)$ from its absolute magnitude, and the luminosity function at M is then

$$\Phi(M) = \frac{1}{dM} \sum_{j=1}^{n} \frac{1}{V_{\text{m},j}}. \tag{7.45}$$

When there is evolution, bins of absolute magnitude as well as redshift have to be considered. The redshift range of each bin again has to be large enough to keep the number of quasars in each bin at a reasonable level. Let $V(z_1, z_2)$ be the comoving volume of the redshift shell bounded by z_1 and z_2. For a quasar in this shell and with absolute magnitude in the range $(M - dM/2, M + dM/2)$, the accessible volume is $V_{\mathrm{m}} = V(z_1, z_2)$ if $z_2 < z_{\mathrm{m}}$; if $z_{\mathrm{m}} < z_2$, $V_{\mathrm{m}} = V(z_1, z_{\mathrm{m}})$. If $z_1 > z_{\mathrm{m}}$ there is of course no contribution, since all quasars in the redshift shell would be fainter than the completeness limit of the sample. The luminosity function is given by

$$\Phi(M, z) = \frac{1}{dM} \sum_j \frac{1}{V_{\mathrm{m},j}} \qquad (7.46)$$

where the summation now extends over all quasars in the redshift interval (z_1, z_2) and absolute magnitude interval $(M - dM/2, \ M + dM/2)$. The luminosity function can be symbolically written as

$$\Phi(M, z) = \sum_{z_1 < z_j < z_2} \frac{\delta(M - M_j)}{V_{\mathrm{m},j}}, \qquad (7.47)$$

where δ is the Dirac delta function and the summation now extends over all quasars in the redshift interval Δz. The error estimate of the luminosity function is given by (Marshall 1985):

$$\sigma = \left(\sum_j \frac{1}{V_{\mathrm{m}}^2,_{j}} \right)^{1/2}. \qquad (7.48)$$

This method is easily generalized to the case with more than one limiting flux, by taking $V_{\mathrm{m},j}$ to be the minimum of the limiting volumes permitted by the two flux limits, as was done in subsection 7.5.3.

This determination of the luminosity function does not depend upon any assumed parameters (except those needed in specifying the cosmological model and the quasar continuum). But there is averaging over the redshift shells, which is necessary because of the finite size of the sample. When the sample sizes grow large enough to have infinitesimal bin sizes, it is possible to determine Φ directly from Equation 7.43. Luminosity functions determined in finite redshift shells are useful in examining the development of the comoving density over widely separated redshift regions, without making any prior assumption about the form of the evolution function. They also help in choosing models for a maximum likelihood analysis of the kind to be described in subsection 7.6.3.

7.6.2   *Luminosity function from weighted accessible volumes*

Schmidt and Green (1983) have used weighted accessible volumes (see subsection 7.6.1) to determine the local luminosity function from the data using an assumed form of the evolution function. From the definition of the luminosity function, it follows that the number of quasars $N(M)$ per unit absolute magnitude in a complete sample is

$$N(M) = \int_0^{z_{\mathrm{m}}(M)} \Phi(M, z) \, dV(z), \qquad (7.49)$$

with $z_{\rm m}(M)$ the usual limiting redshift. Using Equation 7.9 and the definition of density-weighted volume in Equation 7.40, this reduces to

$$N(M) = \Phi(M)V'_{\rm m}, \tag{7.50}$$

so that

$$\Phi(M) = \frac{N(M)}{V'_{\rm m}} = \sum_j \frac{\delta(M - M_j)}{V'_{\rm m}}, \tag{7.51}$$

where the summation extends over all quasars in the sample. The local luminosity function is therefore determined from the data, if the evolution function is known, and weighted volumes can be calculated. Schmidt and Green assumed that there is evolution in the comoving density of sources, the rate of evolution depending upon the luminosity:

$$\rho(M, z) = \exp\left[k(M_0 - M)\tau(z)\right], \tag{7.52}$$

where $\rho(M, z)$ is the evolution function we introduced in Equation 7.7, with ρ now taken to be a function of the absolute magnitude instead of luminosity. The constants k and M_0 are to be determined from observations and $\tau(z)$ is the look-back time to redshift z in units of the age of the universe (see Section 2.8). This is a case of luminosity-dependent density evolution. Given values of k and M_0, Φ is determined. This cannot be compared with an *observed* local luminosity function since this is not available. The function $\Phi(M, z)$, which is obtained from $\Phi(M)$ by multiplication with the assumed evolution function, agrees by construction with the data for all values of the two constants. Different values of the constants will, however, lead to different predictions for the Mz-region outside the limits of the survey. Schmidt and Green determined the local luminosity function using data from the Bright Quasar Survey (see Section 6.7), with the constants in Equation 7.52 chosen so as to obtain a good match with the surface density obtained in other surveys. They found that $k > 0$, i.e., there was steeper evolution for the more luminous sources. We shall not go into the numerical details of their models, and refer the reader to the original work.

7.6.3 *Luminosity function using a maximum likelihood technique*

This method was first used by Marshall, Avni *et al.* (1983) in the context of quasar luminosity functions. In it the luminosity function is represented by a specific model with a number of parameters. The values of these parameters are determined as those at which the probability for obtaining the observed luminosity and redshift values is maximized. The formulation of the technique given by Marshall, Avni *et al.* is as follows.

The number of quasars in the differential element $dLdz$ is given by

$$n(L, z)\, dL\, dz = \frac{\Omega(L, z)}{4\pi} \Phi(L, z)\, dL\, dV(z), \tag{7.53}$$

where $\Omega(L, z)$ is the solid angle of the sky over which a quasar with L and z could have been observed.[1] This is the mean number that will be observed in repeated trials over the volume element in different regions of space. The numbers actually observed will have a Poisson distribution about this mean, providing that the quasars are distributed at random with no correlation.

Let the region of the Mz-plane covered by the survey be divided into elements $dLdz$ which are so small that the probability that more than one quasar is found in any element is very small. Each element therefore has either one quasar or none at all. Assuming a Poisson distribution with mean number given by Equation 7.53, the probability that an element contains exactly one quasar is $e^{-n(z,L)\,dLdz}n(z, L)\,dLdz$, and the probability that it contains no quasar is $e^{-n(z,L)\,dLdz}dLdz$. Therefore the total probability that there is a quasar at each of the observed values of L and z, and that there is no quasar at any other point of the Mz-region covered by the survey is given by

$$\Gamma = \prod_i e^{-n(L_i,z_i)\,dLdz}n(L_i, z_i)\,dLdz \prod_j e^{-n(L_j,z_j)dLdz} \qquad (7.54)$$

where the first product extends over all the quasars in the sample, while the second extends over all the differential regions with no quasars in them. Introducing the quantity $S = -2\ln\Gamma$ and dropping terms that do not depend on the model parameters, we get

$$S = -2\sum_i \ln\Phi(L_i, z_i) + 2\int dL \int \Omega(L, z)\Phi(L, z)\,dV(z), \qquad (7.55)$$

with the limits of integration covering the Mz-region of the survey.

Given a model luminosity function with a certain number of free parameters, their best-fit values are obtained by maximizing the likelihood function Γ, i.e., minimizing S. Confidence regions around the best-fit parameter values can be obtained by using the method of Lampton, Margon and Bowyer (1976). The best-fit models can be tested for goodness of fit using, say, the two-dimensional Kolmogorov–Smirnov statistic (Peacock 1983, 1985, Press *et al.* 1992), and this provides the confidence level to which a particular model can be accepted.

The maximum likelihood method provides one way in which a luminosity function defined analytically can be fitted to the data. The results depend on which functional form is assumed; this is based on considerations of simplicity and mathematical convenience rather than on fundamental physical considerations. To that extent they are arbitrary, and it is possible that a completely different model could provide a better fit than those that are adopted. However, present quasar samples are large enough to rule out some simple models, such as pure density evolution, that have been used in the past, while simple luminosity evolution models continue to provide an excellent fit.

---

[1] In all previous expressions, we have implicitly assumed for convenience that the solid angle covered by the survey is 4π. The comoving volume element in Equation 2.72 corresponds to such an angle.

The best-fit models are now rather robust, in the sense that the goodness of fit cannot be improved or worsened dramatically by making small changes in the form of the models used. It is convenient in various calculations to have simple analytical models, and it is possible that these will provide some insight into the underlying physics.

7.6.4 *The simultaneous analysis of several complete samples*

The discussion in the earlier sections of this chapter was based on a single complete sample of quasars, spread over a connected region of the sky. Modern quasar surveys often involve searches carried out over non-overlapping fields, each with its own limiting magnitude. The survey of Boyle *et al.* (1988) covers 34 fields with limiting *B* magnitude varying over the range 20.15−20.90 from field to field. Each field provides an independent sample of quasars, to which the techniques derived above could be applied. It is, however, advantageous to combine the separate fields into a single joint sample, as this has the benefit that a greatly increased number of quasars are available for the analysis. Using the fact that quasars in the different fields are drawn from the same parent population, Avni and Bahcall (1980) have shown how a single *coherent* sample, in which each observed quasar is allowed to appear a priori in all the independent fields, may be obtained. We will follow Avni and Bahcall and consider the simple case of two independent samples. The generalization to a higher number of regions will be obvious.

Consider two samples A and B that cover solid angles Ω_A and Ω_B, and have N_A and N_B quasars respectively. Let m_A and m_B be the limiting magnitudes of the two samples respectively, with corresponding limiting redshifts $z_A(L)$ and $z_B(L)$ for a quasar with luminosity L. We will assume, without any loss of generality, that $m_A < m_B$, i.e., $z_A(L) < z_B(L)$. A coherent combined sample may now be defined which contains all the $N_T = N_A + N_B$ quasars in the two samples taken together and which spans the solid angle $\Omega_A + \Omega_B$. It is assumed that a quasar in the combined list, with luminosity L, can a priori be found either in region A with redshift $z < z_A(L)$, or in region B with $z < z_B(L)$. The depth of the coherent sample is not uniform over the combined region.

The volume available to a quasar for it to be included in the combined sample (the accessible volume), is defined to be

$$V_a = \frac{\Omega_A}{4\pi} V(z_A) + \frac{\Omega_B}{4\pi} V(z_B). \tag{7.56}$$

The volume enclosed at the redshift of the quasar is defined to be

$$V_e = \frac{\Omega_A}{4\pi} V(z) + \frac{\Omega_B}{4\pi} V(z), \quad z < z_A(L) \tag{7.57}$$

and

$$V_e = \frac{\Omega_A}{4\pi} V(z_A) + \frac{\Omega_B}{4\pi} V(z), \quad z > z_A(L). \tag{7.58}$$

This definition of V_e takes into account the fact that because $z_A(L) < z_B(L)$, a quasar with $z < z_A(L)$ could appear in either of the regions A or B at its actual redshift. For $z > z_A(L)$, the quasar can appear in A up to the maximum allowed redshift of that region, $z_A(L)$, and in B to the redshift $z \leq z_B(L)$. Using a straightforward generalization of the proofs given in subsection 7.5.2, Avni and Bahcall (1980) showed that the formal statistical properties of V_e/V_a are the same as those of V/V_m, so that when there is no evolution, V_e/V_a is uniformly distributed in $[0, 1]$ with mean value $1/2$. When there is evolution, density-weighted volumes V_e' and V_a' can be defined as in Equation 7.40 and V_e'/V_a' has a uniform distribution.

The luminosity function for a combined coherent sample is obtained as in Equation 7.46, but with the combined accessible volume V_a replacing V_m:

$$\Phi(M, z) = \sum_j \frac{1}{V_{a,j}}, \tag{7.59}$$

where the summation extends over all objects in the complete sample in the appropriate absolute magnitude and redshift bin.

Avni and Bahcall have also shown how independent samples that have overlapping regions in the sky may be combined together in the most statistically efficient manner.

7.7   The quasar luminosity function at low redshifts

We have described in Section 6.7 the UVX survey of Boyle, Shanks and Peterson (1988), which spans the redshift range $0.3 < z < 2.2$, and its later extension to $z < 2.9$ using the multicolour technique (Boyle *et al.* 1991). Luminosity functions have been obtained from these surveys using accessible volumes as well as the maximum likelihood method.

In their analysis, Boyle, Shanks and Peterson include data from some of the surveys listed in Table 6.3 other than their own. This improves the statistics at bright magnitudes since the survey itself is rather sparse in this region, because it spanned a relatively small area of the sky owing to its faint limiting magnitude. The absolute magnitude M_B of each quasar was obtained using Equation 2.63 with no corrections made for galactic absorption or the contribution from emission lines, since these were always found to be small ($\lesssim 0.1$ magnitude). The range of the absolute magnitude M_B in the B band was divided into intervals with $\Delta M_B = 1$, and four bins were defined over the redshift range of the sample, $0.3 < z < 2.2$, such that each redshift bin contained approximately the same number of quasars. The luminosity function was obtained using the accessible volume method, described above, for $q_0 = 0.0$ and 0.5. The results are shown in Figure 7.1. It is clear that the shape of the luminosity function is similar in the four bins, with a flat slope at faint absolute magnitudes and a steep slope at bright magnitudes. The break in the slope occurs at increasing luminosity at higher redshift, indicating that luminosity evolution is present, and continues up to the last redshift bin.

The extension of the luminosity function to $z < 2.9$ by Boyle *et al.* (1991) is shown in Figure 7.2. The shape of the function is found to be similar to that of Boyle, Shanks

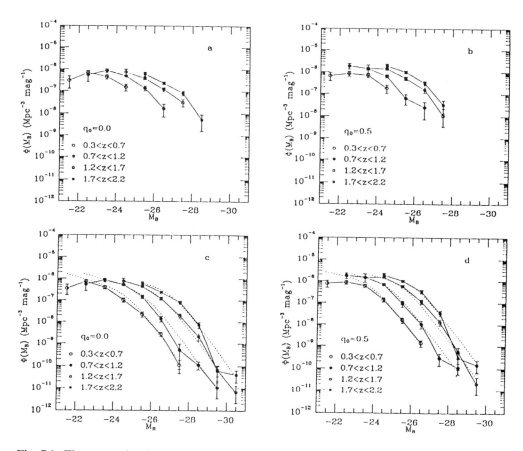

Fig. 7.1. The quasar luminosity function derived using the effective volume method for various redshift bins. See the text for details. Reproduced from Boyle *et al.* (1988), courtesy of the Astronomical Society of the Pacific Conference Series.

and Peterson (1988) but the strong evolution extends only to $z \sim 2.0$. The change in the luminosity function between the bins $1.25 < z < 2.0$ and $2.0 < z < 2.9$ is smaller than the change between lower redshift bins, indicating that the evolution has slowed down.

Boyle, Shanks and Peterson also obtained useful analytical models for the luminosity function using the maximum likelihood method. Taking a hint from the numerical results from the accessible volume method, the luminosity function was represented by the smooth two-power-law function

$$\Phi(M_B, z) = \frac{\Phi^*}{10^{0.4[M_B - M_B(z)](\alpha+1)} + 10^{0.4[M_B - M_B(z)](\beta+1)}}. \tag{7.60}$$

The characteristic luminosity at which the break occurs is a function of redshift, which

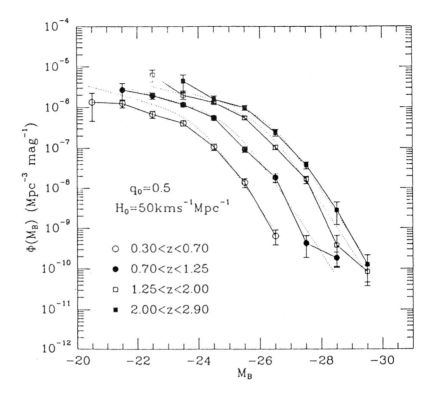

Fig. 7.2. The quasar luminosity function derived by using the effective volume method for various redshift bins extending to $z = 2.9$. See the text for details. Reproduced from Boyle *et al.* (1991).

provides luminosity evolution while maintaining the shape of the spectrum. The form of the redshift dependence of the break luminosity is taken to be

$$M_B(z) = M_B^* - 2.5 k_L \log(1 + z). \tag{7.61}$$

The shape of the luminosity function becomes clear when it is viewed as a function of absolute luminosity using Equation 7.2. Then we get

$$\Phi(L, z) \propto L^\beta, \quad L \ll L(z); \quad \Phi(L, z) \propto L^\alpha, \quad L \gg L(z). \tag{7.62}$$

The different slopes in the power law at low and high luminosity are now obvious. The evolution of the luminosity at which the break occurs is

$$L_B(z) \propto (1 + z)^{k_L}. \tag{7.63}$$

The luminosity function contains four free parameters: α, β, M_B^* and k_L. Φ^* is determined by normalizing the integration of the luminosity function over the Mz-region of the survey to the total number of quasars in the sample. The maximum likelihood values of the parameters obtained using the method described in subsection 7.6.3 are shown in Table 7.1 for $q_0 = 0.0$ and 0.5.

Table 7.1. *Best-fit parameters for the luminosity function models of Boyle et al. (1988)*

q_0	α	β	M_B^*	k_L	Φ^*
0.0	−3.84	−1.61	−23.10	3.34	2.3×10^{-7}
0.5	−3.79	−1.44	−22.42	3.15	1.0×10^{-6}

Table 7.2. *Best-fit parameters for luminosity function models that include a cutoff in the evolution at redshift z_{cut} (Boyle et al. 1991)*

q_0	α	β	M_B^*	k_L	z_{cut}	Φ^*
0.1	−3.8	−1.6	−22.6	3.55	2.1	3.5×10^{-7}
0.5	−3.9	−1.5	−22.4	3.45	1.9	6.5×10^{-7}

Boyle, Shanks and Peterson have tried variants of the form of the luminosity function in Equation 7.60, such as replacing the $\log(1+z)$ term in Equation 7.61 with the look-back time $\tau(z)$ and including density evolution in addition to the luminosity evolution. For each model, best-fit values were obtained and it was found that none of the other variants produced a better fit than the model described above. The density evolution allowed was found to be negligible.

In the extension of their model to include high redshift quasars, Boyle *et al.* (1991) assumed that the luminosity evolution in Equation 7.61 occurs only for $z < z_{cut}$, with

$$M_B(z) = M_B(z_{cut}), \quad z > z_{cut}. \tag{7.64}$$

The cutoff redshift z_{cut} at which evolution stops is an additional variable to be obtained from the maximum likelihood analysis. The best-fit parameters are shown in Table 7.2. The form of the luminosity function for different redshifts is shown in Figure 7.3. The surface density as a function of apparent magnitude predicted by this luminosity function, and the observed surface densities, are shown in Figure 6.7.

The simple picture of pure luminosity evolution, which is based on quasar samples with $z > 0.3$, may not be valid at low redshifts. Köhler *et al.* have derived the luminosity function for their complete sample of quasars from the Hamburg/ESO survey, which contains a substantial number of objects with bright magnitudes and low redshifts. They find that for quasars with $0.07 < z < 0.3$ and $-26 \leq M_B \leq -18$, the differential luminosity can be represented by a single power law, $\Phi(L) \propto L^\alpha$, with $\alpha = -2.10 \pm 0.08$. The luminosity function is independent of the redshift for $z \leq 0.3$. The space densities obtained for the Hamburg/ESO sample are much higher than those of the BQS, which is the only other survey with a substantial number of low redshift quasars. The higher space densities of course result from the higher surface densities discussed in subsection 6.7.2. The discrepancy with the BQS is partly due to the inclusion of objects

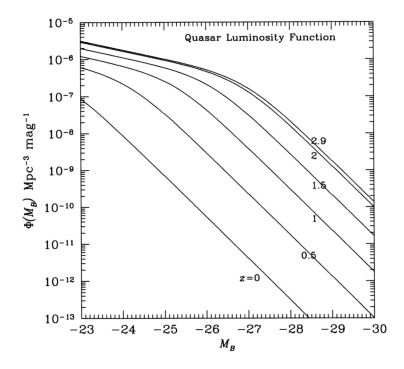

Fig. 7.3. The analytic form of the luminosity function of Boyle *et al.* (1991). See text for details.

with non-stellar appearance in the Hamburg/ESO survey. But this effect appears to be small, since a luminosity function restricted to $z < 0.2$, which omits most objects with extended morphology, has substantially the same normalization as this luminosity function, which includes objects up to $z = 0.3$.

The slope of the luminosity function for the Hamburg/ESO sample is intermediate to the flat slope for low luminosities, and the steep slope for high luminosities, obtained for the luminosity function of quasars with $z > 0.3$. If the results of Köhler *et al.* (1997) are confirmed with larger samples of low redshift quasars, then it would mean that the pure luminosity evolution models are not applicable, since the shape of the luminosity function changes from a single power law, in the local universe, to a broken power law at high redshift. An alternative would be to show that the existing high redshift surveys are substantially incomplete at the highest luminosities. If many more highly luminous quasars were to be found, it may well be possible to describe the luminosity functions at high redshift by a single power law, making the shape of the luminosity function the same at all redshifts. But if a single power law is applicable all over, then one cannot distinguish between the pure luminosity and pure density patterns of evolution, as discussed in subsection 7.3.3. These uncertainties will only be clarified when more substantial samples covering wide ranges of redshift and luminosity become available.

Table 7.3. *Best-fit parameters for the luminosity function models of Warren* et al. *(1995). The Hubble constant used is* $H_0 = 75 \, \text{km sec}^{-1} \, \text{Mpc}^{-1}$

q_0	α	β	M_C^*	k_L	$\log \Phi^*$
0.1	−5.18	−2.11	−14.65	10.33	−1.55
0.5	−5.05	−2.06	−13.21	10.13	−0.99

7.8 The quasar luminosity function at high redshifts

We have seen in Section 6.2 that high redshift quasars are most abundantly found in multicolour and slitless spectroscopic surveys, and it is these samples which are used in the derivation of the luminosity function at high redshifts.

Warren, Hewett and Osmer (1995) found 85 quasars with $z \geq 2.2$ in their multicolour survey (see subsection 6.2.2) and for each quasar assessed the probability p of finding a quasar like it that appears in the survey area. The probability p gives a measure of the incompleteness of the survey; the contribution of each quasar to the luminosity function is weighted by a factor $1/p$ to allow for the missing quasars. Combining their sample with other quasars from the literature, Warren, Hewett and Osmer obtain a total sample of 100 quasars with $z \geq 2.2$. These authors have also used complete samples of quasars with $2.0 \leq z \leq 2.2$ from the LBQS and Boyle *et al.* surveys to obtain the luminosity function in a redshift range overlapping with the range covered by the low redshift luminosity functions in Section 7.7.

Warren, Hewett and Osmer found that the luminosity function in the range $2.0 \leq z \leq 3.5$ is well described by a double power law form with pure luminosity evolution, similar to the one in Equation 7.60:

$$\Phi(M_C, z) = \frac{\Phi^*}{10^{0.4[M_C - M_C(z)](\alpha+1)} + 10^{0.4[M_C - M_C^*](\beta+1)}}; \qquad (7.65)$$

here M_C is the rest frame continuum absolute magnitude at $1216\,\text{Å}$. It is convenient to use this rest frame ultraviolet magnitude because at the high redshifts of quasars in the survey, it is this region that is mainly used in the observed bands. The major difference between the models in Equations 7.60 and 7.65 is that in the latter the luminosity evolution is restricted to quasars with absolute magnitude brighter than the characteristic value M_C^* and fainter quasars are assumed not to evolve. The function $M_C(z)$ is given by

$$M_C(z) = M_C^* - 1.086 k_L \tau, \qquad (7.66)$$

where τ is the look-back time. The constants in Equations 7.65 and 7.66 are determined by using a maximum likelihood technique as described in Section 7.7. The best-fit values are given in Table 7.3 for values of deceleration parameter $q_0 = 0.1$ and 0.5 and the Hubble constant used in the calculation is $H_0 = 75\,\text{km}\,\text{sec}^{-1}\,\text{Mpc}^{-1}$. The luminosity function is expressed in units of magnitude$^{-1}\,\text{Mpc}^{-3}$.

Using two-dimensional Kolmogorov–Smirnov tests, it is found that the best-fit models provide a satisfactory fit to the distribution of points in the $M_C z$-plane when the fit includes only quasars in the redshift range $2.2 \leq z \leq 3.3$. The significance of the fit declines rapidly when quasars in the range $3.0 \leq z \leq 4.5$ are included and Warren *et al.* interpret this to mean that positive luminosity evolution ceases at $z \sim 3.3$. The number of quasars expected in the survey with $3.5 \leq z \leq 4.5$ under the assumption that the luminosity function ceases to evolve at $z = 3.3$ and remains constant thereafter is 49 for $q_0 = 0.1$ and 52 for $q_0 = 0.5$. However, the observed number of quasars with $z \geq 3.5$ is eight, so that there is a decline by a factor ~ 6 compared to the number expected under the assumption of a non-evolving luminosity function for $z \geq 3.3$. Allowing for Poisson fluctuation in the observed number, Warren *et al.* estimate that the decline is by a factor > 3 at the 95 per cent confidence level. It should be noted that this result is based on a small number of quasars (eight) with $z > 3.5$. Also the detection probability p associated with some of the high redshift quasars is as small as ~ 0.1, which leads to large corrections for incompleteness. The results therefore need to be treated with some caution.

The Palomar Transit Grism Survey (PTGS) by Schneider, Schmidt and Gunn (1994) contains 90 quasars in the redshift range $2.75 < z < 4.75$, selected on the basis of their emission lines (primarily their Ly α emission). The results of the survey have been used by Schmidt, Schneider and Gunn (1995) to derive a luminosity function for high redshift quasars. They find that the mean value of the ratio of the effective volume of each quasar to its accessible volume $\langle V_e/V_a \rangle = 0.377 + 0.026$ (see subsection 7.6.4 for the meaning of these quantities). The ratio is < 0.5 at the 4σ level. Assuming that the sample is complete this means that there must be a decrease in the comoving density of quasars with increasing redshift.

The luminosity range of the Schmidt, Schneider and Gunn sample is smaller by a factor 10 and the authors found that it was not possible to fit anything more elaborate than a single power law to the data. The integral form of the luminosity function obtained by them is given by

$$\log \Phi(< M_B, z) = 2.165 - 0.43(z - 3) + 0.748(M_B + 26), \qquad (7.67)$$

which corresponds to the luminosity function per unit absolute magnitude range

$$\Phi(M_B) = 1.376 \times 10^{23 + 0.748 M_B - 0.43 z}\,\text{Gpc}^{-3}. \qquad (7.68)$$

In terms of luminosity this gives $\Phi(L_B, z) \propto L_B^{-1.87} \times 10^{-0.43 z}$. The functional form is that of pure density evolution, but with the comoving density decreasing towards higher redshifts. The luminosity function has been obtained assuming $H_0 = 50\,\text{km}\,\text{sec}^{-1}\,\text{Mpc}^{-1}$ and $q_0 = 0.5$ and is valid only in the redshift and absolute magnitude ranges $2.7 \leq z \leq 4.7$ and $-27.5 < M_B < -25.5$. Schmidt, Schneider and Gunn have made a detailed comparison of their luminosity function with that of Warren, Hewett and Osmer (1995) and find that the two are in good agreement.

It is clear from the above exercises that luminosity functions are not yet understood in a truly cosmological context. The functional forms specifying luminosity or density evolution are put in by hand with the intention of fitting the data in the simplest possible manner. There is at the moment no astrophysical theory of the origin and evolution of quasars in an evolving universe to lead us to these functional forms. Moreover, the models as available now may be subject to extensive revision, if the results at low redshifts, described towards the end of Section 7.7, are correct. It would help to have physical models, because then it will be possible to compare specific model predictions with the observed surface and space densities.

7.9   The luminosity function of Seyfert galaxies

Seyfert galaxies are characterized by bright nuclei that show a non-thermal continuum and strong emission lines. The classical examples of this type were found by Markarian and Seyfert in their searches for galaxies with various peculiarities (see the introduction to Chapter 6), but now we have Seyfert galaxies that have been discovered in the identification of X-ray, infrared or radio sources. As in the case of quasars, a different primary selection technique leads to different sectors of the multivariate luminosity function, and care has to be exercised in drawing inferences about the whole population from any one subset.

A galaxy is classified as a Seyfert from its emission line properties, and a selection technique which addresses this directly is bound to lead to a more complete sample than any other technique. Huchra and Burg (1992) selected a sample of Seyfert galaxies in this manner and though the number of identified Seyferts is small, the sample is important because of its unbiased nature. This sample of Seyferts is obtained from the CFA redshift survey; this consists of 2399 galaxies, in the merged Zwicky–Nilson catalogue (see Huchra and Burg for references), that have $m_{Zw} < 14.5$, where the Zwicky magnitude m_{Zw} is very nearly equal to the photographic magnitude, and which are in the part of the sky defined by (1) either declination $\delta \geq 0$ deg and Galactic latitude $b \geq 40$ deg, or (2) $\delta \geq -25$ deg and $b \leq -30$ deg. More than 98 per cent of the sample have been observed spectroscopically in the range 4600–7000 Å. A galaxy is classified as Seyfert 1 , Seyfert 2 or Liner on the basis of its emission lines. The emission line galaxies identified in this manner include 26 Seyfert 1 galaxies, 23 Seyfert 2 galaxies and 33 Liners. There is also one quasar and a few other objects.

It would be most natural to use nuclear luminosities and magnitudes in the determination of the space density, but the nuclear quantities are not directly measurable since the underlying galaxy can contribute substantially even when a small aperture around the nuclear region is used to measure the magnitude. It is therefore necessary to use other techniques, e.g. the colour-given method, which depends on the colour distributions of the nuclear and galactic (stellar) emissions to separate the two components. This introduces uncertainties in the estimated nuclear magnitudes (see e.g. Cheng *et al.* 1985). Moreover, the completeness of the primary identification is based on total magnitudes. Huchra and Burg have therefore obtained the luminosity function of the integrated magnitude, which includes the galactic as well as the nuclear emission. This must be taken into account when comparing the results from this survey with other luminosity functions from the literature.

Seyfert luminosity functions have been estimated by several groups, including Cheng *et al.* (1985), who have used all confirmed Seyfert 1 and Seyfert 1.5 galaxies that appeared in the first nine Markarian lists. They obtained the space density as a function of nuclear magnitudes, which were estimated in two different ways. The luminosity function that they obtained was found to be consistent with the quasar luminosity function at the faint end, as well as the luminosity function determined using X-ray selected AGN.[1]

Applying the V/V_m technique, Huchra and Burg found that $\langle V/V_m \rangle \simeq 0.5$ for the Seyferts, which shows that the sample is not seriously incomplete. The redshift of all the galaxies is relatively small, and cosmological evolution can be neglected. The (local) luminosity function is then easily determined by again using the V/V_m technique as described in Section 7.5, and is shown in Figure 7.4. Nuclear magnitudes are usually estimated from the total magnitude. The number of galaxies in the sample is, however, too small for the parameters of a model luminosity function to be determined.

Huchra and Burg find that ~ 1 per cent of all the galaxies in their sample are of the Seyfert 1 type, while another ~ 1 per cent are of the Seyfert 2 type. The percentage of AGN in the sample increases if the most luminous galaxies are considered, with ~ 20 per cent of the galaxies with absolute blue magnitude $M_B < -21$ having an AGN. Integrating to the faint limit of $M_B = -18$, the relative space density of Seyfert 2 to Seyfert 1 galaxies is found to be 2.3 ± 0.7.

The use of total luminosities, instead of the more appropriate nuclear luminosities, can seriously affect the shape of the luminosity function at lower luminosities. Using a small sample of just seven galaxies, with carefully determined nuclear luminosities, Köhler *et al.* (1997) obtained a luminosity function for Seyfert 1 nuclei which merges smoothly into the luminosity function of low redshift quasars (see Section 7.7). The entire spread of objects over eight magnitudes can be represented by a single-power-law luminosity function $\Phi(L) \propto L^\alpha$ with $\alpha = -2.17 \pm 0.06$. Using total instead of nuclear luminosities produces a sharp upturn in the Seyfert 1 luminosity function at $M_B \simeq -20$, owing to a substantial contribution from the host galaxy to the total luminosity. These results are based on a small number of objects and need to be confirmed with much larger samples. A smooth luminosity function which goes from Seyfert 1 nuclear luminosities to high luminosity quasars would support the notion that Seyfert galaxies and quasars are basically the same, and differ only in the proportion of light coming from the active nucleus and the host galaxy.

---

[1] It would be interesting to make a detailed comparison between quasar and Seyfert luminosity functions using recent data, to see whether the luminosity evolution observed for quasars extrapolates to the less luminous objects.

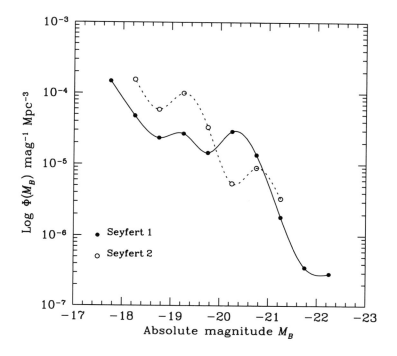

Fig. 7.4. The luminosity function for Seyfert 1 and Seyfert 2 galaxies as determined by Huchra and Burg (1992). The lines joining the points are spline fits provided to aid the eye.

7.10 Evolutionary scenarios

We have seen above, from the inferred space density of quasars as a function of luminosity and redshift, that the luminosity of the quasar population as a whole increases with redshift up to $z \simeq 2$, after which the evolution slows down, and there could even be a decline in luminosity towards higher redshift. This behaviour of the population can be reproduced by different patterns of activity in individual quasars. These have been discussed most clearly by Cavaliere and Padovani (1989).

Let us assume that quasar luminosity is the result of accretion of matter onto a black hole with an efficiency η for converting the accreted mass into energy.[1] If the mass of the black hole M is wholly due to accretion, one has

$$M = \frac{1}{\eta c^2} \int L(t)\, dt, \tag{7.69}$$

where the integral indicates the total radiant energy emitted by the quasar over the

---

[1] See Chapter 5 for a discussion of the standard black hole model of quasars.

period of accretion. Using Equation 5.28 we have $L_{Edd} = Mc^2/t_{Edd}$ and therefore

$$\frac{L}{L_{Edd}} = \frac{\eta t_{Edd} L}{\int L(t)\,dt}, \tag{7.70}$$

where L is the bolometric luminosity at the current epoch. Examination of this equation allows three basic activity patterns to be studied.

In the first scenario, it is assumed that luminosity evolution refers to the dimming of individual quasars. All quasars dim continuously on the same time scale t_L following a short formation phase at $z \simeq 2$. Assuming exponential dimming for simplicity, the integral reduces to $L_{max}t_L$. The epoch at $z = 2$ from which quasars have been dimming corresponds to a look-back time of 5.4 Gyr for a $q_0 = 0$ universe (see Equation 2.79). For the luminosity evolution of Boyle *et al.* (1988), the dimming from redshift z is by a factor $(1 + z)^{3.5}$, and therefore at the present epoch

$$\frac{L}{L_{Edd}} = 2 \times 10^{-4} \left(\frac{\eta}{0.1} \right). \tag{7.71}$$

This increases to at most $2 \times 10^{-3}(\eta/0.1)$ if factors such as the effect of evolution on t_L and the possibility of no evolution at the faint end of the luminosity function are taken into account. In this scenario the quasar phenomenon is limited to a very small fraction of all galaxies. The dimmed counterparts could be Seyfert galaxies, but then the local space density of Seyferts would be required to be consistent with the space density of high redshift quasars. The observational situation here is not clear because of the uncertainties in the Seyfert luminosity function and the evolution of the quasar luminosity function at the faint end.

Another alternative is where all galaxies have a single, short event of quasar-like activity, corresponding mainly to the black hole formation phase. If during the episode the accretion rate is $\dot{M} \sim \eta^{-1}\dot{M}_{Edd}$, then

$$\frac{L}{L_{Edd}} \sim \frac{\eta \dot{M} c^2}{\dot{M}_{Edd} c^2} \sim \frac{\eta (\eta^{-1} \dot{M}_{Edd})}{\dot{M}_{Edd}} \sim 1. \tag{7.72}$$

If, further, the luminosity (i.e., the mass of the black hole formed in the episode) depends inversely on the cosmological epoch, then the episodes will on average present the appearance of luminosity evolution, with $L \sim L_{Edd}$ at all redshifts. This scenario requires that galaxies with $L \gtrsim L_*$, where L_* is the characteristic luminosity of the Schechter luminosity function of galaxies, would all have passed through a phase of quasar activity. Since the average luminosity increases with redshift, at any epoch the inactive galaxies harbour nuclei that are more massive on the average than those of the active galaxies.

Intermediate to the two extreme alternatives above is the scenario where quasar activity occurs intermittently in each of a set of galaxies. The duration of each outburst is short compared to the time scale over which the luminosity of the quasar population is found to evolve. If the total duration of the activity is t_e and L_e is the average luminosity, then from Equation 7.70 it follows that

$$\frac{L}{L_{Edd}} = \frac{\eta t_{Edd} L}{t_e L_e}.$$ (7.73)

For fast average dimming, $L/L_e \sim 10^{-2}$ and assuming that the total active time t_e is short relative to the evolutionary time scale, one has locally

$$\frac{L}{L_{Edd}} \sim 5 \times 10^{-2} \left(\frac{\eta}{0.1} \right) \left(\frac{t_e}{10^{-2} t_{Edd}} \right)^{-1}.$$ (7.74)

The ratio L/L_{Edd} increases from this local value to $\gtrsim 1$ at $z \simeq 2$. The number of host galaxies required in this scenario scales with L/L_{Edd} and for the value 5×10^{-2} approaches the same number as in the scenario with $L \sim L_{Edd}$.

Examining the observational data available to them, Cavaliere and Padovani (1989) have estimated that $L/L_{Edd} \sim 5 \times 10^{-2}$ (it is assumed here that $H_0 = 50$ km sec^{-1} Mpc^{-1}). In spite of the considerable uncertainty in the determination of this ratio, the nominal observed value rules out the model in which quasars are to be found in a very small fraction of galaxies and are subject to continuous dimming. The model with $L \sim L_{Edd}$ is also ruled out, which leaves the scenario involving recurrent episodes of quasar activity.

If quasar activity is long lived, it is limited to the small fraction of all galaxies which are specially suitable for sustaining it. In the standard picture, these galaxies would have a high-mass black hole at their centres. The situation is quite different for the other patterns we have discussed. In these cases a considerable fraction of all galaxies would have passed through single or recurrent episodes of activity. A supermassive black hole is therefore expected in many galaxies, with the presently inactive bright galaxies having more-massive black holes than the presently active ones. These alternatives need to be examined further, both theoretically as well as observationally, because of their important implications for quasar and galaxy formation and evolution.

Small and Blandford (1992) considered a model that is somewhat different from the scenarios described above. This model involves the quick formation of a black hole in newly formed galaxies, with the hole radiating at the Eddington limit. At later epochs, when the supply of fuel is no longer plentiful, the holes accrete only intermittently at an average rate that is the same for all quasars and is given by $\dot{M} \propto M^{-1.5} t^{-6.7}$, t being the cosmological time. The number of relic black holes per decade of mass is $\propto M^{-0.4}$, for black hole masses in the range $3 \times 10^7 \lesssim M/M_\odot \lesssim 3 \times 10^9$. In this model all sufficiently massive galaxies pass through a quasar phase, the mass of the black hole formed being a monotonic function of the host galaxy mass. It is therefore possible to test the model by comparing the local density of galaxies having known central masses with the density of high redshift quasars. Haehnelt and Rees (1993) have developed a model for the evolution of the luminosity function, using a superposition of ~ 100 generations of quasars with lifetime $\sim 10^8$ yr and an evolutionary time scale for the population of $\sim 10^9$ yr. In this model quasars are assumed to be the first phase in the formation of a galaxy, in the potential well of a dark-matter halo. The number of newly forming dark-matter haloes at successive cosmological epochs is estimated using the Press–Schechter formalism in the cold dark matter (CDM) scenario. The luminosity functions calculated from this model are in good agreement with observation.

7.11   Bivariate radio–optical luminosity function

We have seen in Section 7.2 that the luminosity function is in general dependent on the luminosity in various bands, the redshift and other properties of the quasars or AGN under consideration. Owing to the limited number of objects present in complete samples, the luminosity functions considered in the literature have been generally restricted to two luminosities, such as optical and radio or optical and X-ray, and the redshift. We will consider in this section some often used forms of bivariate radio and optical luminosity functions.

Radio-loud quasars and AGN are discovered as the optical counterparts of bright radio sources from radio surveys. Alternatively, sensitive radio observations of optically discovered objects lead to estimates of their radio luminosity. When the two luminosities are known for a class of objects, the bivariate luminosity function $\Phi(L_{op}, L_R, z)$ is defined in the same manner as the optical luminosity function in Equation 7.1, with L_{op} and L_R the optical and radio luminosities respectively, at suitable wavelengths. In the absence of a theory which determines the functional form, one looks for the simplest possibilities that are consistent with the data.

Suppose one assumes that the optical and radio luminosities are uncorrelated. In that case the two will be independently distributed, and we can separate the luminosity function into the form

$$\Phi(L_{op}, L_R, z) = \Phi(L_{op}, z)\Psi(L_R), \tag{7.75}$$

where

$$\int_0^\infty \Psi(L_R) = 1. \tag{7.76}$$

The integration extends over the interval $(0, \infty)$ in only a formal sense, of course, and cutoffs have to be applied to avoid divergence at the endpoints. It is possible to consider more complex forms, with Ψ depending on redshift and so on, but these are not warranted by the quality of the data. The fraction of quasars with radio luminosity greater than L_R is given by the integral form of the radio luminosity function,

$$I(> L_R) = \int_{L_R}^\infty \Psi(L_R') \, dL_R'. \tag{7.77}$$

It was argued by Maarten Schmidt (1970) that this form of the luminosity function is not tenable when applied to quasars. Schmidt based his conclusion on a comparison of the redshift distribution of two sets of objects: (1) optically selected quasars with magnitude $m_V \simeq 18$, from a sample due to Sandage and Lyuten (1967) and (2) radio-bright 3 CR quasars with similar magnitude. The number of optically selected quasars with redshift and magnitude in small ranges around z and m_V can be obtained from Equation 7.75 by integrating over all values of L_R, so that

$$n(m_V, z) \, dm_V \, dz = \Phi(L_{\text{op}}, z) \, dL_{\text{op}} \, dz, \tag{7.78}$$

where dL_{op} is the luminosity interval corresponding to dm_V at z. The radio quasars, however, are from a complete survey and are all brighter than some radio flux limit $F_{R,\text{lim}}$. Therefore at a redshift z only quasars with luminosity $L_R > L_{R,\text{lim}}$ are selected, where the limiting flux and luminosity are related by Equation 2.61. The number of radio quasars in the given magnitude and redshift range is therefore

$$n(m_V, > F_{R,\text{lim}}, z) \, dm_V \, dz = \Phi(L_{\text{op}}, z) \, dL_{\text{op}} \, dz \int_{L_{R,\text{lim}}}^{\infty} \Psi(L_R') \, dL_R'. \tag{7.79}$$

The lower limit of integration in Equation 7.79 increases with redshift, and for the steep power law type of luminosity functions of the kind one normally encounters, there is a progressive reduction in the number of quasars at higher redshift because only those with radio luminosity brighter than the limit are included in the complete sample. The two samples therefore should not have similar redshift distributions.

To account for the similarity in the observed redshift distribution, Schmidt proposed that the mean values of the optical and radio luminosities are correlated, so that quasars brighter in the optical region are also radio bright. In this case a simple representation of the bivariate luminosity function is provided by

$$\Phi(L_{\text{op}}, L_R, z) = \Phi(L_{\text{op}}, z)\Psi(R), \tag{7.80}$$

where $R = L_R/L_{\text{op}}$ and the area under the function $\Psi(R)$ is normalized to unity between the the limits R_{min} and R_{max}, with the latter usually taken to be ∞. The fraction of objects with ratio greater than R is given by

$$G(> R) = \int_R^{\infty} \Psi(R') \, dR'. \tag{7.81}$$

The distribution of the radio luminosity is now provided by a universal function with only the ratio of the two luminosities as the independent variable. The radio luminosity of quasars with optical luminosity L_{op} ranges from $R_{\text{min}}L_{\text{op}}$ to $R_{\text{max}}L_{\text{op}}$ and has the mean value $\langle L_R \rangle = L_{\text{op}} \langle R \rangle$. The quasars clearly become more luminous in the radio band as their optical luminosity increases. In this formulation the distribution of redshift for radio sources with magnitude m_V is given by

$$n(m_V, > F_{R,\text{lim}}, z) \, dm_V \, dz = \Phi(L_{\text{op}}, z) \, dL_{\text{op}} \, dz \, G(> R_{\text{lim}}), \tag{7.82}$$

where

$$R_{\text{lim}} = \frac{L_{R,\text{lim}}}{L_{\text{op}}} = \left(\frac{F_{R,\text{lim}}}{F_{\text{op}}} \right) (1 + z)^{(\alpha_R - \alpha_{\text{op}})}, \tag{7.83}$$

α_R and α_{op} are the radio and optical spectral indices respectively of the assumed power law continua and Equation 2.61 is used. Now, α_{op} and α_R are both in the range ~ 0.5–1, and therefore R_{lim} is approximately constant. It follows that the ratio $n(m_V, z)/n(m_V, > F_{R,\lim}, z)$ is also approximately constant. Factorization of the luminosity function as in Equation 7.80 therefore leads to similar observed redshift distributions for optical and radio quasars, as found by Schmidt, providing of course they come from the same parent population. In Chapter 9 we will consider radio luminosity functions evaluated using the two forms considered here.

7.12   Concluding remarks

We have considered in this chapter luminosity functions primarily in the optical and radio bands. We shall defer discussion of luminosity functions in other bands, such as the X-ray band, to subsequent chapters, since we have yet to develop the required terminology and concepts.

8　The continuum

8.1　Introduction

We shall consider in this chapter some characteristic properties, such as the spectral shape, variability and polarization, of the continuum radiation from quasars and AGN. The continuum has been observed from at least $\sim 10^8$ Hz in the radio region to $\sim 10^{27}$ Hz, which corresponds to extremely high energy γ-ray photons. This vast range spans many different physical processes and the emitting region can vary in size from the $\sim$ Mpc scale of extended radio emission to the scale of $\sim 10^2$ light seconds on which high energy radiation is emitted. Our aim in this chapter will be to provide an overview. Some of the regions of the spectrum, for example the radio and X-ray regions, will be discussed in detail in other chapters, and we will dwell upon them only very briefly here.

8.2　Power law continuum

The continuum spectra of quasars and AGN can, over limited frequency ranges, be represented by simple power law forms. The luminosity as a function of frequency $L(v)$ is then given by

$$L(v) = Av^{-\alpha}, \tag{8.1}$$

where A and the power law index are constant. The luminosity emitted between frequencies v_1 and v_2 is then

$$L(v_1, v_2) = \frac{A}{1 - \alpha} \left(v_2^{1-\alpha} - v_1^{1-\alpha} \right), \quad \alpha \neq 1 \tag{8.2}$$

and

$$L(v_1, v_2) = A \log \left(\frac{v_2}{v_1} \right), \quad \alpha = 1. \tag{8.3}$$

For a power law form, the luminosity per unit frequency can decrease by several orders of magnitude as the frequency increases over its wide range. However, the energy emitted in a finite frequency interval over some part of the spectrum such as the radio, optical or X-ray, changes relatively slowly because of the increase in bandwidth when going from a low frequency region to a higher frequency one. When $\alpha = 1$, the

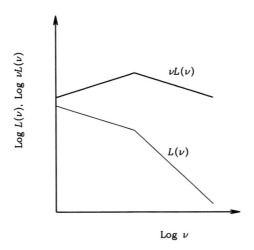

Fig. 8.1. Luminosity $L(v)$ and the energy emitted per decade of frequency $vL(v)$ for a power law form with spectral index $\alpha = 0.5$ at low frequencies and $\alpha = 1$ at high frequencies.

luminosity in the frequency interval $(v, 10v)$, which corresponds to the unit logarithmic interval $\Delta(\log v) = 1$, is given by

$$L(v, 10v) = A = vL(v). \tag{8.4}$$

When $\alpha \neq 1$, $vL(v)$ still is approximately equal to the luminosity in $(v, 10v)$. It is therefore useful to consider the quantity $vL(v)$ rather than $L(v)$ in depicting continuum shapes. In Figure 8.1 we have shown $L(v)$ and $vL(v)$ for a power law spectrum in which the index steepens from $\alpha = 0.5$ at low frequencies to 1.0 at high frequencies. Most of the energy in the spectrum is radiated around the break frequency. For $\alpha = 1$ equal energy is emitted in every decade of the frequency.

On the basis of the observed power law forms, the polarization found at some wavelengths and the compact sizes of the emitting regions, which followed from the observed variability, it was concluded that the continuum radiation of all quasars and AGN must have its origin in non-thermal processes such as synchrotron radiation or Compton scattering. The conclusion that closely related non-thermal processes are responsible for the emission was supported by the observed correlations between the luminosities in widely separated parts of the spectrum, e.g. the infrared and X-ray regions. However, major revisions in this simple picture have become necessary as more detailed observations have become available over the last decade or so.

Observation of the spectrum over a wide frequency range requires the use of many different kinds of telescope and technique. In regions of the spectrum to which the atmosphere is not transparent observations can be made only with satellite-borne instruments.[1] Over the last decade or so, there has been a marked improvement in the sensitivity of ground-based as well as satellite-based instruments, and data in some regions of the spectrum such as the far infrared have become available for the first time for large samples of quasars and AGN.

From the data presently available, it appears that there are basically two kinds of continuum shape, as follows.

(1) Blazars, which include quasars and BL Lacs with bright radio cores and high degrees of polarization and variability, indicating bulk relativistic motion in these cores (see Section 9.9), have non-thermal continuum radiation from the radio to the ultraviolet band, possibly produced by synchrotron emission. The non-thermal emission continues in the high energy region, where it can be affected by the effects of inverse Compton scattering and pair production.

(2) In radio-quiet quasars, Seyfert galaxies and even broad line radio galaxies (BLRG) the dominant emission is in the $\sim 0.01-1\,\mu$m region. It is of thermal origin and is produced in an accretion disk or as free–free optically thin emission. Emission in the $\sim 1-100\,\mu$m region, which is the second most dominant component, is produced as thermal emission from dust with a range of temperatures. Emission in the radio region, when detectable, has a steep spectrum and is of non-thermal origin. X-ray emission is often detectable in these objects and could have a strong non-thermal component. The observed emission can be affected by the orientation of the object relative to the observer.

In the following sections we shall consider the continuum in greater detail. Recent excellent reviews can be found in Bregman (1990, 1994), in various articles in the *Proceedings of the IAU Symposium 159* (CB94) and in the works cited in the chapters on radio and X-ray emission.

8.3 Thermally dominated objects

Thermally dominated quasars and AGN are usually radio-quiet, but only in the sense that they do not appear in radio surveys with high flux limits. A fraction of them ($\sim 10-20$ per cent) are actually radio-loud at the level of a few mJy, while others remain radio-quiet even when probed at the μJy level (see Section 9.8). The 'radio-quiet' objects that have detectable radio emission often have radio luminosity higher than that of normal galaxies. The upper limits on the radio flux from many undetected objects also place them above normal galaxies in radio luminosity. Nevertheless, the radio power emitted is small relative to their total emitted power. Detected objects usually have steep radio spectra indicative of optically thin synchrotron emission. The extrapolation of the radio spectrum to infrared wavelengths lies far below the actual

---

[1] Balloon and rocket-borne instruments can also be used when it is sufficient to attain a limited height. The total duration available for observations is rather limited in these cases.

observed flux, indicating that there are other processes that become dominant in that region.

In the infrared region, the emitted power per unit logarithmic interval $\nu L(\nu)$ rises sharply between 10 and 100 μm and peaks somewhere in this range. The continuum in the 2–300 μm region is known as the *infrared bump* and typically contains a third of the total detected (i.e., bolometric) luminosity of the source, though this fraction can be as high as ∼ 90 per cent. A near-infrared bump, always centred at 58 THz (5.2 μm) and carrying ≲ 40 per cent of the 2.5–10 μm luminosity has been detected by Edelson and Malkan (1986) in a sample of quasars and Seyfert 1 galaxies. The position of this bump is, however, model dependent. There is a local minimum in $\nu L(\nu)$ at ∼ 1 μm, which is most obvious in highly luminous radio-quiet quasars and in Seyfert 1 galaxies.

Shortwards of 1 μm the continuum rises into the optical and ultraviolet, with a prominent peak in $\nu L(\nu)$, called the *big blue bump* or the *UV excess*, occurring between ∼ 0.3 μm and ∼ 10 nm (∼ 10^{15}–3×10^{16} Hz). In some objects the peak power occurs in the 1200–100 Å region, while in others $\nu L(\nu)$ continues to rise into the extreme ultraviolet. Data in the ultraviolet region were obtained primarily from the *international ultraviolet explorer (IUE)* satellite.[1] Moreover, in the extreme ultraviolet range (EUV) between the Lyman limit of 912 Å and ∼ 100 Å, the opacity of the Galactic interstellar medium is generally very high, and observations can only be made through specific low opacity windows. The continuum in this region is responsible for ionization of the broad line and narrow line emitting regions, and therefore some idea of its shape can be obtained from the study of emission lines. It is also possible to observe it more directly in high redshift quasars, but here again the continuum is subject to absorption both within the quasar and within the intervening matter. The radiation can again be observed for wavelengths ≲ 100 Å (125 eV), i.e., in the soft X-ray band, with X-ray detectors aboard satellites. The observed flux here indicates that the peak of the blue bump can occur in the extreme ultraviolet region with the bump providing much of the bolometric luminosity.

Direct observations of AGN and quasars in the extreme ultraviolet (EUV) have been made with the wide field camera (WFC) on *ROSAT*, which was sensitive in this region, and the dedicated *extreme ultraviolet explorer (EUVE)* satellite, which was launched in 1992. An all-sky survey was carried out with the WFC, in which seven AGN were found. Another all-sky survey by the *EUVE* has also produced seven AGN. The list has been increased to a few dozen using pointed observations, serendipitous discoveries and coincidences between *EUVE* all-sky source positions, and known X-ray positions of extragalactic sources. EUV sources are generally found in the directions of low optical depth windows in the galactic interstellar medium, with neutral hydrogen column density $N_{\mathrm{H}} \lesssim 10^{20}$ cm^{-2}. Firm detections have only been made in the shortest

---

[1] This was launched in 1978 and carried a 45 cm Cassegrain telescope with two spectrographs. It was used, until it was shut off in 1996, in the observation of ultraviolet spectra in the wavelength range 1200–3200 Å and had no sensitivity shortwards of 1200 Å.

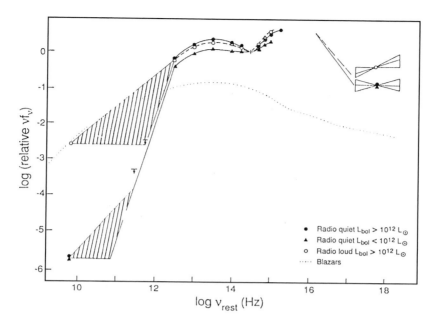

Fig. 8.2. Average continuum distributions of radio-quiet and radio-loud quasars from the sample of Sanders *et al.* (1989) which is a subset of the BQS. Each data point represents the mean over the corresponding sample. The hatched region in the radio band indicates the spread of radio spectral indices for each sample. Note that νf_ν, where f_ν is the flux density, is plotted as the ordinate. The infrared bump and big blue bump are clearly seen. Reproduced from Sanders *et al.* (1989).

wavelength bandpass, which goes from $\sim 60\,\text{Å}$ to $\sim 170\,\text{Å}$. The EUV observations will provide an important link between the ultraviolet and soft X-ray regions. A review of observations with the *EUVE* and some preliminary interpretations of the sparse data may be found in BM96.

Sanders *et al.* (1989) have discussed continuum observations, in the $\sim 0.3\,\text{nm}-6\,\text{cm}$ ($5\,\text{GHz}-10^{19}\,\text{Hz}$) range, of 105 quasars from the Palomar Bright Quasar (BQS) survey. They find that the shapes are remarkably similar, except in a few cases such as 3C 273 that have flat radio spectra and can be classified as blazars, discussed below. Average continuum distributions of the 70 quasars from the sample that were detected at more than one wavelength between 10 and 100 μm are shown in Figure 8.2, separately for the radio-loud and radio-quiet varieties.[1] The average spectrum clearly shows the infrared and big blue bumps, which are also apparent in individual spectra.

These workers also obtained the bolometric luminosity for those quasars in their

---

[1] The BQS quasars are all optically selected, and most of the radio-loud subset have radio luminosities generally much below those of the highly luminous radio selected kind (see Section 9.8). The latter usually have blazar-like properties, because they are discovered mostly in high frequency surveys which are dominated by compact flat radio spectrum sources.

sample for which far-infrared data is available. Using a constant value of $\nu L(\nu)$ in the range $2 \times 10^{15} - 3 \times 10^{16}$ Hz, they find that the bolometric luminosity ranges from $\sim 10^{11} L_\odot$ to $\sim 2 \times 10^{14} L_\odot$. The strengths of the infrared and big blue bump do not show any correlation to the bolometric luminosity. The ratio of the bump luminosities L_{IR}/L_{UV} has a spread of a factor ~ 10 and $\langle L_{IR}/L_{UV} \rangle = 0.4$ with a spread of 0.15.

In contrast to the small spread in the ratio of the infrared and ultraviolet luminosities, the ratio of the radio to the $1\,\mu$m luminosity has a spread of a factor $\sim 10^5$. This has a clearly bimodal distribution, with one peak around

$$\nu L(\nu)(5\,\text{GHz})/\nu L(\nu)(1\,\mu\text{m}) = 3 \times 10^{-6}, \qquad (8.5)$$

corresponding to the radio-quiet objects, and another peak around 3×10^{-3} corresponding to the radio-loud objects. The bimodal distribution is similar to that found for the ratio of radio to optical luminosity in Section 9.8. Sanders *et al.* define the boundary between radio-loud and radio-quiet quasars at

$$\nu L(\nu)(5\,\text{GHz})/\nu L(\nu)(1\,\mu\text{m}) = 10^{-4}. \qquad (8.6)$$

The two kinds of quasar have very similar continuum spectra for wavelengths ~ 1 mm, as can be seen from Figure 8.2. In the radio region both the radio-quiet and radio-loud variety have a range of spectral indices (see Section 9.8). In the case of the radio-quiet quasars, extrapolation of the radio spectrum into the infrared region underestimates the observed flux at $100\,\mu$m by a factor of $\sim 10^2 - 10^6$, depending on the radio spectral index. For the radio-loud quasars, the extrapolation meets the $100\,\mu$m observed flux only for those with the flattest radio spectral indices, while the steep spectrum objects again miss the observed point by a large margin. The radio and infrared emissions therefore appear to be decoupled, except in the case of the flat spectrum radio-loud objects.

Beyond the extreme ultraviolet is the soft X-ray region at ~ 0.1 keV and then there is the X-ray region, followed by the γ-ray region. The spectra up to a few keV are of power law form, with radio-loud quasars having flatter spectra than the radio-quiet ones. Beyond this range and up to a few tens of keV the spectra have an approximate power law shape with a canonical spectral index of ~ 0.7. The simple power law is often modified by X-ray absorption at low energy, a soft X-ray excess and the effects of pair creation due to photon–photon interactions in the most compact sources. In addition there are modifications to the spectrum, including emission lines, that are interpreted to be the consequences of reflection by cold matter. Observations with COMPTEL have shown that Seyferts and radio-quiet quasars are not strong emitters above some hundreds of keV, while the radio-loud blazars often show high energy γ-ray emission. We will discuss these regions at length in Sections 11.1 and 11.7 and will not consider them further here.

Riding on the big blue bump is a subsidiary '3000 Å bump', which extends from ~ 4000 to 1800 Å. This is made up of contributions from (1) the Balmer continuum, (2) a pseudocontinuum formed from higher order Balmer lines that blend into each other and contribute at longer wavelengths than the Balmer continuum limit of 3646 Å

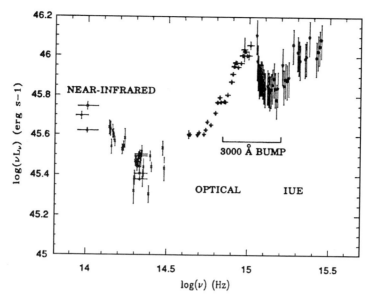

Fig. 8.3. The continuum of the quasar PG 1116+215 with strong emission lines removed. The 3000 Å bump is indicated. Reproduced from M. Elvis, CFA preprint 2846.

and (3) a pseudocontinuum due to a forest of iron lines, mainly Fe II, with the contribution occurring in the ~ 1800–3500 Å region. The Balmer contribution occurs in the ~ 2700–3800 Å region, and where this begins to taper off, the Fe II contribution becomes important, with the blend of the two contributions producing the 3000 Å bump. Such a bump can be clearly seen in Figure 8.3.

While the big blue bump is very common in quasar continua, there are a few examples of quasars in which such a bump is not evident (see Figure 8.4). In a study of bump properties in a sample of quasars, McDowell *et al.* (1989) found that five out of 31 quasars had bumps that were unusually weak or possibly absent. Defining the bump strength in terms of the logarithmic ratio ('colour')

$$C = \log \left[\frac{L(0.1-0.2\,\mu\mathrm{m})}{L(1-2\,\mu\mathrm{m})} \right], \tag{8.7}$$

McDowell *et al.* found that most of the quasars in their sample have strong ultraviolet bumps with colours bluer (i.e., greater) than 0.4, while the weak-bump quasars, which were initially picked out visually, all have $C < 0.15$. The weak bumps could be explained by the reddening of objects with normal bumps for an extinction $E(B - V)$ of about 0.1 to 0.2 magnitude. This would imply neutral hydrogen column densities $N_{\mathrm{H}} > 10^{21}\,\mathrm{cm}^{-2}$. However, such column densities are not allowed in the spectral fits of the IPC data on these objects.[1] Other possible explanations are variability, which

---

[1] See Section 11.1 for a discussion on neutral hydrogen column densities in the context of X-ray spectra.

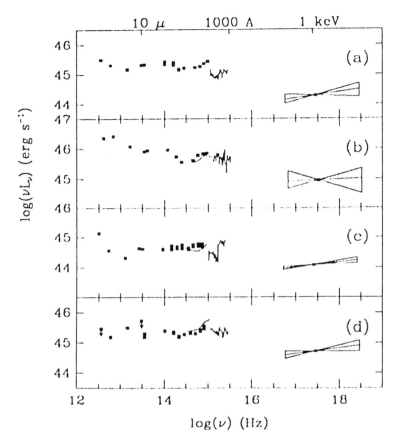

Fig. 8.4. The continuum distributions in the rest frame of the weak-bump quasars (*a*) PHL 909, (*b*) 3C 48, (*c*) Mrk 205, (*d*) PHL 1657. The 3000 Å (10^{15} Hz) bump can be seen. Reproduced from McDowell *et al.* (1989).

would make all quasars spend some time in a weak-bump state, or that the bump occurs in the soft X-ray region, which is observationally inaccessible, or that the bump may be intrinsically weak relative to a power law continuum which may be present, making it difficult to detect the bump.

8.4   Continuum radiation from blazars

In Section 9.9 we will define blazars as objects that have strong radio continua, with a flat or inverted spectrum, high polarization, rapid variability and continuum emission with a smooth spectrum from the radio to the ultraviolet. When emission lines are absent or weak the blazar is called a BL Lac while objects with strong emission lines are optically violent variables (OVV) or flat spectrum radio quasars (FSRQ).

The continuum radiation from blazars has a smooth form from the radio to the infrared to the optical. The luminosity per decade of frequency, $\nu L(\nu)$, increases from the radio region and generally peaks somewhere in the submillimetre to infrared regions. The continuum spectra of four blazars from Bregman *et al.* (1990) are shown in Figure 8.5. The continuum beyond the peak can fall off gradually, as in 3C 345, or more sharply, as in the case of the other three objects in the figure. In the case of X-ray selected BL Lac objects, however, the peak can occur at higher frequencies, in the ultraviolet to soft X-ray energy range (see Section 11.5). Recent observations of blazars by the γ-ray detectors on COMPTEL have extended the continuum to much higher energy (see Section 11.7).

At X-ray energies blazars have an approximately power law spectrum, which is generally flatter than the X-ray spectra of radio-quiet quasars. It has been found that a significant fraction of blazars are γ-ray emitters at GeV energies and that the power in the γ-ray region can exceed by an order of magnitude or more the peak power at lower frequencies. In the case of Mrk 421, photons with energy in the TeV range have been detected. The continuum of the quasar 3C 279 from $\sim 10^{10}$ Hz to $\sim 10^{24}$ Hz is shown in Figure 8.6. The continuum in the X-ray region is discussed in detail in Section 11.1, while that in the γ-ray region in Section 11.7.

The continuum of blazars lacks thermal features such as the infrared and big blue bumps which are present in radio-quiet quasars. The radio cores of blazars are bright and rapidly variable, i.e., their brightness temperatures are very high ($\sim 10^{12}$) and the flux is polarized. All these facts point to a synchrotron origin for the emission at radio wavelengths. The brightness temperature in the infrared and optical regions is $\gtrsim 10^6$ K, which is much higher than the black body temperature that would be associated with emission peaks at these wavelengths. The emission is therefore non-thermal and the continuity of the spectrum between the radio and ultraviolet wavelengths points to a synchrotron origin for the emission in all these regions. The emission at higher energies could also be due to the same process, but this would require the emitting electrons to have very high energies. The high energy photons can, however, be straightforwardly produced, from inverse Compton scattering of the synchrotron photons by the electrons that produced them in the first place, i.e., synchrotron self-Compton emission. This boosts photons of energy ϵ to $\sim \gamma^2 \epsilon$. Superluminal motion has been observed in a number of blazars. This can be interpreted in terms of relativistic bulk motion of the synchrotron-emitting plasma. Such motion is also required if the rapid variability is not to lead to excessively high brightness temperatures as well as to enable the sources to remain transparent to γ-ray photons. These matters have been discussed in detail in Chapters 3 and 4.

8.5   Origin of the big blue bump

We have seen above that the big blue bump (BBB) contains a significant, or even dominating, fraction of the overall energy emitted by AGN or by quasars that are not

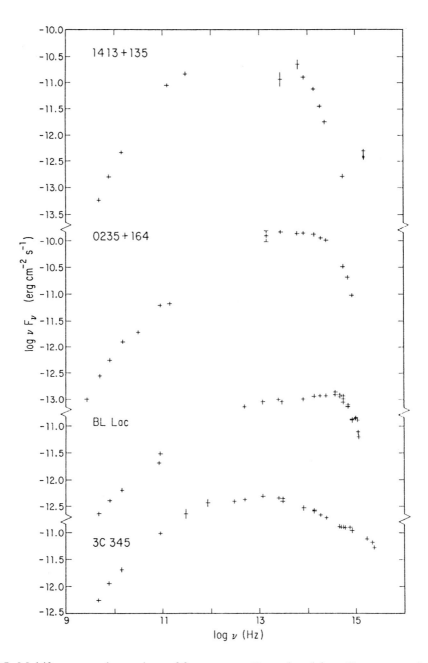

Fig. 8.5. Multifrequency observations of four quasars. Reproduced from Bregman *et al.* (1990).

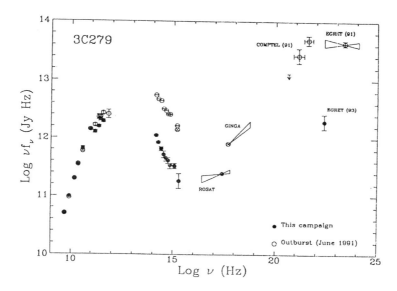

Fig. 8.6. Continuum spectrum of the quasar 3C 279 from the radio to the γ-ray region. Reproduced from Maraschi *et al.* (1994).

classified as blazars. The magnitude of the energy in the bump is an indication that it has a primary origin, in the sense that it is not high energy radiation that has been reprocessed to lower energy through thermalization. The observed continuum flattens from a steep power law with index ∼ 1 in the infrared (IR) region to a flatter power law in the optical/UV region. It was first pointed out by Shields (1978) that if the IR power law extends to higher energies, and is subtracted from the UV flux, then the residual has the form $F(\nu) \sim \nu^{1/3}$, which is remarkably similar to the shape of the continuum expected from a geometrically thin but optically thick accretion disk (see Section 5.5). This has led to the development of detailed models for the emission of the BBB from a putative accretion disk. While these models provide excellent fits to the continuum shape, there are various shortcomings associated with the picture, which has led to the development of an alternative class of models in which the BBB is interpreted as being produced by optically thin thermal bremsstrahlung (free–free) emission. We will now consider some salient points of these models.

8.5.1 *Accretion disk models*

After the suggestion by Shields that the BBB is due to emission from an accretion disk, detailed models were developed and applied to the broad band continuum spectrum from the IR to the UV by Malkan and Sargent (1982) and Malkan (1983). In this approach, it was assumed that a power law derived from infrared data extends over the entire range of the continuum. The BBB was assumed to be generated from a standard geometrically thin but optically thick accretion disk. Balmer continuum emission as well as blended Fe II line emission were also taken into account.

The flux $F(r)$ as a function of the distance r from the disk centre was obtained by using relativistic radial structure equations. It was assumed that the radial heat loss was negligible, so that all the energy was radiated from the surface, the functional form at radius r being a Planckian of temperature $[F(r)/\sigma]^{1/4}$, where σ is Stefan's constant. The spectrum produced by a face-on disk was obtained by integration and has the form $F(v) \propto v^2$ at the lowest frequencies, becoming $\propto v^{1/3}$ and then falling off exponentially at the high frequency end, after passing through a flat part (see subsection 5.5.1).

Calculations were done for the Schwarzschild geometry as well as for the Kerr geometry of a rapidly rotating black hole with angular-momentum-to-mass ratio $a/M_B = 0.998$. The flux produced from the innermost regions was corrected for general relativistic transfer effects, which included gravitational redshift and focusing; these make the finally emergent spectrum much redder than it is at the surface of the accretion disk.

The temperature distribution in the accretion disk depends on the mass of the central black hole M_B and the accretion rate $\dot{M}$. The fitting procedure determines these parameters to within ~ 20 per cent and they are not sensitively dependent on small errors in the data, but depend on the various assumptions that go into the model.

1 General relativistic effects   Improved models that also take into account relativistic effects as a function of the disk inclination angle, as well as Doppler shifts, have been considered by Sun and Malkan (1989). The inclination effects make the emergent spectrum harder, an edge-on disk having a spectrum that is about five times harder than that of the same disk viewed face on. These model spectra, together with a power law Balmer continuum emission, blended Fe II line emission and the host galaxy starlight contribution have been fitted by Sun and Malkan to the IR to UV continuum for a sample of 60 AGN and quasars. There are just two effective parameters that can be varied to obtain the best fit. Sun and Malkan assumed a range of inclination angles for each object in their sample and obtained the corresponding best-fit black hole mass and accretion rate. The disk luminosity as a function of frequency is obtained by integration over the disk after the fit. Sun and Malkan find that the observed spectra are well fitted by their models with sub-Eddington accretion rates. Low luminosity Seyfert galaxies have fitted accretion rates that are just a few per cent of their Eddington rate, while luminous quasars have accretion rates close to the limit. The fitted black hole masses for Seyferts are smaller than those for luminous quasars.

It is assumed in the standard disk models that the disk is optically thick, with opacity provided by thermal absorption and electron scattering. In the inner region of the disk, where the temperature is high enough for the production of BBB radiation, the opacity is dominated by scattering. As a consequence of this, the emergent flux should be significantly polarized when the disk is viewed at high inclination, with the degree of polarization being 11.7 per cent for an edge-on disk and 2.3 per cent for a disk with axis inclined at 60 deg to the line of sight, which is the median inclination for a randomly oriented disk (Gnedin and Silant'ev 1978, as quoted in Antonucci 1984). However, the observed degree of polarization in non-blazar objects is $\lesssim 0$–2 per cent, with an average of ~ 0.6 per cent, much of which is due to local interstellar polarization (Angel and Stockman 1980). This is inconsistent with the predicted polarization from the thin accretion disk model. When an axis is defined by the presence of the extended radio structure associated with a quasar, it is found that the projected polarization vector is approximately aligned with the radio axis, while the model's prediction is for alignment perpendicular to the radio axis, if the radio jets are oriented roughly normal to the disk plane.

The discrepancy between the model's prediction of the polarization and actual observation can be removed in several ways. It has been suggested by Coleman and Shields (1990) that the disk may have ripples on its surface, so that the whole disk is never together viewed at a high inclination angle. Even within the standard disk, detailed calculations by Laor, Netzer and Piran (1990), which take into account general relativistic and opacity effects, show that the degree of polarization is considerably reduced over predictions of the more simplistic calculations.

In this model, in the range $\sim 10^{14}$–3×10^{14} Hz, the polarization is reduced because here the opacity is provided by thermal absorption rather than scattering, while at high frequencies the reduction occurs due to general relativistic effects. In the frequency range 3×10^{14}–3×10^{15} Hz, the absorption decreases as the frequency increases, which leads to a steady increase in the polarization. When the Lyman limit is reached (see below), there is a increase in the absorption opacity due to the onset of bound–free transitions, leading to decreased polarization. Beyond the edge the absorption again decreases, leading to an increase in the polarization, until it reaches the purely relativistic values when absorption becomes negligible beyond $\sim 10^{16}$ Hz. The edge appears as a broad feature due to Doppler and relativistic effects. This behaviour is shown in Figure 8.7.

The degree of polarization varies from negligible values for very small angles of inclination (i.e., a face-on disk) to ~ 1 per cent for an inclination of 60 deg to ~ 2 per cent for a nearly edge-on disk. The relativistic depolarization is small for a non-rotating black hole. The model predicts an increase in polarization with frequency that is similar to the observed increase in going from the R to the B filters. However, the polarization direction in the models is perpendicular to the disk rotation axis, which is in conflict with the results for radio sources mentioned above. In the optical region, there is no observed change in the plane of polarization with frequency, which is again consistent with the predictions of the model.

2 The Lyman edge During propagation through the disk in the vertical direction, radiation shortwards of the Lyman limit at 912 Å is absorbed due to bound–free transitions in hydrogen. Photons with energies higher than the Lyman limit of 13.6 eV, produced deep in the atmosphere, are confronted with a large optical depth to bound–free absorption and those that escape are from the upper layers. Photons with energies < 13.6 eV are not affected by the bound–free transition so that the optical depth is

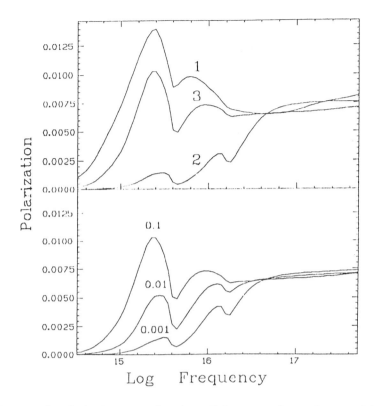

Fig. 8.7. Degree of polarization as a function of frequency for various model parameters, for an accretion disk around a rotating black hole with mass $M = 10^9 M_\odot$, accretion rate $\dot{M} = \dot{M}_{\text{Edd}} = 0.3$, $a/M = 0.9982$, where a is the black hole angular momentum per unit mass and the disk inclination to the line of sight is given by $\cos i = 0.5$. The upper panel shows the polarization for different viscosity laws, for viscosity parameter $\alpha = 0.1$. The lower panel shows the polarization for a fixed viscosity law and values of α as indicated on the curves. Reproduced from Laor *et al.* (1990).

small below the Lyman limit. When there is a temperature gradient in the atmosphere, the escaping higher energy photons come from the upper region where the temperature, and hence the emissivity, is relatively low. However, the lower energy photons, which face no absorption, can come from regions of higher temperature and emissivity. As a result a Lyman edge should be seen at $(1 + z_{\text{em}})916\,\text{Å}$, where z_{em} is the emission line redshift of the object. The amount of reduction in the flux across the edge depends on the temperature gradient, and there is no discontinuity if the temperature is constant.

A Lyman edge at the quasar redshift can also be produced by clouds which give rise to broad emission lines in quasars and AGN, or by intervening absorbers between the source and the observer, in which case the edge is at the absorber redshift. When the absorption above the Lyman edge is produced by a cold intervening cloud, it is said to be *complete*. The edge produced by thermal matter has a different appearance because of the gradual change in opacity and emissivity across the atmosphere and is said to be *partial*. Moreover, the rotation of the disk smears the edge by Doppler broadening, which also affects the absorption lines produced in the disk.

A sample of 11 quasars at redshift ~ 3 was observed by Antonucci, Kinney and Ford (1989) in order to detect the Lyman disk signature of accretion disks. Owing to the high redshift, the Lyman edge is expected in the optical region and is observable from the Earth. Of the sample, four quasars showed no Lyman edge at all at the redshift of the object, while five of the objects either had a complete edge or a partial edge with absorption lines that were too narrow to have originated in a rotating disk. Only two out of the 11 quasars had partial edges and features that could have arisen in a rotating disk. A search for the Lyman edge in low to intermediate redshift quasars observed by the *International Ultraviolet Explorer* (*IUE*) satellite was made by Koratkar, Kinney and Bohlin (1992), who again found that only seven out of the 53 quasars had edges which could possibly be attributed to accretion disks. These results show that the edge is not as ubiquitous and deep as would be expected from simple accretion disk theory. However, there are observational as well as theoretical reasons that can make the edge difficult to detect.

At high redshift the density of lines of the Lyα forest produced by intervening neutral hydrogen clouds is high and it is difficult to separate the absorption lines from the continuum, and therefore to see whether there is an edge present. *IUE* data allows access to the edge region in low redshift quasars which are less affected by intervening clouds, but only a limited signal-to-noise ratio is available in the data. On the theoretical side, the reasoning that an observable edge should be present is based on application of radiative transfer theory developed in the context of stellar atmospheres. In the case of an accretion disk, emission around the Lyman edge arises predominantly in regions with temperature $\sim 5\times10^4$ K, and the opacity is dominated by electron scattering. The vertical structure of the disk in these circumstances can be quite different from the results of stellar atmosphere calculations made assuming local thermodynamic equilibrium. Reliable estimates of the drop in the strength of the continuum at the Lyman edge applicable to disk atmospheres are therefore not available, and it may not be right to expect large absorption at the edge. Any edge effect can also be reduced by external heating of the atmosphere with X-rays, which can reduce the temperature gradient, or even reverse it, producing an emission edge. Since the UV radiation arises in regions close to the black hole, general relativistic effects have to be taken into account, and these shift the energy of the observed edge to shorter wavelengths and broaden it (Laor 1990). As a result of all these possibilities, the Lyman edges produced by accretion disks would become difficult to observe, without contradicting a disk origin for the BBB.

O'Brien, Gondhalekar and Wilson (1988) derived spectral indices in the rest frame 1215–1990 Å range for 68 quasars using *IUE* data. They find slopes of ~ 1 at low luminosities and ~ 0.4 at high luminosities. In the thin accretion disk picture it is expected that higher luminosity objects would have higher mass black holes and therefore cooler disks, i.e., steeper slopes. This poses a difficulty for the accretion disk model, but there are uncertainties including a large scatter in the data. Further difficulties for the accretion disk model that arise from variability studies will be discussed in Section 8.7.

8.5.2   *Free–free emission*

It has been proposed by Barvainis (1993) that free–free or thermal bremsstrahlung emission (see Section 4.5) is responsible for the BBB. Barvainis invokes free–free emission because of the several difficulties faced by the accretion disk model which we have discussed above. The good fit to the data that the accretion disk model provides depends on the combination of a power law and an accretion disk spectrum. The former is used to fit the infrared (IR) region and extrapolates into the optical and ultraviolet (UV) region, while the latter dominates the blue and UV part. However, we will see in Section 8.3 that in non-blazar types of object, the infrared band is very likely to be dominated by dust emission. In this case the IR emission falls off exponentially at lower wavelengths, and an underlying power law in the optical and UV region cannot be assumed. This leads to poor fits of the accretion disk models to the data and other sources for the BBB need to be explored.

Free–free emission has the exponential shape $I(v) \propto \exp(-hv/kT)$, but in the temperature range $\sim 10^5$–10^6 K can provide the observed approximate power law slopes in the optical, UV and soft X-ray regions. The shape of the spectrum as it emerges from the medium is dependent on the *effective optical depth* τ_* defined in Equation 4.34. For $\tau_* \ll 1$ the spectral shape has the same form as the free–free emission, while for $\tau_* \gg 1$ the spectrum has the black body or modified black body form depending on whether the optical depth to scattering is significant. When τ_{sc} is large, Compton scattering can be important, the Comptonization parameter defined in Equation 4.37 is $\gtrsim 1$ and photon energies are altered substantially. This helps in reducing a Lyman emission edge that might otherwise be present.

Bravainis assumes that the medium that gives rise to the free–free emission is in the form of many small clouds with total covering factor < 1, which will allow radiation emerging from any one cloud to escape from the region. He has considered two cases.

In the first case, it is assumed that $\tau_* < 1$, $\tau_{sc} > 1$. Each cloud is effectively optically thin to absorption but there are many scattering events before a photon escapes from the cloud in which it is created. This helps in smearing the edge, which otherwise would become an embarrassment. The size over which the clouds are distributed is large. The overall variability time scale is set by the radius of this size and is not increased by the high optical depth of the individual cloud. If the clouds are taken to be distributed in a disk- or torus-like configuration, the scattering between clouds can produce a small degree of linear polarization parallel to the projected minor axis if the shape is inclined to the line of sight. This is consistent with observation.

In the second case, it is assumed that $\tau_{sc} < \tau_{ab} < 1$. In this case each cloud is thin to both scattering and absorption in the optical and UV. If the covering factor of the clouds is < 1, the brightness temperature is $\simeq \tau_{ab} T$. The high brightness temperature observed in some sources can be explained by taking $T \sim 10^6$ K. This model predicts significant Lyman edges, which can only be smeared by invoking scattering in a hot intercloud medium, since each cloud itself is optically thin. This medium has to be distributed non-spherically so that it does not scatter the observed soft X-rays, which would result in variability times scales longer than observed.

While the free–free model can explain many of the observations, detailed modelling needs to be done to test its viability for producing the observed spectrum and the variability in the optical to soft X-ray region.

8.6 The infrared continuum

The infrared continuum in quasars and AGN can be of non-thermal origin or it can be due to thermal emission from dust grains. It was long ago suggested by Rees *et al.* (1969) that infrared radiation from Seyfert galaxies in the 2.2–22 μm band was produced by dust grains that were heated by the optical and UV emission from the nucleus. However, continuity between the radio and infrared regions found in radio-loud objects, and an observed power law form in the far infrared region, seemed to favour a non-thermal origin. But there were indications that, in Seyfert 2 galaxies at least, a significant thermal component was present (see Bregman 1990 for an account of the early work). Analysis of the *IRAS*[1] data has led to some understanding of the relative importance of the thermal and non-thermal components in different kinds of object.

An unbiased sample of 48 Seyfert galaxies, which were selected on the basis of their emission line properties, was studied in the 1.2 μm–1.3 mm region by Edelson, Malkan and Rieke (1987). Most of the galaxies in the sample were detected in the infrared, but none at 1.3 mm. It was found that the 2.2–25 μm power law index was, on the average, steepest for Seyfert 2 galaxies, with mean value $\langle \alpha_{2.2-25\,\mu m} \rangle = 1.56$, and flattest for optically selected quasars, with $\langle \alpha_{2.2-25\,\mu m} \rangle = 1.09$. The power law index for ~ 70 per cent of Seyfert 1 galaxies was flat, as in the case of quasars, while the rest had steep spectra. The mean value for the whole Seyfert 1 sample was $\langle \alpha_{2.2-25\,\mu m} \rangle = 1.15$. About half the Seyfert 1 galaxies and all the Seyfert 2 galaxies were found to have extended far-infrared emission. About half the galaxies had sharp low frequency turnovers at ~ 80 μm. Edelson *et al.* concluded from these observations that the infrared emission from all Seyfert 2 galaxies, and from Seyfert 1 galaxies that have Seyfert-2-like spectra, is due to dust at a temperature $T \sim 50$ K. The infrared spectra of flat spectrum Seyfert 1 galaxies and quasars were taken to be dominated by non-thermal emission, the low frequency turnover being due to synchrotron self-absorption, which implies (see Equation 3.56) that the source size is ~ 100 Schwarzschild radii.

---

[1] The Infrared Astronomical Satellite was launched in 1983 and carried a 0.57 m-aperture infrared telescope. It surveyed 96 per cent of the sky for point sources at 12, 25, 60 and 100 microns. *IRAS* has detected $\sim 250\,000$ sources in the survey, and was also used in pointing observations.

Carleton *et al.* (1987) studied a sample of hard X-ray selected Seyfert 1 galaxies from the X-ray to the radio band, with the aim of separating thermal and non-thermal contributions to the $1-100\,\mu m$ region of the spectrum. They assume that all Seyfert 1 galaxies have a similar underlying form, with varying amounts of dust absorption and reemission. The intrinsic power law in the $1-100\,\mu m$ region is taken to be $L(v) \propto v^{-1}$, the normalization being fixed by the point with the lowest flux in this region. Any departure from the simple power law is taken to be due to the effects of dust. A number of correlations between luminosities in different bands are studied, and these substantiate the adopted model. In particular, it is found that objects that are significantly reddened have an infrared excess, which is consistent with the assumption that the optical and UV radiation is absorbed by dust and reemitted. The dust temperature is determined by a balance between absorption and emission. Carleton *et al.* determine the radial dependence of the temperature from the simple prescription of Davidson and Netzer (1979):

$$T = 10^{-6} \left(\frac{L_{\text{eff}}}{r^2} \right)^{1/5}, \tag{8.8}$$

where the luminosity is in units of erg sec^{-1}, the distance r is in parsec and it is assumed that the dust emissivity is $\propto \lambda^{-1}$. Dust that is responsible for emission at 3.5, 20 and 80 microns respectively has characteristic temperatures 1000 K, 180 K and 45 K respectively. Assuming a typical value of $5 \times 10^{44}\,\text{erg sec}^{-1}$ for the luminosity, which follows from the observed distribution, the distances of the dust having peak emission at the three wavelengths are 0.7 pc, 50 pc and 1600 pc respectively. The smallest radius here is comparable to the size of the broad-line-emitting region, the size of 50 pc is that of the narrow-line-emitting region, while the largest size is similar to that of the inner part of a galactic disk.

On the basis of their observations, discussed in subsection 8.3, Sanders *et al.* (1989) have proposed that the $2\,\mu m - 1\,\text{mm}$ region is dominated by thermal emission in all but the blazar class of object. In their model, emission in the $0.5-5\,\mu m$ region is produced by heated gas and dust in the outer region of the accretion disk, and in an inner disk made of molecular gas and located within a few parsec of the nucleus. The radiation in the $5\,\mu m-1\,\text{mm}$ region is produced by heated dust in a warped disk that extends from a few parsec to a few kiloparsec from the nucleus. The warping increases the effective covering factor of the disk beyond a few parsec from the nucleus, which helps in producing the observed far-infrared to ultraviolet luminosity ratio. Different parts of the infrared continuum are produced at different distances from the nucleus because of the radial dependence of the temperature, which is obtained in the same manner as Equation 8.8 above but using a more sophisticated emissivity function and allowing for a finite optical depth. A dust origin for the infrared radiation has also been considered by McAlary and Rieke (1988).

The local minimum in the infrared $vL(v)$ that always occurs at $\sim 1\,\mu m$ finds a natural explanation in the dust model. The radiation in this region arises at ~ 0.1 pc from the nucleus, where the temperature of the dust can rise above ~ 2000 K. At these high temperatures the grains sublimate, which reduces the opacity and therefore the radiative efficiency of the disk. Because of the retained energy the temperature of this part of the disk rises to a new equilibrium value of $\gtrsim 10^4$ K, at which the cooling is dominated by free–free emission. Gas in the intermediate temperature range $2000 \lesssim T \lesssim 10\,000$ K is thermally unstable, and is therefore not expected to be present in substantial quantities. This leads to the observed minimum at $\sim 1\,\mu$m.

The spectra of radio-quiet quasars and AGN show a sharp down-turn around $100\,\mu$m and the submillimetre spectral index $\alpha_{\rm sm}$ in the $100\,\mu$m–1.3 mm region is rather steep, with some sources having $\alpha_{\rm sm} > 2.5$. The spectral index in the submillimetre region can be an important factor in distinguishing between non-thermal and thermal emission. In the non-thermal case, a turnover below some frequency can be naturally interpreted as due to synchrotron self-absorption. We have seen in subsection 3.5.1 that for a power law distribution of electrons, self-absorption with optical depth exceeding unity leads to a spectrum with $L(v) \propto v^{2.5}$. This is, however, true only in a homogeneous emission region, while for an inhomogeneous region, with the self-absorption setting in at different frequencies in different parts, the net result is a spectrum that is flatter than in the homogeneous case. On the basis of these arguments, a spectrum with slope steeper than 2.5 would therefore rule out a synchrotron self-absorption origin. One has to be cautious, however, because it is possible to have electron energy distributions more complex than a simple power law that lead to a steeper spectrum in the self-absorbed case over limited frequency intervals (see subsection 3.5.1).

The submillimetre spectral index $\alpha_{\rm sm}$ can be determined by combining $100\,\mu$m *IRAS* data with millimetre data, but the results can then be uncertain because of (1) the wide *IRAS* beam, which includes extended emission not related to the nucleus and (2) uncertainty about the wavelength at which the turnover occurs, which means that a two-point spectral index determined from the fluxes at $100\,\mu$m and 1.3 mm may be an underestimate or overestimate of the spectral index in the millimetre region. Hughes *et al.* (1993) have circumvented these problems by directly observing the spectrum in the submillimetre region, using the 15 m James Clerk Maxwell telescope, of 10 *IRAS* selected radio-quiet quasars. Three of the quasars were detected and were found to have very steep spectra, with $\langle \alpha_{\rm sm} \rangle = 3.75 \pm 0.48$. The spectral indices are significantly different from the value of 2.5, and even the steeper self-absorbed values can be ruled out for the following reason: as per the models of Schlickeiser *et al.* (1991), a spectral index steeper than three is produced only for wavelengths greater than ~ 10 times the turnover wavelength. Since the latter is $\sim 100\,\mu$m, the steep part should occur for $\lambda \gtrsim 1$ mm, with $\alpha > 5$ reached only beyond 5 mm. Since observationally the spectra are steep even in the submillimetre region, synchrotron self-absorption cannot be the emission mechanism.

When a medium is optically thick and the radiation reaches equilibrium with matter at some temperature T, its form is in general given by the Planck function corresponding to that temperature. For photon energies $< hv/kT$ the Rayleigh–Jeans form applies and the spectrum is $\propto v^2$. However, for long wavelengths in the $100\,\mu$m – 1 mm region, the absorption coefficient is small and the optical depth can be less than unity. The radiation at these wavelengths therefore does not acquire the Planck (Rayleigh–Jeans) form, and the spectrum depends only on the emission function and can be steeper than v^2. Dust with temperature in the range 15–60 K would produce the required steep spectrum with turnover in the narrow range 60–200 μm as observed (Hughes *et al.* 1993). The infrared continuum in radio-quiet quasars is therefore dominated by dust emission, which is also the case for half the Seyfert 1 galaxies and all Seyfert 2 galaxies as discussed above. Radio-loud quasars can have substantial contributions from thermal as well as non-thermal processes, as in the case of 3C 273, while the infrared emission of blazars is dominated by the non-thermal component.

Variability in the infrared can be an important indicator of the origin of the emission. Synchrotron self-absorption occurs in compact sources, and for a turnover at 100 μm the source size is of the order of light days. Variability on the time scale of days is therefore expected at all infrared wavelengths. In the dust emission case we have seen that the models require a range of temperatures, the highest temperature being close to the nucleus. Therefore, the shorter the wavelength of the radiation the closer to the nucleus it originates. Variability on the scale of months to years is therefore expected around 1 μm, while the 100 μm radiation would vary only over a time scale of thousands of years. The observed lack of infrared variability in the *IRAS* band is therefore consistent with dust emission.

8.7   Variability

Variability studies are very important in identifying the physical processes and the size of the region in which the radiation in a given wavelength range is produced. If variability is observed on a time scale of Δt_{var} in the source frame, then the radiation must be produced in a region with size constrained by

$$R \lesssim c\Delta t_{var}. \tag{8.9}$$

If the source is bigger than this limit, then different parts of the source would not be causally connected, so that they would not be varying in phase with each other. This would lead to greatly reduced amplitude for the variation. When different variability time scales are found in a source at the same frequency, the most rapid variability is taken to be indicative of the source size, after allowing for redshift and relativistic beaming effects. The slower variations are likely to be due to slow changes in the source structure and other effects such as heating or cooling.

When this simple relation was applied to extragalactic radio sources, it was found that the source sizes turned out to be so small that the sources would be very quickly quenched by the Compton catastrophe. The situation was saved by invoking relativistic beaming of the emitting region (see below and subsection 3.8.4), which resulted in the introduction of new concepts related to beaming in the study of AGN and quasars at radio as well as other wavelengths. More generally, the limit on source sizes that followed from variability helped in assembling the standard picture of an active nucleus as a very compact region with a supermassive black hole at the centre, which releases gravitational energy through accreted matter.

The pattern of variability in different wavelength regions can have important implications for the modelling of the radiation processes. If the variability in two bands shows the same pattern as a function of time, with the variations occurring in phase, then one can conclude that the radiation in the two bands is very likely to be produced by the same mechanism and in the same geometric region. However, if the variation in one band systematically lags behind that in the other, then propagation from one region to another is involved. Multiband variability studies therefore complement the information that can be obtained from single-epoch broad band spectra of the kind described above. These campaigns are, however, difficult to organize as they involve the simultaneous observation of sources, spread over long periods of time, using a number of terrestrial and space-borne telescopes.

In order to determine the different variability time scales that may be present even in a single region of wavelengths, it is necessary to observe a source for long periods of time, with a sampling interval much shorter than the shortest time scales expected. In reality, however, barring a few exceptional cases, the flux from a given source is sampled at long, non-uniform intervals, from different locations and using different techniques. Different sources from an ensemble may have even been observed over different time intervals, when they are not in a single field of view of the telescope and detector being used. Variability studies therefore require the use of sophisticated techniques for studying correlations between different wavelength regions, and the results can be ambiguous because of the paucity of data. We will see below that, in spite of these limitations, much progress has been made in understanding the physics and geometry of the emitting region with the help of variability studies.

When a source has been sampled at several epochs, whether the source is truly variable can be determined by comparing the distribution of flux at the different epochs with a model in which the flux is assumed to be constant (see e.g. Edelson 1992). If F_i are the measured fluxes, then the mean flux, and root mean square dispersion as a fraction of this mean, are given by

$$\langle F \rangle = \frac{1}{N} \sum_{i=1}^{N} F_i \tag{8.10}$$

and

$$\sigma = \frac{1}{\langle F \rangle} \sqrt{\frac{1}{N-1} \sum_{i=1}^{N} (F_i - \langle F \rangle)^2} \tag{8.11}$$

respectively, where N is the number of samples. The goodness-of-fit parameter for testing the model of constant flux is taken to be the reduced chi-square value

$$\chi_v^2 = \frac{1}{N} \sum_{i=1}^{N} \left(\frac{F_i - \langle F \rangle}{\sigma_i} \right)^2, \tag{8.12}$$

where σ_i is the dispersion in the individual measurements. $\chi_v^2 > 1$ indicates that the assumption of constant flux is questionable, and the confidence with which the hypothesis of constant flux can be rejected follows from the value of χ_v^2.

Blazars as a class are found to have rapid and large amplitude variability, which distinguishes them from other quasars and AGN. Blazars have been found to vary at all wavelengths from the radio to X-rays and at all time scales from days or less to years (Impey and Neugebauer 1988). It is found that in general the amplitude of variability increases with frequency, while the variability time scale Δt_{var} decreases with frequency, which is consistent with the fact that the higher energy radiation arises in more compact regions. The dependence of Δt_{var} on the frequency of observation v is given approximately by $\Delta t_{\mathrm{var}} \propto v^{-1/2}$ (Bregman 1990). We will review in the following subsection the variability observations, at different frequencies, of blazars as well as of other quasars and AGN.

8.7.1   *X-ray variability*

The shortest time scales for variability have been found at high energies, but the time scales accessible at any wavelength depend upon the frequency of observation and the signal-to-noise ratio. At X-ray wavelengths, the best data is available from *EXOSAT*, which had good sensitivity and a highly eccentric orbit, and therefore reduced occultation of sources by the Earth. This allowed the continuous observation of a source for several days at a time. At γ-ray energies, however, the number of γ-ray photons received by even the most sensitive instruments is so small that variations on time scales shorter than a day cannot be detected. For these reasons, when the shortest time scale observed is at the limits of an instrument one may expect even shorter time scales at that wavelength to be found.

In the X-ray region, a sky survey made with the *ARIEL* satellite showed that a significant fraction of Seyfert 1 and narrow emission line galaxies (NELG) had large amplitude variability on a time scale of days (Marshall *et al.* 1980). However, analysis of *HEAO 1* data led Tennant and Mushotzky (1983) to the conclusion that large amplitude, short term variability on time scales less than 12 hours was not common amongst active galaxies. Out of a sample of 38 AGN, they found that only NGC 6814 varied on time scales less than 10^4 sec. Marshall *et al.* (1983) also found an example of rapid variability in the X-ray emission from the Seyfert 1 galaxy NGC 4051.

The capabilities of *EXOSAT* were particularly well suited to the study of time variations and a somewhat clearer picture is now available. Contrary to the earlier observations, X-ray flux variations have been found on different time scales, and it has been shown than short term variability is common to AGN (McHardy 1990). Grandi *et al.* (1992) have analysed *EXOSAT* archival observations to study the variability from a sample of 30 AGN, including Seyfert 1 galaxies and NELG (BL Lacs were specifically excluded). They found that long term variability, on time scales from days to months, is a very common phenomenon. It was found in 95 per cent of the sources in the ME detector (1–8 keV) and 71 per cent of the cases in the LE detector (0.05–2 keV). Flux variations up to a factor of 30 were seen, with a trend towards the X-ray spectrum becoming softer (i.e., more steep as a function of energy) with increasing flux. Rapid variability, on time scales from minutes to hours, was found in the ME detector for 12 out of the 29 AGN on which data was available. Short term variability therefore appears to be less common than long term variability, but this could be due to poor statistics. The amplitude of short term variability was often found to have $\delta I/I < 1$.

In the blazar class, Giommi *et al.* (1990) have studied the variability of a sample of 36 BL Lac objects from the *EXOSAT* archives. Strong variability was detected in all but two cases, with the variation reaching a factor of ~ 30 in one case. However, the largest amplitude variations were rare, and the observed distribution of amplitude indicates that the sources spend about half their time within a factor ~ 2 of a quiescent level. The variability was usually more pronounced in the higher energy band. The spectrum becomes harder with increasing flux, which is opposite to the behaviour found in AGN. Large amplitude flux variations on a time scale of hours were found in four objects, and four other objects showed variability on similar time scales but with smaller amplitude ($\delta I/I \simeq 0.2$–0.4). No long term trends in the variability of any object were found, which indicates that the flux variability is due to events lasting over hours and days, above a quiescent level that remains constant over a period of years.

The variability in the X-ray band is always found to be aperiodic. There was much excitement when periodic variability was observed in the low luminosity Seyfert 1 galaxy NGC 6814 (see the references in Mushotzky, Done and Pounds 1993). However, observations with *ROSAT* (Madejski *et al.* 1994) showed that the periodicity was due to another X-ray source ~ 37 arcmin away, just outside the field of view, which had contaminated the flux from NGC 6814.

8.7.2 *Ultraviolet variability*

A large sample of blazars observed with the *IUE* has been analyzed for ultraviolet (UV) variability by Edelson (1992; references to earlier work are listed there). Data are available in the short wavelength range 1150–2000 Å and the longer wavelength range 2200–3200 Å. Out of a large sample of ~ 500 blazars, more than five observations per source in a given wavelength range were available for 14 sources. Using the χ^2_ν technique described above, Edelson found that all these sources were variable over the time scale for which data were available, which ranged from months for some sources to a few years for the others. In all six cases with data in both wavelength ranges, it was found that the variations were stronger at the shorter wavelengths.

Edelson found that the UV variability, as defined by σ in Equation 8.11, was correlated with optical polarization as well as with the observed ultraviolet luminosity (see Figure 8.8). The positive correlation between variability and observed luminosity follows naturally if the UV emission in blazars is relativistically beamed like the radio emission. Beaming contracts the observed time scales and enhances the observed

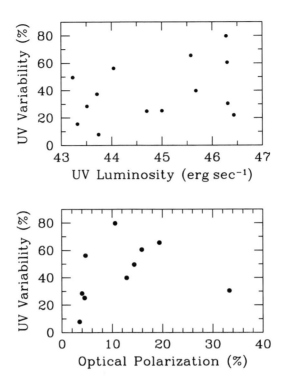

Fig. 8.8. Plot of ultraviolet variability amplitude against ultraviolet luminosity (upper panel), and against degree of optical polarization (lower panel), for blazars. The data is taken from Edelson (1992).

luminosities and the degree of polarization over their rest frame values (see Section 3.8); this leads to the positive correlation observed. It is found that for the two blazars that have the best data, the UV power law spectral index, defined between the shorter and longer wavelength ranges mentioned above, shows no correlation with the ultraviolet flux. This indicates that the emission in the UV range consists of a single component that varies uniformly over the observed wavelength range, preserving the shape of the spectrum (at the level of accuracy reached by the present data).

Edelson, Krolik and Pike (1990) have considered the ultraviolet variability of a sample of 27 Seyfert 1 galaxies from the Centre for Astrophysics (CFA) redshift survey. Though 24 of the 27 galaxies in the sample have been observed by the *IUE*, only a much smaller sample has data at a sufficient number of epochs for variability studies. All galaxies with adequate data are found to be variable. The variability amplitude is *negatively* correlated with the UV luminosity i.e., the more luminous objects show less variation. This is opposite to the correlation observed in the case of blazars, and points to a different mechanism for the generation of the UV radiation. Such a negative correlation is consistent with origin of the radiation in an accretion disk, or some other isotropically emitting source: the more luminous such a source is, the longer would be the light crossing time for it so that it varies slowly, leading to a smaller variability amplitude. For the three galaxies in the sample that have been observed the most often, it was found that the power law spectral index α_{uv} showed a clear correlation with the luminosity, in the sense that the spectrum becomes flatter (harder) with increasing luminosity. This suggests that there are two components to the UV emission, a harder one that varies more rapidly and a softer component that varies less. The observed spectral change is unlikely to be caused by absorption at the lower energy, since line ratios that would be affected by such reddening remain unchanged.

When several AGN which that been observed contemporaneously in the optical band and with the *IUE* are examined (see Clavel 1994 for references), it is found that the fluxes in the optical and UV bands vary in phase, with no measurable delays. Cross-correlation studies show that in the case of the Seyfert 1 galaxy NGC 5548, the delay between the variations is $\lesssim 2$ days (see Figure 8.9), while in the case of NGC 3783 the delay is $\lesssim 1$ day. This simultaneous variation of the optical and UV fluxes goes against the predictions of accretion disk models, in which the region that emits the optical radiation is about seven times larger than the UV emitting region. Since in the disk models the two layers can communicate only at sound speed, the optical variation should lag behind the UV variation, which is produced much closer to the central source, by several years.

From simultaneous UV and 2–10 keV X-ray data for NGC 5548, Clavel *et al.* (1992) found that the UV flux is directly proportional to the X-ray flux and that the delay between the variations in the two bands is $\lesssim 6$ days (see Figure 8.10). Such correlations have also been observed in a few other Seyfert 1 galaxies, and can be interpreted in the framework of the reflection models for the X-ray emission that we have considered in detail in subsection 4.6.5. In these models the observed X-ray spectrum of AGN includes a component that consists of X-rays reflected from a slab of cold matter. A major fraction of the X-ray flux incident on the slab is absorbed, heating it to $\sim 10^5$ K, so that the slab emits in the UV region of the spectrum. If the absorption and reflection from the slab are practically instantaneous, it is to be expected that variations in the two bands will be correlated. A difficulty with this model is that there does not always seem to be enough energy in the X-rays to produce the observed UV flux by heating of the slab. In NGC 5548 and NGC 4151 large outbursts of X-ray flux have been observed during which the X-ray flux does not grow sufficiently to power the UV.

Difficulties for generation of the UV emission in a reflecting disk have further increased as a result of the monitoring of NGC 5548 in the 0.2–2 keV band with the PSPC on *ROSAT* (see Section 10.2), for a period of about a month when the source was at its historical maximum in the X-ray flux (Done *et al.* 1995). It was found that fitting the data required an excess in the 0.1–0.4 keV soft band over the extrapolation of an overall power law with energy spectral index $\alpha_x \simeq 1$, which was compatible with results from *GINGA*. The X-ray fluxes in the soft band as well as in the hard band 1–2.5 keV were found to be highly variable, but the soft excess varied independently of the hard flux. This separate variability was particularly evident when a flare in the

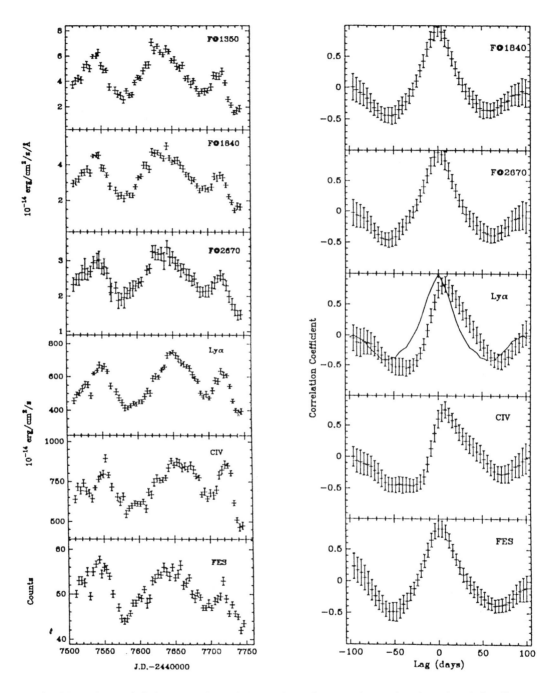

Fig. 8.9. Left panel: light curves for NGC 5548 in various continuum bands and emission lines. Right panel: cross-correlation of the light curves with the flux at 1350 Å. Reproduced from Clavel *et al.* 1991.

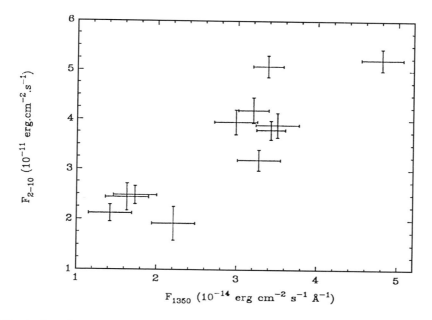

Fig. 8.10. A plot of 2–10 keV flux against the continuum flux at 1350 Å for NGC 5548. Reproduced from Clavel *et al.* (1992).

soft band lasting for about eight days was found to have no corresponding flare in the hard band. Such independent variability is not expected if the soft X-ray flux is generated by heating of a reflecting disk as discussed above, or for that matter in any other reprocessing model, such as the free–free emission model of Barvainis (1993) which we have discussed separately.

8.7.3 *Optical variability*

At optical wavelengths, variability has been observed on time scales as short as days to as long as years. Optically violently variable quasars (OVV) show variation by greater than a magnitude over a few days, while some quasars show no variability at all at the 10 per cent (0.1 magnitude) level. Observations in the optical region are usually available only over a limited number of epochs, spaced non-uniformly, which makes it difficult to apply the usual methods of time series analysis to the study of the variability time scales. When detailed monitoring is available, as in the case of the OVV quasar 3C 446 (Bregman *et al.* 1988), 'flickering activity' on a time scale of days to weeks as well as major outbursts that last for months to years have been observed. Variations over time scales as short as 30 min have been observed in the degree and angle of polarization in the source BL Lac (Moore, Schmidt and West 1987).

There have been some indications in the literature that the degree of optical variability in quasars is dependent on the luminosity and redshift. As the samples used in these studies are generally flux limited, only objects brighter than some limiting flux are included in the sample. Since the surface density of objects increases towards fainter magnitudes, most objects in the sample tend to be at the lowest fluxes. A consequence is that objects that are at high redshift are found to be the most luminous, while those at low redshift are the least luminous. This correlation between luminosity and redshift, which is inherent to the way in which samples are built up, makes it difficult to distinguish between the dependence of some property such as variability, on luminosity and redshift. Even within this limitation, there have been conflicting claims in the literature on how the degree of variability depends on the two parameters. Some groups claim that the more luminous or higher redshift quasars show less variation, while others find no such dependence, or even that higher redshift quasars have greater variability (see Hook *et al.* 1994 for references).

Hook *et al.* (1994) have investigated in detail the variability properties of a sample of 290 quasars, identified in different optical surveys, at the south Galactic pole (SGP). Photographic plate material taken in 12 exposures over the period from 1975 to 1991 is available for this sample. Hook *et al.* define a variability index σ_v by

$$\sigma_v = \frac{1}{N_e - 1} \sqrt{\frac{\pi}{2}} \sum_{i=1}^{N_e} |M_i - \langle m \rangle|, \qquad (8.13)$$

where N_e is the number of epochs over which the quasar was visible on the plates, m_i is the magnitude at epoch i and $\langle m \rangle$ is the mean magnitude. Hook *et al.* find that this index of variability is significantly correlated with absolute magnitude, with the more luminous quasars having lower variability. A correlation with redshift is indicated, but is weaker than that with luminosity. The data can also be used to examine how the amplitude of variation Δm depends upon the interval Δt over which the variation is observed. It is found that there is a functional relationship of the form

$$\Delta m = [0.155 + 0.023(M_B + 25.7)]\Delta t_r^{0.18}, \qquad (8.14)$$

where $\Delta t_r = \Delta t/(1+z)$ is the rest frame variability, with Δt the observed interval over which the variation Δm occurs. An immediate consequence of this dependence is that a model in which the quasar is assumed to be made up of a large number of randomly varying subunits (the 'Christmas tree' model) is ruled out, as in this case it follows from Poisson statistics that the luminosity dependence is given by $\Delta m \propto L_B^{-0.5} \propto 10^{0.2M_B}$.

8.7.4   *Infrared variability*

Variability in the far infrared of a sample of active galaxies and quasars has been studied by Edelson and Malkan (1987). In pointed *IRAS* observations of a sample of 20 sources, they found that the far-infrared fluxes of three blazars were found to vary up to a factor of ~ 2 on time scales of a few months. None of the non-blazar quasars or Seyfert galaxies in the sample showed significant variation. Impey and Neugebauer (1988) have observed a large sample of blazars with *IRAS* and found that 10 out of

24 blazars with multiple observations show variability, with amplitude typically half of the amplitude of optical variability. The time scales tested are in the 3–9 month range, which was dictated by observational constraints. The blazars found to be variable are mostly those that have been found to be very active at other wavelengths. The sizes based on the observed variability time scale in blazars require brightness temperatures $\gtrsim 10^6$ K, which implies a non-thermal origin for the infrared radiation. However, the observed lack of variability on the time scale of months in quasars that are not blazars, and in Seyfert galaxies, is consistent with the hypothesis that the far-infrared radiation is thermal emission from dust in an extended region (see Section 8.6).

At near infrared wavelengths, 1–10 μm, Neugebauer *et al.* (1989) have studied 108 quasars from the Bright Quasar Survey over an interval of about six years. They found that there was a good probability that half the quasars had varied over the period, while the rest showed no signs of variability. The amplitude of variation was generally ~ 0.5 magnitude, with maximum observed amplitude $\lesssim 1$ magnitude. A majority of quasars in the sample that have flat spectrum radio sources had a high probability of having varied over the time interval covered. The infrared colours remained essentially constant during the variation.

Comparing optical and near infrared variations, it is found that the two occur together or with a delay of one day at most. Between optical and 10 μm variations time delays amounting to days have been seen, while delays between optical and far-infrared variations can extend to weeks.

8.7.5 *Radio variability*

In the radio domain, variations are observed in compact sources on time scales of weeks to months, the shorter time scales being observed at the higher frequencies. The time scales are of the order of years in the 0.1–1 GHz range, months to years in the 1–10 GHz range and weeks to months in the 40–100 GHz range (see references in Bregman 1990). Even on these time scales, the small source sizes implied by the variations lead to brightness temperatures $\gtrsim 10^{12}$ K, which would lead to a catastrophic inverse Compton cooling of the energetic electrons. The situation is saved by appealing to relativistic beaming, which reduces the brightness temperature in the rest frame of the source by a large factor $\sim \gamma^3$ (see subsection 3.8.4). Short term variability on an intraday ($\lesssim 1$ day) time scale has also been observed (see Qian *et al.* 1991, Krichbaum, Quirrenbach and Witzel 1992 and references given there). It was usual to attribute this variability to propagation effects involving refractive interstellar scintillation, but the standard interstellar medium can produce fluctuations of only a few per cent amplitude. Higher amplitude flux variations on the intraday time scale can be produced by regions of enhanced density in clouds and filaments in the Solar neighbourhood.

To examine the cause of the intraday radio flux variability, large samples of compact extragalactic radio sources have been observed in a quasi-simultaneous fashion at different radio frequencies as well as in the optical band. It has been found in such observations that in some sources the intraday variation of the radio fluxes at different frequencies is anticorrelated with variations in the degree of polarization. The optical light curves also show variations that are related to changes in the radio flux. In particular, it was observed that in the BL Lac object 0716+71, the optical flux showed variations on a time scale of 1 day for the first week of observations, after which there was a transition to slower variations with time scale of about one week and reduced flux (Wagner and Witzel 1992). Similar behaviour with a much smaller amplitude of variation was observed in the source at radio wavelengths. While it is possible to explain the polarization variation within the framework of radio interstellar scintillation, the optical variations cannot be produced by propagation effects. Since the variations in the optical flux appear to be related to radio variations, it seems natural that the latter too should be produced intrinsically to the source. The small source sizes implied by the observed rapid variations then lead to very high brightness temperatures.

In the case of the quasar 0917+624 at redshift $z = 0.44$, it has been found (Qian *et al.* 1991) that the short term variability time scale at 5 GHz is one day, with amplitude 0.2 Jy in the flux variation. For $H_0 = 100 \, \text{km sec}^{-1} \, \text{Mpc}^{-1}$ and $q_0 = 0.5$, using Equation 3.111 this gives a variability brightness temperature of $T_{\text{var}} \simeq 2 \times 10^{18}$ K, which is far in excess of the limit for the onset of the Compton catastrophe. However, this is the observed brightness temperature, while from Equation 3.114 the rest frame brightness temperature $T' = \mathscr{D}^3 T_{\text{var}}$. For beaming very close to the line of sight, $T' \sim \gamma^3 T_{\text{var}} \lesssim 10^{12}$ K for $\gamma \simeq 100$. The value of γ required to reduce the brightness temperature to a non-catastrophic level is therefore an order of magnitude greater than the values inferred for most sources from superluminal motion and other data, as described in subsection 9.5.2. Qian *et al.* are able to avoid the need for an excessively large Lorentz factor by explaining the observation in terms of a shocked jet model, which is an extension of the model used to describe the observed long term variations and milliarcsecond scale structures of radio sources. The model requires $\gamma \simeq 10$, which is in the usual range. Other models of the intrinsic variability are also possible and are discussed by Krichbaum *et al.* (1993) and Qian *et al.*

Variability in the centimetre and in the millimetre regions of the radio band is found to be well correlated, with outbursts at the higher frequencies occurring earlier than outbursts at the lower frequencies. The delay is weeks to months between 20 and 100 GHz, while it is months to years between 5 and 15 GHz. The optical variability and radio variability have been found to be correlated, with radio variations occurring usually about a year after the corresponding optical variations. In some case the delay has been found to be two months or less. The delay can be interpreted in terms of the dilution of the emitting plasma as it travels outwards from the central source. A very compact emitting region would be opaque at radio frequencies owing to synchrotron self-absorption. As the emitting region moved outwards it would get bigger and become transparent to progressively lower frequencies as per the observations (see Bregman 1990 for references). But the detailed variability profiles in the radio and optical regions are different, indicating that the radio emitting plasma cannot simply be taken to be an expanded version of the plasma emitting in the optical region. The differences arise because there is significant reprocessing of the plasma between its emitting in the optical and in the radio regions, and different physical processes are active in the two regions (Bregman and Hufnagel 1989).

9   Radio properties

9.1   Introduction

About 10 per cent of giant elliptical galaxies and quasars are *radio loud*, which means that they have a radio luminosity of $\sim 10^{41}$–10^{46} erg sec^{-1} in a band extending from $\sim 10^2$ MHz to ~ 10 GHz. This corresponds to a 5 GHz luminosity $\gtrsim 10^{25}$ W Hz^{-1} and, for the observed range of redshift, a 5 GHz flux $\gtrsim 10^2$ mJy.[1] The ratio of the radio luminosity to the optical luminosity in these objects is $\gtrsim 10$. The criteria for separating radio-loud objects from radio-quiet objects are to some extent subjective; they identify a somewhat vague boundary on the brighter side of which rather spectacular manifestations of radio emission from active galaxies become evident.

Radio objects that do not belong to the radio-loud class are *not* radio quiet. Our Galaxy has a radio luminosity of $\sim 10^{37}$ erg sec^{-1}, while Seyfert galaxies show a whole range of radio luminosity up to the boundary of the radio-loud class. The non-thermal emission in normal galaxies can be traced to their centres, supernovae, active stars, energetic particles in the magnetic field of the interstellar medium (ISM) and so on. In the more luminous Seyfert galaxies, the radio emission originates in the supernovae in starburst regions, though in the most luminous cases it is possible that mechanisms similar to those in radio-loud galaxies are in operation.

The radio properties of radio-loud quasars, which are in the minority in the whole quasar population, are quite distinct from those of radio-quiet quasars. As we will see in greater detail below, the ratio R of radio to optical emission in quasars shows a bimodal distribution, with the radio-loud objects clustered around $R \sim 10^3$, and the others around $R \sim 1$. A small fraction of quasars do not show radio emission even at the level of a few μJy. Considering their high luminosity in all other bands, the relative radio-quietness of the majority of quasars is rather surprising. Since all quasars, by definition, have a central active region that produces energetic particles, perhaps in the form of beams, which can lead to radio emission, there must be mechanisms operating that in some manner quench the radio emission in most cases; not much is known about these at the present time.

Radio galaxies and quasars are similar in many respects but different in others. It has been realized through studies in the radio as well as other domains that many of the differences could arise as a result of different orientations, relative to the observer's

---

[1] 1 Jy $= 10^{-26}$ W m^{-2} Hz.

line of sight, of basically similar sources. Some of the differences due to orientation can also be highly exaggerated by bulk relativistic motion in nuclear regions and beyond; this also leads to apparent superluminal motion.

9.2 Radio morphology

The radio sources associated with radio galaxies and quasars come in a variety of morphological structures and sizes. At one extreme of the range are unresolved flat spectrum compact sources coincident with the optical nucleus; at the other end are complex structures, many hundreds of kiloparsec in extent, with a halo, lobes, a compact nucleus and jet-like features connecting the nucleus to the extended regions.

The first double-lobed structure was discovered by Jennison and Das Gupta (1953) in their observation of Cygnus A, which is the brightest extragalactic radio source in the sky. In the following decade it was found from interferometric observations that the double-lobed structure was very common amongst extragalactic sources. Subsequent observations at higher frequencies in the GHz range, which provided better angular resolution than at lower frequencies, revealed that a compact flat spectrum nuclear source was often associated with extended radio structures.

The development of large radio telescope arrays, which make use of Earth rotational aperture synthesis and sophisticated image restoration techniques, made it possible to obtain two-dimensional intensity and polarization maps. The most sensitive such instrument is the Very Large Array (VLA), located in New Mexico. This consists of an array of 27 dish antennae, each 25 m in diameter. The array is Y-shaped, and the distances between the dishes can be varied, to provide four main configurations, A, B, C and D. The maximum distance between the dishes, 36 km, occurs in the A configuration, which therefore has the highest resolution ($\sim$ 1 arcsec at 5 GHz). In the D configuration the antennae are spread over just 1 km. Using the VLA it is routinely possible to make high sensitivity radio maps with an angular resolution of $\lesssim$ 1 arcsecond. Resolution better than 50 milliarcsec at 5 GHz can obtained with the Jodrell Bank MERLIN,[1] which is an array of radio telescopes spread over Great Britain, with separations up to 217 km. Resolutions of $\sim$ 1 milliarcsec can be obtained using the technique of very long baseline interferometry (VLBI). This uses antennae spread over different continents; the baseline is therefore thousands of kilometres long. The signals from the antennae are independently recorded and correlated later. The Very Long Baseline Array (VLBA) consists of 10 dish antennae, each of 25 metre diameter, spread from Hawaii to the Virgin Islands. The baseline has been extended greatly by the launch of a 10 m dish on the Japanese satellite *HALCA*, whose orbit reaches up to 21 000 km from the Earth.

We will first consider radio structure on the arcsecond scale and discuss VLBI observations in a later section.

---

[1] The multi-element radio-linked interferometer network.

9.2.1 *Primary morphological features*

The primary features in a radio galaxy or quasar are the *core, lobes, hotspots* and *jets.* Not all of these features occur in all sources, and quite often the morphology is too complex to provide an unambiguous separation into these parts. It is, however, convenient to view a radio source as being built up from them and consider any complex or ill-defined features to be produced from disturbances in the source itself, or in interaction with the ambient intergalactic environment. We will now briefly describe these features and develop some additional facts about them in later sections. More-detailed descriptions of radio morphology with examples and references can be found in Miley (1980), Bridle and Perley (1984), Perley (1989), Muxlow and Garrington (1991) and Laing (1993).

Cores These are compact components, unresolved when observed with angular resolution $\gtrsim 0.1$ arcsec, and coinciding with the nucleus of the associated optical object. They usually have flat power law[1] or complex spectra, which points to synchrotron self-absorption, as explained in subsection 3.5.1. When cores are examined with VLBI, they can be resolved into subcomponents, often consisting of an unresolved flat spectrum core and a jet-like structure that may comprise more than one knot. There are also compact steep spectrum sources and compact doubles, which we will discuss in Section 9.5.

Cores are best detected at $\sim$ GHz frequencies because they often have flat spectra, while the more extended components have steep spectra. The flux from the extended sources drops sharply at high frequency, while that from the compact components remains relatively high. The latter is therefore preferentially detected in flux-limited high frequency surveys. Another factor is that the angular resolution is proportional to the wavelength, so that a smaller beam size is available at high frequency. As the beam size decreases, the extended emission per beam goes down, while the emission from the compact source remains virtually unchanged as long as the source is unresolved. This again allows relatively weak cores to be detected at high frequency. Cores are found in almost all radio quasars and in ~ 80 per cent of all radio galaxies. It is safe to assume that they will be found in all cases when observations are made with sufficient sensitivity and dynamic range.

The contribution of the core to the total radio luminosity of a source varies from under one per cent in some sources to almost 100 per cent in some quasars. Sources of the latter type dominate high frequency surveys, while a survey made at a few hundred megahertz has sources with a wider range of the core to total radio luminosity ratio. We will see later that the wide range of the relative contribution of a core can be interpreted in terms of a beaming model which unifies radio galaxies and quasars into a single population.

---

[1] By convention a power law radio spectrum is said to be flat when the spectral index $\alpha \leq 0.5$ and steep otherwise.

Lobes   These are extended regions of radio emission. There are often two lobes approximately symmetrically placed on opposite sides of the galaxy or quasar. The lobes sometimes contain regions of enhanced emission called hotspots (see below), which are collinear with the central source. The overall extent of the radio structure, to the ends of discernible lobes, can be several hundred kiloparsec and in some extreme cases up to a few megaparsec. The radio galaxy 3C 236 has an overall size of $\sim 4\,\mathrm{Mpc}$. The lobes have power law radio spectra with $\alpha > 0.5$ and the emission is partially linearly polarized, which suggests it could be an optically thin synchrotron source.

The lines joining the extended structure to the central source (radio or optical) are sometimes bent back, so that the opening angle is less than 180 deg. In the case of *narrow-angle-tail* sources, the tail angle is very small and the source appears to have a relatively bright head with an extended tail. A typical example is shown in Figure 9.1. These sources are of the FR-I type (this terminology is described below) associated with galaxies in clusters, with the ram pressure of the cluster gas sweeping back the radio structure as the host galaxy moves in it. In the *wide-angle-tail* type the extended structure is C-shaped, with properties intermediate between narrow-angle-tail and FR-II sources described below. An example is shown in Figure 9.2.

The lobes often show rotational symmetry and have Z- or S-shaped structures. These shapes are most naturally interpreted as being due to the precession of the axes of the jets which carry energy from the central source to the extended regions.

The luminosities of the two lobes in a typical radio source are usually quite comparable and differ at most by a factor of ~ 2 in most cases. This also applies to the distance of lobe extremities from the nucleus. In some quasars like 3C 273, an extended structure occurs on only one side of the nucleus; quantitatively this means that any extended structure on the opposite side is $\lesssim 1/10$ as bright as the observed structure. One of the reasons for one-sidedness could be that the double structure is oriented close to the line of sight, so that the structures on either side of the nucleus appear to be merged together.

Jets   These are narrow features that link the compact core to the outer regions. A jet can be interpreted as radio emission from beams that transport energy from the AGN to the extended regions. Radio jets occur on parsec as well as kiloparsec scales and can be smooth or knotty. Jets are said to be two-sided when they are seen on both sides of the central source. In the more luminous radio galaxies and in all quasars jets are only one-sided, while those in the less luminous radio galaxies are two-sided. The kiloparsec scale jets have a power law spectral index $\alpha \sim 0.6$. We shall consider jets in detail in Section 9.4.

Hotspots   These are intensity maxima located towards the outer extremities of the lobes of highly luminous sources. When the lobes are observed with insufficient resolution, hotspots give them the appearance of being edge brightened (see below). Hotspots typically have a linear size of $\sim 1\,\mathrm{kpc}$ and a power law spectrum which is steep (i.e., in the range 0.5–1) but generally flatter than the spectral index of the lobe as

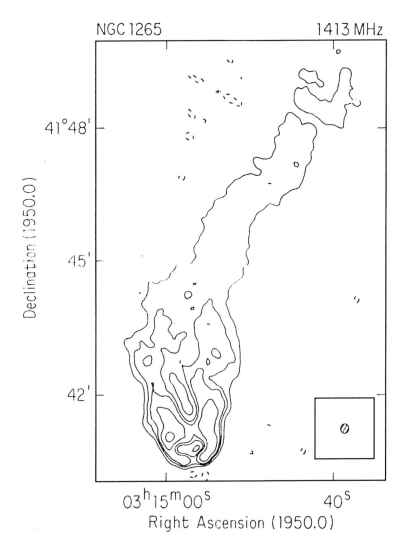

Fig. 9.1. Very Large Array image of the narrow-angle-tail source NGC 1265 at 1428 MHz. The resolution is 12.9×11.4 arcsec^2 in position angle -4 deg. The contours indicate flux density levels of -10, 10, 30, 50, 100, 200 and 300 mJy per beam. The beam shape is shown as a hatched ellipse in the small panel. Reproduced from O'Dea and Owen (1986).

a whole. This is consistent with the interpretation of a hotspot as the place at which a jet from the nucleus hits the ambient medium, producing a shock in which bulk kinetic energy of the beam is converted to random motion. The energetic particles diffuse from the hotspots to the lobes, providing a continuous supply of energy.

Hotspots are not always present, while in some cases there is more than one intensity maximum in a lobe. Jets too can be composed of a number of knots that appear as

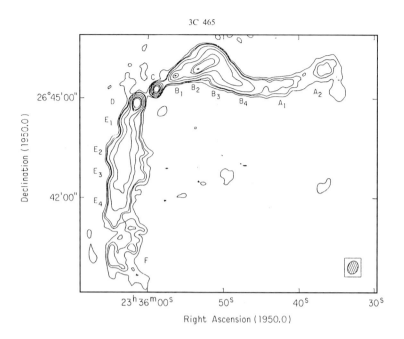

Fig. 9.2. VLA map of 3C 465 made at 1480 MHz. The beam width is 9.5×12 arcsec^2. The contours indicate flux density levels of -5, 5, 10, 20, 40, 80, 100 and 150 mJy per beam. Reproduced from Eilek *et al.* (1984).

intensity enhancements and, in a complex structure, it is difficult to distinguish between these knots and genuine hotspots, yet the two have quite different physical significance. In the literature it is common to refer to primary and secondary hotspots when there is more than one intensity maximum. Recently Bridle *et al.* (1994) have provided a useful empirical definition of hotspots as follows. In a source with no detected jet, a hotspot must (a) be the brightest feature in a lobe, (b) have a surface brightness that is more than four times greater than that of the surroundings and (c) have a linear full width at half maximum that is no more than five per cent of the largest diameter of the source. When a jet is detected, a further condition is added: (d) the hotspot must be further from the nucleus than the end of the jet. The termination of the jet itself is defined by (d1) its disappearance, (d2) an abrupt change in direction or (d3) decollimation by more than a factor of 2. The condition (d) is provided so that hotspots can be distinguished from jet knots. This definition due to Bridle *et al.* ensures that there can at most be one hotspot per lobe. Some examples of radio galaxies with hotspots are given in the following section.

Apart from the main features of radio morphology that we have discussed here, there are several others, such as plumes, tails, bridges and haloes, which appear in the literature. Some of these have been discussed in Muxlow and Garrington (1991).

9.3   Fanaroff–Riley classification

It was first noticed by B.L. Fanaroff and J.M. Riley (1974) that the relative positions of regions of high and low surface brightness in the lobes of extragalactic radio sources are correlated with their radio luminosity. This conclusion was based on a set of 57 radio galaxies and quasars, from the complete 3CR catalogue, which were clearly resolved at 1.4 GHz or 5 GHz into two or more components. Fanaroff and Riley divided this sample into two classes using the ratio R_{FR} of the distance between the regions of highest surface brightness on opposite sides of the central galaxy or quasar, to the total extent of the source up to the lowest brightness contour in the map. Sources with $R_{FR} < 0.5$ were placed in Class I and sources with $R_{FR} > 0.5$ in Class II. It was found that nearly all sources with luminosity

$$L(178 \, \text{MHz}) \lesssim 2 \times 10^{25} h_{100}^{-2} \, \text{W Hz}^{-1} \, \text{str}^{-1}, \tag{9.1}$$

were of Class I while the brighter sources were nearly all of Class II. The luminosity boundary between them is not very sharp, and there is some overlap in the luminosities of sources classified as FR-I or FR-II on the basis of their structures. For a spectral index of $\alpha \simeq 1$ the dividing luminosity at 5 GHz is

$$L(5 \, \text{GHz}) \lesssim 7 \times 10^{23} h_{100}^{-2} \, \text{W Hz}^{-1} \, \text{str}^{-1}. \tag{9.2}$$

At high frequencies the luminosity overlap between the two classes can be as much as two orders of magnitude.

Various properties of sources in the two classes are different, which is indicative of a direct link between luminosity and the way in which energy is transported from the central region and converted to radio emission in the outer parts. We will now provide a somewhat more detailed description of the Fanaroff–Riley classes.

Fanaroff–Riley Class I (FR-I)   Sources in this class have their low brightness regions further from the central galaxy or quasar than their high brightness regions (see Figure 9.3). The sources become fainter as one approaches the outer extremities of the lobes and the spectra here are the steepest, indicating that the radiating particles have aged the most. Jets are detected in 80 per cent of FR-I galaxies. A jet can begin as one-sided close to the core, but beyond a few kiloparsec it becomes two-sided and continuous, with an opening angle $\gtrsim 8$ deg that varies along its length. Along the jet the component of the magnetic field in the plane of the sky is at first parallel to the jet axis, but soon becomes aligned predominantly perpendicular to the axis (see Figure 9.10).

FR-I sources are associated with bright, large galaxies (D or cD) that have a flatter light distribution than an average elliptical galaxy and are often located in rich clusters with extreme X-ray emitting gas (Owen and Laing 1989, Prestage and Peacock 1988). As the galaxy moves through the cluster the gas can sweep back and distort the radio structure through ram pressure, which explains why narrow-angle-tail or wide-angle-tail sources, say, appear to be derived from the FR-I class of objects.

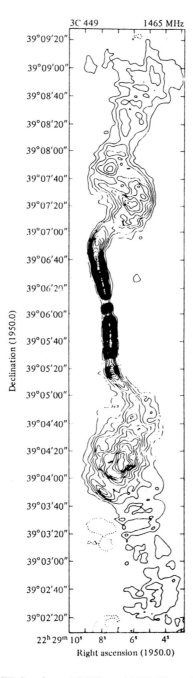

Fig. 9.3. VLA map of the FR-I galaxy 3C 449 at 1465 MHz, with angular resolution 4.8 × 3.4 arcsec². The peak flux is 22.2 mJy per beam, with contours drawn at 5 per cent intervals, beginning with the −5 per cent contour. Reproduced from Perley, Willis and Scott (1979).

A typical FR-I galaxy is shown in Figure 9.3 (Perley, Willis and Scott 1979). This is the radio source 3C 449, which is optically identified with a galaxy of type cDE4 at a redshift of 0.0181, so that 1 arcsec corresponds to $255h_{100}^{-1}$ pc. There are twin jets that are straight for ~ 30 arcsec from the core, after which they deviate towards the west and terminate into diffuse lobes. These jets and outer lobes are mirror symmetric about an axis through the core. The jets are generally smooth in appearance, but higher resolution observations show knots on a smooth ridge of emission, the southern jet being more knotty than the northern one. Within ~ 10 arcsec of the nucleus, the surface brightness of the jets is much reduced. The jets widen at a non-uniform rate close to the core, with the greatest expansion occurring where the jets are faintest. Beyond ~ 10 arcsec from the nucleus the opening angle is constant at ~ 7 deg. The emission from the jets is highly polarized, the average polarization over the jets being ~ 30 per cent, and the projected magnetic field is perpendicular to the jet axis.

Fanaroff–Riley Class II (FR-II)   This class comprises luminous radio sources with hotspots in their lobes at distances from the centre which are such that $R_{FR} > 0.5$. These sources are called edge-darkened, which was particularly apt terminology when the angular resolution and dynamic range used in observing the classical sources was not always good enough to reveal the hotspots as distinct structures. In keeping with the overall high luminosity of this type of source, the cores and jets in them are also brighter than those in FR-I galaxies in absolute terms; but relative to the lobes these features are much fainter in FR-II galaxies. Jets are detected in < 10 per cent of luminous radio galaxies, but in nearly all quasars. The jets have small opening angles (< 4 deg) and are knotty; the jet magnetic field is predominantly parallel to the jet axis except in the knots, where the perpendicular component is dominant. An example of an FR-II source is shown in Figure 9.4, which is a VLA map of the radio quasar 3C 47 made by Bridle *et al.* (1994). The most striking feature of the jets in the FR-II class is that they are often one-sided, as is clearly seen in Figure 9.4. Jet one-sidedness occurs at large (kpc) scales as well as in the milliarcsecond jets which are found in compact cores through VLBI observations and which will be discussed further in Section 9.5. The feature A in the jetted lobe is a hotspot, while feature H on the unjetted side looks like one, but does not qualify for being a hotspot according to the criteria of Bridle *et al.* (1994) listed in Section 9.2.

FR-II sources are generally associated with galaxies that appear normal, except that they have nuclear and extended emission line regions. The galaxies are giant ellipticals, but not first-ranked cluster galaxies. Their average absolute magnitude $\langle M_R \rangle = -19.9$ ($H_0 = 100$ km sec^{-1} Mpc^{-1}) is close to the characteristic value M_* of the Schechter galaxy luminosity function (see e.g. MB81, p. 352) at which the density of galaxies shows an exponential turnover. The environment of FR-II sources does not show enhanced galaxy clustering over the environment of randomly chosen elliptical galaxies (Owen and Laing 1989, Prestage and Peacock 1988). Owing to the large differences in the nature of the host galaxies and the environments of the FR-I and FR-II sources, it is possible that they are intrinsically different types of source not related to each

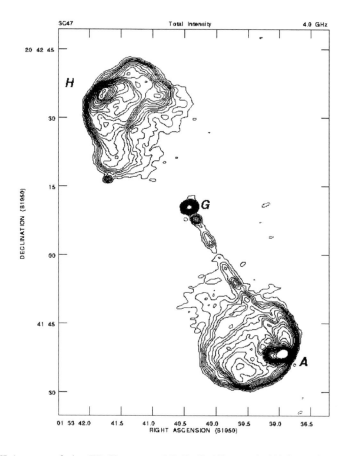

Fig. 9.4. A VLA map of the FR-II quasar 3C47 (Bridle *et al.* 1994) made at 4.9 GHz with $1.45 \times 1.13\,\mathrm{arcsec}^2$ resolution. G is the core, A the jetted hotspot. H does not meet the hotspot criteria of Bridle *et al.* The figure was kindly provided by Alan Bridle.

other through an evolutionary sequence (but see the remark towards the end of this section).

   Host galaxies   The host galaxies of radio sources have the general appearance of ellipticals, but high dynamic range charge coupled device (CCD) images reveal morphologies that are rather complex and disturbed. A detailed study of bright radio galaxies from the Molonglo Reference Catalogue (MRC) by A. Mahabal, A. Kembhavi and P. McCarthy (Mahabal 1998) has shown that in nearly every case there is evidence of a significant departure from the simple structure expected in an elliptical. One can see the presence of disk-like structures, spiral arms and on occasion multiple nuclei. These features can be seen in direct images after careful image processing, but are more readily evident in two-colour images. The intensity profiles of normal elliptical galaxies in most cases are well fitted by de Vaucouleurs' law given in Equation 9.7 below. But

Fig. 9.5. CCD image in the *B* filter of the MRC radio galaxy 1226-252 taken with the 1 m telescope of the Carnegie Observatory at Las Campanas. The image was kindly provided by A. Mahabal.

the profiles of radio galaxies in several cases show significant departures from such a law, and the presence of a significant disk component with exponential intensity distribution can be inferred. Radio galaxy images also show the presence of dust and the signs of interaction with neighbouring galaxies. We show in Figure 9.5 a *B* filter image of the MRC radio galaxy 1222-252. Structure on different scales is evident in the direct image, and the disk-to-bulge luminosity ratio inferred for this galaxy is 2.5.

Bivariate classification   The Fanaroff–Riley classification of a galaxy depends on its radio luminosity: most galaxies that are brighter than $2 \times 10^{25} h_{100}^{-2}$ W Hz^{-1} str^{-1} at 178 MHz are of the FR-II type, while less luminous galaxies belong mainly to the FR-I class. The division between the two classes becomes sharper if the distribution as a function of radio luminosity as well as of absolute optical magnitude is taken into account. This can be seen in Figure 9.6, which shows the distribution of FR-I and FR-II radio galaxies as a function of their optical and radio luminosity (Owen and Ledlow 1994). The two types are seen to be distributed over a wide range of optical luminosity. The brighter radio galaxies tend to be of the FR-II type, but the dividing point along the radio axis is not sharp. However, there seems to be a fairly clear diagonal division of the types, which shows that the radio luminosity at the division between the classes increases with optical luminosity. The diagonal boundary is given approximately by $L_{\mathrm{R}} \propto L_{\mathrm{op}}^2$. When the two-dimensional distribution of points is projected onto the radio luminosity axis, the two types remain separated but with some mixing at the boundary. The dependence of the dividing radio luminosity on the optical luminosity is a reflection of the form of the bivariate luminosity function (see Section 7.11).

Physical distinction   The jets in FR-II galaxies are generally smooth, often one-sided and end in hotspots in well-separated lobes. However, jets in FR-I galaxies are two-sided, and the radio structures are often distorted and plume-like. The smooth

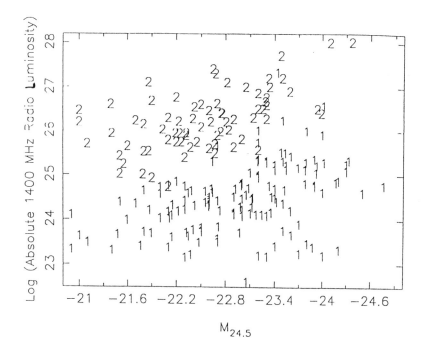

Fig. 9.6. Distribution of FR-I and FR-II radio galaxies as a function of their 1400 MHz radio luminosity and absolute *B* magnitude. Reproduced from Owen and Ledlow (1994) in BDQ94.

nature of FR-II jets is thought to be indicative of highly supersonic flows, while FR-I jets are thought to be subsonic, which makes them amenable to distortions in interaction with the ambient medium. There are two possibilities for this difference in jet speeds: (1) the jets in all radio galaxies and quasars are produced in similar central engines, always emerge with supersonic speeds from the engine and are slowed down to subsonic speeds when there is sufficient interaction; (2) the engines that power FR-I and FR-II sources are different in nature, and produce subsonic and supersonic jets respectively. Models based on both these assumptions have been proposed, but the development is suggestive only, and the theoretical concepts involved as well as observational consequences remain to be explored.

A model of the first kind has been proposed by De Young (1993), who has made an earlier version of Figure 9.6 as the basis. He notes that for a given radio luminosity in the figure, there is an optical luminosity limit that separates the two types of source. Sources fainter than this limit are FR-II, while those brighter than it are FR-I. Since the transition occurs at a fixed radio luminosity, De Young concludes that there is no major change in the engine that produces the radio emission, and the difference between the two types must be an environmental effect.[1]

---

[1] It may be more appropriate here to consider the core radio luminosity, but this is not known in all cases, and where known it is at least approximately proportional to the extended radio luminosity.

De Young suggests that jets in FR-I galaxies are decelerated a short distance outside the production region. Thus, in the short distance before significant deceleration begins, the jets interact very little with the matter, and have low luminosity, which explains the gap often found between the nucleus and the base of jets in FR-I sources (see Section 9.4). After the deceleration the jets must proceed relatively unimpeded over large distances, as otherwise the outflow would cease completely. The deceleration can be produced by transfer of momentum from the jet to a dense ambient gas if the Reynolds number is very large, which is likely to be the case for reasonable jet dimensions and speeds. The dense ambient medium can be produced by the inflow of gas into the central region, owing to stellar mass loss, flows set up by interactions, cooling flows etc.

It is expected that active star formation will take place owing to the enhanced density of the central region, and also the action of the jet on it. This should make the central regions of FR-I galaxies bluer than those in FR-II galaxies. Testing such a prediction requires good signal-to-noise images in two or more colours of a number of radio galaxies, and a detailed study along these lines is yet to be made.

The second possibility, that the differences between the FR-I and FR-II galaxies are due to qualitative differences in the properties of the central engine, has been considered by Baum, Zirbel and O'Dea (1995). They base their conjecture on a detailed study of the correlations between radio luminosity, emission line luminosity and host galaxy magnitude of a large sample of FR-I and FR-II galaxies; this spans 10 orders of magnitude in luminosity and contains a number of galaxies of the two types that overlap in luminosity.

The principal differences found in the two types by Baum *et al.* are as follows.

(1) At the same host galaxy absolute magnitude or radio luminosity, FR-II galaxies produce about an order of magnitude more optical line emission than the FR-I galaxies. FR-II galaxies are orders of magnitude brighter in line emission than radio-quiet galaxies of the same optical magnitude, while FR-I and radio-quiet galaxies of the same magnitude have comparable line emission.

(2) The emission line luminosities of FR-I galaxies are correlated with their absolute magnitude. Such a correlation is not seen in the case of FR-II galaxies.

(3) FR-II galaxies produce substantially more line emission than FR-I galaxies of the same total or core radio luminosity.

(4) The emission line luminosity in both types of galaxy is correlated with total and core radio luminosities, but the regression line for each type has a different slope and intercept.

(5) There is a strong correlation between core and total radio luminosities for both galaxy types. There is continuity in the distribution of the two galaxy types in the $\log(L_{\mathrm{rc}}/L_{\mathrm{r,ext}})$ log L_{R}-plane, where L_{rc} and $L_{\mathrm{r,ext}}$ are the core and extended radio luminosity at 408 MHz and $L_{\mathrm{R}} = L_{\mathrm{rc}} + L_{\mathrm{r,ext}}$. The regression lines fitted separately to FR-I and FR-II galaxies in this plane are not significantly different.

The observed distribution of the [O III]/Hβ line ratio in FR-I galaxies is similar to the distribution for radio-quiet ellipticals and cooling flow galaxies. Baum *et al.* suggest that the emission lines here are of the low ionization type, which are different from the high ionization lines produced owing to ionizing radiation from the nucleus in Seyfert and other active galaxies. Since the line luminosity is also correlated with the host galaxy optical luminosity, Baum *et al.* conclude that the line emission in FR-I galaxies could be produced by processes in the host galaxy. In contrast to this, available evidence suggests that in FR-II galaxies the lines are produced as a result of an ionizing continuum from the nucleus. This points to a possible important difference between the central engines of FR-I and FR-II galaxies: the engines in the former produce far less ionizing radiation and funnel a higher fraction of their total energy output into the kinetic energy of the jets than FR-II galaxies.

Baum *et al.* furthermore suggest that FR-I sources are produced when the accretion rate onto the central black hole is low, and the black hole has relatively less angular momentum; the FR-II sources arise when the accretion rate is high and the hole spins more rapidly. The different degrees of black hole spin make a difference to the nature of the jets produced, leading to subsonic jets when the spin is low and supersonic jets when it is high. As we have mentioned above, these different jet properties can lead to different levels of interaction with the ambient medium, creating different radio morphologies. Baum *et al.* also suggest that a high accretion rate could decline with time, causing an FR-II galaxy to evolve into an FR-I type. The correlations that have been used in arriving at this picture, the conclusions drawn from them and the theoretical conjectures all need thorough study before the scenario can be accepted as plausible. We have seen above that the environments of the two types of radio galaxy are likely to be different. This seems to argue against the two types having an evolutionary connection. But it is possible that at least some of the FR-I galaxies began as FR-II galaxies in the dense environments of rich clusters and relatively quickly evolved to the FR-I state (Hill and Lilly 1991).

9.4 Jets

Many radio galaxies and quasars contain *jets* that link the central component to the outer regions, which could be several hundred kiloparsec away. Jets also occur on parsec scales, with small angular extent accessible only to VLBI observations. In the beaming models that are now the 'standard' in the interpretation of observed radio structures, the jets are conduits for the transport of energetic particles from the nucleus to the extended radio structures. The radio luminosity of a typical jet is only a small fraction of the total radio luminosity, which means that the energy is transported in the form of bulk kinetic energy of the particles, the observed radio emission from jets being indicative of inefficiency during the transport. The energy in the beam is randomized when it meets the ambient medium at the hotspots, leading to synchrotron radio emission.

An extragalactic jet-like feature was first noticed at optical wavelengths in the galaxy M87 (NGC 4486) by Curtis (1918) and was later discussed by Baade and Minkowski

(1954). At radio wavelengths Hazard, Mackay and Shimmins (1963) observed a jet-like feature in 3C 273, and Miley *et al.* (1970) discovered that one of the two small radio components in the radio source Virgo A coincided with the nucleus of M87, while the other was associated with the tip of the optical jet. Further examples of jets were discovered using the Cambridge one-mile and five-kilometre telescopes (3C 266, Northover 1973; 3C 219, Turland 1975). Observations of radio galaxies and quasars with high sensitivity, dynamic range and subarcsecond angular resolution which have become routinely possible with the development of the VLA, have led to the discovery of a large number of jets with varied and often highly complex morphology.

On the theoretical side, beams of energetic particles were suggested in the context of AGN by Morrison (1969), while Rees (1971) proposed that radio lobes may be energized by beams of low frequency electromagnetic radiation. Jets viewed as narrow beams of energetic charged particles were proposed by Blandford and Rees (1974) and Scheuer (1974a) and proved to be very useful in interpreting the jets and hotspots that were then being discovered.

The need for some mechanism to transport energy continuously to the outer regions became most obvious with the discovery of hotspots in the radio galaxy Cygnus A by Hargrave and Ryle (1974). The two hotspots in this source are at a distance of $\sim 49 h_{100}^{-1}$ kpc and $\sim 42.5 h_{100}^{-1}$ kpc from the nucleus, so that the (average) light travel time to a hotspot is $\sim 135 h_{100}^{-1}$ Myr. The minimum energy in the hotspots obtained using the arguments in Section 3.6 is $U_{\mathrm{min,hs}} \sim 3.5 \times 10^{57}$ erg sec^{-1}, and the corresponding equipartition magnetic field is $B_{\mathrm{eq}} \sim 3.2 \times 10^{-4}$ G. The 'half-life' of a synchrotron electron which emits in this field at 5 GHz (which was one of the frequencies used in the observations) is obtained from Equation 3.6 as $t_{1/2} \sim 3 \times 10^4$ yr. If the electrons remain in the hotspot for longer than $t_{1/2}$, the energy loss should lead to a break in the spectrum at frequencies < 5 GHz; this is, however, not observed. The observed (projected) size of each hotspot is $\sim 1.1 h_{100}^{-1} \times 1.5 h_{100}^{-1}$ kpc^2, which gives a light travel time of $\sim 3 \times 10^3$ yr across it. The emitting relativistic particles therefore quickly diffuse out of the hotspots and if they are continuously replenished in some manner a break in the spectrum will not appear. The energy replenishment is required at the rate of $\sim U_{\mathrm{min,hs}}/\tau \sim 10^{45}$–$10^{46}$ erg sec^{-1}, where τ is the particle lifetime in the hotspot. The particles leaving the hotspot can supply the minimum energy requirement of the lobes, which is $\sim 3 \times 10^{59}$ erg, if the process continues for 10^6–10^7 yr. The energy replenishment in Cygnus A can be provided by the jets which have now been observed but which were not apparent to Hargrave and Ryle. They were nevertheless able to argue that models based on a continuous flow of energy provided a more satisfactory explanation of the observations than any of the other models popular at that time.

9.4.1   *Large scale jets: definition and incidence*

While radio jets in some AGN are unmistakable, in spite of being relatively faint, in other cases they appear to be made up of a number of discrete knots and can also be highly distorted. It is therefore not always easy to distinguish a jet from other structures, and it is necessary to have a definition that will introduce some uniformity in features which are identified as jets. In the definition now standard, introduced by Bridle and Perley (1984), a feature is termed as a jet when it is: (1) at least four times as long as it is wide; (2) separable at high resolution from other standard structures either spatially or by brightness contrast; (3) aligned with the compact radio core when it is closest to it.

We show some examples of kiloparsec scale jets in Figures 9.7 to 9.9. Cygnus A (3C 405), discussed above and shown in Figure 9.7 (Carilli and Barthel 1996), is one of the most luminous radio galaxies known and has been studied in great detail because of its large angular size and flux density. It is optically identified with a cD galaxy at redshift 0.0562 and is located in a luminous X-ray cluster. A very long and narrow jet extends from the core into the lobe in the northwest and there is a faint counterjet to the southeast. Hargrave and Ryle's argument based on synchrotron ageing in the hotspots of Cygnus A provided support to the beaming models, but the large scale jet itself was first reported much later by Perley, Dreher and Cowan (1984). Although the jet has very high luminosity, it has only a small fraction (~ 0.1 per cent) of the total source flux density and high dynamic range observations are required to study it in detail. The features in the jet and counterjet have been described by Carilli and Barthel (1996). The jet is straight to a distance of ~ 35 arcsec from the core, after which it first bends towards the south and then back to the north. It is made up of a series of regularly spaced knots, which are ~ 7 arcsec apart. In a high-pass filtered image, which enhances the small scale structure, the knots appear as oblique ridges oriented at ~ 7 deg to the core. The opening angle derived for the jet is ~ 1.6 deg, but the morphology is consistent with regions of larger opening angle ~ 5 deg and a constant-width jet in between. The counterjet is straight to ~ 20 arcsec from the core, after which it bends gradually by ~ 20 deg to the south and terminates at the hotspot in the lobe towards which it is heading. On the parsec scale there is a prominent jet and a faint counterjet.

Figure 9.8 shows another example of a prominent jet in the bright quasar 3C 175 from Bridle *et al.* (1994). The jet in this case is one-sided, like all quasar radio jets. It emanates from the compact core at M, and is straight until knot I, after which it curves southward toward feature D. The hotspot, identified as per the definition of Bridle *et al.* given above, is indicated by A. The jet expands steadily between knots L and H, beyond which the transverse profile is asymmetric, the intensity decreasing more rapidly towards the northwest. The lobe in the northeast has a hotspot at O, even though there is no sign of a counterjet.

An example of a jet that is not as prominent as in the previous two cases is shown in Figure 9.9. This is again a quasar from the sample of Bridle *et al.* The data shown in the figure was obtained in the B configuration with angular resolution 1.1 arcsec. The source is a widely spaced double, with lobes placed symmetrically about the central feature E; however, the northern lobe has ~ 16 times the flux density of the southern lobe. The feature D satisfies the conditions for being a jet, especially when observed

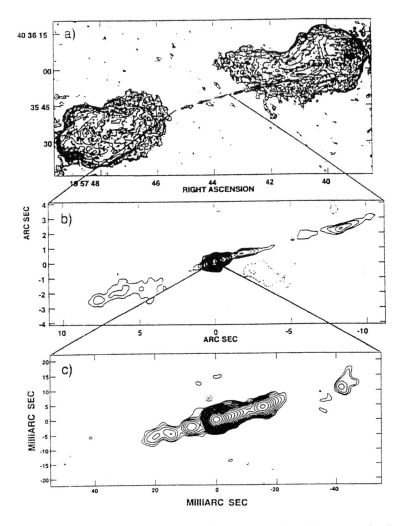

Fig. 9.7. The jets in Cygnus A from parsec to kiloparsec scales. The images in the top and middle panels are from VLA observations at 0.5 arcsec angular resolution, while those in the bottom panel are from VLBI observations at 3 milliarcsec resolution. Reproduced from Carilli and Barthel (1996).

at 0.35 arcsec angular resolution (see Bridle *et al.* 1994). There are complex emission features at A, B and C, of which B is tentatively identified as the hotspot. Feature F, which lies on the side opposite to the jet, shows, at 0.35 arcsec resolution, a compact feature and an extension to the southeast. F is directly opposite the peak in feature D, and Bridle *et al.* consider it to be a counterjet candidate, since its location and features are suggestive but do not satisfy the conditions for its being a jet. G, H and I are subsidiary peaks in the southern lobe; of these, H is the most compact, but it does not have sufficient brightness contrast to be termed a hotspot.

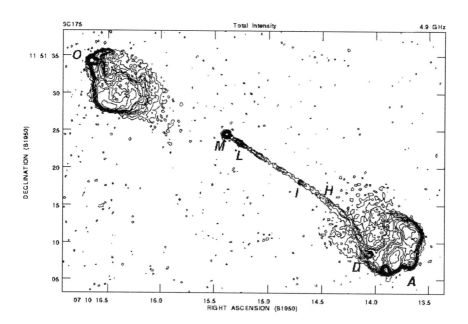

Fig. 9.8. A VLA map of the quasar 3C 175. The angular resolution is 0.75 arcsec FWHM. M is the core, A and O are the hotspots. Other labelled features are described in the text. There is no candidate counterjet. The figure was kindly provided by Alan Bridle.

The general features of radio-source models involving jets are consistent with observation, and jets occur in extragalactic radio sources of all types. This means that jets should be ubiquitous to extragalactic radio sources and are probably an indispensable component. However, jet luminosity, in absolute terms as well as relative to the total radio luminosity of the source, shows a large spread, so that it can be difficult to detect jets in many sources. Apart from source-to-source variation, jet morphology, nature and detectability depends upon the type of source under consideration. A list of 125 jets and their properties was provided by Bridle and Perley (1984), and a more extensive list with 276 jets has been compiled by Liu and Xie (1992). The number of known jets has gone up considerably since the important review by Bridle and Perley, and many observations have been made with improved dynamic range and angular resolution, but the general trends mentioned by them have remained largely unchanged.

Jets occur in ~ 80 per cent of weak (FR-I) radio galaxies and are generally two-sided, as in Figure 9.3. They are detected in only a small fraction of highly luminous radio galaxies. Recent observations by Fernini *et al.* (1993) of a sample of five FR-II radio galaxies with luminosity

$$L(5\,\mathrm{GHz}) \gtrsim 4.5 \times 10^{26} h_{100}^{-2}\,\mathrm{W\,Hz^{-1}\,str^{-1}} \tag{9.3}$$

led to the definite detection of a radio jet in only one case, which is consistent with

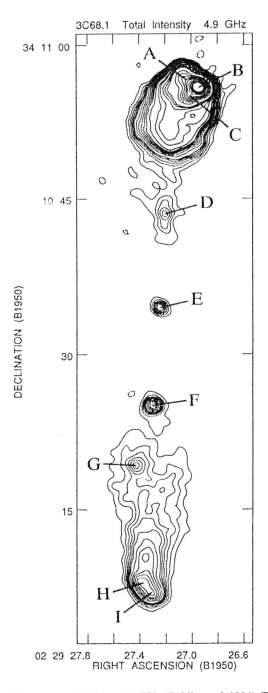

Fig. 9.9. A VLA map of the quasar 3C 68.1 at 4.9 GHz (Bridle *et al.* 1994). The angular resolution is 1.1 arcsec FWHM. E is the core and B the hotspot. A jet is defined by the features D and C along with the ridge that joins them. F defines the candidate counterjet. Other labelled features are described in the text. The figure is reproduced from Bridle *et al.* (1994).

the ~ 10 per cent detectability found by Bridle and Perley (1984). In the observations
of 30 FR-II galaxies that are far less luminous than the galaxies used by Fernini *et al.*
(1993), Black (1993) has found radio jets in ~ 57 per cent of the sample. This detection
rate is intermediate between that for FR-I galaxies and that for highly luminous FR-II
galaxies. Jets are found in a large fraction of radio quasars (40–70 per cent as quoted
in Bridle and Perley 1984). Recently Bridle *et al.* (1994) have found jets in all the 12
luminous quasars they observed with the VLA with high dynamic range and angular
resolution. Jets in FR-II radio galaxies are often one-sided, and those in quasars have
always been found to be one-sided, but Bridle *et al.* have been able to identify *candidate*
counterjets in some cases (see below).

9.4.2   Large scale jet properties

The nature of large scale radio jets has been reviewed in detail by Bridle and Perley
(1984), Muxlow and Garrington (1991) and Laing (1993). We will now sketch some of
their important observational properties.

Radio spectra   Large scale jets have steep radio spectra, with spectral index in
the interval 0.5–0.9 in most cases. The spectral index is nearly constant along the jet. In
the well-studied jet of M87 (Biretta and Meisenheimer 1993), the large scale structure
has the spectral index $\alpha \simeq 0.5$ throughout its length, with some evidence of steepening
in the weaker knots and in inter-knot regions. Beyond the jet the spectrum steepens
appreciably, with $\alpha \simeq 0.65$.

Magnetic field and polarization   The radio emission from jets at centimetre
wavelengths shows polarization to a level of up to ~ 40 per cent, with values exceeding
~ 50 per cent found in some regions. We have seen in Section 3.4 that the plane of
polarization is normal to the projection of the magnetic field on the sky, $\mathbf{B}_a$. The high
level of polarization means that $\mathbf{B}_a$ is well ordered in jets and it can be mapped using
multifrequency polarization measurements.

It is found that in sources with

$$L(178 \text{ MHz}) \lesssim 2 \times 10^{25} h_{100}^{-2} \text{ W Hz}^{-1} \text{ str}^{-1}, \tag{9.4}$$

i.e., in FR-I sources where the jets are predominantly two-sided, close to the core the
field is parallel to the jet axis and is indicated by $B_\parallel$. After ~ 10 per cent of the jet
length is covered the field becomes perpendicular to the axis and is indicated by $B_\perp$.
Sometimes the field in such jets is in the $B_\perp$ configuration on the axis, but becomes $B_\parallel$
at the periphery. Such a field is indicated by $B_{\perp-\parallel}$. In luminous sources with one-sided
jets (FR-II), the projected magnetic field has the $B_\parallel$ configuration over the whole length
of the jet. The field configuration can change where a jet bends, inside a knot along a
jet or within a hotspot at the end of a jet.

The projected magnetic field configuration in the twin jets of the FR-I radio galaxy

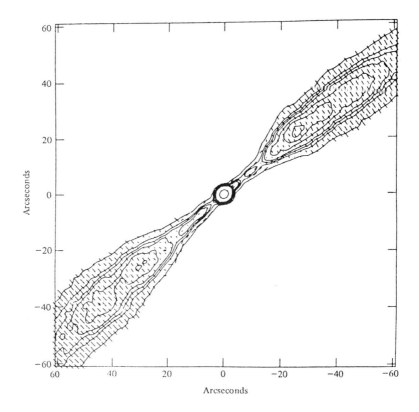

Fig. 9.10. VLA map of the FR-I galaxy IC 4296. The short lines represent the projected magnetic field, with a length of 1 arcsec indicating 22.1 per cent polarization. The field is seen to change configuration from $B_\parallel$ close to the nucleus to $B_\perp$ at $\sim$ 15 arcsec from it. Reproduced from Killeen, Bicknell and Ekers (1986).

IC 4296 is shown in Figure 9.10 (Killeen, Bicknell and Ekers 1986). The field, indicated by the short lines in the figure, is seen to be $B_\parallel$ close to the nucleus but changes to $B_\perp$ at $\sim$ 15 arcsec ($2.6 h_{100}^{-1}$ kpc) from the core. At this point the jet also flares substantially, the rate of change of opening angle increasing from < 0.1 to ~ 0.45. At 120 arcsec from the core the field changes to $B_{\perp-\parallel}$ (this region is not shown in the figure).

An example of the field configuration in a highly luminous FR-II quasar is shown in Figure 9.11, which is a polarization map of the quasar 3C 334, obtained from VLA A- and B-array observations by Bridle *et al.* (1994). The short lines in the figure indicate the degree of polarization and the orientation of the electric field vector **E** (i.e., the plane of polarization). In this convention the projected magnetic field is everywhere perpendicular to the lines, which is different from the convention in Figure 9.10. This jet is one-sided, the magnetic field being $B_\parallel$. The degree of linear polarization decreases from $\sim$ 40 per cent near the base of the jet to 20–30 per cent near the outer regions. Some features show polarization > 70 per cent.

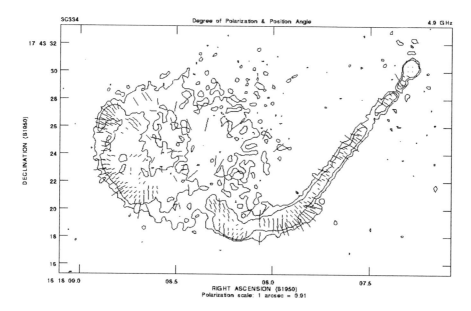

Fig. 9.11. A VLA map of the quasar 3C 334, showing the polarization of the electric field. The short lines indicate the projected electric field **E**, a length of 1 arcsec indicating 91 per cent polarization. The projected magnetic field at any point is perpendicular to the electric field. The magnetic field in the jet is seen to be $B_{\parallel}$, even where it bends. Figure kindly provided by Alan Bridle.

Size and curvature Jets at the two extremes of the radio-loud population, i.e., weak FR-I galaxies and highly luminous core-dominated quasars, generally have short lengths, only ~ 10 per cent of sources having jets longer than 40 kpc. Jets in sources of intermediate luminosity are longer, with over 50 per cent having lengths exceeding 40 kpc (Bridle and Perley 1984). In relativistic beaming models, core-dominated sources have one of their jets beamed towards the observer, which leads to a perceived foreshortening of the jet. But in the case of weak sources the jets are not expected to be biased in their orientation, so that the short lengths are likely to be intrinsic. Jets tend to be curved in the case of weak galaxies as well as that of the most powerful ones. In the former the curvature leads to a head–tail structure and is likely to be due to the ram pressure of the gas in the cluster environment in which FR-I galaxies tend to be located (see the discussion in Section 9.3). In the case of the core-dominated sources, any small curvature in the jet path will be exaggerated if the jet is indeed beamed towards the observer.

Collimation The extent to which jets are collimated has an important bearing on jet physics. Collimation is parameterized by the rate of spreading with distance from the core; this is given by $d\Phi/d\Theta$, where Φ is the FWHM of the jet, obtained after deconvolution to account for the point spread function, and Θ is the angular separation from the core.

In weak galaxies jets start off being well collimated, with $d\Phi/d\Theta < 0.1$, but flare with $d\Phi/d\Theta \sim 0.25$–0.6 (Bridle and Perley 1984) at distances of 1–10 kpc from the core (see Figure 9.10). At greater distances the jets may recollimate. Jets in highly luminous radio galaxies and quasars are much better collimated, with little or or no systematic expansion in many cases. The hotspots found in powerful galaxies subtend small angles at the core ($\lesssim 1$ deg), which is consistent with little jet spreading.

Sidedness   One of the most interesting properties of large scale jets is that they tend to be one-sided in powerful sources, even when the rest of the large scale structure shows reasonable symmetry in size and luminosity on either side of the core. Bridle and Perley (1984) have quantified jet sidedness by introducing a parameter $\mathscr{S}$ that is the ratio of the intensities of the brighter and fainter jets measured at the same distance from the core. Jets are said to be *one-sided* when $\mathscr{S} \gtrsim 4$, and *two-sided* when $\mathscr{S} < 4$ everywhere. There is no physical significance to the limiting value $\mathscr{S} = 4$; Bridle and Perley chose it as a reasonable point of separation, which divided the sample of jets available to them into two more or less equally numerous groups.

We have seen that the jets in weak (FR-I) galaxies are generally two-sided. However, they often begin as one-sided close to the core, but after a few kiloparsec become two-sided. The one-sidedness is confined to less than ~ 10 per cent of their length. Jets in the more powerful radio galaxies (FR-II) are often one-sided, but several examples of two-sided jets in powerful radio galaxies are known, an example being Cygnus A. Such *counterjets* usually have only a few per cent of the luminosity of the main jet. Jets in quasars are always one-sided. In their very sensitive and high angular resolution observations of a set of 12 highly luminous FR-II quasars, Bridle *et al.* (1994) failed to detect a single counterjet. Candidate counterjets have however been identified in several cases, on the basis of features found at positions which are thought to be natural for the course of a putative counterjet (see the discussion on Figure 9.9 given above). Even if these candidates are confirmed as jets, it is clear that there is a real asymmetry in jet luminosity in quasars and luminous radio galaxies. We will discuss possible reasons for this asymmetry in Section 9.6.

9.5   Compact sources and jets

Compact radio cores are found in a large majority of radio galaxies and in all quasars. In the former the core luminosity ranges from a fraction of 1 per cent to ~ 10 per cent, while in quasars the core contribution can be almost the whole radio luminosity of the source. While cores are (by definiton) not resolved on the arcsecond scale, they do have discernible structure when observed on the ~ 1 milliarcsec scale using VLBI. Dominant cores can be broadly classified into the following types.

Core-jet sources   These have a strong flat spectrum core, unresolved at the ~ 1 milliarcsec scale, and in most cases a one-sided steep spectrum jet on the parsec scale.

Compact steep spectrum (CSS) sources   These have small apparent linear sizes $\lesssim 15h_{100}^{-1}$ kpc, and yet they have steep radio spectra, with spectral index $\alpha > 0.5$, which shows that they are not core-dominated. CSS sources can be found in radio galaxies as well as quasars, and form 15–30 per cent of radio sources in catalogues (see Fanti *et al.* 1990, and Dallacasa *et al.* 1993 for reviews). Bright jets are common in CSS sources and have properties similar to those in the larger-sized radio galaxies and quasars. A majority of the jets are one-sided (Spencer and Akujor 1992) and superluminal motion has been observed in some cases. Observations of nearby CSS sources show good correlation between their complex radio structure and optical emission line gas, suggesting strong interactions between jets feeding the radio source and the gas in the ambient medium. The small size of the CSS sources could be explained if (1) these are young sources (or old sources restarting after a period of quiescence) which will later grow to a larger size or (2) the jets are trapped by a dense gas.

Compact double sources   These were first identified as a class by Phillips and Mutel (1982) and are dominated by two separate regions of emission, of comparable flux density, on the parsec scale. They usually have low variability and polarization, and lack lobe emission and superluminal motion. Some of the compact doubles in fact have been found to have a linear triple structure (see Conway *et al.* 1992).

Pearson and Readhead (1988) have carried out the morphological classification of 45 detected parsec scale sources in a sample of 65 objects. They found 25 could be classified as very compact or core-jet type, seven as CSS sources and six as compact doubles; six remained unidentified. The cores of lobe-dominated sources are weak and therefore difficult to study on a small scale, but the examples that are known show a core-jet structure (see Pearson 1990 for references).

9.5.1 *Parsec scale radio jets*

Parsec scale jets were first discovered in the late 1970s when it became possible to make VLBI images with a resolution of ~ 1 milliarcsec using closure phase techniques. While the first maps had only limited dynamic range, the best VLBI images available now reach the quality obtainable with the VLA.

Parsec scale jets present themselves as a series of knots that extend more or less in the direction of the kiloparsec scale jet (when the latter is detected). Jets that have been observed with a high dynamic range show continuous emission with knots superposed on it. The knots can define a linear or a curved structure. The small and large scale jets are usually aligned, but large curvature is sometimes found. New components are observed to arise close to the unresolved core and the components tend to be more extended away from the core.

The apparent speed at which the knots separate from the core often exceeds the speed of light, and such motion is said to be superluminal. Different components in a given jet can move with different velocities and can accelerate or decelerate as they

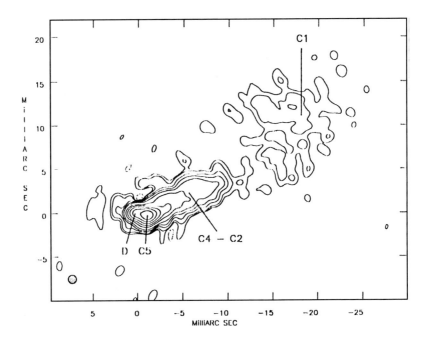

Fig. 9.12. A VLBI image of 3C 345 at 4990 MHz with 1 milliarcsec resolution. The beam shape is shown by the small circle in the lower left hand corner. An unresolved core is at D. The other features are described in the text. Reproduced from Unwin and Wehrle (1992).

move out, and successive components need not follow the same track. Superluminal sources always expand, i.e., the distance between a knot and the core always increases. The complexity of knot motion implies that we are probably not seeing the physical velocity of individual components, but the phase velocity of brightness peaks in jets.

An excellent example of the core-jet type of compact source is the quasar 3C 345. A VLBI image of this quasar made at 5 GHz with 1 milliarcsec resolution (Unwin and Wehrle 1992) is shown in Figure 9.12. The core at D is unresolved at the highest frequencies and has been determined to be stationary with reference to the more distant compact quasar NRAO 512, with a fractional uncertainty of two parts in 10^8 in the angular coordinates. The component C5 emerged from the core in 1984; the older components lie to the west of it. Components C2–C4, which were distinct in images made at earlier epochs, are now difficult to separate, either because the jet has changed over several years or because the earlier data failed to detect the faint emission from the region between the components. The apparent speed of the components relative to the core is in the range 1.4–9.5 ch_{100}^{-1} ($q_0 = 0.5$). The speed appears to increase continuously with separation from the core, which could be because of acceleration or because the direction of the motion relative to the line of sight is changing. On the smallest scales, successive components do not appear to follow the same trajectory. No milliarcsecond counterjet has so far been detected in this source.

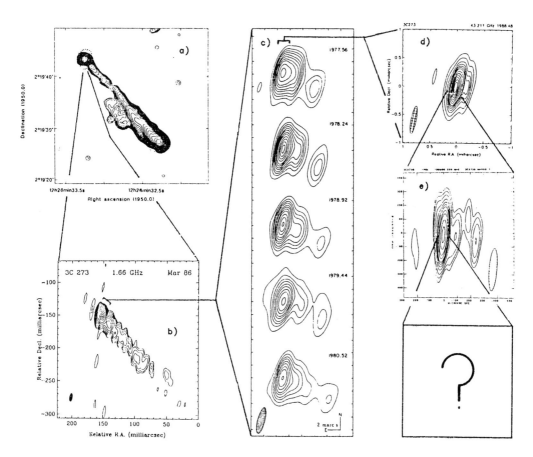

Fig. 9.13. A composite of images of the quasar 3C 273 made at different angular resolutions. Top left, Merlin image at 408 MHz at a resolution of 1 arcsec (Davis, Muxlow and Conway 1985) showing a core and jet. Bottom left, VLBI image at 1.66 GHz with 4 milliarcsec resolution (Unwin and Davis 1988). Middle, VLBI images at 5 GHz with 1 milliarcsec resolution made at various epochs (Pearson and Readhead 1987). The change in the position of some of the features is obvious. Top right, VLBI image at 43 GHz and 0.1 milliarcsec resolution (Krichbaum *et al.* 1993). Middle right, VLBI image at 100 GHz using a beam of 280×50 microarcsec^2 FWHM (Baath *et al.* 1991). The unresolved core at 43 GHz is seen to be made up of a number of features at the higher resolution. Reproduced from Schilizzi (1992).

Another well-studied quasar is 3C 273, which has a redshift of 0.158 and was the first object in which superluminal motion was detected (Whitney *et al.* 1971, Cohen *et al.* 1971). A composite made by Schilizzi (1992) with images at various frequencies is shown in Figure 9.13. When observed with ~ 1 arcsec resolution (Davis, Muxlow and Conway 1985) the quasar is found to have a core-jet structure with a bright flat spectrum core and a jet that extends to 22 arcsec ($\sim 40 h_{100}^{-1}$ kpc). Furthermore, it extends beyond an optical jet at the same position angle and ends in a bright head. There is no counterjet detected at 408 MHz and the observed jet-head to counterjet-head brightness ratio is > 5500 : 1. The ridge of the jet shows a wiggle whose wavelength decreases by a factor of ∼ six along the length of the jet. On the VLBI scale there is a jet that curves through 20 deg in the first 10 milliarcsec and then extends continuously for ∼ 150 milliarcsec, closely aligned with the large scale jet. A number of knots in the jet move away from the core (which is assumed to be stationary) at superluminal speeds in the range $5-7h_{100}^{-1}$ kpc. When observed at 5 and 1.67 GHz the superluminal motion is seen to extend to at least 2.5 milliarcsec. Superluminal components appear to emerge from the core at the time of increasing flux at millimetre wavelengths. There are wiggles in the small scale jet similar to those at larger scales. There is no parsec scale counterjet detected on scales of a few tens of parsec at 1.67 GHz for a dynamic range of ∼ 1000. Structure on scales smaller than 300 microarcsec are seen at 100 GHz. This corresponds to a linear size of $1.5 \times 10^{18} h_{100}^{-1}$ cm.

9.5.2   *Superluminal motion*

We have studied in the previous section some examples of apparent superluminal motion and have seen in Section 3.8 how it can be interpreted in terms of relativistic bulk motion in the source. There are several mechanisms which can give rise to apparent component speeds that exceed the speed of light c (see Blandford, Mckee and Rees 1977 for a review) and some of these could very well be operating in some cases. However, with the availability of extensive data on superluminal motion, jet sidedness and related phenomena, relativistic motion has emerged as the clear favourite in interpretation, even though little direct evidence is available of actual motion in the source at relativistic speeds.

Superluminal motion was observed soon after the development of VLBI techniques (see Pearson and Zensus 1987 for a brief historical review). The first results, on 3C 273 by Cohen *et al.* (1971) and on 3C 279 by Whitney *et al.* (1971) were obtained using model fitting to the brightness distributions because of the limited data available. The data were consistent with models that involved expansion of the source at transverse speeds exceeding c. However, this interpretation was not the only one consistent with the data, and the brightness changes could also be understood in terms of intensity fluctuations in a complex stationary system. With the availability of better coverage due to an increased number of telescopes in the VLBI array and the development of techniques to recover phase information, it is now possible to make actual maps of the brightness distribution of the sources. The separation of components with time can be unambiguously determined on these images and superluminal motion has been detected in a large number of sources, as we shall see below.

Apparent superluminal motion is common in the nuclei of quasars and BL Lacs and has also been observed in radio galaxies. It is found especially often in sources with core-jet morphology and has been detected in compact steep spectrum sources, but not in compact doubles. Superluminal motion has been found in at least four out of eight flat spectrum sources with 5 GHz flux greater than 1 Jy studied by Witzel *et al.* (1988). Cohen (1990) has detected superluminal motion in nine out of 11 of the brightest sources from a sample of 10.7 GHz strong, core-dominated sources that are variable. While these numbers are small, they clearly indicate the ubiquity of the phenomenon.

A list of 66 extragalactic sources for which multi-epoch VLBI observations are available has been made by Vermeulen and Cohen (1994). Many of the sources in the list contain more than one component with measured proper motion. While this list does not constitute a homogeneous, complete sample of sources with proper motion measurements, it is a comprehensive presentation of the data available in the literature in 1994. The list contains seven radio galaxies, 11 BL Lacs, 47 quasars and one empty field (NRAO 150). Amongst the quasars are 25 selected at high radio frequency; the emission in these cases is dominated by the flat spectrum cores. In the beaming model the core emission is amplified due to relativistic motion, so that a flux-limited sample will preferentially contain sources that are beamed towards the observer. There are 13 quasars in the sample which were selected at low frequency and have dominant lobe emission that is expected to be independent of any relativistic beaming effects. There is no bias towards brighter cores, i.e., jets directed towards the observer, in these cases. The jets will therefore be oriented at random relative to an observer's line of sight, and in the framework of the relativistic beaming model one expects to see slower transverse motion in them. There are nine quasars that have compact morphologies, with spectra which peak in the GHz range.

The distribution of the apparent transverse speed β_a for the different classes is shown in Figure 9.14. The GHz-peaked quasars do not show significant proper motion. Core selected quasars have higher values of β_a than lobe-selected quasars, which is consistent with the former having jets beamed towards the observer. BL Lac objects appear to have lower apparent speeds than core-dominated quasars as a class, but the number in each bin is too small for a meaningful statistical comparison. Lower speeds in a BL Lac object would be consistent with its being beamed very close to the line of sight, which reduces the apparent speed because then the angle of inclination is less than the angle $\sin^{-1}(1/\gamma)$ at which the maximum apparent speed occurs. The small angle, however, leads to enhancement of the flux and rapid variability, making a *blazar* out of the object. The calculated speeds used in the diagram of course depend on the value of q_0 assumed, but it is only the distribution of the core-dominated quasars that is significantly affected when q_0 is changed. The reason is that these quasars have the highest redshifts in the sample.

The μz-diagram   We show in Figure 9.15 the joint distribution of the proper motion and redshift for the sample of 66 sources. We have followed Vermeulen and Cohen (1994) in including only the brightest component from multicomponent sources. It is clear that the proper motion decreases with increasing redshift and there is a reasonably well-defined envelope which is the locus of the maximum proper motion as a function of redshift. This can be readily understood in the bulk relativistic motion model. We have seen in subsection 3.8.1 that the maximum apparent speed of

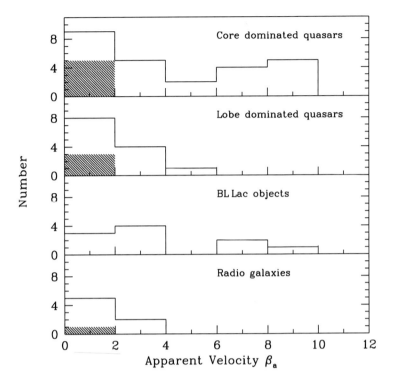

Fig. 9.14. The distribution of observed proper motion in core- and lobe-dominated quasars, BL Lac objects and radio galaxies. The shaded areas indicate objects with upper limits and negative values of β_a that are significantly different from $\beta_a = 0$. The values $q_0 = 0.5$ and $h_{100} = 1$ are used. After Vermeulen and Cohen (1994).

a blob moving with Lorentz factor γ is $\beta_a = \gamma\beta$ and it occurs when the direction of motion makes an angle $\psi = \arcsin \gamma^{-1}$ with the line of sight. Using Equation 3.89, this corresponds to a maximum proper motion

$$\mu_{\text{max}} = \frac{2.12 \times 10^{-2} q_0^2 \gamma h_{100} \beta (1 + z)}{q_0 z + (q_0 - 1)(\sqrt{1 + 2zq_0} - 1)} \qquad (9.5)$$

at redshift z. The curve for $\gamma h_{100} = 9, q_0 = 0$ is shown in Figure 9.15, and is seen to provide a reasonable envelope to the observed distribution of μ. The curve for $\gamma h_{100} = 6$ is also shown. In practice there will be a distribution of Lorentz factors, which may even depend on the luminosity and other properties of the sources, and therefore the envelope is not expected to be sharply defined. Cohen *et al.* (1988) have considered predictions for the envelope shape made by various models that seek to explain superluminal motion without invoking relativistic bulk motion. They find that the local quasar theory of Terrell (1966, see Chapter 15), in which quasars are shot out from the Galactic centre at one time and the fastest quasars (i.e., those with the highest redshift) are furthest away, leads to an envelope shape that conforms reasonably with

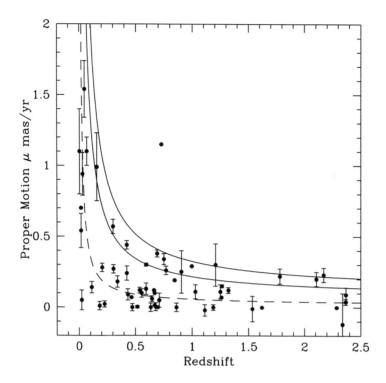

Fig. 9.15. The distribution of observed proper motion in the μz-plane. In the cases of sources with multiple components, the motion of the brightest component is used. The two solid lines are maximum proper motion curves for $\gamma h_{100} = 9$ and 6 respectively. The broken curve is the 'median line' described in the text. After Vermeulen and Cohen (1994).

the observed maximum proper motion. The other models all make predictions that are at variance with the observations, or require rather low ($h_{100} < 0.4$) or high ($h_{100} > 3$) values of the Hubble constant.

Cohen *et al.* (1988) have used the distribution of points in the μz-diagram to argue that the sources must be predominantly oriented close to the line of sight of the observer. If the sources were oriented at random, the probability of having the line of sight angle $> \psi$ would be $p(> \psi) \propto 1 - \cos \psi$, so that half the sources would be expected to have $\psi > 60$ deg. The broken line in Figure 9.15 is the locus of the proper motion for sources with $\gamma = 4$, $h_{100} = 1$ and $\psi = 60$ deg, with the apparent speed obtained from Equation 3.91. Assuming that all sources have Lorentz factor $\gamma = 4$ and are randomly oriented, those with $\psi > 60$ deg should have a proper motion below this line. Sources with $\psi < 60$ deg, except those in a small cone around the line of sight, should be above the line. It is seen from the figure that in fact many more sources are on or above the 'median line' than below it. The position of the line does not change much for $\gamma > 4$, and for smaller values of γ the lines are lower than the one for $\gamma = 4$.

Reducing the value of h_{100} or q_0 again lowers the line. The asymmetric distribution of points around the 'median line' therefore is rather robust, which means that the motion of the sources is preferentially oriented towards the observer. Cohen *et al.* attribute this to Doppler boosting of the flux, which would make the cores brighter at small angles and lead to such sources being favoured in flux-limited surveys.

When the direction of motion is transverse to the line of sight, i.e., the angle made with the line of sight is $\psi = 90\,\mathrm{deg}$, the apparent transverse speed coincides with the true speed: $\beta_a = \beta$. As ψ decreases the apparent speed increases until a maximum is reached at $\sin\psi = 1/\gamma$. For smaller angles, β_a decreases to 0, which occurs when the motion is directed along the line of sight. For highly relativistic motion the angle at which the maximum occurs is very small, so that for all directions of motion other than those in a small cone around the line of sight, we have $\beta_a \geq \beta$. The proper motion $\mu_\perp(z)$ corresponding to transverse motion can be obtained from Equation 3.89 by simply setting $\beta_a = \beta$ on the left hand side. Assuming that all sources have the same Lorentz factor, the ratio of the limiting proper motion for sources at redshifts z_1 and z_2 is (Pelletier and Rolland 1989)

$$\frac{\mu_\perp(z_1)}{\mu_\perp(z_2)} = \left(\frac{1+z_1}{1+z_2}\right)\frac{q_0 z_2 + (q_0-1)(\sqrt{1+2z_2 q_0}-1)}{q_0 z_1 + (q_0-1)(\sqrt{1+2z_1 q_0}-1)}. \tag{9.6}$$

The ratio on the left hand side of this equation is observationally determined, so in principle it can be used to obtain q_0, independently of the distance scale problems encountered in the usual methods for its determination. The question now is how does one identify the sources moving transversely to the line of sight? Since most sources have $\beta \geq \beta_a$, this can be done by choosing sources along the minimum proper motion envelope in the μz-diagram. For this of course it is necessary to have a sufficient number of sources available at all redshifts so that the envelope is robustly defined. However, there are only a few sources at high redshifts, while at low redshifts the right hand side of the above equation is not sensitively dependent on the value of q_0. Another difficulty is the assumption that β has the same value for all sources. Consider a source with two-sided jets that move away from a stationary core, the ratio of the speeds of the advancing and receding components being $\lesssim 2$. Such a source cannot be oriented close to the line of sight, and for large angles, the mean of the apparent speeds of the two components of the source is within a few per cent of c if $\gamma \geq 5$ (Pelletier and Rolland 1989). One can therefore set $\beta_a = \beta \simeq 1$ and the cosmological parameters are determined with reasonable accuracy even if the sources are not exactly transverse and the value of β varies from one source to another.

The $\beta_a z$-diagram A plot of apparent transverse speed against redshift also shows an upper envelope defined mostly by core-dominated quasars. For speeds obtained with $q_0 = 0.5$ the envelope is flat, while for $q_0 = 0.05$ it rises somewhat with redshift. This result is uncertain because of the rather sparse data at the present time, but with more points could be used to choose between values of q_0.

Beaming models Vermeulen and Cohen (1994) have considered detailed beaming model fits to the data using the apparent transverse speeds described above. Only
in the case of core selected quasars does the sample have a sufficient number of objects
for one to be able to distinguish even broadly between different possibilities. For a
random distribution of jet speeds relative to the observer, the distribution of β_a is given
by Equation 3.94. If the core fluxes are Doppler boosted, the core selected sample will
have an excess of sources that are beamed towards the observer, relative to a sample
which has random orientation. There will therefore be an excess of high β_a over the
prediction of Equation 3.94. The distribution will depend on the Lorentz factor γ_b of
the bulk relativistic motion as well as on the Lorentz factor γ_p of the pattern speed,
which determines the apparent transverse speed (we have not distinguished between the
two Lorentz factors in most of the above discussion, and have simply used a single γ).
Vermeulen and Cohen have derived the distribution of β_a, allowing for Doppler boosting and a difference between the pattern speed and bulk motion. Comparing the model
distribution with the observed data, they find that if all core selected quasars have the
same Lorentz factor, then the pattern speed must either be slower, with the ratio of
pattern speed to bulk motion speed $r \simeq 0.5$, or considerably faster, $r \simeq 5$, than the speed
of bulk motion. However, there is another fit with $r = 1$ but with a broad distribution
of γ_b. Many intermediate combinations are possible, but *not* $r = 1$ with a single γ_b.

9.6   One-sidedness of jets

We have seen in Subsections 9.4.1 and 9.5.1 that large scale as well as small scale jets
tend to be one-sided in powerful radio galaxies (mainly FR-II), while they are always
one-sided in quasars, except for the candidate counterjets found by Bridle *et al.* (1994).
A plot of the counterjet prominence, which is defined as the ratio of the integrated flux
density in a counterjet to that in the extended source, including the jet and counterjet,
is shown in Figure 9.16 (Bridle 1990). It is seen that the flux from most counterjets in
sources with $L(1465\,\text{MHz}) > 10^{26} h_{100}^{-2}\,\text{W Hz}^{-1}$ is only a small percentage of the total
extended flux, which makes the counterjets difficult to detect given the limited dynamic
range. Even with highly sensitive observations it is not easy to detect faint jets against
the background of the lobe emission, which shows complex small scale structure at
faint levels. Jets in sources that do not show classical FR-I or FR-II structures and
have a complex morphology tend to be very prominent. Bridle (1990) has pointed out
that counterjets in such sources do not show an enhanced prominence.
     The standard model of a quasar (see Chapter 5) or AGN involves a supermassive
black hole which accretes matter through an accretion disk, energy being released
as radiation and in the form of twin beams of relativistic charged particles. Such a
model has no natural mechanism for the beam to be one-sided, and therefore different
possibilities have to be considered. The three basic possibilities are as follows.

- *Asymmetric energy transport.* In this case, the beam is taken to be inherently
  one-sided, with much less energy transported on the *dark side* than on the jetted

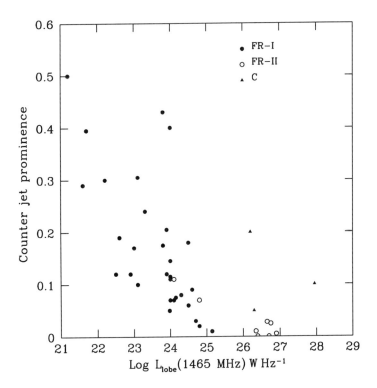

Fig. 9.16. Counterjet prominence shown as a function of lobe power in FR-I and FR-II galaxies, and sources with complex morphology. After Bridle (1990).

side. The model requires a mechanism for shutting off the beam on one side of the nucleus; this goes beyond the prescription of the simple standard model. However, the two-sided lobe structure that is present in most extended sources requires that the beam flips from one side to the other periodically.

- *Asymmetric dissipation of energy.* Here it is assumed that energy leaves the quasar or AGN in the form of a twin beam and that the dissipation of energy is asymmetric in the two directions. Higher radio power is therefore produced along the bright jet than along the dark one, but similar quantities of energy are transported to the opposite lobes. The asymmetric dissipation can occur because of differences in the properties of the beams, such as the number of relativistic particles or the magnitude and configuration of the magnetic field. Asymmetry can also come about because of differences in the ambient medium on the two sides, which would lead to differing energy dissipation in the interaction between the beams and the medium. Since the one-sidedness of parsec scale and kiloparsec scale jets is correlated, any asymmetry in the medium has to extend from the innermost regions close to the nucleus to the furthest reaches.

- *Relativistic bulk motion.* We saw in subsection 3.8.3 that the observed flux from matter moving relativistically very close to the line of sight of an observer can be very substantially boosted. The ratio of the flux from a jet beamed towards an observer to that beamed away is shown in Figure 3.8, and even for modest values of the Lorentz factor and a beaming angle $\lesssim 20\,\mathrm{deg}$ the ratio can be $\gtrsim 10^3$. Owing to the limited dynamic range of observations, jets will appear to be one-sided in these cases, unless a special effort is made to detect them.

Models that involve asymmetric energy transport or flip-flops from one side to another are virtually ruled out because of the existence of compact hotspots on the unjetted side in some sources. The synchrotron lifetime in the hotspots is shorter than the light travel time from the nucleus to the hotspot, which means that there must be a continuous replenishment of the electron energy. The radio galaxy M87 is classified as an FR-I source, but has a large scale one-sided jet (see Biretta 1993 for a detailed description of the jet and a review of previous work). There is no jet observed on the opposite side. However, a hotspot has been discovered at optical wavelengths at a distance 2 kpc from the nucleus on the unjetted side. The spot is located precisely where the end of a counterjet, similar in size and structure to the observed jet, would be located (Sparks *et al.* 1992, Stiavelli *et al.* 1992). The optical radiation has polarization characteristics similar to those of the radio emission from the site, and is very likely to be synchrotron emission. The lifetime of the synchrotron electrons emitting the optical radiation is $\lesssim 1600\,\mathrm{yr}$, which is significantly less than the light travel time to the spot. Even granting the uncertainty in the estimation of the synchrotron lifetime, its short value implies that the hotspot must have been last energized in the relatively recent past. If the one-sidedness of the jet is intrinsic, the beam must have flipped to the other side $\lesssim 2000\,\mathrm{yr}$ ago, which is not consistent with the large size of the jet visible on the other side. It seems likely therefore that a counterjet continuously feeds the hotspot, but is not seen because of either asymmetric dissipation or beaming effects.

In their detailed VLA study of jets in 12 radio quasars, Bridle *et al.* (1994) have obtained a number of correlations that have implications for the origin of one-sidedness. The correlations are based on a small sample and therefore the formal statistical significance of the results is not high; nevertheless the results, many of which are supported by data gathered over the years by a number of workers, provide valuable insight into the circumstances.

In the sample of Bridle *et al.* (1994), small scale and large scale features have the same sidedness. A correlation is found between the prominence[1] of the milliarcsecond features and that of straight segments of jets. An interpretation of this is that flows on both scales are relativistic and are simultaneously beamed towards the observer. In the beaming model the slope of the correlation depends on the Lorentz factors of the beams on the two scales, and Bridle *et al.* have argued that the low observed slope implies that

---

[1] The prominence of a feature is defined as the ratio of the integrated flux density in it to the integrated flux density in a suitably defined lobe.

the beam has slowed down from the parsec to the kiloparsec scale, $\gamma_j < \gamma_c$ and the jet Lorentz factor $\gamma_j \simeq 2$. In asymmetric dissipation models a correlation between features on the small and large scales requires that the mechanism that produces the asymmetry must operate on all scales. One way of doing this would be to have an asymmetry in the magnetic fields or the spectrum of relativistic particles on the parsec scale, and to propagate this to larger scales. Neither the theoretical models nor the observational data are good enough at the present time to be able to distinguish between these possibilities.

Bridle *et al.* have identified counterjet candidates in seven quasars (see subsection 9.4.1). They find that the prominence of the candidates is enhanced significantly by jet bending and that there are no counterjet candidates opposite long straight segments of jets. There are suggestions in the data that a jet bends more readily further from the central source and that its ability to form hotspots decreases if there is abrupt bending. In the beaming models these features can be explained by invoking a velocity structures across the jet, so that the Lorentz factors are different at different distances from the centre of the jet, and by allowing the outer layer to decelerate as it interacts with the environment. The data can also be accommodated in asymmetric dissipation models, and in fact what is required is a combination of beaming with a difference in the beam and environmental properties on the two sides. But this introduces a large number of parameters and it is again not possible to distinguish between different models.

9.6.1   *Depolarization asymmetry*

It was noticed by Laing (1988), Garrington *et al.* (1988) and Garrington, Conway and Leahy (1991) that there is an asymmetry in the polarization properties of the lobes of bright FR-II sources with one-sided large scale jets. It is found in such sources that depolarization with increasing wavelength is much higher in the lobe on the side away from the one-sided jet. This is seen clearly in the polarization maps of the quasar 3C 133 shown in Figure 9.17. The figure at the top is the polarization map at 5 GHz, which shows polarization in both the lobes. The one-sided jet (not clearly visible in the polarization maps shown) points to the left of the figure. In the 1.5 GHz map below, it is clearly seen that the degree of depolarization in the lobe not on the same side as the jet is very much lower than on the jetted side. This effect has been observed in a number of bright sources with one-sided jets. Garrington, Conway, and Leahy (1991) have considered the polarization properties of 69 FR-II sources, which consisted of 19 radio galaxies and 50 quasars, having one-sided jets. They find that 49 of these sources show greater depolarization on the side away from the observed jet. The asymmetry is found to be similar in radio galaxies and quasars.

The Laing–Garrington effect can be interpreted straightforwardly if the one-sidedness were due to relativistic beaming. In this case the jet would be beamed in a direction towards and close to the line of sight of the observer, while the counterjet would be beamed away, and the properties of the radio structure on the two sides would be

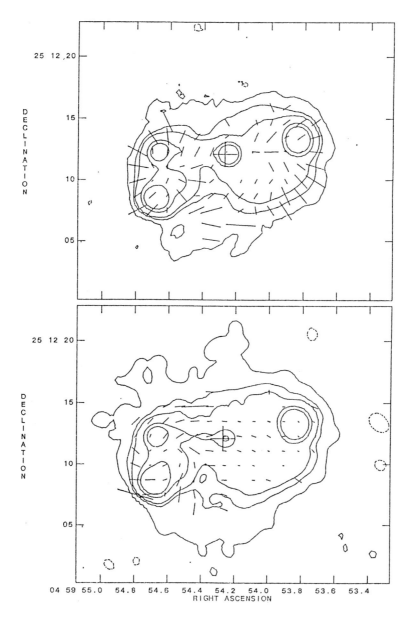

Fig. 9.17. Polarization maps of the quasar 3C 133: upper panel, at 5 GHz; lower panel, at 1.5 GHz. The lines show the direction of the electric vector and have lengths proportional to the fractional polarization. Reproduced from Scheuer (1987).

largely similar. The asymmetry could then be understood if the depolarization were due to differential Faraday rotation in a magneto-ionic screen surrounding the radio source (see subsection 3.4.1. The radio emission from the lobe on the side of the invisible jet, which points away from the observer, would have to traverse a greater distance through the medium, and therefore undergo greater depolarization. This interpretation requires that the matter in the large scale jets moves at relativistic speeds, and that the line of the jets is close to the observer's line of sight. If the jets are more or less transverse to the line of sight, the path lengths traversed are similar for the two lobes and there is no marked asymmetry.

If the jet one-sidedness is intrinsic, then the jet can be oriented at random relative to the observer. In this case the depolarization asymmetry cannot be attributed to different path lengths through a medium surrounding the radio source. A possibility then is that the depolarization is internal to the lobes, with the asymmetry arising from different lobe properties. Consider a medium, with uniform thermal electron density n and uniform magnetic field B, in the form of a slab of thickness L. Let ψ be the inclination of the magnetic field to the line of sight. The Faraday depth (see subsection 3.4.1) at a distance x from the edge towards the observer is $nBx \cos \psi$. The distance varies from 0 to L over the thickness of the slab, so that the mean value of the Faraday depth is $\Phi = 0.5nBL \cos \psi$ and the dispersion about this mean is $\Delta = |\Phi| / \sqrt{3}$. Garrington, Conway and Leahy (1991) have shown that this simple relation between the dispersion and the mean of the Faraday depth is preserved when the magnetic field is tangled on a length scale $\ll L$, except for a small change in the coefficient. A correlation between Φ and Δ should therefore be seen in the observations if the polarization is mainly internal. A complication here is that Faraday rotation due to our Galaxy dominates the measured values of Φ. But the Galactic contribution does not change much over the angular scale of a radio source, and can be eliminated by considering correlations between the differences $|\Phi_1 - \Phi_2|$ and $|\Delta_1 - \Delta_2|$ for the two components of each source. Garrington and Conway find that there is little or no correlation between the two quantities, and that there are many sources in which $|\Phi_1 - \Phi_2|$ exceeds $|\Delta_1 - \Delta_2|$. This behaviour makes it unlikely that the depolarization is internal, and the interpretation of external depolarization, which is consistent with the beaming hypothesis, is favoured.

It has been found that in radio galaxies the lobe closer to the nucleus shows stronger depolarization (see Laing 1993 for a brief review and references). Most galaxies used in these studies do not have jets so there is no orientation bias due to relativistic beaming. It has been found that the closer lobe often occurs on the side of the nucleus that shows more extended narrow line emission. The asymmetry in the distance of the lobes from the nucleus, and the depolarization asymmetry are therefore related to the different ambient matter densities on the two sides. In the unification scheme proposed by Barthel (1989) and others and discussed in Chapter 12, it is argued that radio galaxies and quasars come from a single population, with the former oriented closer to the plane of the sky than the quasars. In the case of radio galaxies the lobe depolarization asymmetry due to different environments dominates, while in the case of quasars, which have jets oriented close to the line of sight, the differences due to the different Faraday depths traversed through a medium drive the observed asymmetry.

9.6.2 *The linear size of beamed sources*

If apparent superluminal motion or flux enhancement leading to jet one-sidedness is due to relativistic beaming, the bulk flow has to be directed close to the line of sight. If the the direction of motion is aligned with the major axis of the radio source, then beamed sources should appear to be foreshortened relative to their unbeamed counterparts owing to projection effects.

Hough and Readhead (1989) have defined a complete sample of 28 radio quasars with double lobes. This is a subset, selected at 178 MHz, of all quasars from the flux-limited 3CR survey. At this low frequency the radio emission is not dominated by any beamed flux that may be present and the sample is expected to be free from orientation bias. We have shown in Figure 9.18 the distribution of the linear sizes of these sources, obtained from their largest angular sizes assuming $h_{100} = 1, q_0 = 0.5$. Superluminal motion has been observed in five of these source (Vermeulen and Cohen 1994), and these are indicated in Figure 9.18 by the hatching. Three of the five sources are amongst the largest sources in the sample. Now if the direction of relativistic motion in these sources is the same as the axis of the large scale structure, the observed size is an underestimate of the true linear size because of projection. The angle of inclination ψ that minimizes the bulk-flow Lorentz factor required to produce the observed value of β_a is $\psi = \mathrm{cosec}^{-1}\sqrt{1 + \beta_a^2}$ (see Section 3.8). It is conventional to use this angle to deproject the observed linear size of the source l_{obs} to obtain the intrinsic linear size $l = l_{obs}/\sin\psi$. The sizes obtained in this manner for the five sources are indicated in the lower panel in Figure 9.18 with left-inclined hatching. Also shown are the deprojected sizes of the five quasars with superluminal motion from Barthel *et al.* (1986), and the *observed* sizes of the two largest radio quasars known, 4C 34.47, which is $560h_{100}^{-1}$ kpc in extent, and 4C 74.46, which extends to $800h_{100}^{-1}$ kpc.

It is clear that in many observed cases the deprojected sizes turn out to be much larger than the sizes obtained in the complete sample of double-lobed quasars, which is expected to be free of any orientation bias. This is particularly the case with the two largest quasars. Superluminal motion has been observed in 4C 34.47 by Barthel *et al.* (1989) with $\beta_a = 2.5$, which gives $\psi = 21.8$ deg when the Lorentz factor of the flow is minimized. The corresponding deprojected size is 1.5 Mpc. In the case of 4C 74.26, a one-sided jet of length $\gtrsim 6h_{100}^{-1}$ pc has been observed (Pearson *et al.* 1992). This small scale jet is well aligned with a large jet of length $200h_{100}^{-1}$ kpc. The jet-to-counterjet flux ratio is ≥ 50, which leads to a line-of-sight inclination angle $\lesssim 45$ deg when Equation 3.107 is used.

The probability that a randomly oriented quasar is beamed within ψ of the line of sight is $1 - \cos\psi$. For 4C 34.47, with an angle of inclination of 21.8 deg, this is seven per cent. An alternative route to the angle of inclination is to maximize the

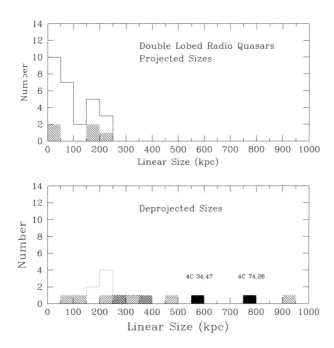

Fig. 9.18. Upper panel: linear size distribution for 3CR quasars from the sample of Hough and Readhead (1989). The hatched areas indicate the five quasars in which superluminal motion is detected. Lower panel: deprojected sizes for the five superluminal quasars in the upper panel are shown with left-inclined hatching. The right-inclined hatching indicates superluminal quasars from Barthel *et al.* (1986). The dotted lines show deprojected linear sizes of the Wardle and Potash (1984) quasars. The observed sizes of two of the largest quasars known are also indicated.

Lorentz factor, i.e., to set $\beta = 1$. In this case the angle has twice the value it does when the Lorentz factor is minimized. The unprojected length is then 800 kpc and the probability of beaming within the angle is 27 per cent. It follows that if the sources are oriented at random, then we should be seeing the many unbeamed counterparts of the superluminal sources. In the case where there are intrinsically symmetric jets on either side of the central source, for beaming directions within ∼ 15 deg of the plane of the sky, the observed jet-to-counterjet flux ratios will be < 4, i.e., two-sided jets should be seen. As we have noted in subsection 9.4.2, not a single case of a quasar with two-sided jets is known.

Wardle and Potash (1984) have found large scale one-sided jets in each of the eight largest sources from a complete sample of 4C quasars. The catalogue was selected at 178 MHz and is again expected to be free of orientation bias and has at least 32 double sources. If the sources all have the same intrinsic size, then the eight largest are distributed between the plane of the sky, $\psi = 90$ deg, and $\psi = \cos^{-1}(8/32) = 75.5$ deg. The median angle, which corresponds to the four brightest sources, is 82.8 deg. Wardle and Potash find that the angles do not change much for reasonable distributions of intrinsic linear sizes. On the beaming hypothesis, the ratio of jet-to-counterjet flux is given by Equation 3.106. For $\psi = 75.5$ deg, an assumed average jet radio spectral index $\alpha = 0.75$ and the extreme speed $\beta = 1$, the ratio is $F_{\rm in}/F_{\rm out} = 4$, which is an *upper limit* since the bulk flow speed will always be less than what we have assumed. If the median angle of 82.8 deg is used, the ratio obtained is 2. However, all eight sources have observed *lower limits* on the flux ratio that are *larger* than the upper limit obtained for the median angle, while four sources have flux ratios larger than the maximum upper limit of 4.1. The observed flux ratios therefore are not consistent with the beaming model if the sources in the complete sample are indeed oriented at random, with the largest sources in the plane of the sky. The flux ratio can be increased if the quasars are beamed close to the line of sight, but in that case the intrinsic sizes of the sources would again be very large. Assuming that the flux ratio is given by the observed lower limit and $\beta = 1$, we have used Equation 3.106 to obtain the angle of inclination ψ and to deproject the observed linear size. The resulting lower limits are shown by the dotted line in Figure 9.18. The lower limits on the linear sizes are again found to be at the upper end of the observed distribution.

Some caution has to be exercised in comparing the sizes of quasars from different samples, as the intrinsic distribution of sizes could depend upon quasar luminosity and redshift. Browne (1987) has considered the linear size distribution of a sample of quasars that show superluminal motion, with a control sample. The latter is constructed by choosing quasars, with extended radio structure, each of which have luminosity and redshift close to those of some quasar from the superluminal sample. Two quasars are chosen in this manner for each superluminal one. Comparison shows that the differences in size distribution are sensitively dependent on the value of the Hubble constant used and the angle used in deprojection.

The arguments above show that quasars with very large linear sizes and two-sided jets are required by the simple beaming models and the assumption of random orientation. It is a major challenge to the beaming models to be able to explain the absence of such quasars from the observed samples. There are several ways (which are somewhat contrived) in which excessive numbers of large quasars can be avoided without having to give up the beaming model altogether. We have seen in subsection 9.5.1 that misalignments have been observed between parsec scale and kiloparsec scale jets. If large misalignments are present, the large scale axis could be close to the plane of the sky even when the central source is beamed towards the observer, contradicting the prediction of many large sized quasars. It is also possible that beams are ejected in a wide cone from the central source, and we see only the part close to the line of sight. This would allow the large scale structure to be oriented at random even when the observed parsec scale structure is Doppler boosted. The wide beam, however, would have to be collimated to produce the large scale jets. An explanation that appears to be more natural than any of these is due to Barthel (1989), who has suggested that *all* radio quasars are beamed towards the observer, luminous FR-II galaxies being their unbeamed counterparts. In this model, which we will discuss further in Chapter 12, the missing quasars with large linear sizes would therefore simply be radio galaxies.

9.6.3   *The evidence for relativistic motion*

We have seen that several observed characteristics of quasars and other AGN can be readily explained by invoking relativistic motion of the jet fluid on the parsec scale. Some of these are: (1) apparent superluminal motion; (2) jet one-sidedness on parsec scales; (3) rapid variability, which leads to superluminal flux variations and very high brightness temperatures; and (4) the strong jet bending seen in some sources. We have already discussed these phenomena in some detail and seen how the introduction of relativistic bulk motion provides a satisfactory explanation for them. There are also other lines of evidence in support of such motion, including (5) the observation that superluminal speeds tend to be higher in core-dominated sources than in lobe-dominated sources and (6) the distribution of superluminal sources in the μz-plane (see subsection 9.5.2).

None of these observations of course *proves* that the small scale jets are indeed relativistic. The superluminal motion could be produced in a variety of different ways, while the one-sided nature could be intrinsic. The problems due to rapid variability arise because of the high radiation energy density that follows from the small source size inferred from the variability, and the high luminosity obtained by assuming that the sources are at cosmological distances. If the last assumption is given up, the redshift treated as intrinsic and the source brought much closer to the observer, the problem of the Compton catastrophe no longer exists because of the reduced luminosity. While these other explanations are possible, relativistic motion provides a single, simple and therefore compelling explanation for the observations. It is possible that examples of sources will be found where the other mechanisms make a contribution, but relativistic motion provides the best unifying basis for the explanation of the present observations.

If beaming is to explain the one-sidedness of large scale radio jets, the motion must remain at least mildly relativistic on large scales for flux boosting to be able to produce apparent one-sidedness. For a jet-to-counterjet ratio of 50, the required Lorentz factor is $\gamma \simeq 2$ (i.e., $\beta = 0.57$) for an inclination angle of 45 deg and an assumed radio spectral index of 0.7. We have seen that small and large scale one-sided jets always occur on the same side, which makes it plausible that the same mechanism is responsible on all scales for producing the one-sidedness. Since there is good evidence for relativistic motion on small scales, one must therefore believe it also exists on the large scale.

Proper motion has been detected in the knots in the one-sided jet of M87, from a comparison of observations at different epochs (see Biretta 1993 for a review and references). The angular motion of the knots observed over several years is only a small fraction of the angular resolution of the images and it is therefore necessary to use a two-dimensional cross-correlation technique to measure the motion. Values of β_a in the range ~ 0.1–0.65 have been found for various knots, while superluminal speeds have been observed for condensations in individual knots.

The Laing–Garrington effect of polarization asymmetry in radio lobes (see subsection 9.6.1) is best explained in terms of a model in which one-sided jets are always directed towards the observer. The most natural interpretation of this is in terms of relativistic beaming, and represents the most compelling argument for relativistic motion on large scales.

9.7   The production and collimation of jets

Radio jets transport enormous amounts of energy from the nuclear region to distances which extend sometimes to hundreds of kiloparsec. The jets are very likely to be highly relativistic on the parsec scale, and are expected to be mildly relativistic on much larger scales, at least in the more luminous quasars and radio galaxies. How these jets are produced and collimated over many orders of magnitude in length scale are issues fundamental to the physics of quasars and AGN. But only limited progress has been made in understanding jet physics, because of the lack of detailed structural information on the very small spatial scales on which the jets are produced, as well as the complexity of the physical processes involved in the production and collimation. Close to the central engine a general relativistic treatment is necessary. Even without this complication, taking into account the magnetohydrodynamic flow and radiation transport can be very difficult. As a result, simple models which address only some of the issues have been considered, and there is uncertainty about even the basic mechanisms. Reviews of these topics, as well as references to the original literature, may be found in H91 and RM93. Excellent summaries of the issues involved have been provided by Begelman, Blandford, and Rees (1984), Blandford (1990) and Camenzind (1993).

In the twin-exhaust model first proposed by Blandford and Rees (1974), two channels propagate in opposite directions from the nucleus. The jets, assumed to be made up of ultrarelativistic plasma, are subsonic close to the nucleus, but pass through a nozzle where the cross-sectional area is a minimum, and the flow is trans-sonic. Beyond the nozzle the flow is supersonic, and though the cross-sectional area increases, the angle which is subtended at the nucleus decreases, and well-collimated jets can be produced. Since observations require that the collimation has to be produced on a scale $\lesssim 1$ pc, the gas density and pressure required are high, and the X-ray emission from the hot gas would exceed the observed upper limits, when the power carried by the jets is high. This model for collimation can therefore work only for jets of low power $\lesssim 10^{43}$ erg sec^{-1}.

We have mentioned in Section 5.6 that jets could naturally form in the funnels associated with radiation supported tori (see Figure 5.4). However, in spite of the Eddington or super-Eddington luminosities encountered here, the jets can be accelerated only to mildly relativistic speeds, with Lorentz factor $\gamma \lesssim 3$, if they are composed of a pair plasma. Even smaller values of γ are obtained for a normal plasma of electrons and ions. This radiative acceleration model may not apply to sources which are observed to radiate at sub-Eddington rates. Moreover, radiation-supported tori could be dynamically unstable, which puts in question their long term existence.

It is possible to accelerate jets in a class of models in which magnetic fields are an important ingredient. These models are mainly based on geometrically thin, magnetized accretion disks, and are generalizations of the models of spherically symmetric stellar winds. The accretion disk centrifugally flings out gas, which remains tied to the magnetic field lines. The inertia of the gas causes the magnetic field lines to be bent backwards, creating a toroidal component. This collimates the flow of the plasma. In these magnetohydrodynamic (MHD) flows, there are three critical points, along a flow line, at which the gas becomes sonic (these points correspond to the three types of wave modes which exist in MHD). After the fluid passes through the third sonic point, the magnetic field becomes mostly toroidal, and about half the energy and angular momentum flux in the field are converted to mechanical energy and angular momentum. In the process the jets become collimated, whether they are relativistic or non-relativistic. In these models there is a correlation between the accretion rate and the speed of jets on the parsec scale, so that quasar jets have highly relativistic speeds while jets in FR-I galaxies are non-relativistic (Camenzind 1993). A possible difficulty with this model is that the magnetically collimated plasma is unstable, and particularly so to non-axisymmetric perturbations. Further work is necessary to verify that the model is viable.

Numerical simulations of jets have for long been used to test theoretical models, questions of stability and so on. The quality and scope of simulations has improved vastly with improvements in computer speed, available memory and software. Three-dimensional simulations of MHD jets can now be attempted, and these will go a long way in clarifying complex jet physics. Reviews on recent progress in the field can be found in RM93 and CW97.

9.8   Radio emission from optical quasars

While radio-loud quasars and galaxies have spectacular radio structures and properties, they constitute only a minority: by any reasonable criterion, only ~ 10 per cent of the total population of quasars is radio-loud. The criterion can be in terms of the observed radio flux, or the radio luminosity or the ratio of the radio and optical luminosities. When a sample of optically selected quasars is surveyed for radio emission, it has been observed that in ~ 10 per cent of the cases, radio flux at a level $\gtrsim 10$ mJy is detected at 5 GHz. Most of the detections are of quasars that have already been found in radio surveys, and the detection rate goes up only slowly below a flux level of 100 mJy.

A flux of 10 mJy corresponds to a rest frame 5 GHz luminosity of about $3 \times 10^{22} h_{100}^{-2}$ W Hz^{-1} at a redshift of $z = 0.05$ and at $z = 1$ the luminosity is about $10^{25} h_{100}^{-2}$ W Hz^{-1}. This is well above the fiducial luminosity which is used to separate radio sources into FR-I and FR-II classes (see Section 9.3). When sources over a large range of redshifts are considered, a criterion of radio-loudness based on the observed flux cannot therefore be very useful. In terms of luminosity, it is conventional (Kellermann *et al.* 1989) to say that quasars with $L(5\,\mathrm{GHz}) > 2.5 \times 10^{24} h_{100}^{-2}$ are radio-loud and classify the rest as radio-quiet. In terms of the ratio $R = L_\mathrm{R}/L_\mathrm{op}$, the boundary is conventionally taken to be at $R = 10$.

Table 9.1. *Radio surveys of optically selected quasars*

Radio survey[a]	Optical sample[b]	Magnitude range[c]	Number	Radio flux limit[d] (mJy)	Detections
1		$B = 15–17$	22	0.6	9
2	10	$m_{1475} \sim 17–19$	122	10	12
3		$m_{4500} = 17–21.5$	228	2	19
4	11	$m_{1475} \sim 17–19$	70	10	10
5		$B < 18.25$	22	0.15	7
6	BQS	$B < 16.16$	92	0.15	75
7	10	$m_{1475} < 20$	108	0.80	9
8	BAL	$B = 17–18.6$	38	0.4	8
9	LBQS	$B = 16–18.4$	124	0.255	24

[a] Radio survey: 1, Condon *et al.* 1980; 2, Smith and Wright 1980; Condon *et al.* 1980a; 3, Sramek and Weedman 1980; 4, Stritmatter *et al.* 1980; 5, Marshall 1987; 6, Kellermann *et al.* 1989; 7, Miller *et al.* (1990); 8, Stocke *et al.* 1992; 9, Visnovsky *et al.* 1992.

[b] Optical sample: 10, Osmer and Smith (1980); 11, Lewis *et al.* 1979. BQS, Bright Quasar Survey (Schmidt and Green 1983); LBQS, Large Bright Quasar Survey (Hewett *et al.* 1995); BAL, broad absorption line quasars.

[c] For magnitude other than *B*, the wavelength in Å at which the continuum magnitude is measured is indicated.

[d] The radio flux is at 5 GHz for all surveys except 9, where it is at 8.4 GHz.

There have been many systematic radio observations of optically selected quasars (see Table 9.1 for a brief description and references). High sensitivity radio investigation of the quasars and AGN from the Palomar Bright Quasar Survey (BQS, see subsection 6.7.1) has been carried out by Kellermann *et al.* (1989, 1994) with the VLA to a limiting sensitivity of 0.2 mJy at 5 GHz, which corresponds to $R = 0.1$ for the BQS.

Kellermann *et al.* have detected above the limiting flux 82 per cent of the 92 quasars that form the complete BQS sample. The distribution of the total flux density at 5 GHz measured with the VLA D configuration, with an angular resolution of $\sim$ 18 arcsec is shown in Figure 9.19. There appear to be two populations, with $\sim$ 20 per cent of the quasars clustered around 0.5 Jy and the rest increasing slowly in number until $\sim$ 0.5 mJy. The trend does not continue at lower fluxes, and there are 16 undetected quasars, which have an estimated average flux of $8 \pm 17\,\mu$Jy. Even at the highest redshift the radio luminosity of the undetected quasars is a factor $\lesssim$ 5 below the boundary between radio-loud and radio-quiet quasars, but these are not the least luminous ones in the sample. For example, the nearest non-detection, which is at $z = 0.087$, has a 5 GHz upper limit of luminosity $10^{22}\,\mathrm{W\,Hz^{-1}}$. This is higher than the measured luminosity of several nearer quasars. Many of the radio-quiet quasars have luminosity (or its upper limit) substantially higher than the nuclear-region luminosities of spiral and elliptical or lenticular galaxies, which are usually $\lesssim 10^{22}\,\mathrm{W\,Hz^{-1}}$ (see reviews by Wrobel 1991 and van der Hulst 1991). The host galaxies of these quasars therefore

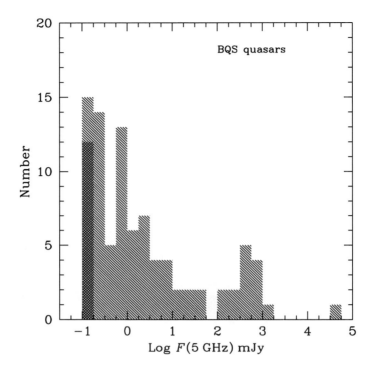

Fig. 9.19. The distribution of 5 GHz radio flux for BQS quasars. The cross-hatched regions indicate radio flux upper limits.

harbour sources of radio emission, such as starburst regions, in excess of what normal galaxies have.

The distribution of the 5 GHz radio luminosity of the BQS quasars is shown in Figure 9.20 and the distribution of the ratio R of the radio luminosity at 5 GHz to the optical luminosity at 4400 Å is shown in Figure 9.21 (Kellermann *et al.* 1994). Also shown in the lower panel of each of these figures is the corresponding distribution for quasars from the catalogue of radio sources brighter than 1 Jy by Kuhr *et al.* (1981). The BQS quasars identified as radio-loud have a luminosity distributed around $\sim 2\times10^{26}$ W Hz^{-1}, and a ratio R around ~ 100. It is clear from the figure that radio selected quasars have radio luminosity a factor ~ 100 higher than the radio luminosity of the radio-loud component of the optically selected quasars and there is little overlap between the populations. Most of the radio-quiet population has $0.1 < R < 1$ over a range of $\sim 10^4$ in optical luminosity.

Many of the radio-loud BQS quasars are resolved by the VLA A configuration (angular resolution 0.25 arcsec). These show structures with features similar to those found in higher luminosity radio selected quasars, which are edge brightened. Five of the radio-loud ones have apparent projected diameter $> 0.5h_{100}^{-1}$ Mpc. The radio

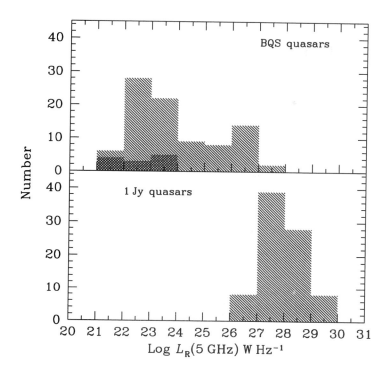

Fig. 9.20. Upper panel: the distribution of the 5 GHz radio luminosity of the BQS quasars. The cross-hatched area indicates upper limits. Lower panel: the corresponding distribution for quasars from the 1 Jy radio catalogue of Kuhr *et al.* (1981) (the data for the lower panel is adapted from Kellermann *et al.* 1989).

structures of the radio-quiet population are mostly unresolved, with just four cases showing resolved linear structure ranging up to $\sim 300 h_{100}^{-1}$ kpc in size.

9.8.1 *Radio luminosity function of optical quasars*

We have seen in Section 7.11 that there are two simple types of bivariate luminosity function which address the distribution of optical and radio luminosity. In the first, the optical and radio luminosities are considered to be independently distributed, in which case the form in Equation 7.75 applies. In the second, the optical and radio luminosities are taken as correlated, so that optically bright quasars also have a high radio flux, in which case the form in Equation 7.80 can be used. The BQS sample of quasars was selected without any reference to their radio properties and has subsequently been observed with sufficient sensitivity that all but a small fraction of the quasars have been detected. The distribution of $L(5\,\mathrm{GHz})$ and R in this sample is therefore known and the radio part of the luminosity function can be determined simply by counting

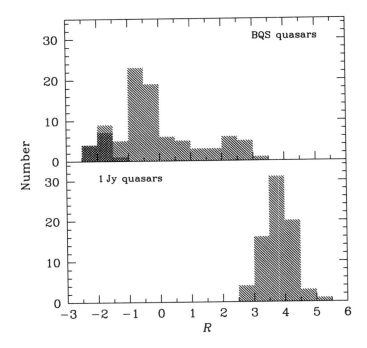

Fig. 9.21. Upper panel: the distribution of the ratio R of the 5 GHz radio luminosity to the 4400 Å optical luminosity for the BQS quasars. The cross-hatched area indicates upper limits. Lower panel: the corresponding distribution for quasars from the 1 Jy radio catalogue of Kuhr *et al.* (1981) (the data for the lower panel is adapted from Kellermann and Pauliny-Toth 1981).

the numbers of objects in different bins in Figures 9.20 and 9.21. The resultant integral luminosity functions $I(> L_R)$ and $G(> R)$ defined in Equations 7.77 and 7.81 are shown in Figures 9.22 and 9.23 respectively (Kellermann *et al.* 1989). These luminosity functions simply represent the distributions as observed, and are in principle applicable only to the optical and radio luminosity range covered by the BQS. Since there is no analytic form provided in the present treatment, there is no rule available for extrapolation of the functions to other luminosities. In particular, radio selected quasars are not covered by the present treatment, since their L_R and R values have very little overlap with the BQS distributions.

Kellermann *et al.* (1989) have used the luminosity functions to predict the number of radio detections expected at different limiting radio-flux levels for subsets of the BQS that cover different magnitude intervals. They find that both forms of the luminosity function are in reasonable agreement with observation. This also applies to predictions made about the other optical quasar surveys in Table 9.1. The present data therefore cannot distinguish between the luminosity functions. Miller *et al.* (1990) carried out VLA observations at 5 GHz, to a limiting flux of ~ 0.6 mJy, of a sample of optically selected quasars with a continuum magnitude at 1475 Å brighter than 20. The quasars

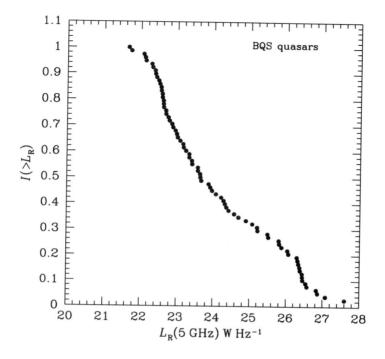

Fig. 9.22. The normalized integral radio luminosity function $I(> L_R)$. The data is taken from Kellermann *et al.* (1994).

were restricted to the redshift range $1.8 < z < 2.5$. They found that the distribution of radio luminosity is distinctly bimodal: the nine detected radio-loud quasars in their sample have $L(5\,\text{GHz}) > 2 \times 10^{24} h_{100}^{-2}\,\text{W Hz}^{-1}$, while the radio upper limits are all with $L(5\,\text{GHz}) < 2 \times 10^{23} h_{100}^{-2}\,\text{W Hz}^{-1}$. Quasars with absolute magnitude $M_B > -22.5$ are predominantly radio-quiet. There is no evidence for a correlation between optical and radio luminosity and the two appear to be independently distributed. The radio-loud quasars in this sample are mostly compact with flat radio spectra. Because of the small number of detections, it is not clear whether the differences in the radio properties of the LBQS and BQS are primarily due to the higher redshifts or the higher luminosities of the LBQS.

9.8.2 *Radio-loud and radio-quiet quasars*

It appears from Figure 9.21 that the ratio R of radio to optical luminosity has a bimodal distribution, optically selected quasars having distinctly lower R values than radio selected quasars. The distribution of R has been obtained from observations of radio selected quasars, which are very bright in the radio, and optical quasars which have low radio fluxes. It is possible that when objects with intermediate and faint radio

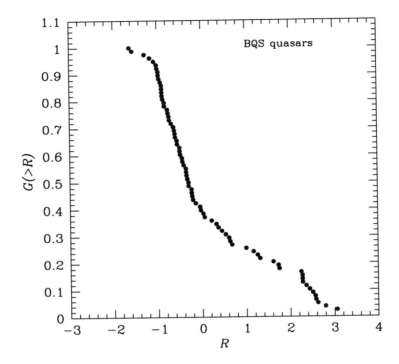

Fig. 9.23.   The normalized integral radio luminosity function $G(> R)$.

flux are identified, and the sample grows in size, the distribution may be found to be continuous. But as matters stand, bimodality seems to be evident, pointing to different physical mechanisms operating in the two kinds of object.

Radio quasars often have very large radio structures, and even where the sources are compact, high dynamic range observations do reveal faint large scale extensions. There is ample evidence that the radio lobes are powered by jets which carry energy to the extended parts. In this aspect radio quasars and radio galaxies appear to be similar, except that the jets in quasars are perhaps beamed close to the observer's line of sight. The Scheuer–Readhead model (1979) attempted to unify radio-loud and radio-quiet quasars through beaming, but later observations showed that any such attempt would not be viable. Unification appears to be restricted to radio quasars and radio galaxies. This points to special circumstances in these objects that allow the production and collimation of radio jets which power large radio structures. It is not clear what these circumstances are, why they arise only in a small fraction of objects and why a majority of quasars and AGN are radio-quiet, even though they are highly active at other wavelengths.

We have seen in earlier chapters that the emission from quasars which are not blazars could be attributed to thermal emission, generated perhaps in an accretion disk. Radio emission in highly luminous radio objects can be looked upon as an additional effect, which contributes power law emission not only in the radio band but in other regions of the spectrum as well. It seemed possible at one time that the power law component is present in all quasars, and that the radio emission is not observed because it is absorbed, owing to, say, synchrotron self-absorption or thermal absorption. This does not seem realistic, however, because the absorption would have to take place over a very wide range of wavelengths, and the properties and covering factor of the absorber are not compatible with the models of the emission line regions.

Seyfert galaxies appear to be the lower luminosity analogues of radio-quiet quasars. They often have radio emission, but this is in general at lower luminosity levels than the radio galaxies, and shows well-ordered jet and lobe structure in only a few cases such as 3C 120 (Walker, Benson and Unwin 1987). In other cases the radio emission is diffusely spread out, and does not show the core–jet–lobes structure typical to radio galaxies.

While the radio-jet-producing mechanism may be absent in most quasars and Seyfert galaxies, they do have radio luminosities which are far in excess of normal galaxies. This radio emission is plausibly due to thermal and non-thermal emission associated with regions of rapid star formation. The starbursts could act in conjunction with an AGN (see e.g. Perry 1994 for models of this sort), or they could even act all by themselves. It has been argued that the emission from young stars and supernovae remnants in starbursts in nuclear regions of galaxies could reproduce most of the observed properties of, at least, the low luminosity AGN (see e.g. Terlevich 1992). It is not clear at the moment how powerful an AGN has to be for it to manifest the canonical radio galaxy or radio quasar properties, and whether in fact there are two different kinds of radio emitting objects. In general one expects that there is a mix of thermal and non-thermal emission from an AGN and from surrounding starburst regions, with the different components dominating in different circumstances.

9.9 BL Lac objects

BL Lac objects share many of the properties of quasars, such as a non-thermal continuum and variability, but are lacking in strong spectroscopic lines. The first object of the class was discovered when the 'variable star' BL Lacertae was identified with the radio source VRO 42.22.01, which had variable radio flux density and linear polarization. The rotation measure of the plane of polarization (see subsection 3.4.1) in the radio source indicated, however, that the source was extragalactic, and the unusual nature of BL Lac was established when the optical continuum was found to be featureless (i.e., with no emission or absorption lines), variable and linearly polarized to a relatively high degree (Visvanathan 1969). Soon, further examples of sources similar to BL Lac, such as OJ 287 and AP Lib, were found and the class of 'BL Lac' sources was established (Stritmatter *et al.* 1972).

Around a hundred BL Lac objects are now known, the early identifications being nearly all based on radio surveys. X-ray surveys are also proving to be very useful in finding objects of the class, and the X-ray selected objects have properties, such as luminosity, somewhat different from the radio selected variety. It is difficult to find BL Lacs in optical surveys since they do not always have ultraviolet excess and lack strong emission lines. A few BL Lacs have been discovered from their colours and variability, but complete samples are only available from optical identifications of radio and X-ray surveys. The classic references are the *Proceedings of the Pittsburgh Conference on BL Lac Objects* (W78), and the reviews by Stein, O'Dell and Stritmatter (1976) and Angel and Stockman (1980). More recent developments may be found in MMU89, VV92 and Urry and Padovani (1995), the last dealing extensively with matters connected with the unification of BL Lacs and other kinds of AGN.

Characteristics   BL Lac objects have several characteristic properties which distinguish them as a class. (1) They usually have flat ($\alpha_R < 0.5$) or inverted radio spectra. (2) The spectra steepen in the infrared or optical region as a consequence of which many BL Lacs have little or no ultraviolet excess. (3) There is rapid and large amplitude flux variation in the radio, optical and X-ray bands. (4) The radio and optical emissions show strong and variable linear polarization. (5) The continuum is smooth and featureless, and unlike in quasars and all other kinds of AGN, emission lines are entirely absent or are very weak. Absorption lines are also sometimes seen.

Polarization   In the optical band, BL Lac objects typically show several percent linear polarization, which is higher than for the average quasar. The degree of polarization is as high as 30 per cent in some objects. The polarization is in general highly variable on a time scale of about one day, both in degree and position angle, and is largely wavelength dependent. Properties in the infrared are found to be similar to those in the optical band. Polarization in the optical and infrared are generally found to be similar in strength as well as position angle, but cases of substantial rotation of the position angle in passing from one band to another are known. A few objects with strong polarization do not show detectable variability; the degree and position angle remains essentially constant. In some objects the high variability seen in one epoch is no longer found in another, the object having settled down to a state of nearly constant high or low polarization (see Angel and Stockman 1980 for a detailed discussion). The polarization is generally weaker at radio wavelengths and for most objects there is no correlation between the position angles in the optical and radio regions.

Spectral lines   Many BL Lac objects show only weak emission and absorption lines. The equivalent width of any emission lines that may be present is necessarily small in the discovery observations, since otherwise the object would not be taken to be a BL Lac. As the continuum flux varies the equivalent width changes too, and one might expect it to become large when the flux is at its lowest. However, the equivalent widths seldom exceed $\sim 5\,\text{Å}$ (see Stickel *et al.* 1991 for some exceptions and references to them). Broad as well as narrow emission lines from the nuclear region have been seen. Emission features that can be ascribed to a galaxy in which the BL Lac is embedded are also seen.

Stickel *et al.* (1991) have identified a complete radio-flux-limited sample of BL Lac objects with radio flux > 1 Jy at 5 GHz, flat or inverted radio spectra with $\alpha_R < 0.5$, optical magnitude $V < 20$ and rest frame equivalent width of the brightest line emission < 5 Å. Spectra of these objects in the range 4000 Å $\lesssim \lambda \lesssim 7500$ Å are available. The sample contains 34 BL Lacs and, in spite of its modest size, is very useful in studying various distributions and luminosity functions and for comparison with X-ray selected samples.

Weak emission lines have been detected in ~ 75 per cent of the 1 Jy sample. The emission lines most often observed are the broad line Mg II $\lambda 2798$ Å in the case of high redshift quasars and the narrow forbidden line [O III] $\lambda 5007$ Å in the case of low redshift objects (for $z > 0.5$ this line passes beyond the wavelength range for which spectroscopy is available). Sometimes a weak narrow [O II] $\lambda 3727$ Å line is also seen. The distribution of the luminosities of the [O III] and Mg II lines for the 1 Jy sample of BL Lacs as well as for some quasars is shown in Figure 9.24. The narrow [O III] line is seen to have considerably lower luminosity in BL Lacs than in quasars. In the case of the Mg II line, while the BL Lacs have on the whole lower luminosities, there is some overlap and the brightest BL Lac line luminosities are well within the distribution for quasars. It is not clear what the situation with the Mg II line is for the low redshift objects, and therefore one cannot say whether the overlap with quasars at high redshift but not at low redshift is a consequence of some redshift-dependent effect.

Absorption lines are also seen in some BL Lac spectra. In some cases the absorption lines are like those from elliptical galaxies and arise in the host galaxy of the BL Lac object (see below). In this case the redshift of the absorption lines z_{abs} is the same as the emission line redshift z_{em}, if the latter has been measured from the emission spectrum. Host galaxy absorption lines are occasionally found even when no emission line is to be seen. There are cases where $z_{abs} < z_{em}$ and the absorption lines can be traced to intervening cosmological matter, as for high redshift quasars.

Host galaxies and environment   Host galaxies have been observed around $\gtrsim 20$ BL Lac objects at $z \leq 0.2$. Examining the details of the galaxy at all but the smallest redshifts has been difficult because of low surface brightness and contamination by the bright nucleus. Study of the galaxy morphology from direct imaging has become easier with the use of CCDs and modern image processing techniques, which allow the effects of the nucleus (the BL Lac) to be subtracted from the extended structure.

From direct imaging the host galaxies appear to be elliptical in nature. The surface brightness profiles of elliptical galaxies are usually well fitted by de Vaucouleurs' law

$$I(r) \propto 10^{-3.33(r/r_e)^{1/4}}, \tag{9.7}$$

where $I(r)$ is the surface brightness at a distance r from the centre along the major

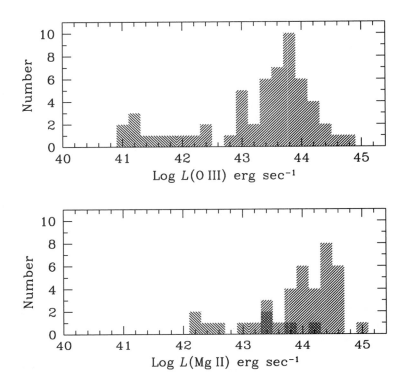

Fig. 9.24. The distribution of the luminosities of [O III] (upper panel) and Mg II (lower panel) for the 1 Jy sample of BL Lacs and some quasars. BL Lacs are indicated by left-inclined hatching and quasars by right-inclined hatching. The data is taken from Stickel *et al.* (1991).

axis and r_e is the effective radius within which half the total light from the galaxy is contained. Stickel *et al.* have fitted the intensity profiles of the hosts of seven $z < 0.2$ BL Lacs from their sample. They use a law similar to the one in Equation 9.7, but with the $1/4$ power replaced by β, the value of which is to be obtained from the fit. The disk component of spiral galaxies has a profile with $\beta = 1$. Stickel *et al.* find that in all but one case β is much less than unity, which they interpret as an indication that the host galaxy is elliptical. The mean absolute magnitude of the host galaxies, after taking into account K-correction and Galactic absorption, is found to be $\langle M_V \rangle = -22.9 \pm 0.3$, which is comparable to the values for host galaxies of other BL Lac objects (Ulrich 1989) and brightest cluster galaxies.

An excellent example of a fit by de Vaucouleurs' law is provided by the host galaxy of the BL Lac PKS 0548-322 at redshift 0.069, which has been observed at subarcsecond resolution with the ESO 3.5 m New Technology Telescope (Falomo, Pesce and Treves 1995). The observed surface profile can be fitted very well with a combination of de Vaucouleurs' law and a point source at the centre, the flattening due to the point

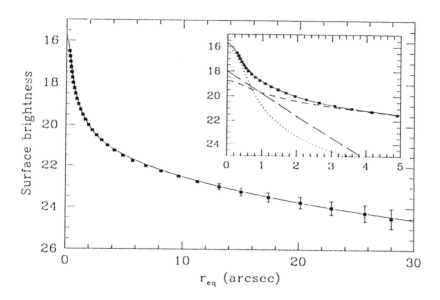

Fig. 9.25. *R* band surface brightness profile of the BL Lac PKS 0548-322. The inset shows the inner 5 arcsec. The squares are data points, while the solid line is the fitted model, which includes a bulge with a de Vaucouleurs' law profile (short-broken line) and an exponential disk (long-broken line). The model galaxy was convolved with a point spread function, shown as the dotted line, prior to the fitting. Reproduced from Falomo, Pesce and Treves (1995).

spread function being taken into account. The fit is satisfactory everywhere except inside $r \sim 1.5$ arcsec, but it is improved with the inclusion of a small faint disk component with characteristic radius 0.6 arcsec. The galaxy profile and the fit are shown in Figure 9.25.

It has been possible to obtain the spectrum of the host galaxy in several cases, though this is again difficult because of the need to do careful sky subtraction and to avoid contamination from the nucleus. The spectra show clear evidence of absorption lines (see Ulrich *et al.* 1975 for an early example) and have the general form corresponding to elliptical galaxies.

The regions around several of the nearer BL Lacs have been searched for the presence of companion galaxies, and a statistically significant excess of galaxies has been found. Stickel *et al.* have obtained spectroscopic data for galaxies in the fields of some BL Lacs in their sample. They found that, in the the case of three of the four low redshift BL Lacs from this set, one to three neighbouring galaxies at the same redshift are present, indicating the presence of a group or small cluster of galaxies. In the case of the BL Lac PKS 0548-322 mentioned above, a surrounding rich cluster at redshift ~ 0.07 has been found. All galaxies in the cluster which have been observed spectroscopically have been found to be within ~ 200 kpc of the BL Lac. The environment of BL Lac objects therefore appear to be similar to those of FR-I galaxies (see Section 9.3).

Radio morphology   High dynamic range VLA observations are now available for a number of BL Lac objects. The number of BL Lacs in any single well-defined sample is still rather too small to enable one to come to statistically significant conclusions, but nevertheless is still possible to get a fair idea of the range of possibilities.

Laurent-Muehleisen *et al.* (1993) have observed a sample of 15 X-ray selected BL Lacs (XBL), which were discovered in the identification of X-ray sources discovered in the *HEAO 1* Large Area Sky Survey. The BL Lacs in the sample had redshifts in the range $0.028 < z < 0.698$. It was found that 80 per cent of the sources showed extended morphologies and were of the one-sided, head–tail, double-lobed or amorphous type. Perlman and Stocke (1993) have found similar morphologies for another X-ray selected sample of BL Lacs from the Extended Medium Sensitivity Survey (EMSS, see Section 6.4), and for radio selected BL Lacs including some from the 1 Jy sample of Stickel *et al.* (1991)

The mean extended radio power of BL Lac objects is found to be similar to that of FR-I radio galaxies. Amongst the BL Lacs on the whole there are fewer double sources and more amorphous sources than are found in samples of FR-I galaxies. Superluminal motion has been found in several BL Lacs (see Section 3.8) and they exhibit high variability. All these factors are consistent with BL Lac objects being FR-I radio galaxies viewed end on, which would make the apparent sizes smaller and the structures amorphous. A detailed comparison (see e.g. Laurent-Muehleisen *et al.* 1993) shows that core radio power increases from FR-I galaxies to X-ray selected BL Lacs (XBL) to radio-selected BL Lacs (RBL) and that the ratio of core-to-extended radio power also increases in the same direction. From comparison of X-ray selected and radio selected BL Lac populations, it has been observed that the XBL have lesser variability in total and polarized light. They have a lesser degree of polarization and the polarization position angle often remains constant. These facts all suggest that the RBL have a smaller beaming angle than the XBL. In a unified picture, FR-I radio galaxies form the parent population of these beamed objects. We shall see in Section 12.5 how the BL Lac luminosity function can be derived from the effects of relativistic beaming on luminosity function for FR-I radio galaxies.

Though the data is broadly consistent with BL Lac objects being beamed FR-I galaxies, there are several examples of BL Lac objects with FR-II type radio morphology. Laurent-Muehleisen *et al.* find that radio selected BL Lacs can have greater extended radio luminosities than typical FR-I galaxies, and two X-ray selected BL Lacs have sufficient radio luminosity to belong to the FR-II class. A sample of 17 BL Lacs has been observed at 5 GHz with the VLA by Kollgaard *et al.* (1992). They find that of the 10 sources which have extended structures, three have morphology and polarization properties (see Section 9.3) that clearly put them in the FR-II class. Of the remaining seven, three are definitely of the FR-I type, while the rest have luminosities close to the dividing line and complex morphologies, which makes it difficult to classify them.

We will see in Section 12.4 that radio-loud quasars can be taken to be highly beamed versions of FR-II radio galaxies. Since BL Lacs too are highly beamed objects, the possibility arises that the more luminous BL Lacs are an extension of the quasar population with low-equivalent-width emission lines. However, the polarization properties of quasars and BL Lacs are different on the milliarcsecond scale. VLBI observations (Gabuzda *et al.* 1992, 1994) show that (1) the degree of polarization in unresolved VLBI cores of BL Lacs is 2–5 per cent, which is higher than the 0–2 per cent found in quasars, (2) the projected magnetic field in BL Lacs is oriented perpendicularly to the jet direction, while in quasars it is parallel to the jet. The numbers presently available are too small to look for systematic differences, on the VLBI and VLA scales, between BL Lacs with different kinds of radio structure, radio galaxies and radio-loud quasars.

9.9.1 *Blazars*

Apart from the absence of strong emission lines, the properties of BL Lac objects are very similar to those of flat spectrum radio-loud quasars, which are highly variable and polarized. The radiation from objects with these properties is largely of non-thermal origin and appears to be relativistically beamed, making the objects distinct in their properties from radio-quiet quasars and related objects. It was suggested by E. Spiegel at the Pittsburgh Conference on BL Lac objects in 1978 that objects with such properties be grouped together into the single class of blazars. An object is defined as a *blazar* (see e.g. Bregman 1990) if it has the following properties: (1) smooth continuum emission from the infrared to the ultraviolet band; (2) high optical polarization ($\gtrsim 3$ per cent); (3) rapid optical variability on a time scale of $\lesssim 1$ day; (4) a strong and variable radio continuum. When an object of this class has weak emission lines compared to radio-quiet quasars, it is called a BL Lac; when the emission lines are strong, the object is said to be an *optically violent variable* (OVV) quasar.

Gravitational microlensing Because of the similarities between some of the properties of BL Lac objects and flat spectrum radio quasars (FSRQ), it has been proposed by Ostriker and Vietri (1990) that BL Lacs are gravitationally microlensed FSRQ. This idea is based on the observed short variability time scales in blazars, which imply that the regions in which the continuum is produced have small angular sizes. Ordinary stars in intervening galaxies along the line of sight to the blazar can gravitationally focus and amplify the continuum radiation, i.e., microlens it (see Section 14.6). Since the line emitting region is much larger in size, it is not significantly affected by the lensing, and therefore the lensed FSRQ will have reduced equivalent line widths and the observed properties of BL Lac objects. This would mean that what are believed to be the host galaxies of BL Lacs are in many cases intervening galaxies. BL Lac redshifts are often taken to be the absorption redshifts of their host galaxies; if these are in fact lensing galaxies between the BL Lac and the observer, then the BL Lac redshifts, and therefore their luminosity, would be overestimated.

This is an elegant model, which unifies blazars with strong and weak lines, but there are a number of arguments against it, which have been summarized in some detail by

Urry and Padovani (1995). These include (1) difficulties faced in obtaining the required amplification factors, which are ~ 10, in numerical simulations, (2) the fact that BL Lacs are observed to be at the centres of their host galaxies, while it would be possible for microlensed FSRQ to be quite off the centre of the lensing galaxy, and (3) the observed equality between emission and absorption redshifts in several BL Lacs, which indicates that in a substantial fraction of cases the measured absorption redshift is of the host galaxy rather than an intervening one. The observed differences in the polarization properties of BL Lacs, discussed above, also constitute a serious argument against all BL Lac objects being microlensed quasars, since the lensing should not change polarization properties.

9.9.2   *The luminosity function of BL Lac objects*

Stickel *et al.* (1991) have determined the luminosity function of the BL Lacs in their 1 Jy complete sample, using the techniques described in subsection 7.6.2. The mean value of V/V_m for the sample is $\langle V/V_m \rangle = 0.6 \pm 0.05$ ($H_0 = 50\,\mathrm{km\,sec^{-1}\,Mpc^{-1}}, q_0 = 0$), which is consistent with the value 0.5 for no evolution within 2σ. However, interpreting the mean value obtained as probably indicative of evolution, Stickel *et al.* consider luminosity evolution of the form

$$L_R(z) = L_R(0)\exp\left[\frac{\tau(z)}{\tau_R}\right], \tag{9.8}$$

where $\tau(z)$ is the look-back time and τ_R is the evolutionary time scale in units of the Hubble time. τ_R is determined by finding that value which makes the distribution of the weighted accessible volume V'/V'_m uniform between 0 and 1. The best-fit value obtained in this fashion is $\tau_R = 0.36$, with the associated 1σ interval (0.24, 0.59). The local luminosity function can be obtained from weighted accessible volumes as in Equation 7.51, and fitting a simple power law form to it gives

$$\Phi_R(L_R) = 2.0\times10^{50}\,L_R^{-2.53\pm0.15}\,\mathrm{Gpc^{-3}}L_R^{-1}, \tag{9.9}$$

where L_R is the 5 GHz luminosity in units of $\mathrm{erg\,sec^{-1}\,Hz^{-1}}$ and the 1σ error on the power law index is 0.15. The index and normalization are also affected by the size of the luminosity bins used, and the values in Equation 9.9 correspond to a bin size of $\Delta(\log L_R) = 0.3$. It follows from the luminosity function that the local number density of BL Lac objects with 5 GHz luminosity in the range $6\times10^{31}\,\mathrm{erg\,sec^{-1}\,Hz^{-1}}$ to $3.4\times10^{34}\,\mathrm{erg\,sec^{-1}\,Hz^{-1}}$ is $40\,\mathrm{Gpc^{-3}}$. If the luminosity function is obtained assuming that there is no evolution, then

$$\Phi_R(L_R)dL_R = 8.9\times10^{38}L_R^{-2.18\pm0.13}\,\mathrm{Gpc^{-3}}dL_R \tag{9.10}$$

and the local density of BL Lacs having luminosities in the range 6.6×10^{31} to $1.5\times10^{35}\,\mathrm{erg\,sec^{-1}\,Hz^{-1}}$ is $31\,\mathrm{Gpc^{-3}}$.

Wolter *et al.* (1994; also see this paper for references to earlier papers with similar results), have obtained the X-ray luminosity function for a sample of 30 X-ray selected BL Lacs from the extended medium sensitivity survey (EMSS). These were identified on the basis of weak emission lines (equivalent width < 5 Å) and the presence of a non-thermal continuum. Deriving the luminosity function here requires the use of the techniques applicable to coherent samples described in subsection 7.6.4, since the EMSS X-ray fields are disjoint, with a different flux limit for each field. It is found that $\langle V_e/V_m \rangle = 0.36 \pm 0.05$ ($H_0 = 50$ km sec^{-1} Mpc^{-1}, $q_0 = 0$), which shows that there is negative evolution with a significance of $\sim 3\sigma$. Wolter *et al.* parameterize the evolution of the luminosity function as in Equation 9.8, with τ_x for the evolutionary time scale, and, following the technique described above, obtain $\tau_x = -0.14$, with the $1, \sigma$ range $(-0.23, -0.06)$. The local luminosity function is given by

$$\Phi_X(L_x)dL_x = 1.52 \times 10^9 L_x^{-1.62 \pm 0.13} \, \text{Gpc}^{-3} dL_x, \quad (9.11)$$

where L_x is the luminosity at 1 keV in units of erg sec^{-1} Hz^{-1}.

A plot of X-ray luminosity against radio luminosity for the RBL shows that the two are well correlated, with $L_x \propto L_R^{0.97}$. The radio luminosities in Equation 9.10 can be transformed to X-ray luminosities using this relation, and the result can be used to compare the relative number of radio and X-ray selected BL Lacs at the same X-ray luminosity. This procedure can in fact be used to compare the relative abundances of the RBL and XBL for luminosities in any band. Wolter *et al.* found in this manner that the XBL greatly outnumber the RBL, the excess being a factor of ~ 100 for high X-ray luminosities ($\sim 10^{46}$ erg sec^{-1}). This is, however, the ratio of the *model-dependent* local space densities; the ratio of the *observed* surface densities of objects can be quite different, especially because the RBL show positive luminosity evolution while the XBL show negative evolution. The surface densities of objects to some flux level are obtained by integrating the luminosity function over the redshift, and the ratio of surface densities can be quite different from the ratio of the space densities. A direct comparison of the radio and X-ray counts of BL Lac objects (e.g. Urry, Padovani and Stickel 1991) shows that over the same flux range the surface density of the XBL is about an order of magnitude higher than that of the RBL. This deduction is supported by results of the unification scheme that identifies BL Lac objects with highly beamed FR-I galaxies (see Section 12.5). It should be noted that the deductions about the evolution and density ratios are all based on relatively small samples and could be the result of selection effects rather than any real difference in the populations. Firmer results will have to await the identification of larger complete samples of BL Lac objects.

Apart from their different densities, there are other significant differences in the properties of the XBL and RBL. The latter are on average found to have higher polarization, variability and radio core dominance. While the XBL and RBL have similar X-ray luminosities, the former are found to have substantially lower radio luminosities. The two classes also have different broad band spectra (see Section 11.5).

The properties of BL Lac objects are indicative of the effects of relativistic beaming, as first suggested by Blandford and Rees (1978), and in this picture the RBL appear to be subject to the effects of beaming to a greater extent than the XBL. Indeed, the beamed luminosity-function models discussed in Section 12.5 lead to Lorentz factors $\langle \gamma_r \rangle \simeq 7$ for the RBL and $\langle \gamma_x \rangle \simeq 3$–4 for the XBL. This can be interpreted in terms of an accelerating-jet model (Ghisellini and Maraschi 1989), since the X-rays, because of the observed rapid variability, are expected to arise from the most compact regions while the radio emission arises in more extended regions. If the relativistic jet in which the beamed emission is produced accelerates outwards, the behaviour of the Lorentz factor surmised from the models is expected. The higher value of γ for radio emission implies that the RBL have to be viewed closer to the beam axis for them to appear in flux-limited surveys. In this model RBL are expected to be rarer than XBL, because the latter will be observable over a wider angle, which is consistent with observation. In a variant of the beaming model, the actual collimation of the beam increases in the outward direction while the bulk Lorentz factor remains the same (Celotti *et al.* 1993). This makes it less likely that the observer's line of sight intercepts the radio emitting part of the beam than the X-ray emitting one. In a completely different approach, Padovani and Giommi (1995) explain the differences between the RBL and XBL on the basis of differences in the broad band spectral shapes applying to a single population of BL Lacs (see Section 11.5). In this picture, X-ray selection chooses those objects whose luminosities peak in the UV or X-ray region and is biased against objects for which the peak is in the IR or optical region. It is the radio selected objects that are picked up in an unbiased fashion, and even though the X-ray selected objects appear to be more numerous, they are intrinsically rarer. We shall discuss some of these matters further in Section 11.5.

10 X-ray emission

10.1 Introduction

X-ray emission has been found to be a universal characteristic of all AGN and quasars. The X-ray luminosity of these objects is in the range $\sim 10^{42}$–10^{48} erg sec^{-1} and accounts for a considerable fraction of their bolometric luminosity. The X-ray flux shows large amplitude variability on a time scale of days, hours and in some cases hundreds of seconds. This implies that the X-ray emission has its origin very close to the central objects. For energies $\gtrsim 1$ keV the X-ray spectra appear to have a simple power law shape, which is modified at lower energies owing to absorption at the source. Improved energy resolution has shown the presence of emission and absorption features, which prove to be important diagnostics of conditions close to the active nucleus. At energies below a few keV, AGN and quasars contribute a significant fraction of the observed X-ray background (XRB). The observed spectra of AGN and quasars in the 2–10 keV range appear to be different from the spectrum of the X-ray background and a way has to be found to reconcile AGN spectra with the XRB, if the AGN indeed contribute a substantial fraction to it at high energy.

In the following sections we will discuss some important X-ray missions, elementary steps used in the analysis of data, X-ray surveys, luminosity functions and broad band properties of AGN and quasars. We will consider X-ray spectra in Chapter 11.

10.2 X-ray observatories

X-rays are highly absorbed by the Earth's atmosphere and it is necessary to reach high altitudes in order to observe X-rays from celestial sources. The first observations in the X-ray band were made from instruments on rockets and balloons. Sounding rockets reached altitudes greater than 100 km, and X-rays in the range ~ 0.25–10 keV could be observed from them with Geiger counters. These flights were of short duration, typically only a few minutes, and only the brightest sources could be observed. The balloons went up to about 40 km and only X-rays with energy $\gtrsim 40$ keV could be observed from them. But balloons can carry heavier payloads in the form of scintillation detectors and remain afloat for hours. Rocket and balloon observations complemented each other and led to many important discoveries that included the detection of X-rays from the Sun, SCO X-1, the Crab nebula and the active galaxy M87 and of the diffuse X-ray background (see Bradt *et al.* 1992 for a brief review and references). In spite of the proliferation of X-ray satellites, rockets and balloon platforms continue to be in use in X-ray observations.

The first artificial satellite dedicated to X-ray astronomy, *UHURU*,[1] was launched in 1970, and has been followed by a succession of satellite missions. We will provide a brief description of the missions most important to the study of AGN, with the aim of acquainting the reader with the names and properties of the detectors that most often occur in the literature. Further information about missions in orbit before 1992 may be obtained in the review of Bradt *et al.* and the references to the original literature provided there and in the book by Charles and Seward (CS95).

UHURU   This satellite carried proportional counters with a total area of $840\,cm^2$ and was sensitive to X-rays in the 2–20 keV range. It provided a view of X-ray sources in the whole sky to the level of 1 milliCrab. The detectors were sensitive enough to observe X-rays from some Seyfert galaxies and clusters of galaxies, as well as from numerous galactic X-ray sources.

SAS 3   Launched in 1975, the third *small astronomy satellite* was a spinning satellite with controllable spin rate. It could be used for pointed observations for up to $\sim 30\,min$. It carried a proportional counter array, a system for the study of the Galactic soft X-ray background and a modulation collimator system to determine source positions to an accuracy of $\sim 1'$. *SAS 3* made several important discoveries to do with Galactic X-ray sources and identified the first X-ray selected quasar.

HEAO 1   The first *high energy astronomy observatory* satellite was launched by NASA in 1977 and was a scanning mission. Great-circles of the sky were scanned every $\sim 30\,min$ with the spin axis always pointing towards the Sun. Consequently the entire sky was covered once every six months and almost three such coverages were completed over the life of the mission (17 months). The time for which a source could be observed varied from just a few days over the entire mission for a source situated close to the ecliptic, to nearly continuous scans for sources near the celestial poles. An object was in the field of view typically for about six days, which provided a total exposure time of $\sim 2000\,sec$. The satellite also had a limited degree of pointing capability.

The cosmic X-ray experiment (A2) on *HEAO 1* was designed primarily to study the XRB from $\sim 0.25\,keV$ to $\sim 50\,keV$ and consisted of a proportional counter array with total area $\sim 4000\,cm^2$ and with energy resolution ~ 16 per cent at 6 keV. The A2 experiment provided a broad band spectrum of the XRB in the three to 50 keV range, broad band spectra from many AGN and an improved AGN luminosity function.

The *EINSTEIN* observatory (*HEAO 2*)   The *EINSTEIN* observatory carried a fully imaging X-ray telescope based on grazing incidence focusing optics (see Giacconi

[1] This means *freedom* in Swahili, the national language of Kenya. *UHURU* was launched on the anniversary of Kenya's Independence Day.

et al. 1979 for a lucid description of the telescope and detectors). The effective mirror area was $400\,cm^2$ at $0.25\,keV$ and $200\,cm^2$ at $2\,keV$. The telescope had an angular resolution of a few arcseconds, a field of view up to 1 deg, moderate resolution and sensitivity about three orders of magnitude greater than previously achieved. The focusing capability meant that images of extended objects could be obtained; it also increased the sensitivity to point-like sources, because of the reduced background due to the small beam width. The observations were made in the pointing mode with the target object kept in the field of view for long periods of time. The telescope was provided with a focal-plane transport assembly which could position any one of four X-ray instruments at the telescope focus. The detectors consisted of an imaging proportional counter (IPC), a high resolution imager (HRI), a solid state spectrometer (SSS), a focal plane crystal spectrometer (FPCS) and a monitoring proportional counter (MPC) mounted on the outside of the telescope and aligned within 1 arcmin of the optical axis.

The IPC was a position-sensitive proportional counter and had a field of view of $\sim 1\,deg^2$ and a moderate angular resolution of ~ 1 arcmin. It had a peak effective area of $\sim 100\,cm^2$ (in combination with the mirror), and was sensitive in the ~ 0.1–$4\,keV$ energy range. The energy resolution of the IPC was a modest ~ 100 per cent at $1\,keV$, but its relatively wide band-pass allowed assumed spectral shapes to be fitted to counts observed in different channels, when the source was on-axis and a sufficient number of counts was available. In many cases it was possible only to determine the total flux in the IPC band.

The HRI was a digital X-ray camera with high angular resolution but no intrinsic energy resolution. The angular resolution was a few arcsecond over a field of view of ~ 25 arcmin diameter and the effective area varied from $\sim 20\,cm^2$ at $\sim 0.25\,keV$ to $\sim 5\,cm^2$ at $2\,keV$. Using the high angular resolution of this detector it was possible to determine accurately the positions of X-ray sources, which considerably facilitated their optical identification.

The MPC was a collimated proportional counter filled with argon and sealed with a $38\,\mu m$ window made of beryllium foil. It had an effective area of $667\,cm^2$ and no focusing ability. The MPC was sensitive in the range 1.2–$20\,keV$ and had an energy resolution of ~ 20 per cent at $6\,keV$.

EXOSAT Launched in 1983, the *European X-ray astronomy satellite* was a three-axis satellite that operated in the pointing mode. It had a highly eccentric orbit with period of 90 hr. This enabled up to several days of uninterrupted observation of a source. The major instruments here were medium energy (ME) proportional counters and a low energy (LE) telescope. The ME counters had a total area of $1600\,cm^2$, a field of view of 0.75 deg and were sensitive in the range 1–$50\,keV$. The LE was sensitive in the range 0.05–$2\,keV$ with a peak effective area of $10\,cm^2$, and had an angular resolution of 20 arcsec on the axis of the telescope. The LE telescope and ME array were used to obtain the broad band spectra of a large number of AGN and to establish that short term variability in the X-ray region is a common characteristic of these objects.

GRANAT[1]   This satellite was launched by the erstwhile Soviet Union in 1989 to image the sky at high energies. It is in a highly elliptical orbit with a period of four days, which allows for long uninterrupted viewing of sources. It has two telescopes: Sigma, to map selected regions of the sky in the 35—1300 keV range, and the ART-P, which is coaligned with the other telescope and is sensitive in the $4 - 60$ keV range. It also has a all-sky monitor for observing persistent sources and localizing transients, as well as the capability to detect γ-ray bursts.

GINGA[2]   This was the third in a series of Japanese X-ray astronomy satellites and was launched in 1987. It was again a three-axis stabilized satellite that worked in the pointing mode, providing long exposures. The primary instrument was a large area counter (LAC), which consisted of eight multicell proportional counters with a total area of ~ 4000 cm^2 and covering the energy range 1.5—37 keV (Turner *et al.* 1989). The energy resolution was $\propto E^{-0.5}$ throughout the energy range and was ~ 18 per cent at 6 keV. The LAC was used to obtain 2—20 keV spectra of quasars and Seyferts, and the observations established that emission and absorption features of iron were common in Seyfert 1 galaxies.

ROSAT   The *Röntgen-Satellit* is a joint German, US and British project. The three-axis stabilized satellite in low Earth orbit was launched in June 1990 and carries an imaging X-ray telescope (XRT) and a wide field camera (WFC). The WFC is an autonomous telescope with micro-channel plate detectors and is sensitive in the range ~ 20 eV—0.2 keV. The XRT has a field of view of 2 deg diameter and geometrical area 1141 cm^2. There are two focal plane instruments mounted on a carousel in the focal plane of the XRT. These are a multiwire position sensitive proportional counter (PSPC), which is similar to the *EINSTEIN* IPC, and the high resolution imager (HRI), which is basically identical to the *EINSTEIN* HRI.

The PSPC-XRT combination is sensitive over the range ~ 0.1—2.4 keV, with effective area larger than that of the IPC at low energies, which makes it more sensitive to the soft part of the X-ray spectrum. The effective area is ~ 200 cm^2 at 1 keV with energy resolution $\propto E^{-0.95}$ and 43 per cent at 0.93 keV. The on-axis spatial resolution corresponds to ~ 25 arcsec at ~ 1 keV in the focal plane. The HRI is a position-sensitive micro-channel plate detector with no intrinsic energy resolution in its range of ~ 0.1 keV to 2.4 keV. It has a field of view of 38 arcmin diameter and on-axis angular resolution of ~ 1.7 arcmin.

ROSAT carried out an all-sky survey in the first six months of its scientific operation. This was the first-ever imaging survey of the X-ray sky and produced $\sim 60\,000$ sources; their optical identification and further study are stupendous tasks. The satellite has also been used extensively in the pointing mode in the study of individual sources and in making deep surveys over small areas.

[1] *GRANAT* means pomegranate in Russian.
[2] *GINGA* means galaxy in Japanese.

ASCA[1]   The *advanced satellite for cosmology and astrophysics*, which is the fourth Japanese X-ray astronomy satellite, was launched in 1993. It is three-axis stabilized and has pointing accuracy of $\sim$ 3 arcsec with a stability of better than 10 arcsec. Orientation of the spacecraft is limited by the constraint that the direction of the Solar paddles must be within 30 deg of the Sun, which limits the amount of sky observable at any one time. The entire sky is covered every six months.

ASCA carries four identical grazing-incidence X-ray telescopes, each with an imaging spectrometer in its focal plane (Tanaka *et al.* 1994). Each of the telescopes has a modest angular resolution with a half-power diameter of $\sim$ 3 arcmin, but the images have a sharp core, which makes it possible for sources that are of $\sim$ 1 arcmin size to be resolved.

The focal plane instruments are two CCD cameras (solid-state imaging spectrometer, SIS), and two gas scintillation imaging proportional counters (gas imaging spectrometer, GIS). The SIS has an energy resolution of $\sim$ 2 per cent at 6 keV and 5 per cent at 1.5 keV and is sensitive at low energies up to 0.25 keV. It has a field of view of 20 arcmin $\times$ 20 arcmin. The GIS has a resolution of $\sim$ 7 per cent at 6 keV and $\sim$ 14 per cent at 1.5 keV. It has no sensitivity below $\sim$ 1 keV but has higher detection efficiency than the SIS above $\sim$ 3 keV. The field of view of the GIS is circular with a diameter of 50 arcmin. All four detectors operate simultaneously so that data from each one of them is separately available all the time.

AXAF   The *advanced X-ray astronomy facility*, to be launched in 1998, will consist of two payloads to be launched separately: *AXAF-I* will be devoted to imaging, while *AXAF-S* will be for spectroscopy. Compared to the instruments on *EINSTEIN*, the telescopes will have four times the collecting area, the angular resolution will be an order of magnitude better and the imaging sensitivity will be two orders of magnitude higher.

10.3   Flux and spectrum

Consider a source with unabsorbed flux density $F(E)$ at the Earth. The net count produced by the source in an X-ray detector in the observed energy range (E_1, E_2) is given by

$$C(E_1, E_2) = T \int_{E_1}^{E_2} \frac{F(E)}{E} \exp[-\sigma(E)N_H]A(E)\,dE, \qquad (10.1)$$

where T is the exposure time, $\sigma(E)$ is the atomic absorption cross-section per hydrogen atom for the intervening matter between the source and the detector, N_H is the equivalent hydrogen column density and $A(E)$ is the effective area of the detector at energy E. The effective area takes into account the available geometric area as well as the efficiency of the detector as a function of energy. The exponential term includes

---

[1] *ASCA* is derived from Asuka, which means a flying bird in Japanese.

absorption due to intervening matter in the source as well as in the interstellar medium in the Galaxy. Using Equation 2.61 we have $F(E) = (1 + z)L(E(1 + z))/(4\pi D_L^2)$, where D_L is the luminosity distance. Using this relation and its version integrated over the range (E_1, E_2) in Equation 10.1 gives a useful relation between the observed count in a given band and the X-ray flux in it:

$$F(E_1, E_2) = \frac{C(E_1, E_2)}{T} \left(\frac{\int_{E_1}^{E_2} L(E(1 + z))\, dE}{\int_{E_1}^{E_2} \frac{L(E(1+z))}{E} \exp[-\sigma(E)N_H]A(E)\, dE} \right). \qquad (10.2)$$

This is the flux from the target that would be received at the Earth in the absence of any absorption, i.e., the luminosity obtained from it is the energy emitted by the source per unit time in its own frame, in the blueshifted band $[(1 + z)E_1, (1 + z)E_2]$.

The flux corresponding to some observed count rate depends on the shape of the spectrum and the intervening absorption. In the case of quasars and AGN it is usual to consider the simple power law form $L(E) \propto E^{-\alpha_x}$, in which case the count-to-flux relation is specified when the power law slope α_x and N_H are given. Given the flux corresponding to a particular value of α_x, it is not straightforward to convert to another value of the power law index; it is necessary to perform the integrations in Equation 10.2. When the spectrum is more complex than a simple power law, the relation between flux and count rate depends on the redshift of the source.

Absorption   When absorption intrinsic to the source is negligible, as in the case of highly luminous quasars, it is only the photoelectric absorption in the ISM of the Galaxy that matters. An X-ray photon can dislodge an electron from an atom by the photoelectric effect only when the photon energy is at least as much as the binding energy of the electron. For the hydrogen atom in the ground state this is 13.6 eV. The photoelectric absorption cross-section as a function of energy for a complex atom shows a series of *absorption edges*, which correspond to the binding energies of the various electron energy levels. At energy E greater than a particular edge the cross-section is approximately $\propto E^{-3}$ until the next absorption edge is reached. Photoelectric ejection from the 1s shell contributes most to the cross-section and produces the *K edge*.

The total absorption cross-section of a medium depends upon the cross-sections of the atoms, of different elements and in various ionization states, present in it and their abundances. An effective absorption cross-section for the ISM was obtained by Morrison and McCammon (1983) using the best atomic cross-section and cosmic abundance data available to them. Morrison and McCammon have provided the coefficients of piecewise polynomial fits to their numerical results and it is very convenient to use these in calculations. The total cross-section per hydrogen atom obtained using these coefficients is shown in Figure 10.1. The optical depth for a medium of the assumed composition can be obtained simply by specifying a hydrogen column density N_H. We show in Figure 10.2 the optical depth corresponding to three different values of N_H. A convenient approximate formula that describes the energy dependence of the optical

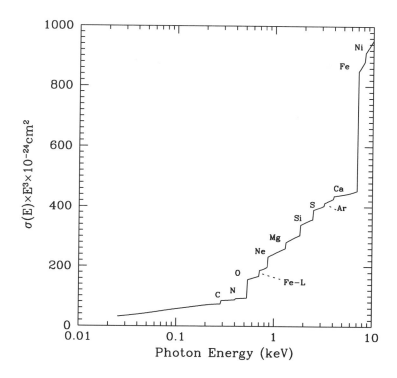

Fig. 10.1. The total photoelectric absorption cross-section of Morrison and McCammon (1983) as a function of energy, obtained by using piecewise polynomial fits. It is assumed that all elements are in the gas phase and in neutral atomic form. The absorption edges due to different elements are indicated.

depth τ is

$$\tau = 2.04 \left(\frac{N_H}{10^{22}\,\text{cm}^{-2}} \right) \left(\frac{E}{1\,\text{keV}} \right)^{-2.4}. \tag{10.3}$$

The optical depth generated from this formula for $N_H = 1.0 \times 10^{22}\,\text{cm}^{-2}$ is shown as a dotted line in Figure 10.2. The Galactic column density of neutral N_H corresponding to the position of a source can be obtained for declination $\delta > -40\,\text{deg}$ from the Bell Laboratories H I Survey by Stark *et al.* (1992), or from Burstein and Heiles (1982).

Determination of the spectrum When energy calibration of the channels of the detector is available and the source is strong enough to produce a sufficient number of counts over the exposure, Equation 10.1 can be used to determine spectral parameters. If the width ΔE of a channel in which counts have accumulated is small compared to the energy corresponding to that channel, i.e., $\Delta E / E \ll 1$, then the integration in Equation 10.1 can be dropped over the channel and the flux in it is $F(E) = C(E)E \exp[\sigma(EN_H)]/[A(E)T]$. The spectrum of the source is then directly

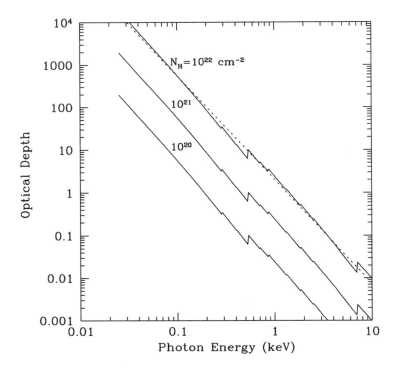

Fig. 10.2. The optical depth obtained from the cross-section of Morrison and McCammon (1983) for different values of the hydrogen column density. The absorption edges are obvious. The dotted line shows the optical depth as given by the approximation described in the text.

determined from the observed count in different channels. X-ray detectors, however, have only modest energy resolution and it is necessary to integrate the flux over the individual channels using Equation 10.1 in order to relate it to the observed count rate. The relation between the flux and count rate therefore depends on the unknown shape of the spectrum. One gets over this by assuming some spectral form with a small number of free parameters, and using the known detector response to predict the counts expected in the various channels. The best-fit parameters are then determined by minimizing

$$\chi^2 = \sum_i \left(\frac{C_i - C_{\mathrm{mod},i}}{\sigma_i} \right)^2, \qquad (10.4)$$

where C_i is the observed count in channel i, σ_i is the estimated uncertainty in it and $C_{\mathrm{mod},i}$ is the count predicted by the model.

In the best circumstances it is possible to show that a spectrum of one particular functional form provides a much better fit than other forms, while in other cases it is not possible to choose between different alternatives. When the detector has no energy resolution, or it is uncalibrated or the count rate is low, one can only determine the total flux in the detector corresponding to some assumed form. If it is believed that there is no absorption in the source, the hydrogen column density N_H used in relating count rate and flux is simply the Galactic value corresponding to the position of the source in the sky. However, if absorption over and above the Galactic value is suspected to exist, then N_H has to be determined through the fitting procedure, assuming some functional form, say the one in Equation 10.3, for the absorption cross-section.

In the case of quasars and AGN, power law spectra provided good fits to the data from missions before *GINGA*. We will see later that even in the case of the most recent data, power law fits are used as a first approximation, relative to which residues are examined. The power law spectrum from a source is modified due to absorption in the Galactic ISM, and a significant number of AGN show intrinsic absorption above the Galactic value in the direction of the source. The power law is represented by $F(E) = AE^{-\alpha_x}$, where A is a normalization constant equal to the flux at 1 keV. The free parameters α_x and N_H are determined through the fit.

When an emission line is present, such as the Fe lines now known to be common to Seyfert spectra (see Section 11.2), the lines can be taken to have a Gaussian shape, in which case the model spectrum is given by (see e.g. Nandra and Pounds 1994)

$$F(E) = AE^{-\alpha_x} + \frac{I_1}{\sqrt{2\pi}\sigma_1} \exp\left[\frac{-(E - E_1)^2}{2\sigma_1^2}\right], \tag{10.5}$$

where E_1 is the central energy of the line, σ_1 its width and I_1 is the integrated line flux. When an absorption edge is suspected to be present, an exponential absorption term is included in the function. For example, for a power law with an iron K-absorption edge (Nandra and Pounds 1994) the spectral form is

$$F(E) = AE^{-\alpha_x} \exp(-\sigma_{Fe} N_{HFe}). \tag{10.6}$$

Here $N_{HFe} = N_H \times$(Fe abundance), the Fe K-absorption cross-section is given by

$$\sigma_{Fe} = 0, \qquad E < E_{th} \tag{10.7}$$
$$\sigma_{Fe} \propto E^{-\delta}, \qquad E \geq E_{th} \tag{10.8}$$

and σ_{th} is the cross-section at the threshold energy E_{th}. The threshold cross-section is a function of ionization state, while $\delta \sim 3.1$ and varies only slowly with ionization (Nandra and Pounds 1994).

X-ray upper limits The counts observed from a target include background counts due to the diffuse X-ray emission and instrumental background. These have to be subtracted from the total count to estimate the net count from the target, which would be related to the flux from it. The background from a point source can be estimated from the counts in an annulus around the target with a large enough radius for significant contamination by counts from the target to be avoided (other sources in the frame also have to be avoided). In the case of the *EINSTEIN* IPC detector, background maps can be created for a source using standard software. The background is not smooth and has its own fluctuations. When the net number of counts detected at some position in an image is small, it could be contributed by a faint actual source, or simply be the result of a fluctuation in the background. It is necessary to choose a criterion that will distinguish between the two, so that the number of spurious detections is limited to some acceptable level. This is done by setting a threshold which the net count from a source must exceed for it to be accepted as a detection. It is usual to set the threshold to three times the standard deviation σ in the background. When the available count is not enough to have produced a detection, a 3σ upper limit to the count rate from the source is taken to be determined. A number of quasars that were observed with *EINSTEIN* have only this type of upper limit on their X-ray flux. Elaborate statistical techniques have been developed to include such upper limits in the study of the X-ray properties of quasars (see Section 10.6).

Optical-to-X-ray spectral index   The ratio of X-ray and optical luminosities of a source is often described in terms of an optical-to-X-ray spectral index (Tananbaum *et al.* 1979)

$$\alpha_{ox} = -\frac{\log(L_x/L_{op})}{\log(v_x/v_{op})},$$
(10.9)

where L_x is the X-ray luminosity at some appropriate frequency v_x and L_{op} the optical luminosity at v_{op}. α_{ox} serves as an effective power law index connecting the optical and X-ray regions, and provides a convenient parameterization of the ratio of the luminosity in these bands. Using α_{ox} does *not* imply that an assumption regarding the shape of the continuum between the optical and X-ray band is being made. If the luminosities are taken at 2 keV and 2500 Å, as is often the case while analyzing *EINSTEIN* data, the denominator in Equation 10.9 has the value 2.605. An effective radio-to-optical spectral index α_{ro}, which connects the luminosities at 5 GHz and 2500 Å, and a radio-to-X-ray index α_{rx}, which connects 5 GHz and 2 keV luminosities, can be defined analogously to Equation 10.9.

10.4   X-ray surveys

We have seen in Section 6.4 how surveys for X-ray selected quasars and AGN can be carried out, either using very deep exposures covering small areas of the sky, or by using a large number of fields originally devoted to the observation of targets independent of the X-ray survey. In the latter case the sources found are said to be serendipitous. We will now consider some mainly recent surveys of the two kinds.

10.4.1   *HEAO 1 A2 high latitude survey*

This was a complete X-ray survey of 8.2 steradian of the sky with Galactic latitude $|b| > 20$ deg to a limiting sensitivity of $\sim 3.1 \times 10^{-11}$ erg cm^{-2} sec^{-1} in the 2–10 keV band

(Piccinotti *et al.* 1982). A total of 85 sources were detected, of which 61 were identified with extragalactic objects. Thirty of these were clusters of galaxies, and the rest were mainly Seyfert 1 and Seyfert 2 galaxies, with the exception of one quasar, four BL Lacs and one normal galaxy. The extragalactic surface density of sources was found to be consistent with the Euclidean relation $N(> F) \propto F^{-3/2}$ as is expected from a nearby non-evolving population. The luminosity function derived for the Seyferts showed that for $3.0 \times 10^{42} < L_x < 1.5 \times 10^{45}$ erg sec^{-1}, Seyfert galaxies contributed ~ 20 per cent of the 2–10 keV X-ray background.

10.4.2 *Extended Medium Sensitivity Survey*

The Extended Medium Sensitivity Survey (EMSS, Gioia *et al.* 1990) made with the *EINSTEIN* observatory consists of 835 serendipitous X-ray sources found in IPC fields originally devoted to observations independent of the EMSS. The EMSS is an extension of the earlier medium sensitivity survey and contains about eight times as many sources. All the fields in the data set were reprocessed using new detection algorithms and background determination which became available long after the processing related to the original observations was done. IPC fields with high galactic latitude ($|b| > 20$ deg) were selected, to avoid regions with high values of N_H and regions with a large number of stars per unit area, as these would contaminate the survey, which is designed to detect extragalactic sources. Fields with bright or extended X-ray sources and those that included groups of objects were avoided, the former because they made background subtraction difficult and the latter so that sources related to each other, and therefore not truly serendipitous, were not included in the complete sample. The exposures of the fields used ranged from ~ 800 to $\sim 40\,000$ sec, with limiting 0.3–3.5 keV X-ray flux in the range $\sim 5 \times 10^{-14}$ to $\sim 3 \times 10^{-12}$ erg cm^{-2} sec^{-1}. For Seyferts, quasars and BL Lacs a power law spectrum with energy index $\alpha_x = 1$ was used. 1435 IPC fields chosen in this manner were searched, and a source was accepted as part of the sample if the significance of detection was $\geq$ four times the rms noise. Avoiding a region of 5 arcmin diameter around the target at the centre, the area searched per field was ~ 0.6 deg^2.

Of the 835 sources in the EMSS, 804 have spectroscopic optical identifications and 25.8 per cent of all sources are found to be Galactic stars. The extragalactic component consists of normal galaxies (2.1 per cent), cooling flow galaxies (0.6 per cent), clusters of galaxies (12.2 per cent), BL Lacs (4.3 per cent) and Seyferts and quasars (51.1 per cent), while 3.9 per cent of the survey sample remain unidentified (Stocke *et al.* 1991). The Seyferts and quasars together span a wide range of luminosity, from $\sim 10^{42}$ to 10^{47} erg sec^{-1}, and have redshifts extending to > 2.

The integral surface density of extragalactic sources with 0.3–3.5 keV flux $> 7 \times 10^{-14}$ is $7.7^{+0.6}_{-0.5}$ deg^{-2}. The best-fit integral source count relation is

$$N(> F) = 140.6 F^{-1.48} \text{ deg}^{-2}, \tag{10.10}$$

where F is the 0.3–3.5 keV flux in units of 10^{-14} erg cm^{-2} sec^{-1} and the 1σ range of the power law index is given by 1.48 ± 0.05. Because of the near complete optical identification of sources the EMSS can be used to obtain luminosity functions of the extragalactic components that are present in sufficient numbers. These will be considered in Section 10.5.

10.4.3 *The EINSTEIN Extended Deep Survey*

The *EINSTEIN* observatory Extended Deep Survey (EDS, Primini *et al.* 1991) consists of a series of 10 deep IPC exposures in six fields, with 34 follow-up exposures with the HRI, which has much better spatial resolution. As in the case of the EMSS, five of the fields were chosen to have galactic latitude $|b| > 20$ deg, low N_H and the absence of bright X-ray sources, while one field was chosen so that it satisfied these criteria and covered an area that had previously been optically surveyed for quasars. The IPC energy band used was 0.8–3.5 keV and a power law spectral index $\alpha_x = 0.5$ was assumed for all sources. X-ray sources detected in the central 32 arcmin $\times$ 32 arcmin region were accepted as part of the sample, with the detection threshold set at 4.5 times the noise. The HRI fields were analyzed independently, but the detections were used only to reduce the positional uncertainties of sources already in the IPC sample.

A total of 25 sources were detected in the ~ 2.3 deg covered by the survey above a limiting flux of 3.68×10^{-14} erg cm^{-2} sec^{-1}. Application of the V/V_m test (see Section 7.5) shows that incompleteness, i.e., the percentage of missed sources which satisfy the selection criteria, is not significant above a flux of 4.5×10^{-14} erg cm^{-2} sec^{-1}. Of the 25 sources, six have been optically identified with Galactic stars, nine with quasars, two with other extragalactic objects and eight remain optically unidentified. An upper limit on the X-ray-to-optical flux ratio has been obtained for the unidentified sources, assuming that they have optical magnitude $V > 20$. On the basis of these it has been surmised that the unidentified sources are extragalactic.

The number of detected sources and the range of their X-ray flux is not large enough to permit a statistically significant fit to a power law surface density distribution, as was done with the EMSS. Primini *et al.* therefore assume that $N(> F) \propto F^{-3/2}$ and determine the normalization from the observations. The result is

$$N(> F) = 91.7F^{-1.5} \text{ deg}^{-2}, \tag{10.11}$$

with the 68 per cent confidence range for the normalization given by the interval $(81.3, 129.2)$. The predictions of the EDS are consistent with the earlier results of the *EINSTEIN* deep survey over a smaller number of fields.

10.4.4 *ROSAT Deep Survey*

The *ROSAT* position sensitive proportional counter (PSPC) deep survey (Hasinger *et al.* 1993; see the *Erratum* in Hasinger *et al.* 1994) of the sky in the region of the Lockmann Hole is the most sensitive soft X-ray survey that has been made up to the time of writing. The equivalent neutral hydrogen column density in this region, $N_H = 5.7 \times 10^{19}$ cm^{-2}, is the lowest known in the sky (and hence the name). The Lockmann Hole is ideally suited for a survey with the PSPC detector: this is highly sensitive at low energies where the flux is seriously attenuated for any significant column density because of the high photoabsorption cross-section at soft X-ray energies.

The point spread function (PSF) of the X-ray telescope depends on the off-axis angle. On the axis it is a circular Gaussian with full width at half maximum (FWHM) of ~ 25 arcsec. The FWHM increases with off-axis angle, which means that the background counts included in its area increase too, leading to a reduced signal-to-noise ratio for an off-axis source. The increase in the FWHM also increases the probability of source confusion, i.e., the mis-identification of two faint sources with angular separation comparable to the FWHM as one bright source. Beyond $20'$, the PSF becomes asymmetric and the circular support structure of the PSPC entrance window casts a shadow on the image (this is called vignetting). To avoid these effects, only sources with off-axis angle < 15.5 arcmin were used, i.e., the usable area in each field was the central 0.21 deg^2.

The total usable deep exposure amounted to 143.7 kilosec and there were 661 detections in the central region. The flux limit reached was 2.5×10^{-15} erg cm^{-2} sec^{-1} in the 0.5–2 keV band. In addition to the deep field, the central 15.5 arcmin of 26 other PSPC fields with exposure of shorter duration, < 8000 sec, from other pointing programmes were also used so that a significant number of brighter sources would be included in the survey. A statistically complete sample of 661 sources was found having 0.5–2 keV flux $> 2.5 \times 10^{-15}$ erg cm^{-2} sec^{-1}, with a range of ~ 200 in the observed fluxes. After allowing for corrections due to source confusion and the variation of the PSF across the field of view, this amounts to a surface density of 413 deg^{-2} above the limiting flux.

The differential source count, i.e., the surface density of discrete sources per unit flux range can be represented by a combination of two power law functions, with

$$N(F) = 228F^{-2.72}, \quad F > 2.66, \tag{10.12}$$
$$N(F) = 111F^{-1.94}, \quad F \leq 2.66, \tag{10.13}$$

where F is the 0.5–2 keV flux in units of 10^{-14} erg cm^{-2} sec^{-1}. The 90 per cent confidence ranges of the slopes β_f and β_b of the faint and bright parts are 2.72 ± 0.27 and 1.94 ± 0.19 respectively. The range of the normalization N_f of the faint part is 111 ± 10 and continuity of the source count at the boundary $F_b = 2.66$ requires that the normalization of the bright part is $N_b = N_f \times F_b^{\beta_f - \beta_b}$. The slope at the bright end is consistent at the 90 per cent level with the Euclidean differential count slope of 2.5 (see Section 7.4), but there is significant flattening of the slope at the faint end. A single power law model is excluded at the 99.9 per cent confidence level. The flattening of the slope at low flux levels has been observed in other *ROSAT* surveys, and was earlier noticed in the fluctuation analysis of IPC deep fields by Hamilton and Helfand (1987) and Barcons and Fabian (1990).

Even when a source is too faint to be detected above the sky background, it contributes to the total background flux measured over an area. The total contribution to the background by such faint sources can fluctuate from one area to another, because of the statistical (Poisson) fluctuation in the number of sources 'uniformly' distributed in the sky. The distribution of the fluctuations in the intensity, called the $P(D)$ distribution, can be used to estimate the differential source count to a flux level of about an order of magnitude fainter than the discrete source detection limit (Condon 1974, Scheuer 1974b; see Barcons and Fabian 1990 for a more recent discussion). Using this technique, Hasinger *et al.* (1993) have obtained

$$N(F) = 116F^{-1.8}, \tag{10.14}$$

with the 90 per cent ranges of the slope and normalization given by 1.80 ± 0.08 and 116 ± 10 respectively. The departure of the slope from the Euclidean value at the faint end is indicative of evolution and/or the effects of space–time curvature. However, it is possible that at very faint levels a population of Euclidean sources, say starburst galaxies, begin to dominate the source count. Hasinger *et al.* find from the fluctuation analysis that a resteepening of the slope to the Euclidean value is possible only below a flux $F_e \leq 0.2$, with the normalization of the source count given by $N_e = N_f \times F_e^{2.5 - \beta_f} = 45$.

The deep survey counts are shown in Figure 10.3 along with results from the fluctuation analysis, the *EINSTEIN* deep survey and the EMSS. Implications of the counts for the contribution of the discrete sources to the X-ray background will be discussed in Section 11.6.

10.4.5 *Other ROSAT surveys*

Two fields from the Durham UVX quasar survey (see subsection 6.2.1) have been observed with the PSPC by Boyle *et al.* (1993) for 24 500 sec and 27 500 sec respectively. These authors used the central 40 arcmin diameter region from each field and detected 89 sources, of which 66 constitute a complete sample above the 5σ detection threshold. The limiting flux in the 0.5–2 keV band is 6×10^{-15} erg cm^{-2} sec^{-1} (assuming $\alpha_x = 1$). Of the complete sample, 42 have been spectroscopically identified with quasars, and there are seven stars, four galaxies and one BL Lac object. Twelve of the objects in the complete sample remain unidentified. Unambiguous redshifts have been obtained for 85 per cent of the quasars and the mean redshift is ~ 1.5. The surface density obtained from the survey is consistent with the EMSS surface density in the region where the fluxes, transformed to the same energy band, overlap. We shall discuss in Section 10.5 the X-ray luminosity function obtained from this survey.

Branduardi-Raymont *et al.* (1994) have surveyed two regions of low hydrogen column density, $N_H = (6-9) \times 10^{19}$ cm^{-2}, with the PSPC. One region was subjected to a long exposure of 70 kilosec, while the other was surveyed in six shorter exposures, of 13–20 kilosec duration. A region in each field with 12 arcmin diameter was used, and

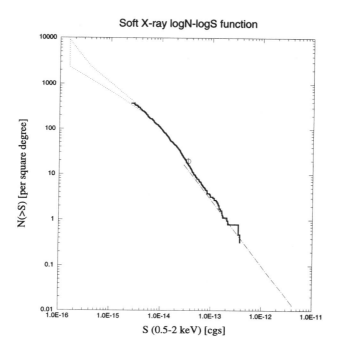

Fig. 10.3. The integral surface density of X-ray sources obtained by Hasinger *et al.* (1993, 1994). The open circle represents the *EINSTEIN* extended deep survey point (Primini *et al.* 1991). The dotted and broken line represents the best fit to the EMSS survey. The dotted area at faint fluxes shows the 90 per cent confidence region obtained from fluctuation analysis of the deepest *ROSAT* field. Figure kindly provided by G. Hasinger.

47 sources were detected in the field with the long exposure and 94 sources in the six other fields. A mean value of $N_{\rm H} = 8 \times 10^{19}\,{\rm cm}^{-2}$ and $\alpha_{\rm x} = 1.1$ was used in converting from counts to flux in the 0.5–2 keV band.

Upon using a Monte Carlo technique to correct for the effects of source confusion and incompleteness, Branduardi-Raymont *et al.* found that a broken power law provides a good fit to the surface density of X-ray sources, with the break at $1.64^{+0.97}_{-0.41} \times 10^{-14}\,{\rm erg\,cm}^{-2}\,{\rm sec}^{-1}$ in the 0.5–2 keV band. This flux as well as the other parameters of the best fit are consistent to within the error limits with the values obtained by Hasinger *et al.* (1993).

Fluctuation analysis of the seven fields of the Branduardi-Raymont *et al.* survey has been performed by Barcons *et al.* (1994). From analysis of the six shorter-exposure fields they find that the surface density inferred from fluctuations is in good agreement with direct source counts from their deep field, as well as the results of Hasinger *et al.* (1993); a flattening of the surface density is observed for 0.5–2 keV flux $\lesssim 2.2 \times 10^{-14}\,{\rm erg\,cm}^{-2}\,{\rm sec}^{-1}$. Fluctuation analysis of the field with the deeper 70 kilosec exposure shows that the surface density $N(F) \propto F^{-\gamma}$ with $\gamma = 1.8^{+0.2}_{-0.1}$ up to a flux level of $7 \times 10^{-16} \, \mathrm{erg \, cm^{-2} \, sec^{-1}}$. There is no evidence of a Euclidean population with a steeper power law at very low fluxes. The surface density of X-ray sources inferred from the fluctuation analysis is $900-1800 \, \mathrm{deg^{-2}}$ for flux $> 7 \times 10^{-16} \, \mathrm{erg \, cm^{-2} \, sec^{-1}}$.

10.5   X-ray luminosity functions

An X-ray luminosity function (XLF) provides the number of objects of some species per unit volume of space as a function of their X-ray luminosity, redshift and possibly additional attributes such as luminosity in other bands. It can be obtained by convolving the optical luminosity function, for the type of object being considered, with the distribution of the ratio of X-ray-to-optical flux (i.e., α_{ox}). How well the XLF can be determined in this manner depends upon how well the distribution of the ratio is known. If the part of the distribution corresponding to faint X-ray fluxes is missed, because some of the sources observed are not detected with the available exposure times, then extrapolating the observed distribution to all sources will lead to an overestimation of their mean X-ray luminosity. We will see in Section 10.6 how the non-detections (i.e., X-ray upper limits) can be incorporated in estimating the observed distribution of the X-ray-to-optical flux ratio in an unbiased fashion.

Another way to determine the XLF is to derive it directly from X-ray surveys that have complete or nearly complete optical identifications. The EMSS (subsection 10.4.2) has 835 X-ray sources, of which 96 per cent have been optically identified and include 427 Seyferts and quasars. The EMSS is complemented by the *ROSAT* survey of Boyle *et al.* (1993) (subsection 10.4.5), which goes a factor of ~ 10 deeper than the EMSS in much the same energy band. The optical identification of the 66 sources in this sample is ~ 85 per cent complete and there are 42 spectroscopically identified quasars. These provide a coverage of the $L_x z$-plane that goes an order of magnitude fainter than the EMSS for $z > 0.5$ (see Figure 10.4). Boyle *et al.* have used the combined EMSS and *ROSAT* samples to derive the XLF for quasars and Seyferts, considered as a single population.

The XLF can be obtained in a parameter-free form by applying the V/V_m technique. However, the sources in the sample are scattered over disjoint areas of the sky and it is therefore necessary to use a generalization of the method, due to Avni and Bahcall (1980) (see subsection 7.6.4). Boyle *et al.* divide the sample into bins of equal width in $\log L_x$ and four equal bins in $\log(1 + z)$ over the redshift interval $0 < z < 3$. The luminosity used is the $0.3-3.5 \, \mathrm{keV}$ luminosity in the *rest frame* of the quasar. To correct from observed to rest frame energies, a two-power-law model with energy spectral index $\alpha_x = 1.3$ for photon energies $E < 1.5 \, \mathrm{keV}$ and $\alpha_x = 1.0$ for $E > 1.5 \, \mathrm{keV}$, is assumed, in order to represent the steepening of the spectrum at soft energies (see Section 11.1). The XLF is obtained separately for deceleration parameter values $q_0 = 0$ and $q_0 = 0.5$. For each bin the comoving number density of objects is determined by summing over the reciprocal of the accessible volume V_a for all the objects in the bin (see Equations 7.46 and 7.56). As in the case of the optical luminosity function for

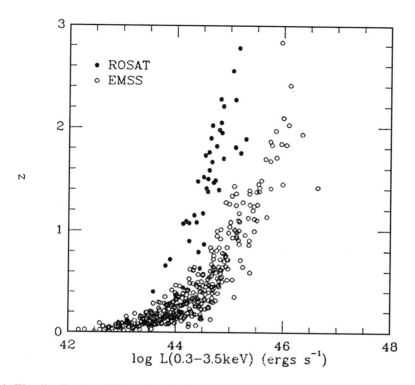

Fig. 10.4. The distribution of X-ray luminosity and redshift in the EMSS and the *ROSAT* survey of Boyle *et al.* (1993). EMSS quasars are shown as open circles, while the *ROSAT* quasars are shown filled. Reproduced from Boyle *et al.* (1993).

quasars (see Section 7.7), the derived XLF in each redshift bin shows curvature, with a flatter slope at low luminosities. The redshift evolution can be seen as a constant change between different redshift bins, consistent with pure luminosity evolution, again as in the optical case. The X-ray luminosity function is shown in Figure 10.5.

Taking a hint from the XLF derived as described above, Boyle *et al.* have applied the maximum likelihood analysis described in subsection 7.6.3 to find the best-fitting values of the parameters of a two-power-law model. The local ($z = 0$) luminosity function in the model is given by

$$\Phi(L_x) = \Phi_x^* L_{x,44}^{-\gamma_1}, \quad L_x < L_x^*(0), \tag{10.15}$$
$$\Phi(L_x) = \Phi_{x1}^* L_{x,44}^{-\gamma_2}, \quad L_x > L_x^*(0), \tag{10.16}$$

with $L_{x,44}$ the X-ray luminosity in units of 10^{44} erg sec^{-1}. The break in the two slopes occurs at a redshift-dependent luminosity given by

$$L_x^*(z) = L_x^*(0)(1 + z)^{k_x}. \tag{10.17}$$

Continuity of the luminosity function at the break luminosity requires that $\Phi_{x1}^* =$

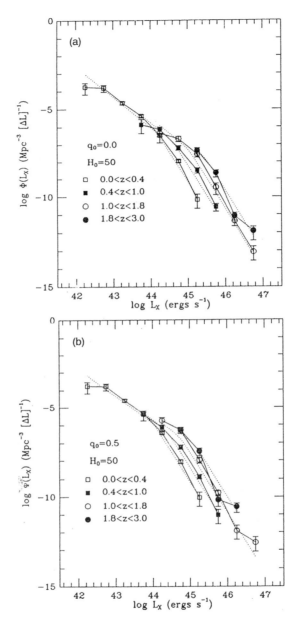

Fig. 10.5. The X-ray luminosity function (XLF) derived using the accessible volume method. The dotted lines in the upper panel show the XLF corresponding to model **D** and in the lower panel to model **G** from Table 10.1. Reproduced from Boyle *et al.* (1993).

Table 10.1. *Best-fit parameter values for X-ray luminosity functions, obtained using the combined ROSAT and EMSS data sets. The models are named as in Boyle* et al. *(1993). The luminosity is given in* $\mathrm{erg\,sec^{-1}}$ *and* Φ_x^* *is given in* $10^{-6}\,\mathrm{Mpc^{-3}}(10^{44}\,\mathrm{erg\,sec^{-1}})^{-1}$. *All models except* **A** *correspond to the broken power law X-ray spectrum, with* $\alpha_x = 1.3$ *for* $E < 1.5\,\mathrm{keV}$ *and* $\alpha_x = 1$ *for* $E > 1.5\,\mathrm{keV}$. P *is the two-dimensional confidence level with which the model can be accepted*

Model	q_0	Evolution function	α_x	γ_1	γ_2	$\log L_x^*(0)$	k	Φ_x^*	Prob. P
A	0.0	power law	1.0	1.67	3.40	43.84	2.75	0.57	0.27
D	0.0	power law	1.3/1	1.71	3.44	43.91	2.80	0.54	0.35
G	0.5	power law	1.3/1	1.55	3.33	43.79	2.50	0.87	0.09
B	0.0	exponential	1.3/1	1.65	3.38	43.62	5.10	0.52	0.01

$L_{x,44}^*(0)\Phi_x^*$. The normalization Φ_x^* can be obtained by integrating the luminosity function over the range of variables and setting the result equal to the total number N of quasars in the sample:

$$\int \int \Phi(L_x, z)\Omega(L_x, z)\, dv(z)\, dL_x = N. \tag{10.18}$$

Here $\Omega(L_x, z)$ is the solid angle covered by the survey as a function of redshift and luminosity. The free parameters are $\gamma 1$, $\gamma 2$, $L_x^*(0)$ and k_x. While the maximum likelihood technique produces a best-fit set of values for the parameters, it does not say how good is the best fit which has been obtained. Boyle *et al.* test the goodness of fit using a two-dimensional analogue of the Kolmogorov–Smirnov test (Peacock 1983, 1985, Fasano and Franceschini 1987; see PTVF92 for a brief description and computer routines). In this test the distribution of sources in the $L_x z$-plane is compared with the predictions of the best-fit model and a statistic is produced from which the significance can be estimated. We have shown in Table 10.1 the parameters derived for some of the models considered in Boyle *et al.* (1993).

The main results emerging from the analysis of Boyle *et al.* are as follows. (1) The two-power-law luminosity function with pure luminosity evolution described by an exponential provides a poor fit to the data for $q_0 = 0$ (model **B** in Table 10.1) as well as for $q_0 = 0.5$; this was also found by Della Ceca *et al.* (1992), from analysis of the EMSS sample. (2) When the evolution function is taken to be of power law form and a power law X-ray continuum with $\alpha_x = 1$ is used, a satisfactory fit to the combined *ROSAT* and EMSS sample is obtained for the $q_0 = 0.5$ cosmology (Model **A**). (3) The fit is improved somewhat when an X-ray spectrum that steepens to $\alpha_x = 1.3$ at $E < 1.5\,\mathrm{keV}$ is used (model **D**). (4) The fit is not as good for the $q_0 = 0.5$ cosmology (model **G**), but improves when the evolution is cut off for $z \gtrsim 2$. Evolution of the break luminosity as an exponential function of the look-back time was also tried but did not provide acceptable fits.

The general trend of results obtained by Boyle *et al.* was already evident in the analysis of the EMSS data in Maccacaro *et al.* (1991) and Della Ceca *et al.* (1992), but there are some small but possibly significant differences in the values of the derived best-fit parameters. The differences in the shape parameters, i.e., in the power law luminosity-function slopes γ_1 and γ_2 and the break luminosity $L_x^*(0)$, can be attributed to the somewhat different analysis used with the EMSS data (see Boyle *et al.* 1993 for a detailed discussion), but there is a difference in the derived values of the power law evolution parameter k that cannot be explained in this fashion. The k value for model **D** is ~ 0.2 in excess of the value obtained in the EMSS analysis, which means that the *ROSAT* sample shows faster evolution than the EMSS sample. The EMSS best-fit models (Della Ceca *et al.* 1992) can be used to predict the redshift distribution expected in the *ROSAT* survey. It is found that the observed distribution contains many more high redshift quasars than are predicted by Della Ceca *et al.* (1992), which is consistent with the faster evolution obtained by Boyle *et al.* using the new *ROSAT* data.

The faster evolution is somewhat surprising, because for $z \gtrsim 1$ the new objects have luminosities about an order of magnitude lower than EMSS objects at the same redshift. Boyle *et al.* have suggested that the faster evolution that they find could be a consequence of the departure of the X-ray spectral shape from a simple power law. Let k be the power law evolution parameter and α_x the X-ray spectral index for a set of quasars. Now suppose that the set of sources is analyzed using an X-ray spectral index equal to 1 instead of α_x. Then it is easy to show, using the relation between flux and luminosity obtained in Equation 2.61 and the expression for the break luminosity $L_x^*(z)$ given above, that an evolution parameter k' will be obtained which is related to k through

$$k = k' + \alpha_x - 1. \tag{10.19}$$

There are indications that at low energies the X-ray spectra of quasars have steeper slopes than the extrapolation of the high energy spectra to lower energies would provide (see Section 11.1). Boyle *et al.* argue that if the slope at energies $E < 1.5\,\text{keV}$ is ~ 1.3 instead of 1, then the difference of ~ 0.2 between the evolution parameters can be understood using Equation 10.19. The EMSS quasars have a mean redshift of ~ 0.3 and therefore it is the steep part of the spectrum that is being probed in the EMSS observations. However, in the *ROSAT* data quasars at higher redshift and therefore higher rest frame energies, with $\alpha_x \simeq 1$, are observed. Boyle *et al.* in fact introduce a broken power law X-ray spectrum into their models to account for the different evolution parameters observed. While there is much evidence for a soft X-ray excess in quasars, there is a great deal of uncertainty about its exact properties, and some caution needs to be exercised in attributing the observed small difference in evolution to it. To obtain reliable models of evolution, further data will be required to test whether any real differences exist, and to examine the reasons for them. In any case, as we have emphasized earlier, these deductions of evolution are not related to any evolution of the physical environment in the expanding universe and are therefore somewhat *ad hoc*.

We have seen in Section 7.7 that the optical luminosity function of quasars has the form of a broken power law with pure luminosity evolution. The close similarity between the shapes and perceived evolution of the optical and X-ray luminosity functions should not come as a surprise. If all quasars had the same ratio of optical-to-X-ray luminosities and the same overall spectral shape, going from one luminosity function to another would simply be a matter of scaling the luminosity, and all properties would remain unchanged. If the ratio of the two luminosities had a universal distribution independent of the other properties of quasars, again the two luminosity functions would be identical, except at the extremities of the luminosity range. It is known from study of the broad band properties of a large number of quasars (see Section 10.6) that their X-ray and optical luminosities show a statistical relation of the form $L_x \propto L_{op}^{0.7\pm0.05}$, i.e., the ratio of the two luminosities is weakly dependent on the optical luminosity (see Equation 10.25 below). Therefore if the X-ray luminosity function is derived from the optical luminosity function, it is to be expected that the two will have the same characteristics overall but with differences in the values of the various parameters. Inspection of Tables 7.1, 7.2 and 10.1 shows that the slopes of the two luminosity functions are similar but the values of the evolution parameters are significantly different, since errors on these parameters are ±0.1 in all cases (see Boyle *et al.* 1993). If the X-ray and optical luminosities are related by $L_x \propto L_{op}^A$, where A is a constant, then it is straightforward to show that the slopes of luminosity functions are related by $\gamma_x - 1 = (\gamma_{opt} - 1)/A$ and the evolution parameters by $k_x = Ak_{opt}$. Comparing the faint- and bright-end slopes and evolution parameters for the corresponding models and taking into account uncertainties in their values, Boyle *et al.* find that $A = 0.88 \pm 0.08$, which is steeper than the power law relation with $A = 0.7 \pm 0.05$ noted above, but not inconsistent with it.

10.6   Broad band X-ray properties

The first quasar to show up as an X-ray source was 3C 273, which was detected in rocket and balloon flights (Bowyer *et al.* 1970). X-ray emission from the Seyfert galaxy NGC 4151 and the radio galaxy Cen A was detected by the *UHURU* observatory (Giacconi *et al.* 1974), which also provided the first X-ray spectra for AGN. It became evident from the detection of a number of Seyfert galaxies by the *ARIEL* satellite (Elvis *et al.* 1978) that X-ray emission was common to these objects. However, the number of detected quasars remained very small until the launch of the *EINSTEIN* observatory, which had the capability to obtain long exposures in fixed directions with imaging detectors.

X-ray samples of a class of objects can be produced in quite different ways.

(1) Known objects belonging to the class that have been discovered in surveys in other bands can be observed in the pointing mode to obtain the X-ray flux from them. A large number of quasars have been observed by *EINSTEIN* in this

manner, principally with the IPC detector. The first such sample was observed by Tananbaum *et al.* (1979), followed amongst others by Ku, Helfand and Lucy (1980), Zamorani *et al.* (1981) and Canizares and White (1989). A large number of Seyferts were observed in the pointing mode by Kriss *et al.* (1980).

(2) X-ray selected quasars and AGN have been discovered in the identifications following deep X-ray surveys. These surveys cover small parts of the sky to very deep levels and most of the sources are close to the limiting flux of the survey. Optical identification is therefore a slow and difficult process. Quasars have been discovered in this manner in *EINSTEIN* (Giacconi *et al.* 1979, Primini *et al.* 1991) and *ROSAT* deep surveys (Boyle *et al.* 1993).

(3) X-ray selected objects have also been discovered serendipitously in the fields centred on other targets (see Section 6.4 for a discussion of this technique). The first serendipitous X-ray selected quasars were reported by Grindlay *et al.* (1980). The largest such sample is the Extended Medium Sensitivity Survey (EMSS) described in subsection 10.4.2. Of the 835 sources in that survey, ~ 51 per cent are quasars and Seyferts, while some of the rest are identified with BL Lac objects. Kriss *et al.* (1982) have produced a list of X-ray selected Seyferts.

There are some interesting variations on this theme of finding certain species of object in exisiting IPC fields. For example, Anderson and Margon (1987) selected a number of suitable IPC fields, surveyed optically the regions of the sky covered by them for quasars, and having identified these, obtained their X-ray flux from the fields in which they appeared. Though there is an element of serendipity in this method as well, the sample produced is optically selected and *not* X-ray selected, because the quasars are first identified using an optical technique and their X-ray flux is then determined. Green *et al.* (1995) have cross-correlated quasars from the Large Bright Quasar Survey (LBQS, see subsection 6.2.3) with fields observed with the *EINSTEIN* IPC and obtained X-ray detections or upper limits for a large number of quasars.

The large samples of optically selected and radio selected quasars with X-ray observations can be used in the study of population properties such as the ubiquity of X-ray emission, the relation between X-ray and other luminosities, the correlation between X-ray emission and special properties like high radio luminosity and so on. The first analyses of this sort were reported by by Tananbaum *et al.* (1979), Ku, Helfand and Lucy (1980) and Zamorani *et al.* (1981). The main conclusions to emerge from these works were as follows.

- A large fraction of quasars are strong X-ray emitters.
- While the ratio of X-ray to optical luminosity L_x/L_{op} (i.e., α_{ox}) has a broad distribution, on average $L_x \propto L_{op}^b$, with $b \lesssim 1$.
- The mean value $\langle \alpha_{ox} \rangle$ for radio-quiet quasars is in the range $1.4 - 1.5$.
- Radio-loud quasars are on average about three times more luminous than radio-quiet quasars with the same optical luminosity, so that their mean value $\langle \alpha_{ox} \rangle \simeq 1.3$.

- The quasar population could contribute up to ∼ 30 per cent of the diffuse X-ray background at a few keV. In order not to exceed the observed background, the steep source count for quasars, applicable at $B \lesssim 20$, would have to flatten at fainter magnitudes.

These early conclusions have been strengthened and put in a more quantitative form with the increased availability of data, application of advanced statistical techniques and better understanding of the astrophysics behind the perceived relationships. The extent and spectral shape of the contribution of quasars to the X-ray background has, however, remained enigmatic, because of the many unknowns involved in the estimation. We shall consider these matters further in the following sections.

10.6.1 *The EINSTEIN database*

A large database of the quasars and Seyfert galaxies observed with the IPC detector on *EINSTEIN* has been produced by Wilkes *et al.* (1994). This consists of X-ray count rates, fluxes and luminosities, or their 3σ upper limits (see Section 10.3) for previously known optically selected or radio selected quasars and Seyfert 1 galaxies that have been targeted for observations by different groups. In order to ensure uniformity, the results included in the database have been obtained by a systematic re-analysis of the observations using the most recent data processing software and calibration. The X-ray flux for each source has been obtained from the observed counts using the formulation presented in Section 10.3 for values of X-ray spectral index α_x equal to 0.0, 0.5, 1.0, 1.5 and 2.0.

The 514 sources included in the database of Wilkes *et al.* were all known quasars or Seyfert 1 galaxies targeted in some programme or other, and are therefore found mostly to lie on-axis in the images (unlike serendipitous sources, which could be anywhere in the field). Taken together the sources do not form a well-defined statistically complete sample, as they were observed by different groups as part of various scientific programmes. But the database includes three complete subsamples consisting of the following. (1) 64 quasars that form an unbiased subset of the Bright Quasar Survey of Schmidt and Green (1983, see subsection 6.7.1). The 64 quasars were selected on the basis of the scheduling constraints of the *EINSTEIN* observatory, and not on their properties. As such they constitute an unbiased subset of the BQS. Of these quasars, 57 have positive X-ray detections while the other nine have X-ray upper limits. (2) 30 quasars from the optical survey of Braccesi *et al.* (see Marshall *et al.* 1984 for a description and references). These form a UVX selected complete sample with magnitude $B < 19.5$. Of these, nine quasars have X-ray detections and the rest have upper limits. This sample is denoted by BF. (3) 33 radio selected 3CR quasars, of which 31 have X-ray detections and two have upper limits. In spite of the heterogeneity of the sample, the database can be used to assess the distribution of various broad band X-ray emission related parameters in the quasar population as a whole and to obtain statistical relations between these and properties in other bands. The distribution of

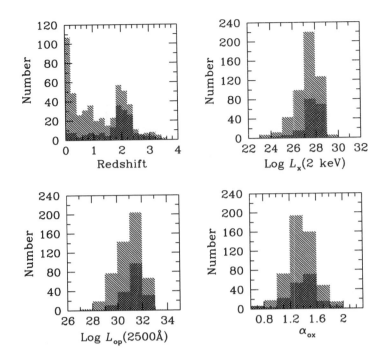

Fig. 10.6. The distribution of redshift, X-ray luminosity at 2 keV in units of $\mathrm{erg\,cm^{-2}\,sec^{-1}\,Hz^{-1}}$, optical luminosity at 2500 Å in the same units and α_{ox} for the sample of Wilkes *et al.* (1994). X-ray upper limits are indicated by the cross-hatching.

redshift, optical and X-ray luminosities and the optical-to-X-ray spectral index α_{ox} (see Section 10.3 for the definition) are shown in Figure 10.6.

The 835 quasars and Seyfert 1 galaxies from the Extended Medium Sensitivity Survey (EMSS, see subsection 10.4.2) form another large database suitable for statistical studies. Sources from the EMSS and from Wilkes *et al.* (1994) cannot, however, be indiscriminately mixed together in such studies, since the two use different primary selection techniques. The EMSS sample, being X-ray selected, is biased towards brighter X-ray fluxes. The sample in the *EINSTEIN* database has no such bias, but the weaker fluxes largely appear as upper limits. The two samples come from different regions of the three-dimensional (L_{op}, L_x, z) distribution, which has to be kept in mind while making any comparisons.

10.6.2 *Relation between X-ray and optical luminosities*

The first X-ray studies of large samples of quasars showed that a significant correlation was present between the X-ray and optical fluxes (see Zamorani *et al.* 1981), indicating a relation between the corresponding luminosities. If the continuum from the optical

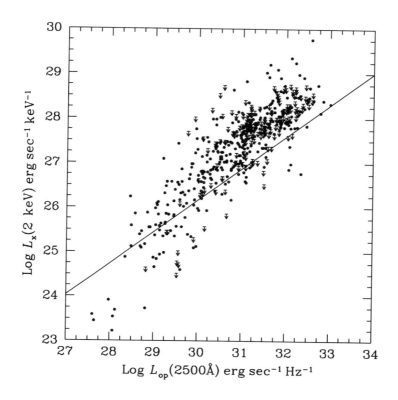

Fig. 10.7. A plot of 2 keV X-ray luminosity against optical luminosity at 2500 Å for the Wilkes *et al.* (1994) sources. The arrows indicate X-ray upper limits. The line shows the fit by Wilkes *et al.* (1994) which takes into account the upper limits.

to X-ray frequencies had the form of a single power law $L(E) = AE^{-\alpha}$, one would have

$$\log L_x = \log L_{op} - \alpha \log \left(\frac{E_x}{E_{op}} \right) - \log A, \qquad (10.20)$$

where E_x and E_{op} are the energies at which the X-ray luminosity L_x and optical luminosity L_{op} are defined. If all quasar spectra had the same power law index α and normalization A, a plot of $\log L_x$ against $\log L_{op}$ would produce a straight line of unit slope with a small scatter about the line, reflecting noise in the measurements. If instead of universally constant values there are distributions of A and α in the quasar population, the plot will show a large scatter, with $\langle \log L_x \rangle \propto \langle \log L_{op} \rangle$.

We have shown in Figure 10.7 a plot of $\log L_x(2\,\text{keV})$ against $L_{op}(2500\,\text{Å})$, for optically selected quasars from Wilkes *et al.* (1994). Radio selected quasars are excluded from this plot because it is possible that a part of their X-ray emission is related to their high radio luminosity, which complicates any relation between L_x and L_{op}. We have taken the luminosities as listed in Wilkes *et al.* (1994), which correspond to $H_0 = 100\,\text{km}\,\text{sec}^{-1}\,\text{Mpc}^{-1}, q_0 = 0, \alpha_x = 0.5$ and $\alpha_{op} = 0.5$, for ease of comparison with the literature. Using other values for these parameters does not change any of the results significantly. The corrections used in obtaining luminosities from fluxes were discussed in Section 6.6 and the way they have been applied to the present sample has been described in detail in Wilkes *et al.* (1994). There is clearly a correlation between the luminosities but two important points have to be taken into account before a regression analysis is carried out and conclusions drawn from it.

Induced correlations   Consider a set of sources that have been observed in two bands, say the X-ray and optical. Let a narrow range only of fluxes be available in each band. Using Equation 2.61 the X-ray luminosity of a source can be written in terms of the observed X-ray flux as

$$\log L_x(2\,\text{keV}) = \log F_x(2\,\text{keV}) - (1 - \alpha_x)\log(1 + z) + 2\log D_L + \log(4\pi), \quad (10.21)$$

where D_L is the luminosity distance, the luminosity is in the rest frame of the source with emitted photon energy $2\,\text{keV}$ and the flux is at observed photon energy $2\,\text{keV}$. A similar expression is obtained for the optical luminosity in terms of the optical flux, with the spectral index α_{op}. Owing to the stretching effect of the redshift-dependent factor, the luminosity can cover a wide range even when the flux is confined to a narrow range. As a result, starting with a group of sources with uncorrelated luminosities it is possible to obtain an apparently highly correlated distribution in the luminosity plane. Such an effect is, of course, most pronounced when the dynamic range (i.e., the ratio of the maximum to the minimum value) of the fluxes is much smaller than the dynamic range of the luminosities. We see from the plot in Figure 10.8 that for optically selected quasars from Wilkes *et al.* the dynamic ranges of flux and luminosity are comparable in both the bands. Any spurious correlation induced by the redshift will therefore be small.

It is apparent from the distribution of points in Figure 10.7 that one should look for a linear relationship of the form

$$\log L_x = a\log L_{op} + b. \quad (10.22)$$

The parameters a and b are estimated from linear regression analysis (see B69 for an excellent introduction). A measure of the dependence of the luminosities on each other is provided by the *linear correlation coefficient r*, which is 0 when there is no correlation and ± 1 when the data points lie along a perfect straight line with positive or negative slope respectively. A statistic t is defined by

$$t = r\sqrt{\frac{N - 2}{1 - r^2}}, \quad (10.23)$$

where N is the number of data points. When the population from which the data points are drawn is uncorrelated, the statistic t has *Student's t-distribution*. If the value of the statistic obtained from the data is t_d, one can obtain from the distribution the probability $P(> t_d)$ that t_d would be exceeded if the data points were drawn at random

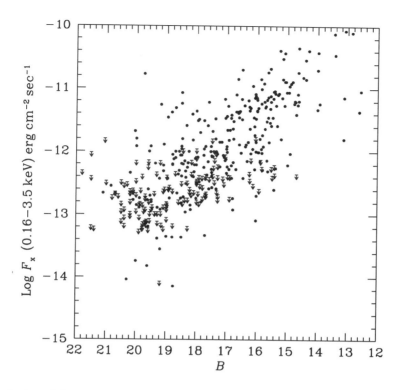

Fig. 10.8. A plot of the X-ray flux at 2 keV against B magnitude for optically selected quasars from Wilkes *et al.* (1994). The arrows indicate X-ray upper limits.

from an uncorrelated population. $(1-P) \times 100$ provides the percentage confidence with which the hypothesis that the two luminosities are correlated can be accepted.

We have seen above that a situation can arise in which an observed correlation between the X-ray and optical luminosities is mainly due to the dependence of each luminosity on the redshift z. Even where the observed correlation is not wholly induced in this manner, it is important to see whether the correlation remains significant when the effect of the third parameter is taken into account. This can be done by evaluating *a partial linear correlation coefficient* as follows. Let $r_{x,op}$, $r_{x,z}$ and $r_{op,z}$ be the correlation coefficients between the pairs $\log L_x$ and $\log L_{op}$, $\log L_x$ and z, and $\log L_{op}$ and z respectively. A partial linear correlation coefficient is now defined by

$$r_{x,op;z} = \frac{r_{x,op}^2 - r_{x,z} r_{op,z}}{\sqrt{1 - r_{x,z}^2}\sqrt{1 - r_{op,z}^2}}. \tag{10.24}$$

The partial correlation coefficient has the same distribution as the ordinary correlation coefficient and therefore the same tests of significance can be applied to it. A statistically significant value for it means at that level the luminosities are correlated even after accounting for their individual dependence on the redshift. However, if $r_{x,op;z}$ is not significant at some acceptable level, one has to be cautious in attributing any observed correlation between $\log L_x$ and $\log L_{op}$ to a physical relationship between the two quantities. When there is no correlation between at least one of the luminosities and redshift, the partial correlation coefficient reduces to the ordinary correlation coefficient.

We have discussed correlation coefficients in the context of luminosities and redshift, but the discussion applies to any pair of variables. The one important requirement, which often goes unstated and untested, is that the variables in question jointly have a Gaussian distribution about their mean values. If this is not true at least approximately, the inferences drawn will not be meaningful. When the variables have non-Gaussian distributions, non-parametric *rank correlation* tests can be applied (see e.g. S88). These provide a correlation coefficient with known statistical distribution which can be used to estimate the significance of the observed rank correlation. However, the rank tests do not provide a fit to data such as the starlight line obtained from linear regression analysis.

The use of X-ray upper limits   We have seen in Section 10.3 that in some cases of pointed X-ray observations of quasars and AGN, the counts detected from the source do not exceed the noise in the background sufficiently for the source to be considered a positive detection. In such a case an upper limit on the X-ray flux is obtained. As indicated in Figures 10.7 and 10.8, about 40 per cent of the sources in Wilkes *et al.* (1994) are upper limits. Note that the upper limits do not cluster around the lowest luminosities observed. Whether a source is a detection or upper limit depends on the counts received from it relative to the background. A luminous source at high redshift would have a low flux and, if it is observed only for a relatively short time, sufficient counts may not be received from it for a detection. The upper limit on the luminosity may nevertheless be quite high.

When the number of upper limits in a sample is much smaller than the number of detections, the former can simply be omitted from the analysis, without seriously affecting the results. However, when the number of upper limits is a considerable fraction of the observed population, as in the case of the Wilkes *et al.* sample, omitting the upper limits biases the sample towards the more luminous objects. This does not necessarily affect regression and correlation analysis, if the upper limits are scattered throughout the luminosity plane, but luminosity functions determined from the sample can be affected.

Techniques for incorporating X-ray upper limits in the analysis were first proposed by Avni *et al.* (1980). The *detections and bounds* (DB) method uses an algorithm that iteratively distributes each upper bound proportionately to the distribution of detections less luminous than the upper limit. The DB method provides the distribution of the ratio L_x/L_{op}, i.e., α_{ox},[1] and linear regression coefficients when upper limits are present (see Avni and Tananbaum 1986). It was later realized that the DB method was similar

---

[1] Note that an upper limit on the X-ray flux corresponds to a lower limit on α_{ox} for a given optical flux.

to one of several statistical techniques developed over the years, which are used in drawing inferences from *censored* data, i.e., data that use upper limits. The techniques go under the general name of *survival analysis* and a brief review of the field may be found in Feigelson (1992).

The straight-line fit to the points in Figure 10.7 obtained by Wilkes *et al.* is

$$\langle \log L_x \rangle = 0.71 \log L_{op} + 4.863, \tag{10.25}$$

where L_x and L_{op} are the rest frame X-ray and optical luminosity in erg sec^{-1} at 2 keV and 2500 Å respectively. It is assumed that there is no dependence of the relationship on redshift, which will be justified by the discussion on α_{ox} below. The 1σ error on the slope is ± 0.05 and on the intercept ± 0.065. It follows from the fit that in the mean $L_x \propto L_{op}^{0.7}$.

The dependence of α_{ox} on optical luminosity and redshift has been studied in detail by Avni and Tananbaum (1986), using the DB technique. They assumed that the functional form of the mean value α_{ox} is

$$\langle \alpha_{ox} \rangle = A_z [\tau(z) - 0.5] + A_{op}(\log L_{op} - 30.50) + A, \tag{10.26}$$

where $\tau(z) = z/(1+z)$ is the look-back time in units of the Hubble time H_0^{-1} for $q_0 = 0$ (see Equation 2.83), and A_z, A_{op} and A are the parameters to be determined. The subtraction terms are introduced so that A reduces to the typical value of α_{ox} for the central values of the look-back time and optical luminosity. It is assumed that α_{ox} has a Gaussian distribution around $\langle \alpha_{ox} \rangle$, with a width independent of luminosity and redshift. Fits were made separately to the BQS and BF complete samples mentioned above as well as to a heterogeneous collection of quasars. Avni and Tananbaum found that the results for all the three samples were consistent with each other. This is to be expected in regression analysis if there are no biases that lead to the systematic deletion of sources from some regions of the three-dimensional space of the data points. It follows that the best results would be obtained using the large data base of 343 optically selected quasars in Wilkes *et al.* (1994).

Following the prescriptions of Avni and Tananbaum, Wilkes *et al.* obtained the best-fit parameters $A_{op} = 0.10 \pm 0.04$, $A_z = 0.006 \pm 0.17$ and $A = 1.54 \pm 0.04$. The best-fit width of the Gaussian distribution of α_{ox} about the mean value is given by $\sigma = 0.25 \pm 0.02$. These results are consistent with those of Avni and Tananbaum, but with a decrease in the size of the contours that define the 90 per cent confidence intervals in planes defined by the fitted parameters A_{op}, A_z and A taken two at a time. The value of A_z is consistent with zero, which shows that the dependence of $\langle \alpha_{ox} \rangle$ on the redshift is not significant. Making a fit after assuming that $\langle \alpha_{ox} \rangle$ depends only on the optical luminosity, Wilkes *et al.* obtained

$$\langle \alpha_{ox} \rangle = 0.11(\log L_{op} - 30.5) + 1.53, \tag{10.27}$$

with 1σ errors of ± 0.02 in both the slope and the intercept. A plot of α_{ox} against $L_{op}(2500 \text{ Å})$ and the fit in Equation 10.27 are shown in Figure 10.9. It is clear that there is a large scatter, but the correlation coefficient is nevertheless significant because

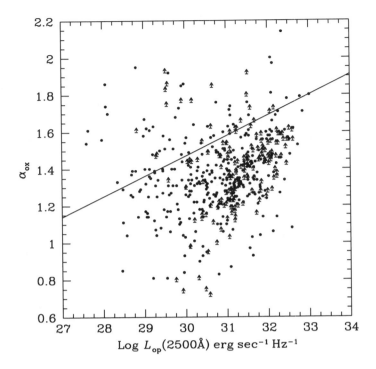

Fig. 10.9. A plot of α_{ox} against optical luminosity at 2500 Å for optically selected quasars from Wilkes *et al.* (1994). The arrows indicate lower limits on α_{ox}. The line shows the linear least squares fit of Wilkes *et al.*

of the large number of points. The relation between the luminosities obtained from Equation 10.27 is $L_x \propto L_{op}^{0.71}$, which agrees with the relation in Equation 10.25. Wilkes *et al.* find that their results do not change significantly when the analysis is carried out with different values of the X-ray spectral index α_x. When radio-quiet and radio-loud quasars from the optically selected sample are treated separately, it is found that for the latter population $\langle\alpha_{ox}\rangle$ shows no dependence on redshift and the same dependence on optical luminosity as in the case of the radio-quiet quasars. However, the normalization constant for the radio-loud subset is higher at $A = 1.53 \pm 0.05$, against the value $A = 1.27 \pm 0.07$ for the radio-quiet objects. This means that for a given optical luminosity, radio-loud quasars have higher X-ray luminosity than the equivalent radio-quiet objects.

   It was mentioned in the introduction to this section that Green *et al.* (1995) have obtained X-ray data on a sample of quasars from the LBQS, using cross correlation between optical positions and the coordinates of *EINSTEIN* IPC fields. Of the 146 quasars in this sample, 29 are X-ray detections at the usual 3σ level, the rest being upper limits. The data have been analyzed using survival analysis as well as X-ray image stacking, which was introduced by Caillault and Helfand (1985) and used in the context of quasars by Anderson and Margon (1987). In this technique, quasars in the sample are first divided into subgroups that depend on properties such as optical luminosity and redshift. The X-ray images of quasars from each subgroup are stacked together, regardless of whether they are X-ray detections or upper limits. These stacked images provide the mean X-ray properties of quasars in each subgroup with a sensitivity that is improved by a factor $\sim \sqrt{N}$, where N is the number of images in a stack. Margon *et al.* find that results obtained from this technique are consistent with those from survival analysis. In particular, the slope of the $\alpha_{ox} - \log L_{op}$ relation is 0.11 ± 0.01, which agrees with the result in Equation 10.27.

The regression analysis presented so far has been based on the assumption that at any optical luminosity and redshift the residuals $r = \langle \alpha_{ox} \rangle - \alpha_{ox}(L_{op}, z)$ have a Gaussian distribution about zero. Avni and Tananbaum (1986) have considered the possibility that the distribution of the residuals may be asymmetric about the mean, and that there could be a population of quasars with very large values of r, i.e., with very low X-ray-to-optical luminosity ratio. They introduced a distribution made up of two half-Gaussians which are joined at their maxima. The distribution is characterized by σ, which is the rms width of the joint distribution, and the ratio $R = \sigma_R/\sigma_L$ of the widths of the two Gaussians. The *X-ray-quiet* population is assumed to constitute a fraction P of optical quasars, situated at some $r = r_{xq} \gg \sigma_R, \sigma_L$.

Using this model of the residuals and the DB technique, Avni and Tananbaum performed a regression analysis on a sample of 154 quasars to find the best-fit values of the parameters in Equation 10.26 and those defining the asymmetric Gaussian. They obtained the values $P = 0, R = 3.3, \sigma = 0.21, A_z = 0.02, A_{op} = 0.09$ and $A = 1.54$. The distribution of the α_{ox} is therefore skewed, with a longer tail at high values of α_{ox} than at low values. Such a skew distribution of the residuals has important implications for the X-ray luminosity function, source count and contribution to the X-ray background derived using an optical luminosity function and the distribution of α_{ox}. The values of the parameters A_z, A_{op} and A are not significantly affected by the assumed shape of the distribution of residuals. The best-fit value $P = 0$ indicates that the fraction of X-ray-quiet optically selected quasars is very small. Avni and Tananbaum in fact find, that treating P as a single parameter, an upper limit of $P \leq 8$ per cent is obtained with 95 per cent confidence. A large majority of all optical quasars must therefore be X-ray loud.

10.6.3 *Relation between X-ray and radio luminosities*

The relation between the X-ray and optical luminosities was readily apparent in the *EINSTEIN* quasar observations by Ku, Helfand and Lucy (1980) and Zamorani *et al.* (1981), but it took much longer to unravel the dependence of the X-ray emission on the radio properties. Ku, Helfand and Lucy and Zamorani *et al.* showed that the X-ray luminosity of radio-loud quasars is on average higher than that of radio-quiet quasars of comparable optical luminosity. The X-ray luminosity is higher by on average only a factor ~ 3 compared to a total spread of luminosity that spans several orders of magnitude, but this difference is unequivocally present. The presence of the radio emission seems therefore to provide an extra channel for the generation of X-rays, and one would expect that at least in those sources where the radio induced X-ray emission dominates, the two should show a strong correlation.

Though no straightforward correlation between the X-ray and radio luminosity was apparent in the data available in the early 1980s, there were indications that the two were closely related. Owen, Helfand and Spangler (1983) found that there was a tight correlation between the X-ray and radio flux densities at 90 GHz in a sample of 25 extragalactic radio sources selected on the basis of their strong emission at millimetre wavelengths. In a study of a complete sample of 33 3CR radio quasars, Tananbaum *et al.* (1983) found that there was no correlation between their X-ray and total radio luminosity at 5 GHz. However, in a subsample of 13 quasars they found that the X-ray emission was correlated with radio emission from a compact core coincident with the central optical object.

One reason for the difficulty in detecting a correlation is that on the one hand, the radio emission associated with a quasar or AGN is spread over different scales, ranging from compact nuclear emission on the parsec scale to jet and lobe emission spread over tens or even hundreds of kiloparsec (see Chapter 9). On the other hand the X-ray emission is restricted to nuclear regions that are unresolved at the ~ 1 arcsec scale, except in some highly luminous nearby sources. Any correlation between the X-ray and nuclear radio luminosities is therefore expected to be obfuscated when the total or lobe radio emission is used. Examining any link between the luminosities therefore requires high resolution radio maps that allow the radio luminosity arising in compact regions to be used.

Kembhavi, Feigelson and Singh (1986) were the first to examine a correlation between X-ray and radio core emission in a large heterogeneous sample of radio selected quasars. They obtained VLA radio maps at 5 and 1.5 GHz of 35 radio selected quasars that had previously been observed with *EINSTEIN*. The 5 GHz data enabled the core emission to be separated from the extended emission with a resolution of ~ 4 arcsec (FWHM). The 1.5 GHz data, which was obtained with ~ 13 arcsec FWHM, was used to obtain the radio spectral index α between the two frequencies (after taking into account the difference in the FWHM at the two frequencies). A flat spectrum ($\alpha < 0.5$, see Section 6.3) indicates that the emission from the VLA core is dominated by self-absorbed radio emission, with any steep spectrum component (which could be resolved with a decrease in the beam width) limited to a small fraction. A steep radio spectral index ($\alpha > 0.5$), however, is indicative of radio emission which could be further resolved.

Kembhavi, Feigelson and Singh assembled a list of 127 radio selected quasars with *EINSTEIN* observations and VLA data from their own observations or from those in the literature with resolution compared to theirs. Browne and Murphy (1987) have added a few quasars to this set and included some improved measurements. This large data set, which includes all radio selected quasars from the lists of Ku, Helfand

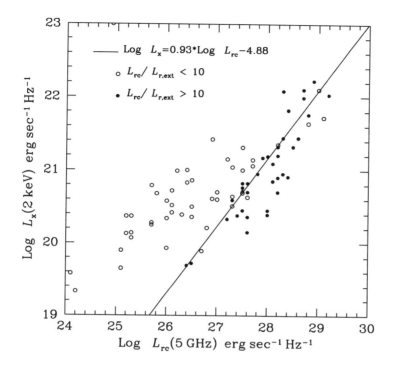

Fig. 10.10. A plot of X-ray luminosity at 2 keV against radio core luminosity at 5 GHz. Radio quasars with $L_{rc}/L_{r,ext} > 10$ are indicated by full circles and and those with $L_{rc}/L_{r,ext} < 10$ with open circles. The straight line shown is obtained by linear regression on points with $L_{rc}/L_{r,ext} > 10$. Data taken from Kembhavi, Feigelson and Singh (1986).

and Lucy (1980), Zamorani *et al.* (1981) and Tananbaum *et al.* (1983), had 11 X-ray flux upper limits as well. These were omitted from the sample and the analysis was performed using only the detections. The relatively small number of upper limits meant that the results were not seriously affected by their exclusion.

Kembhavi, Feigelson and Singh have found from their analysis that the X-ray luminosity of radio selected quasars is tightly correlated with the 5 GHz radio core luminosity L_{rc}, with $L_x \propto L_{rc}^{0.42\pm0.04}$. The correlation is significant at better than the 0.1 per cent level. As explained in subsection 10.6.2 the correlation could in principle arise because of the dependence of each luminosity on the redshift; it is also possible that the radio core luminosity is correlated with the optical luminosity. In this case, since X-ray and optical luminosity are correlated, a false correlation could be induced between the X-ray and radio bands. The extent to which the X-ray–radio correlation is real may be judged by evaluating partial correlation coefficients. It is found that these remain significant at better than the 0.1 per cent level after the effects of redshift and optical luminosity are separately accounted for.

A plot of $\log L_x$ against $\log L_{rc}$ is shown in Figure 10.10. The correlation is very obvious in the figure, but it is also clear that there is some structure in the distribution of points, indicating that there are well-organised subsets. Kembhavi, Feigelson and Singh found that if the sample was divided into subgroups with steep and flat radio spectra, the group with the flattest spectra ($\alpha < 0.3$) showed the best correlation and had the steepest slope, with $L_x \propto L_{rc}^{0.71 \pm 0.07}$.[1] Cores that are unresolved by the VLA but have a steep spectrum have, mixed with the self-absorbed component, a radio component with steep spectrum similar to that found in the extended resolved structures. If this is not related to the X-ray emission, it is expected that its presence dilutes the observed correlation and reduces the slope of the $\log L_x - \log L_{rc}$ fit.

We will see in the next section that the trends observed in Figure 10.10 can be elegantly interpreted in terms of relativistic beaming of the X-ray emission. Worrall *et al.* (1987) have obtained results consistent with those of Kembhavi, Feigelson and Singh (1986). They have also considered combined fits of the X-ray, radio and optical data.

Feigelson, Isobe and Kembhavi (1984) considered the correlation of the X-ray emission with the extended radio emission for resolved quasars from the sample in Kembhavi, Feigelson and Singh. While there is a formal correlation between the two luminosities, the partial correlation coefficients are not significant, showing that the observed effect is spurious. This is consistent with results for 3CR quasars in Tananbaum *et al.* (1983).

10.7   Relativistic beaming of X-ray emission

We have seen in Section 3.8 that the superluminal motion and rapid variability observed at radio wavelengths have led to the concept of relativistic bulk motion in the nuclear regions of quasars and AGN. This motion also leads to flux amplification, which can change the appearance of a source and is the basis of unification schemes (see Chapter 12). Since the X-ray luminosity too arises in compact regions, and is closely related to core radio emission, it is natural to expect that some of it also is relativistically beamed.

At radio wavelengths, VLBI techniques provide the milliarcsecond resolution that is necessary to spot the superluminal motion. Even at the resolution of a few arcseconds routinely available with the VLA, we have seen that it is possible to isolate compact self-absorbed cores, which are affected by beaming if it is present. This leads to the observation of one-sided jets that can be interpreted in terms of beaming. Comparison of the distribution of core luminosity relative to the luminosity of the extended emission, which is thought to be largely free of beaming effects, leads to the determination of beaming model parameters (see Section 12.3). In the case of the X-ray emission the situation is different, as nearly all of this is confined to a region that cannot at present

---

[1] This division into flat and steep radio spectrum quasars is not shown in Figure 10.10. The quasars are differentiated on the basis of the ratio of their radio core luminosity (L_{rc}) to extended radio luminosity ($L_{r,ext}$), as explained in subsection 10.7.2.

be resolved. It is therefore not possible to separate the emission into beamed and unbeamed parts. This makes the modelling more indirect and uncertain.

10.7.1 The Browne–Murphy beaming model

The first attempt at producing a model of X-ray beaming was made by Browne and Murphy (1987), who applied it to the interpretation of the correlation between X-ray and radio emission observed by Kembhavi, Feigelson and Singh. The model was closely related to the radio beaming model of Orr and Browne, which we have described in detail in Section 12.3. Browne and Murphy assumed that the X-ray luminosity L_x is divided into a beamed part L_{xb} and an unbeamed, i.e., isotropic, part L_{xu}:

$$L_x = L_{xb} + L_{xu}. \tag{10.28}$$

In the radio beaming model it was assumed that the ratio R_t of the beamed radio emission, at transverse orientation, to the extended radio emission is a constant. This is equivalent to assuming that the compact (beamed) radio luminosity at transverse orientation is proportional to the extended (unbeamed) part,[1]

$$L_{rc}(90\,\text{deg}) \propto L_{r,\text{ext}}. \tag{10.29}$$

In analogy with the radio case, Browne and Murphy assumed that the beamed X-ray luminosity at transverse orientation is also proportional to the extended radio emission, i.e.,

$$L_{xb}(90\,\text{deg}) = AL_{r,\text{ext}}, \quad A = \text{constant}. \tag{10.30}$$

The beamed luminosity for inclination angle ψ between the beam direction and the line of sight is

$$L_{xb}(\psi) = g_x(\beta, \psi)L_{xb}(90\,\text{deg}), \tag{10.31}$$

where $g_x(\beta, \psi)$ is the *X-ray beaming factor*, given by

$$g_x(\beta, \psi) = \tfrac{1}{2}(1 - \beta\cos\psi)^{-(1+\alpha_x)} + \tfrac{1}{2}(1 + \beta\cos\psi)^{-(1+\alpha_x)}, \tag{10.32}$$

β is the Lorentz factor of the bulk flow (see Section 12.3), $g_x(\beta, 90\,\text{deg})$ equals unity and α_x is the power law spectral index of the beamed X-ray emission. $g_x(\beta, \psi)$ and the radio beaming factor $g_r(\beta, \psi)$ to be introduced in Equation 12.4 differ only to the extent of the power law spectral index used. The ratio of the beamed to unbeamed X-ray luminosity is

$$R_x = \frac{L_{xb}}{L_{xu}} = R_{tx}g_x(\beta, \psi), \quad R_{tx} = \frac{L_{xb}(90\,\text{deg})}{L_{xu}}. \tag{10.33}$$

---

[1] The notation we use here to represent quantities at radio wavelengths is somewhat different from the notation used in other chapters. The changes are necessary to help distinguish between beamed and unbeamed quantities at different wavelengths.

As in the case of the Orr–Browne model, it is assumed that the beamed X-ray part at transverse orientation is proportional to the unbeamed part, i.e.,

$$R_{\text{tx}} = \text{constant.} \tag{10.34}$$

The beaming model considered here is a very simple one, in which it is assumed that the only differences between the radio and X-ray beaming are in the ratio of the intrinsic beamed to unbeamed luminosities and the spectral index. The same bulk flow velocity β is assumed to apply to the two cases. More complicated models in which the degree of beaming is different in the radio and X-ray cases are of course possible. Since the emission at different wavelengths can arise at different points along the jet, which may have different bulk velocities, the effective beaming angle can be quite different. There could even be recollimation of the jet, again changing the degree of beaming. There is some evidence of such differences between the X-ray and radio emission in the case of BL Lac objects (see Section 9.9).

In order to find an expression for the isotropic X-ray component, Browne and Murphy considered the correlation of the X-ray and extended radio emission in sources with $R = L_{\text{rc}}/L_{\text{r,ext}} < 1$. They found that, for these sources, linear regression of $\log L_x$ with $L_{\text{r,ext}}$ produces a straight line with slope ~ 0.5. In these sources the beamed radio emission is not the dominant component, and it is to be expected that the X-ray emission too is not dominated by beaming. Browne and Murphy assumed that the beamed component for the sources is small enough that the observed relation between the luminosities is indicative of the relation between the pure isotropic components. Therefore they adopted the simple law

$$L_{\text{xu}} = B L_{\text{r,ext}}^{0.5}, \quad B = \text{constant.} \tag{10.35}$$

For a given radio quasar with resolved radio structure, the ratio $R = L_{\text{rc}}/L_{\text{r,ext}}$ is known. Introducing this into Equation 12.5 below and using $R_t = 0.024$ (see Section 12.3), the quantity $\beta \cos \psi$ is obtained. This in turn can be used in Equation 10.32 to obtain $g_x(\beta, \psi)$ if α_x is known or some standard value is assumed for it. Once the two constants A and B are fixed, the beamed and unbeamed X-ray parts can be separately determined for each source simply from its extended radio luminosity.

Browne and Murphy determined the two constants by minimizing the quantity $\sum \left[\log \left(L_x / L_x^{\text{obs}} \right) \right]^2$, where L_x and L_x^{obs} are the model and observed X-ray luminosities for a given source and the summation extends over all the sources in the sample. Values obtained from the minimization are $A = 6.89$ and $B = -10.02$. In this model the beamed X-ray emission outshines the isotropic component only for $\psi < 15 \deg$.

10.7.2 *X-ray beaming and radio core luminosity*

There are two problems with the Browne–Murphy model which imply that it needs some modification. (1) For radio-quiet quasars, which constitute the vast majority, L_{rc} and $L_{\text{r,ext}} \simeq 0$, which means from Equations 10.30 and 10.35 that $L_x \simeq 0$. But we have seen that radio-quiet quasars have X-ray luminosities which are comparable to those of radio-loud quasars, and therefore the X-ray emission cannot in general depend on the extended radio luminosity alone. (2) The hypothesis $L_{xu} \propto L_{r,ext}^{0.5}$ is based on the observed slope of the straight-line fit to the data points in the $\log L_x \log L_{r,ext}$-plane. However, when the partial correlation coefficients which take into account the effect of redshift and radio core luminosity are considered, it is found that the significance of the correlation reduces considerably. This is best illustrated for the subset of 26 quasars from Browne and Murphy with $L_{rc}/L_{r,ext} < 0.1$. In these objects having low radio beaming factors, the effect of the beamed X-ray component is expected to be small. The correlation between X-ray luminosity and extended radio luminosity for this sample should therefore be indicative of the relationship between L_{xu} and $L_{r,ext}$. The linear correlation coefficient for this set is 0.69, which is significant at the 99.99 per cent significance level. However, the partial correlation with the effect of redshift suppressed is 0.33, with confidence level reduced to 93.5 per cent. The decrease in significance is most dramatic when the dependence of the X-ray and extended radio luminosities on the radio core luminosity is accounted for. The corresponding partial correlation coefficient is 0.18, with a confidence level of only 78.9 per cent. Clearly the dependence of L_x on $L_{r,ext}$ obtained from the linear correlation cannot be accepted at face value.

Kembhavi (1993) has proposed a beaming model that uses the formalism suggested by Browne and Murphy but with a different scheme for separating the X-ray luminosity into beamed and isotropic parts. It has been seen in subsection 10.6.3 that X-ray luminosity and radio core luminosity are tightly correlated. For the subset of 34 quasars from Browne and Murphy with $L_{rc}/L_{r,ext} > 10$, the best-fit relation is

$$\log L_x = (0.93 \pm 0.09) \log L_{rc} - 4.88. \tag{10.36}$$

The correlation coefficient is 0.87, which is significant at a confidence level > 99.99 per cent. All partial coefficients are also significant at this level. The points used and the best-fit line are shown in Figure 10.10.

Since the fit is dominated by beamed emission, Equation 10.36 is indicative of a relation between the beamed X-ray and radio components. Kembhavi therefore assumed that the relation

$$\log L_{xb} = 0.93 L_{rc} + \log k, \tag{10.37}$$

with k a constant, holds for all radio quasars. From the definition in Equation 10.33 the isotropic X-ray component L_{xu} is given by

$$L_{xu} = \frac{L_{xb}(\psi)}{R_{tx} g_x(\beta, \psi)}, \tag{10.38}$$

so that the total X-ray luminosity is given by

$$L_x = L_{xb} + L_{xu} = L_{xb} \left(1 + \frac{1}{R_{tx} g_x(\beta, \psi)}\right) = k L_{rc}^{0.93} \left(1 + \frac{1}{R_{tx} g_x(\beta, \psi)}\right). \tag{10.39}$$

This expression for the model X-ray luminosity has two constants, k and R_{tx}, to be determined. This is done by minimizing $\sum \left[\log \left(L_{\mathrm{x}}/L_{\mathrm{x}}^{\mathrm{obs}}\right)\right]^2$ as in the case of the Browne and Murphy model. The best-fit values are

$$\log k = -4.81, \quad R_{\mathrm{tx}} = \frac{L_{\mathrm{xb}}(90\,\mathrm{deg})}{L_{\mathrm{xu}}} = 1.3 \times 10^{-2}. \tag{10.40}$$

The intrinsic level of the beamed component is only ~ 1 per cent of the isotropic component, but it can be boosted to dominance when the beaming direction is aligned at an angle $\sim 1/\gamma$ relative to the line of sight, where γ is the bulk Lorentz factor.

Using Kembhavi's beaming model it is possible to relate the two components of the X-ray emission to the radio emission. To this end, a straight line can be fitted to the distribution of $\log g_{\mathrm{x}}(\beta, \psi)$ and $\log g_{\mathrm{r}}(\beta, \psi)$ over the quasar radio sample to obtain

$$\log g_{\mathrm{x}}(\beta, \psi) = 1.23 \log g_{\mathrm{r}}(\beta, \psi) + 0.17. \tag{10.41}$$

From Equation 10.31, a similar equation for the core radio luminosity and the values of the constants k and R_{tx} in Equation 10.40, it follows that

$$L_{\mathrm{xb}}(90\,\mathrm{deg}) = 8.9 \times 10^{-6} L_{\mathrm{rc}}^{0.93} g_{\mathrm{r}}^{-0.31}, \tag{10.42}$$

$$L_{\mathrm{xu}} = 4.6 \times 10^{-4} L_{\mathrm{rc}}^{0.93} g_{\mathrm{r}}^{-1.55}. \tag{10.43}$$

These relations determine the beamed and isotropic components of the X-ray emission from a radio quasar given only its radio properties. The X-ray luminosity determined in this manner is in reasonable agreement with the observed X-ray luminosity.

Using Equation 10.38 and defining

$$R_{\mathrm{xbu}} \simeq \frac{L_{\mathrm{xb}}}{L_{\mathrm{xu}}} = R_{\mathrm{tx}} g_{\mathrm{x}}(\beta, \psi), \tag{10.44}$$

it is easy to show that, for an observed X-ray luminosity $L_{\mathrm{x}}(\mathrm{obs})$,

$$L_{\mathrm{xb}}(\mathrm{obs}) = \frac{L_{\mathrm{x}} R_{\mathrm{xbu}} g_{\mathrm{x}}(\beta, \psi)}{1 + R_{\mathrm{xbu}}}, \quad L_{\mathrm{xu}}(\mathrm{obs}) = \frac{L_{\mathrm{x}}}{1 + R_{\mathrm{xbu}}}. \tag{10.45}$$

For a given radio quasar, which has been resolved into a compact and an extended structure, the beaming factors $g_{\mathrm{r}}(\beta, \psi)$ and $g_{\mathrm{x}}(\beta, \psi)$ are determined as explained above; R_{xbu} follows from the value of R_{tx} in Equation 10.40 and the beamed and isotropic components of the observed X-ray luminosity are then determined from Equation 10.45.

We have shown in Table 10.2 the components obtained in this manner for the radio quasars 3C 273 and 3C 263. It is seen that even though the beamed X-ray component in 3C 273 appears to contribute 90 per cent of the observed X-ray luminosity, the fractional intrinsic contribution, which is

$$\frac{L_{\mathrm{xb}}(90\,\mathrm{deg})}{L_{\mathrm{x}}(\mathrm{obs})} = \frac{L_{\mathrm{xb}}(\psi)}{L_{\mathrm{x}}(\mathrm{obs})} g_{\mathrm{x}}^{-1}, \tag{10.46}$$

is only ~ 0.13 per cent. In 3C 263 the intrinsic beamed component contributes as much as ~ 1.2 per cent of the observed luminosity, while the apparent contribution is only ~ 7 per cent, because the direction of the relativistic motion does not lie close to the line of sight, as is evident from the radio data.

Table 10.2. *The beamed and isotropic X-ray components for two radio quasars*

Source	$L_{rc}/L_{r,ext}$	g_r	g_x	L_{xb}/L_{xu}	L_{xu}/L_x(obs)	L_{xb}/L_x(obs)
3C 273	6.3	263	673	8.75	0.10	0.90
3C 263	0.08	3.3	6.1	0.08	0.93	0.07

The model explains straightforwardly the result of Kembhavi, Feigelson and Singh (1986) that the slope of the $\log L_x - \log L_{rc}$ line is steepest for the quasars with the flattest radio spectra. In these cases the beamed radio component dominates the unbeamed component, and therefore so must the beamed X-ray emission. It follows that $L_x \sim L_{xb}$ and, using Equation 10.37, that $L_x \sim L_{rc}$, which is consistent with the steepening of the slope. In the case of steep spectrum quasars, the extended radio and isotropic X-ray components dominate and these are not related to each other. The X-ray emission now no longer increases when L_r increases (except for the non-dominant beamed emission), and therefore the observed slope of the $\log L_x - \log L_{rc}$ line is significantly < 1, as seen in subsection 10.6.3. The Browne and Murphy model provides a similar explanation. We will see in Section 11.1 that the relatively low equivalent widths of iron lines in some radio quasars, as well as a correlation between X-ray power law index and the radio beaming factor can be understood in terms of the beaming model.

Baker, Hunstead and Brinkmann (1995) have compared the radio properties of an unbiased sample of quasars from the low-frequency-selected Molonglo quasar sample with soft X-ray data from the *ROSAT* All-Sky Survey. As in the previous studies, they find a relationship between the radio core and X-ray luminosities, with somewhat separate trends for steep and flat radio spectrum quasars. They find that highly anisotropic and beamed X-ray components are present, and that the aspect dependence of the beamed component is consistent with relativistic beaming models.

The X-ray beaming model described in this section depends upon parameters determined for the radio beaming model of Orr and Browne (1982), particularly on the value $R_t = 0.024$. A more general treatment, consistent with the more recent unification models discussed in Chapter 12, needs to be developed. The present model simply illustrates how the two components of X-ray emission may be separated. An important point to note is that the model applies specifically to radio-loud quasars: radio-quiet quasars are assumed to lack the beamed component and in their case $L_{xb}/L_{xu} = 0$. They only have the isotropic component, which is not determined by the model.

11 X-ray and gamma-ray spectra

In the previous chapter we considered how the broad band X-ray emission from quasars and AGN is related to the emission from other bands. The results from the large database on broad band fluxes are important in providing some pointers to the processes by which the X-ray emission is produced, but a detailed understanding requires observation of the spectral shape of the continuum, and any emission lines that may be present. Owing to the small collecting area of the X-ray telescopes that have so far been used, and the high energy per photon, the total number of photons received from a source is rather limited. It is therefore possible to obtain reliable spectra only for nearby AGN, which in spite of their moderate luminosity have a relatively large flux, and for the most luminous quasars. In spite of this limitation, spectral shapes have been sufficiently well established for a detailed comparison to be made with the predictions of the models described in Chapter 4. In this chapter we will consider the observation and modelling of the spectra from discrete sources as well as the diffuse background radiation in the X-ray and γ-ray domains.

11.1 X-ray spectra

The X-ray continuum for quasars and AGN in the 2–10 keV energy range can usually be well approximated by a power law. It is found that the spectral index is distributed in a narrow range in the case of AGN, but shows a somewhat wider distribution in the case of quasars. At other energies the spectrum exhibits a more complex structure: at $E \lesssim 1$ keV lower luminosity sources can be seriously affected by absorption along the line of sight over and above that due to the matter in our Galaxy. There can also be a soft X-ray excess, which is difficult to observe because of the drop in sensitivity of the detectors at $E \lesssim 0.1$ keV and the absorption. At energies ≥ 10 keV there could be a flattening of the spectrum. This must be followed by a resteepening at some higher energy so that constraints from γ-ray observations of AGN and the γ-ray background are not violated and divergence in the total energy is avoided.

The characteristic form of the X-ray spectrum was first noticed in observations of the radio galaxy Cen A (NGC 5128) with *UHURU* in the 2–20 keV range (Tucker *et al.* 1973) and with the *OSO* 7 satellite (Winkler and White 1975) in the 1–60 keV range.[1]

---

[1] See Mushotzky, Done and Pounds (1993) for a review of early developments and references.

These observations established that Cen A has a spectrum which can be fitted by a power law with spectral index α_x as flat as ~ 0.4 and that it is very strongly absorbed at $E \lesssim 3\,\mathrm{keV}$ with $N_\mathrm{H} \sim 10^{23}\,\mathrm{cm}^{-2}$. It was not possible to rule out exponential or black body fits to the spectra. Mushotzky *et al.* (1978b) found an Fe emission line at 6.4 keV and an absorption edge at 7.1 keV in Cen A. Mushotzky *et al.* (1978b) found from *OSO 8* observations in the range 2–60 keV that the spectrum of the Seyfert 1 galaxy NGC 4151 was consistent with a power law form having $\alpha_x = 0.42 \pm 0.06$ and $N_\mathrm{H} = (7.5 \pm 0.5) \times 10^{22}\,\mathrm{cm}^{-2}$. Thermal fits were allowed only with $kT > 70\,\mathrm{keV}$, i.e., with the knee in the thermal bremsstrahlung spectrum located outside the observed range. An iron fluorescent line at 6.4 keV was detected at the 2.5σ level.

We will now consider the wealth of spectral data that have become available from a number of missions since these pioneering observations, and its interpretation in terms of the theoretical models developed in Chapter 4.

11.2   X-ray spectra: Seyfert 1 galaxies

We will first consider the continuum spectrum and then emission lines, and follow this with a discussion on recent theoretical models of the spectrum. We will divide the observations on the basis of the satellites from which they were obtained, as a means of organizing the presentation of data acquired by many observers using a bewildering number of detectors, each with its own characteristics leading to different restricted views of the X-ray spectrum.

11.2.1 *Continuum emission*

HEAO 1   Though Seyfert 1 galaxies are in general far less luminous than quasars, a greater X-ray flux is received from them because of their relative proximity. A number of Seyfert 1 galaxies were observed in the 2–10 keV range by the A2 experiment on the *HEAO 1* satellite. It was found that the counts could best be fitted by simple power law spectra $I(E) \propto E^{-\alpha_x}$ modified by Galactic absorption, with α_x in the narrow range 0.7 ± 0.19. The spectral index had a similar range for the small number of broad line radio galaxies (BLRG) and narrow emission line galaxies (NELG) in the sample. α_x was found to be independent of luminosity for

$$3 \times 10^{42} < L_x(2\text{–}10\,\mathrm{keV}) < 3 \times 10^{45}\,\mathrm{erg\,sec}^{-1} \tag{11.1}$$

and independent of the presence or absence of radio emission (see Mushotzky 1984 for a review).

This 'universal' nature of the power law spectrum was considered very interesting, as it had the potential to impose tight constraints on any model used to describe the X-ray spectrum. As an example, we can consider the generation of the power law spectrum from the inverse Compton scattering of low energy photons by highly relativistic electrons with a power law energy distribution. In this case the steady state electron power law index is $p = 2\alpha_x + 1$, which equals 2.4 for $\alpha_x = 0.7$ (see Equation 4.23). This is steeper by one than the power law index Γ at injection into the system of the energetic electrons, because of the energy loss in the scattering (see Equation 3.85). Thus the narrow distribution of α_x implies a narrow distribution of Γ around 1.4, for which there is no compelling physical reason. We will see below that while later observations have confirmed a narrow distribution of α_x in the 2–10 keV range, the overall X-ray spectrum is much more complex, and pair models with reflection (see Section 4.6) have to be invoked in the explanation.

EXOSAT   Broad band X-ray spectra of a large number (46) of X-ray selected emission line galaxies were obtained using the medium energy array (ME) and the low energy imaging telescopes (LE) on *EXOSAT* (Turner and Pounds 1989). The sample contained six Seyfert 2 galaxies or NELG and two quasars, the rest being of Seyfert 1 type. The ME provided counts in the $\sim$ 2–10 keV range with an energy resolution of $51(E/1\,\mathrm{keV})^{-0.5}$ per cent, while the LE data was in the range $\sim$ 0.1–2 keV, with no energy resolution. The low energy data allowed possible intrinsic absorption and soft X-ray excess to be examined. The main results obtained from these observations are as follows.

- X-ray continuum spectra in the 2–10 keV range can be represented by a simple power law, modified by low energy absorption.
- The narrow distribution of the power law energy index α_x found from *HEAO 1* A2 measurements is confirmed. The distribution has mean $\langle \alpha_x \rangle = 0.7$ and dispersion $\sigma_I = 0.17$. This dispersion reflects measurement errors as well as any real spread in the distribution. If both types of error are assumed to have a Gaussian distribution, the intrinsic spread can be estimated to be $\gtrsim 0.15$ at the 90 per cent confidence level.
- The spectral index α_x is independent of the 2–10 keV X-ray luminosity over the four decades of spread.
- About half the sources show low energy absorption exceeding the Galactic value for their position, with intrinsic equivalent hydrogen column density $N_H \simeq 10^{21}$–$10^{23}\,\mathrm{cm}^{-2}$. N_H is found to be highest in the low luminosity AGN. The Seyfert 2 galaxies and NELG are generally found to have higher N_H values than the Seyfert 1 galaxies.
- A soft X-ray excess above the extrapolation of the power law to low energies is seen in about 30 per cent of the sample, including half the sources with low absorption. The excess was previously observed in just a few sources (Arnaud *et al.* 1985, Singh *et al.* 1985). Such an excess has also been seen by Ghosh and Soundararajaperumal (1992).

GINGA   A sample of 27 AGN, consisting of 20 Seyfert 1 and seven NELG, was observed with the large area counter (LAC) (Nandra and Pounds 1994). The LAC is sensitive over the $\sim$ 1.5–37 keV range, with energy resolution $44(E/1\,\mathrm{keV})^{-0.5}$ per cent.

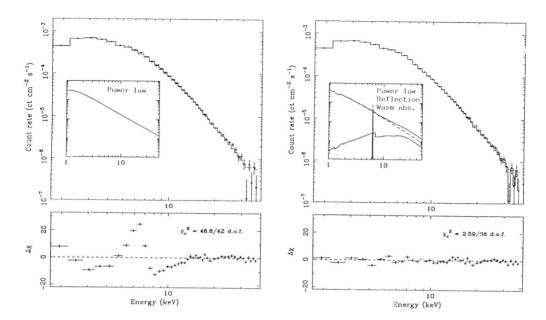

Fig. 11.1. Model fits to the summed data on the 27 quasars in the sample of Nandra and Pounds (1994). The figures on the left show a power law with absorption (inset) fitted to the counts (upper panel). The residual counts after subtracting the power law fit from the data are shown in the left lower panel. An emission line is clearly seen. The panels on the right show a fit that includes a power law, a reflected component, an iron line and an ionized absorber. Reproduced from Nandra and Pounds (1994).

The counts used in the analysis were restricted to the 2–18 keV range, since above these energies the data are of very low statistical weight and do not help in constraining the fits; moreover they can be contaminated by fluorescence lines from the collimator. Fits were made separately in the ranges 2–18 keV and 10–19 keV. In the former case the fits are dominated by counts up to 10 keV, so a comparison with the spectral index obtained from earlier measurements can be made.

The larger collecting area and higher energy resolution of the LAC relative to the earlier detectors allow the spectrum to be examined in much greater detail then was earlier possible. When power law fits were made to the data, it was found that their quality was poor, with large reduced chi-squared values χ_ν^2. The residuals were dominated by clear emission features in the region where iron fluorescence lines are expected. A power law fitted to the summed data for all the 27 AGN in the sample is shown in Figure 11.1, with the line clearly visible. A hard component above 10 keV as well as iron absorption edges are also found in the data.

In spite of these significant residuals, which are obvious because of the quality of the data, Nandra and Pounds have fitted pure power law spectra in the 2–18 keV range, with the usual low energy absorption, to the counts from individual objects.

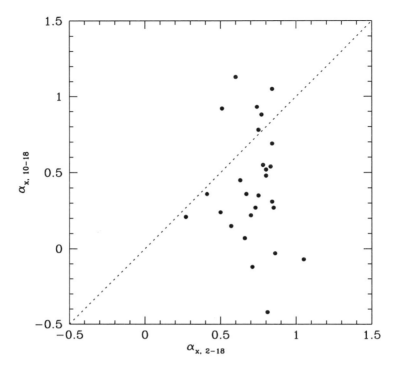

Fig. 11.2.   A plot of $\alpha_{x,10-18}$ against $\alpha_{x,2-18}$. The data is taken from Nandra and Pounds (1994).

This helps in making a simple parameterization of the spectrum and comparing results with earlier work. The spectral slopes obtained in fits that include the other features are not significantly different from these simple power law slopes. The results of these fits are as follows.

- In the range 2–18 keV the population's mean energy spectral index is given by $\langle \alpha_x \rangle = 0.73 \pm 0.05$, with an intrinsic spread of $\sigma_I = 0.15$, which is consistent with the earlier results discussed above.
- The mean spectral index in the range 10–18 keV is $\langle \alpha_{x,10-18} \rangle = 0.45 \pm 0.08$, with an intrinsic spread of $\sigma_I = 0.22$. Both the indices therefore show a small but significant spread that must be taken into account in the modelling.
- A plot of $\alpha_{x,10-18}$ against $\alpha_{x,2-18}$ is shown in Figure 11.2. It appears that in a majority of cases the former has flatter values. The distributions of the spectral indices in the two energy ranges are shown in Figure 11.3. Using the Kolmogorov–Smirnov test for comparing the two distributions, they are found to be different at a confidence level > 99.9 per cent. This indicates that there is a flattening of the spectrum for energies $\gtrsim 10$ keV.
- In about half the cases, the 2–18 keV fits show absorption in excess of the Galactic value, with N_H greater than a small multiple of 10^{21} cm^{-2}. The LAC

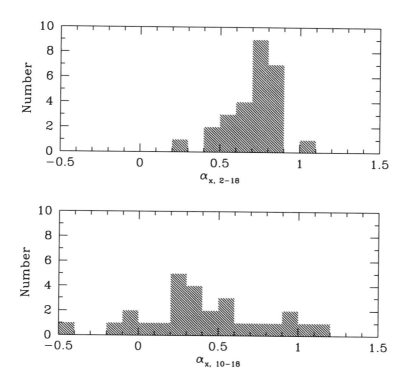

Fig. 11.3. The distributions of $\alpha_{x,2-18}$ and $\alpha_{x,10-18}$. The data is taken from Nandra and Pounds (1994).

detector is in fact not sensitive to absorption below such values, because its energy range is confined to $> 2\,\mathrm{keV}$. The absorption found here is in general greater than in the *EXOSAT* data, even though the latter went down to $\sim$ $0.1\,\mathrm{keV}$, where the effects of absorption are much greater.

We shall consider the non-thermal-pairs modelling of the *GINGA* data after summarizing the situation with iron emission and absorption.

EINSTEIN Kruiper, Urry and Canizares (1990) have studied a sample of 52 Seyfert 1 galaxies, on which *EINSTEIN* IPC data is available. The spectra are well determined for the more luminous objects, and in these cases the spectra are found to be consistent with power laws with energy spectral index $\alpha_x = 0.81$ in the range 0.2–$4.0\,\mathrm{keV}$. For objects where data in higher energy bands is also available, it is found that the soft energy spectral index is systematically steeper. Some objects show the presence of an intrinsic absorbing column density and a few objects show some sign of a soft X-ray excess. Higher luminosity and higher redshift objects are found to have steeper soft X-ray spectra.

ROSAT Walter and Fink (1993) have considered the spectra of 51 Seyfert 1 galaxies detected in the *ROSAT* All-Sky Survey (RASS) by the PSPC detector. The survey was carried out in the scanning mode, along great circles perpendicular to the ecliptic. The period for which a source was observed depended on its ecliptical latitude. Exposures for a given point in the sky varied from a few hundreds of seconds to a few thousands. By comparing the positions of X-ray point sources found in the survey with the positions of known AGN, the X-ray fluxes from these objects as measured by the PSPC have been determined. From these sources Walter and Fink have selected Seyfert 1 galaxies which had more than 300 counts and were observed at least once with the *IUE*.

Walter and Fink found that, in all but two cases, the spectra were well represented by a power law with low energy absorption. The 0.1–2.4 keV energy spectral index was found to be in the range 0.6–2.4, with $\langle \alpha_x \rangle = 1.5$ and $\sigma = 0.48$. In ~ 90 per cent of the galaxies, the best-fit value of N_H was found to be consistent with the corresponding Galactic value and the mean of the difference between the measured and Galactic values was found to be consistent with zero. This shows that the intrinsic absorption along the line of sight is small.

GINGA LAC spectra in the 2–10 keV band are available for 27 of the sources and it is found that these are significantly flatter than the PSPC spectra in 17 cases, indicating that a soft excess may be present in ~ 60 per cent of the sample. The energy resolution of the PSPC is not sufficient to find the shape of the excess, so Walter and Fink have tried different models for the excess, such as a thermal bremsstrahlung spectrum. They have added these model excesses to a power law with the slope fixed at the measured 2–10 keV value and fitted the combination to the data, using Galactic values of N_H. These fits show an excess above the extrapolation of the high energy power law in 90 per cent of the cases. The excesses obtained using different models and bandwidths are found to be well correlated with each other, as well as with the spectral index in the PSPC band. This means that general conclusions about the excess may be considered to be robust within the constraints of the observation, even though its existence is inferred from modelling that involves a number of assumptions.

Walter and Fink have investigated the correlation of the soft X-ray excess with the big blue bump (see subsection 8.3 for a discussion on the latter). They considered the ratio $R_{UV,x} = \nu F(1375\,\text{Å})/\nu F(2\,\text{keV})$, where F is the flux at the indicated point in the spectrum and ν is the corresponding frequency. $\alpha_{UV,x} = \log R_{UV,x}$ is the two-point spectral index between the UV wavelength and 2 keV, where any soft excess is expected to make only a small contribution. It is found that the 0.1–2.4 keV spectral index is significantly correlated with $\alpha_{UV,x}$. When six sources which seem to lie apart from the main distribution are omitted, the correlation improves significantly (the probability of obtaining it by chance goes down from 0.1 per cent to 0.001 per cent). Of the six peculiar sources, five are strongly absorbed by dust in the UV, which explains their departure from the trend. Walter and Fink concluded from the correlation that the big blue bump extends up to soft X-ray energies. The sample of quasars in this study has been selected on the basis of their being bright in the *ROSAT* All-Sky Survey (RASS).

This biases the sample towards objects with steep soft X-ray spectra, as these would have a high count rate in the survey even if their integrated X-ray flux was low. This has to be kept in mind in interpreting the correlations.

11.2.2 *Iron emission and absorption*

We have seen in subsection 4.6.5 that, owing to the relatively high abundance and fluorescence yield of iron, emission and corresponding absorption edges from the element may be expected in AGN spectra beyond ~ 6 keV. An iron absorption edge in the Seyfert galaxy NGC 4151 was observed long ago from the *ARIEL* satellite by Barr *et al.* (1977). The Japanese satellite *TENMA* was able to observe emission lines in NGC 4151 (Matsuoka *et al.* 1986) and Cen A (Wang *et al.* 1986). It was clear from the well-determined line energies in these observations that the emission was due to iron fluorescence from cold matter. Inspired by a suggestion of Guilbert and Rees (1988), Nandra and Pounds (1992) searched for and found iron K_α emission in *EXOSAT* observations of the Seyfert 1 galaxy MCG-6-30-57. However, it was only with observations from *GINGA* that iron emission and absorption features were shown to be common to AGN (Pounds *et al.* 1989, Matsuoka *et al.* 1986, Pounds *et al.* 1990, Nandra, Pounds and Stewart 1990). We shall now consider the results of Nandra and Pounds (1994), who include in their work many of the sources reported earlier and provide a detailed comparison of the observations with theoretical models.

As seen above, power law fits in the 2–18 keV range to the AGN sample of Nandra and Pounds leave residuals at energies where iron emission features are expected. Nandra and Pounds have therefore included an emission line in their fits along with the power law, as in Equation 10.5. They consider separately cases with the line energy (1) fixed at the energy, 6.4 keV, of the Fe I K_α line, (2) fixed at 6.7 keV, which is the energy of the line for Fe XXVI, and (3) a free parameter.[1]

Nandra and Pounds detected lines at > 90 per cent confidence in 25 of the 27 sources. The energy of the line in the free fit peaks around 6.4 keV, with weighted-mean line energy 6.37 ± 0.04 keV. A good fit is obtained with reduced $\chi_\nu^2 = 0.66$,[2] if a constant line energy equal to 6.4 keV is used in all sources. The fit with emission line energy 6.7 keV is at an unacceptable level ($\chi_\nu^2 = 3.56$). The energy of the line is therefore consistent with its being produced by fluorescence in cold iron (Fe I, see subsection 4.6.5). If the line were due to thermal emission from a hot plasma, the iron would be ionized and the line would have energy in the range ~ 6.7–6.9 keV, which is not consistent with the data. The mean equivalent width obtained for the line with energy fixed at 6.4 keV is 140 ± 20 eV. The inclusion of the line does not affect the best-fit power law index appreciably.

---

[1] The energies indicated are in the rest frame of the source. An appropriate redshift factor is applied while making a fit to the observed counts.

[2] Reduced χ_ν^2 is simply the usual χ^2 divided by the number of degrees of freedom. $\chi_\nu^2 < 1$ indicates a good fit.

Though inclusion of the emission line improves the fit considerably, in many cases the fits are still not at a statistically acceptable level, which indicates that there are other features in the spectrum to be explored. Nandra and Pounds have tested for the presence of an iron K-absorption edge, which would be caused by the same matter that produces the fluorescence lines. The energy of the edge, like that of the emission line, depends on the ionization state. Nandra and Pounds have considered power law plus absorption fits, with the edges fixed at (1) 7.1 keV for Fe I, (2) 8.85 keV for Fe XXV (helium-like iron) and (3) an intermediate value of 8 keV. When an improvement over the pure power law fit was seen, a further fit was made with the edge energy taken to be a free parameter. The result of the fit to individual sources was a preferred edge energy of ~ 8 keV, with column densities of a multiple of 10^{23} cm^{-2}. When fits are made that include a power law together with iron emission *and* absorption, similar edge energies and somewhat lower column densities are obtained. The absorption edge energy implies the existence of significant column densities of highly ionized iron along the line of sight.

A warm absorber should make its presence felt at energies below a few keV through absorption, due principally to ionized carbon, nitrogen and oxygen. This region is not accessible to the LAC, but can be studied with the PSPC on *ROSAT*, which is sensitive in the ~ 0.1–2.4 keV region. Pounds *et al.* (1994) have carried out simultaneous PSPC and LAC observations of a small sample of galaxies, the combined data extending over more than two decades in energy. They find that, in every case, the underlying continuum has a power law form, with energy spectral index $\alpha_x = 0.9 \pm 0.1$. Superposed on the power law spectrum are a variety of features, including Fe emission and a soft excess. In one case, NGC 4051, absorption from a substantial column of ionized gas along the line of sight is seen. Mathur, Elvis and Wilkes (1995) have discovered that the X-ray and UV absorption found in two radio-loud quasars, 3C 212 and 3C 351, as well as in the Seyfert galaxy NGC 5548, can be attributed to absorption due to a single warm absorber. In the case of NGC 5548, analysis of X-ray data from *ASCA* and UV data from HST showed that the absorber is highly ionized, has a high column density of $N_{\mathrm{H}} = 3.8 \times 10^{21}$ cm^{-2}, is situated outside the broad line emitting region and is outflowing with a mean velocity of 1200 ± 260 km sec^{-1}.

11.2.3 *Models*

The pre-*GINGA* observations of a power law X-ray spectrum with nearly constant slope were not good enough to distinguish between different possibilities for producing the spectrum through elementary physical processes such as Compton scattering from thermal or non-thermal electrons. In fact the uniformity of the spectra seemed to impose severe restrictions on the parameters that define the input spectra of the low energy photons and energetic electrons. The data from *GINGA* is much richer and the complexity of the X-ray spectrum, including the common occurrence of iron emission and absorption features, shows that a variety of processes must be involved in its production.

The iron lines can be produced either in transmission or in reflection. In the former case, the X-rays pass through neutral or partially ionized gas in the line of sight, causing photoionization followed by fluorescence. In the reflection models, which we have described in subsection 4.6.5, part of the spectrum emerging from the production region interacts with a cold slab of gas, eventually being reflected in the direction of the observer with an altered spectrum, which includes fluorescent lines produced in the slab. The observer sees a combination of the direct and reflected radiation.

In the transmission models, the neutral iron K_α line with observed equivalent width $\sim 150\,\text{eV}$ requires a column density of $\sim 2\times10^{23}\,\text{cm}^{-2}$ if the gas is in the form of a spherical cloud with Solar abundances (Nandra and Pounds 1994). This requirement is not consistent with the much lower column densities, $\sim 10^{21}-10^{22}\,\text{cm}^{-2}$, actually observed. If the gas were highly ionized, its observed column density would be underestimated because of the low absorption it would offer to soft X-ray photons. However, if the ionized component were to dominate, the energy of the iron line would be in the range $\sim 6.5-6.7\,\text{keV}$, which is higher than the line energy found above. An iron line with higher equivalent width could be produced with neutral matter for the observed range of N_H if the abundance of iron were raised to $\sim 10-100$ times its Solar value, relative to the lighter elements that produce soft X-ray absorption. However, the abundant neutral iron would produce a K-absorption edge at $7.1\,\text{keV}$, which is not seen in a majority of the sources, and iron L-shell absorption at $\sim 0.7\,\text{keV}$ would become significant, leading to higher absorption. It appears therefore that transmission models are not very successful in accounting for the observed features.

We saw in subsection 4.6.5 that when X-ray photons are incident on a slab of cold matter, the reflected spectrum is affected by absorption at energies $\lesssim 10\,\text{keV}$, while photons with energy $> m_e c^2/\tau_\text{sc}$ are scattered to lower energy because of Compton recoil. The result is a characteristic hump in the spectrum. When this reflected spectrum is added to the direct part reaching the observer, the net result is a spectrum that is harder than just the incident part. This can account for the flattening of the spectrum seen in the $10-18\,\text{keV}$ window by the *GINGA* LAC. The emission lines seen in the observations can be attributed to fluorescence from cold iron photoionized by X-rays incident on the slab.

Nandra and Pounds (1994) considered a spectrum of the form

$$F(E) = AE^{-\Gamma} + A_\text{ref}\psi(E,\Gamma), \tag{11.2}$$

where the power law on the right represents the spectrum incident on the slab, which is generated in a region through non-thermal or thermal pair processes, and $\Gamma = \alpha_x + 1$ is the photon number index. $\psi(E,\Gamma)$ is the reflected spectrum generated for a face-on slab which subtends a solid angle 2π at the source and on which is incident a power law continuum with $A = 1$. Nandra and Pounds considered only face-on slabs, since the shape of the spectrum is not a very sensitive function of the angle of inclination to the line of sight, and considered the variation in intensity with inclination and solid angle together through the normalization constant A_ref. The reflected spectrum was obtained from the monte-carlo calculations of George and Fabian (1989), which we considered in subsection 4.6.5. It was assumed that the cold matter had Solar abundances and that line broadening, either due to orbital motion of the matter around the centre or due to gravitational redshift, could be neglected.

Nandra and Pounds fitted the model in Equation 11.2 to individual spectra in their sample and found that acceptable fits (defined as those with $\chi_v^2 < 1.4$) were obtained in ~ 80 per cent of the cases. The fits were in general better than those with a simple power law and emission line. The mean value of the ratio of the reflected part to the incident part was $\langle A_{ref}/A \rangle = 0.72^{+0.18}_{-0.12}$, with no significant intrinsic spread. For the spectral indices in the fitted range, the maximum K_α equivalent width attained is ~ 150 eV. Features like the lines and the depth of the absorption edges appear only in the reflected part of the spectrum, and therefore are diluted in the net spectrum.

In those sources that could not be fitted satisfactorily, it was found that the residuals from the observed profile were significant in the 2–4 keV range, where the standard Morrison and McCammon (1983) opacities proved to be inadequate. This has prompted Nandra and Pounds to consider a *warm absorber*, i.e., a highly ionized gas, both for providing the required absorption at soft energy, as well as to account for the iron edges found at energies higher than 7.1 keV, which corresponds to cold iron. Including an iron absorption edge in the fits produces significant improvement but the line energy is not well constrained, though it lies in the range 8–9 keV and N_H values are similar to those obtained for the simpler fits considered above. Nandra and Pounds have also fitted a warm absorber model by introducing an ionization parameter in the fits in addition to those in Equation 11.2. Inclusion of this parameter improves the fits in all those cases where a significant absorption edge has been detected. However, the mean value of the ionization parameter is found to be a factor $\sim 10^2$ higher than the typical value in standard broad line region (BLR) models.

An important result to emerge from the reflection modelling of the *GINGA* data is that the mean value of the energy spectral index of the intrinsic power law continuum obtained from the fits is $\langle \alpha_x \rangle = 0.95 \pm 0.05$. The values of α_x for the individual sources are concentrated around the mean, but there is a real dispersion (i.e., not due to measurement errors) of $\sigma_p = 0.15 \pm 0.04$ around the mean. The spectral index therefore is significantly steeper than the value 0.7 ± 0.15 obtained from fitting power law spectra (with or without emission lines) to the observed counts from the *GINGA* LAC observations, or those from earlier missions. This discrepancy between the 'intrinsic' and the 'observed' spectral index arises because the latter is obtained without taking into account the flattening of the spectrum beyond ~ 10 keV due to reflection and the low-energy absorption due to the warm absorber. The intrinsic energy spectral index is close to the value 1.0 expected from saturated pair cascades (see subsection 4.6.2).

11.3   X-ray spectra: Seyfert 2 galaxies

In the soft X-ray band, a sample of 12 Seyfert 2 galaxies with IPC data has been studied by Kruiper, Urry and Canizares (1990), while at higher energies a sample of 30 narrow line AGN including optically and *IRAS* selected Seyfert 2 galaxies, and narrow line radio galaxies (NLRG) have been observed by *GINGA* (Koyama 1992). These observations show that the distributions of the power law index for Seyfert 2 and Seyfert 1 galaxies are similar, but the former, in nearly every case observed by *GINGA*, shows strong absorption at low energy and iron line emission.

As an example, in the case of Mrk 3, an equivalent hydrogen column density of $N_H = 7 \times 10^{23}$ cm^{-2} is found. The power law energy index is $\alpha_x = 0.3 \pm 0.3$ and, after correction for obscuration, the intrinsic 2–10 keV luminosity is 4×10^{43} erg sec^{-1} (Awaki *et al.* 1990). The intrinsic X-ray properties of the Seyfert 2 galaxy Mrk 3 are therefore very similar to those of Seyfert 1 galaxies, and it may very well be an obscured Seyfert 1 galaxy, which is consistent with the unification model considered in Section 12.6. A large column density has also been observed in the case of the NLRG IC 5063 and it too can be considered to be an obscured broad line radio galaxy (BLRG). However, *IRAS* selected Seyfert 2 galaxies show no significant absorption or iron line emission. Mulchaey, Mushotzky and Weaver (1992) have considered a sample of Seyfert 2 galaxies that were detected in the ultraviolet and for which reliable X-ray spectral fits were available. They again find high column densities in the range 1.6×10^{22}–7×10^{23} cm^{-2}, while the other properties of the sample are similar to those of Seyfert 1 galaxies.

In the unified scheme, Seyfert 2 galaxies are viewed through a molecular torus which obscures the nuclear and broad-line-producing regions. The torus also obscures the soft X-rays, but at the inferred values of N_H, it would be transparent to X-rays with energies exceeding several keV. Seyfert 1 galaxies, on the other hand, are viewed through the opening in the torus, so that the full unobscured X-ray spectrum is seen. This scheme readily explains the low luminosity of Seyfert 2 galaxies observed in the *EINSTEIN* IPC band. However, a difficulty is the observation that while the intrinsic X-ray luminosities of Seyfert 2 galaxies are within the range of Seyfert 1 luminosities, no Seyfert 2 galaxies as bright intrinsically as the brightest Seyfert 1 have been found.

The unification scheme for Seyfert galaxies was inspired by the observation of Seyfert 1 characteristics in the typical Seyfert 2 galaxy NGC 1068. One would expect that the X-ray behaviour of that galaxy would also point to an obscured X-ray emitting region. However, the observed X-ray spectrum of NGC 1068 shows no signs of significant absorption, and the fitted neutral hydrogen column density is with $N_H < 10^{22}$ cm^{-2}. Moreover it has a 2–10 keV X-ray luminosity of $\sim 10^{41}$ erg sec^{-1}, which makes it much fainter intrinsically than a Seyfert 1 galaxy (Koyama *et al.* 1989), and an intense iron line at 6.55 ± 0.1 keV with equivalent width $1.3^{+0.7}_{-0.3}$ keV. The behaviour of NGC 1068 can be understood within the unification scheme if the torus has a column density $\sim 10^{25}$ cm^{-2}, which would prevent any of the 2–10 keV X-ray emission from escaping through it. The observed X-ray luminosity is attributed to the radiation scattered into the line of sight of the observer after escaping through the opening in the torus, which is oriented away from the observer. Koyama *et al.* (1989) have shown that the observations can be explained by a scattering gas with cosmic abundances distributed spherically symmetrically about the nucleus. For a gas sphere of radius ~ 1 pc, the reflection efficiency is estimated to be a few per cent, which would mean that the intrinsic X-ray luminosity of NGC 1068 is comparable to that of a luminous Seyfert 1. The iron line equivalent width predicted by the model is consistent with observation.

11.4   X-ray spectra: quasars

In spite of their large intrinsic X-ray luminosity, it has been difficult to obtain good X-ray spectra of quasars because of their low X-ray flux. While the number of AGN spectra available from *HEAO 1* observations was sufficient to examine general trends, data on quasar spectra remained meagre. Good quality X-ray spectra in the 2–30 keV region were available for only two bright quasars, the radio-loud 3C 273 (Worrall *et al.* 1979) and the radio-quiet 0241+622 (Worrall *et al.* 1980). The former was found to have a flat power law spectrum with $\alpha_x \simeq 0.4$, while the latter had a steeper spectrum with $\alpha_x \simeq 0.9$.

A number of quasar X-ray spectra were observed with the IPC on *EINSTEIN* and, unlike Seyfert galaxies, power law fits produced a wide range of spectral index. This seemed to go against the 'universality' of the power law index of ~ 0.7 found in AGN. It soon became clear that an important cause of the difference was that low energy steep spectrum components not evident to the *HEAO 1* A1 detectors were being observed. The overall quasar spectrum is therefore quite complex, and can be represented by a simple power law form over only limited energy ranges, particularly when the resolution and signal-to-noise ratio are not adequate to observe departures from the simple form. The change in spectral shape over different energy ranges has to be kept in mind while comparing the results from different missions and detectors. A number of quasar spectra have been observed with *EXOSAT* and *GINGA* and more recently with *ROSAT* and *ASCA*. Though the quality of the spectra is not as good as for the AGN because of the low flux, it is clear that there are important spectral differences between the lower luminosity AGN and high luminosity quasars. We will now look at the data obtained from different missions.

EINSTEIN   The IPC on *EINSTEIN* had rather limited energy resolution, ~ 100 per cent at 1 keV, but because of the energy range, which spanned 0.1 to 3.5 keV, it was possible to fit power law spectra to the data with an accuracy better than ± 0.3 for the spectral index α_x. Power law spectral fits for a sample of 33 radio-loud and radio-quiet quasars from the *EINSTEIN* data base were reported by Wilkes and Elvis (1987). This sample was extended to 45 quasars, with new spectral fits, by Shastri *et al.* (1993). The spectral index α_x and the hydrogen column density N_H were the independent variables in these fits. We will now summarize the main deductions from these references. In what follows, a quasar is defined to be radio-loud when the radio-loudness parameter $R_L \equiv \log[F(5\,\mathrm{GHz})/F(B)] \geq 1$, where $F(5\,\mathrm{GHz})$ is the total radio flux density at 5 GHz and $F(B)$ is the optical flux density at the centre of the *B* band. A quasar is radio-quiet when $R_L < 1$.

- The 0.2–3.5 keV spectral index α_x spans the wide range $-0.2, 1.8$.

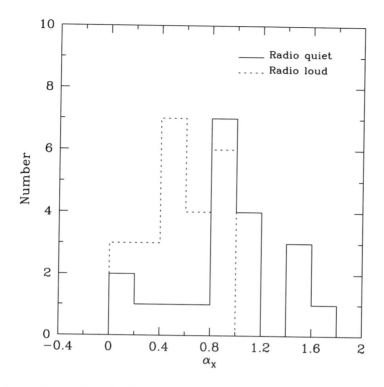

Fig. 11.4. The distribution of α_x for radio-loud and radio-quiet quasars. The data is from Shastri *et al.* (1993).

- Radio-loud quasars in general have flatter X-ray spectra (i.e., smaller values of α_x) than radio-quiet quasars. The distribution of α_x for radio-loud and radio-quiet quasars is shown in Figure 11.4. Statistical tests show that the radio-loud objects have systematically flatter α_x values at a confidence level > 99.95 per cent (Shastri *et al.* 1993). A plot of α_x against R_L is shown in Figure 11.5. It is seen that the distribution of α_x is distinctly different for the two populations. However, the spectral index is not significantly correlated with radio loudness within each population. This is consistent with the division of the quasar population into distinct radio-quiet and radio-loud types that we came across in Section 9.8, and the increased X-ray emission in radio-loud quasars relative to the radio-quiet quasars seen in Section 10.6. The presence of high luminosity radio emission contributes a flat spectrum X-ray component, which is in addition to the component with $\alpha_x \simeq 1$ present in radio-quiet quasars and presumably in the radio-loud population as well.
- In radio-loud quasars, there is a tendency for α_x to be anticorrelated with the beaming parameter R, which is the ratio of the beamed radio flux to the unbeamed radio flux from an extended region (see Equation 12.2), as can be seen

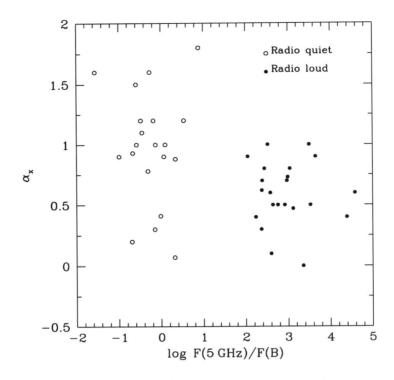

Fig. 11.5. A plot of α_x against the radio-loudness parameter for radio-loud and radio-quiet quasars. The data is from Shastri *et al.* (1993).

in Figure 11.6 (Kembhavi *et al.* 1989, Shastri *et al.* 1993). If the two discrepant points in the figure are dropped, the Spearman rank correlation coefficient is significant at the 99.95 per cent level. The significance is reduced to 90 per cent with the inclusion of the two points. A relationship between α_x and R is expected if the X-ray component related to radio emission is subject to beaming like the core radio emission (see Section 10.7): the higher the value of R, the greater is the contribution of the beamed component, which has a flat X-ray spectral index.

- Wilkes and Elvis found that the best-fit values of N_H never exceeded the Galactic N_H for the position of the quasar, determined from 21 cm radio observations, and were in fact systematically lower. By eliminating other causes such as peculiarities in the structure and composition of the Galactic ISM, Wilkes and Elvis concluded that the low values of N_H obtained in the fit were due to the presence of a soft X-ray excess above the level expected from the single power law flux density at 2 keV.

Canizares and White (1989) used a sample of 71 optically selected and radio selected quasars with $0.1 < z < 3.5$ to obtain composite X-ray spectra for ensembles of

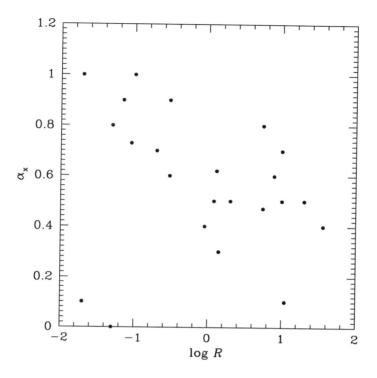

Fig. 11.6. A plot of α_x against the beaming parameter R for radio-loud quasars. The data is from Shastri *et al.* (1993).

different types of object. The procedure followed is effectively equivalent to making a single joint fit to the data for all the quasars in a subset. It helps in finding mean values and confidence intervals for α_x and N_H, including data from quasars which are so faint that their parameters are not individually well constrained by the fits. Canizares and White found that flat radio spectrum quasars have $\alpha_x \sim 0.4$, steep radio spectrum quasars have $\alpha_x \sim 0.7$ and radio-quiet quasars have $\alpha_x \sim 1$–1.4, which is consistent with the results of Wilkes and Elvis (1987). The flat X-ray spectra continue to be obtained even for objects with redshift in the range $2.0 < z < 3.5$. This means that quasar spectra are flat in the ~ 1–$10\,keV$ region, which is redshifted to the IPC band when the redshift is high. To constrain the soft X-ray component further, Masnou *et al.* (1992) analysed the IPC observations of the 14 (five radio-loud) quasars from Wilkes and Elvis (1987) with the highest signal-to-noise ratio.

Data from the monitor proportional counter (MPC) covering 1.2–$10\,keV$ was also used when a sufficient number of counts was available. The entire energy range covered by the detectors was used to fit a broken power law to the data with a fixed break energy of $0.6\,keV$. The Galactic value of N_H was used to account for low energy absorption. It was found that the spectral index above the break energy was consistent with the value obtained by Wilkes and Elvis (1987). Below the break energy a soft excess was found, with the increase over the high energy component being a factor ~ 1–6 at $0.2\,$keV. This corresponds to an increase $\Delta\alpha \simeq 1$ in the spectral index. The strength and energy of the excess observed are such that it would not be detectable beyond $z \gtrsim 0.5$.

EXOSAT   The *EXOSAT* database is a rich source for detailed X-ray study of different classes of object. Comastri *et al.* (1992) studied a sample of 17 quasars, most of which are from the Bright Quasar Survey, on which LE and ME data was available in the *EXOSAT* database. These quasars were originally observed with the *EINSTEIN* IPC, and had a count rate of at least $0.2\,\mathrm{ct\,sec^{-1}}$. Power laws modified by absorption were fitted to the ME data in the range 2–10 keV. It was found that the distribution of the spectral index α_x was significantly different from the distribution for Seyfert 1 galaxies (Turner and Pounds 1989), with $\langle\alpha_x\rangle = 0.89 \pm 0.06$ and a dispersion $\sigma = 0.25$. The quasars have generally higher luminosity than the galaxies, which could be the reason for their steeper spectra, but the steepening is also found in the range of overlapping luminosities. This difference is expected because while the quasars were selected for their flux in the soft IPC band, the AGN were selected in the harder 2–10 keV band. The latter would therefore be biased against steep values of the spectral index. When the power law fits were extended to the low energy LE band, it was found that for about two-thirds of the sample single-power-law fits proved to be adequate, while the rest showed evidence of the existence of a soft excess. The soft excess is not well constrained by the data. It corresponds to power laws with indices in the range 2.7–4.2, or black body spectra with $T = 40$–$80\,$eV. The break energy is in the region 0.35–0.75 keV at the source. The ME spectral index does not show any obvious dependence on X-ray or optical luminosity and radio-loudness, while the 0.1–4 keV spectral index is flatter for the radio-loud objects.

Lawson *et al.* (1992) compared the spectra of a sample of 18 radio-loud quasars and 13 radio-quiet quasars from the *EXOSAT* database. They fitted power law models with Galactic absorption to ME data and find that $\langle\alpha_x\rangle_{RQ} = 0.90 \pm 0.11$ with intrinsic dispersion $\sigma = 0.20$, while $\langle\alpha_x\rangle_{RL} = 0.66 \pm 0.07$, with $\sigma = 0.0$. The mean spectral indices for flat spectrum and steep spectrum quasars are $\langle\alpha_x\rangle_{FS} = 0.62 \pm 0.07$ and $\langle\alpha_x\rangle_{SS} = 0.87 \pm 0.17$ respectively, with the intrinsic dispersion in each case being zero. The dependence of the spectral index on radio-loudness and radio spectral flatness is consistent with the *EINSTEIN* data, but appears to be inconsistent with the absence of any such dependence in the sample of Comastri *et al.* (1992). The difference between the samples is, however, found to be reduced when the distribution of the spectral indices obtained by Comastri *et al.* is treated statistically using the same methods as are adapted by Lawson *et al.* Moreover, all correlations and differences found need to be treated with caution since the sample sizes are quite small.

ROSAT   The PSPC detector on *ROSAT* operates in much the same band as *EINSTEIN*, but it has a significantly improved sensitivity at low energies and higher energy resolution. It is therefore ideally suited to the the study of the soft excess. Laor *et al.* (1994) have observed with the PSPC 10 quasars from the BQS as part of a larger program. They find that spectra of nine quasars are consistent with single-power-law fits in the range 0.15–2 keV, with $\langle \alpha_x \rangle = 1.5 \pm 0.40$. The power law index is determined with statistical error $\sim$ 2–4 per cent, which is an order of magnitude smaller than that attainable with earlier instruments. There is no strong evidence for significant soft excess emission over the power law and deviations from it are $\lesssim$ 30 per cent. PSPC counts, with energy > 0.47 keV, produce better power law fits than counts over the entire range. The higher energy spectral index is flatter than the overall slope in five of the 10 quasars, but the average slopes in the two bands are practically identical. This indicates that if there is a steepening in the X-ray spectrum due to a soft excess, the break must occur for energy considerably higher than 0.5 keV.

PSPC spectra are steeper than IPC spectra, and seem to lack the soft excess found in the latter (as well as by other detectors as we discuss elsewhere). Similar results have also been obtained by Fiore *et al.* (1994) for a sample of six radio-quiet quasars. The missing soft excess is surprising because it has been observed extensively in Seyfert galaxies with *EXOSAT* and in quasars with *EINSTEIN* as well as *EXOSAT*, and the reason for the observed lack of excess is not clearly understood. However, the different responses of the detectors and changes in the 'baseline' over which the excess is measured have to be taken into account in arguments about the presence or absence of an excess.

GINGA Since the most sensitive X-ray detectors operate at below a few keV, very few quasar spectra in the range covered by *HEAO 1* A1 and *EXOSAT* had been available before the launch of *GINGA*. The LAC on that satellite was the first instrument capable of obtaining the spectra of a significant number of quasars in the 2–10 keV range in which much data on AGN was already available. The data on Seyfert 1 galaxies was good enough to reveal exciting features that had a direct bearing on the environment of the X-ray-producing region and on models for the generation of the spectrum. Quasars too have to be studied in the energy range in which these features appear, to see what effect their higher luminosities and redshifts have on the nature of the spectrum.

Williams *et al.* (1992) observed 13 bright quasars with luminosities $\gtrsim 10^{44}$ erg sec^{-1} with the LAC. These authors have fitted power law models together with an emission line at 6.4 keV to the data, with absorption produced by a combination of the Galactic hydrogen column density and intrinsic absorption by cold matter with a column density N_H^{int}. It was found that single-power-law models in all observations gave the best-fit value $N_H^{int} = 0$, and a range of 2–10 keV spectral indices with $\langle \alpha_x \rangle = 0.81$ and standard deviation 0.19, after accounting for the spread due to statistical errors. The spectral index was found to be significantly correlated with the spectral index in the IPC band. There is therefore no evidence in this sample for the flattening of the spectral index between the IPC and LAC bands.

Three of the quasars in the sample, including 3C 273 and E1821+643, that have been independently very well studied (see below), show significant reduction in the χ_v^2 value when an iron emission line is included in the fits. This provides support for the existence of the iron line. In fact, the data are compatible with all the quasars' having iron lines with equivalent widths $\sim 100\,\mathrm{eV}$, which means that the feature could occur as frequently in quasars as in the spectra of Seyfert 1 galaxies. A number of quasars in the sample show poor fits to a pure power law, and a significant improvement is obtained when more complex models are fitted. But the energy resolution available is not good enough to distinguish between models involving a broken power law, partial covering or Compton reflection. The last provides good fits to low luminosity quasars in the sample but not to high luminosity quasars. Since reflection models adequately represent the data in the case of Seyfert 1 galaxies, this could mean that these models fail in the case of high luminosity objects. However, higher resolution observations on a larger number of quasars are needed for the situation to be better understood.

3C 273   The quasar 3C 273 was observed over a period of five years using the LE telescope and ME proportional counters on *EXOSAT* and the LAC on *GINGA* (Turner *et al.* 1990). With these observations the spectrum was covered in the region $0.1–35\,\mathrm{keV}$ and the variation in the continuum spectrum and any soft excess and emission lines present could be studied.

A power law fit with absorption to the ME and LAC data leads to $\alpha_x \simeq 0.5$ and a hydrogen column density consistent with the best Galactic value of $N_\mathrm{H} = 1.8 \times 10^{20}\,\mathrm{cm}^{-2}$ in the direction of the quasar. Extrapolation of this fit to the LE band leaves residuals that indicate a soft X-ray excess. The LAC data is not good enough for model fitting, so a thermal bremsstrahlung spectrum with $T = 0.2\,\mathrm{keV}$ was adopted. The $0.1–2\,\mathrm{keV}$ excess in the various observations ranges from 5 per cent to 20 per cent of the total LE+ME flux in the range and has a luminosity of $\sim 10^{45}\,\mathrm{erg\,sec}^{-1}$. The soft excess shows a variation by a factor ~ 2 over 40 days, which is the shortest available interval between the observations. This corresponds to a size $\lesssim 10^{17}\,\mathrm{cm}$ for the soft X-ray emitting region, but as yet unobserved shorter time scales may be present. The variations in the soft excess are not found to be correlated with variations in the $2–10\,\mathrm{keV}$ flux, which indicates that the two have separate origins. The $2–10\,\mathrm{keV}$ spectral index is significantly flatter than the range of values obtained for Seyfert 1 galaxies.

The spectral index fitted to the $1.5–35\,\mathrm{keV}$ *GINGA* LAC data shows unambiguous signs of variation. It changed from $\alpha_x = 0.53 \pm 0.01$ to 0.42 ± 0.01 over a period of of one year. There is a significant excess in the LAC data over the power law fit near the observed energy of $5\,\mathrm{keV}$. This can be adequately modelled by a narrow line with equivalent width $\sim 50\,\mathrm{eV}$ and luminosity $\sim 5.4 \times 10^{43}\,\mathrm{erg\,sec}^{-1}$ ($H_0 = 50\,\mathrm{km\,sec}^{-1}\,\mathrm{Mpc}^{-1}, q_0 = 0.5, z = 0.158$). The central value of the line energy is below $6.4\,\mathrm{keV}$ in the emitter frame, indicating cold matter, but it is not possible to formally reject emission from highly ionized iron.

We saw in Section 10.7 that the X-ray emission in 3C 273 may be considered to have a component that is relativistically beamed towards the observer. Such a component is naturally associated with the milliarcsecond scale radio jet in the quasar, which is is thought to be moving relativistically, because of its observed one-sidedness and superluminal motion. The spatial scale of the radio jet is much larger than the scale on which the isotropic X-ray emission is produced. The iron emission line in Seyfert 1 galaxies is associated with this isotropic emission. If that is the case with radio quasars such as 3C 273 also, the presence of the beamed component adds to the continuum without adding to the flux in the line, which reduces its equivalent width. A simple prediction of the X-ray beaming model is that the width of X-ray emission lines should be systematically smaller in radio-loud quasars than in radio-quiet quasars, and the equivalent width should decrease with increasing radio beaming factor.

E 1821+643 This is an X-ray selected quasar with 2–10 keV luminosity $\sim 2\times10^{45}$ erg sec^{-1}. It is one of the most luminous radio-quiet quasars in the far infrared, has redshift 0.297 and is possibly located in a cluster of galaxies. *EXOSAT* observations by Warwick *et al.* (1989) have shown that the 2–10 keV spectral index is ~ 0.9 and that the X-ray flux in the 0.2–2 keV band fluctuates by a factor ~ 2 on a time scale of months.

Kii *et al.* (1991) have observed E 1821+643 with the LAC on *GINGA*, and a power law fit with absorption has produced $\alpha_x \sim 0.9$, in agreement with the earlier observations, and $N_H < 10^{22}$ cm^{-2}. The power law fit left residuals around 5 keV and a fit including line emission produced a line centred at 6.7 ± 0.3 keV in the quasar rest frame. The equivalent width of the line is 275 ± 105 eV and it has luminosity $\sim 6\times10^{43}$ erg sec^{-1}, which is ~ 3 per cent of the quasar emission. The range of the line energy means that it can arise from neutral to highly ionized iron. If the quasar is situated in a cluster of galaxies, the hot inter-cluster gas can contribute to the continuum as well as to the line emission. The cluster emission should make the continuum deviate from a pure power law. By modelling the cluster gas and constraining its properties with the observed distribution of counts, Kii *et al.* (1991) concluded that subtracting the cluster contribution makes the line energy consistent with 6.4 keV, i.e., with emission from a cold gas. The line equivalent width is reduced to the 60–380 eV range. The larger equivalent width of this quasar relative to 3C 273 is consistent with the former's being radio-quiet.

11.5 X-ray spectra: BL Lac objects

At soft X-ray energies, observations with the IPC on *EINSTEIN* showed that spectra of BL Lac objects could be represented by a simple power law with $\langle\alpha_x\rangle \simeq 1.1$ (Madejski and Schwartz 1983). The spectral indices were on the whole steeper than those of flat radio spectrum quasars (Worrall and Wilkes 1990), which like the BL Lacs are considered to be highly beamed objects. These early observations were based on small samples and often had large uncertainties in the power law spectral indices obtained as best fits. The situation has improved considerably with the availability of *ROSAT* data on complete radio selected and X-ray selected samples of BL Lac objects.

Urry *et al.* (1996) have observed 32 of the 34 BL Lacs from the 1 Jy radio selected sample due to Stickel *et al.* (1991), which we described in Section 9.9, with the
PSPC on *ROSAT*. Urry *et al.* found that in the 0.1–2 keV band the spectra could
be well represented with a simple power law model after taking into account Galactic absorption. The power law energy index showed a large spread, ranging from
~ 0 to ~ 3, with $\langle \alpha_x(\text{RBL}) \rangle = 1.16$ and dispersion 0.56. The dispersion was found
to be intrinsic, in the sense that it was not possible to describe the data with a
single-power-law spectrum, the observed distribution being attributed solely to measurement errors. Some of the brightest sources showed more complex spectra, with
evidence of spectral flattening towards higher energies in some cases and steepening
in others. Rapid X-ray variability was detected in three cases, with doubling time
scales $\lesssim 1$ week and detected amplitudes as large as ~ 7. A comparison with the
ROSAT All-Sky Survey data showed that the observations were consistent with the
variation of the X-ray emission from all BL Lacs over a time scale of months to
years. The spectral shape was also found to vary in some cases. Similar results were
earlier obtained for a subset of the 1 Jy BL Lacs by Comastri, Molendi and Ghisellini
(1995).

PSPC spectra have also been obtained for 23 of the 30 BL Lacs from the complete
X-ray selected sample from the Extended Medium Sensitivity Survey (see Section 9.9).
For these sources the mean X-ray power law energy index is $\langle \alpha_x(\text{XBL}) \rangle = 1.30$ and
the dispersion is 0.52. Application of the Kolmogorov–Smirnov test shows that the
distributions of α_x for the RBL and XBL in these samples are drawn from the same
parent population at a significance level of 85 per cent. The X-ray spectral indices for
the BL Lac objects are found to be significantly steeper than the indices for a sample
of eight flat spectrum radio quasars (FSRQ), which have mean and dispersion 0.54
and 0.56 respectively (Brunner *et al.* 1994). As mentioned above, Worrall and Wilkes
(1990) had earlier arrived at a similar result from IPC data. They found that the X-ray
spectral indices in the 0.1–3.5 keV range were similar for FSRQ and highly polarized
quasars (HPQ), with $\langle \alpha_x \rangle \simeq 0.5$, while for BL Lac objects (24 RBL and six XBL)
$\langle \alpha_x \rangle \simeq 1.0$.

Urry *et al.* (1996) have obtained the X-ray power law spectral index in the *EINSTEIN*
IPC band 0.3–3.5 keV for 13 RBL from their sample for which IPC data is available,
and find that $\langle \alpha_x(\text{IPC}) \rangle = 1.74$ with dispersion 0.38, compared to the PSPC values of
$\langle \alpha_x(\text{PSPC}) \rangle = 2.10$ with dispersion 0.40 for the same objects. This difference could be
due to a soft X-ray excess in the PSPC band, or a flat component that dominates the
IPC band at the higher energy end. However, the available data is too meagre and the
uncertainties in the IPC indices are too large for any firm statements to be made.

We show in Figure 11.7 a plot of the two-point spectral index α_{ro} against α_{ox} for the
XBL, RBL and FSRQ in the samples used by Urry *et al.* (1996) and described above.
The definitions of these indices are as in Equation 10.9 and the description below it,
with the difference that the radio, optical and X-ray luminosities are now considered
at 5 GHz, 5500 Å and 1 keV respectively, following the conventions in Sambruna *et al.*
(1996) from which we have taken the data. It is clear that of the three kinds of sources

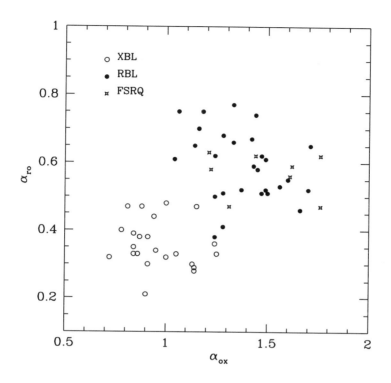

Fig. 11.7. The distribution of X-ray selected (XBL) and radio selected (RBL) BL Lacs, and flat spectrum radio quasars (FSRQ) in the $\alpha_{ox}\alpha_{ro}$-plane. The data is from Sambruna *et al.* (1996).

considered, the XBL have the lowest values of the two-power-law indices i.e., they have the flattest spectra in going from the radio to the optical to the X-ray regions. The RBL occupy a region of the plane quite separate from the XBL and have steeper spectra. The FSRQ are mixed up with the RBL. While there seems to be an overall correlation between the points, some caution is required in interpretation, since the XBL and RBL form distinct groups. Within the XBL themselves there is very little correlation, while the RBL appear to have a correlation between themselves that is quite opposite to the one which would formally apply to the whole sample taken together. A plot of this sort was first made for BL Lacs by Ledden and O'Dell (1985).

A variant of Figure 11.7, due to Sambruna *et al.* (1996), in which α_{ro} is plotted against $\alpha_{oxx} = \alpha_{ox} - \alpha_x$ is shown in Figure 11.8. When $\alpha_{oxx} \simeq 0$, the X-ray spectrum in the *ROSAT* band is not too different from the extrapolation of the optical-to-X-ray slope. Positive values of α_{oxx} indicate a spectrum that flattens in the X-ray region (a *concave* spectrum), while positive values indicate steepening (a *convex* spectrum). It is again seen that the XBL are on the whole in a different region of the plane from the RBL and the FSRQ, but now there is more mixing of the XBL and RBL than in Figure 11.7. The XBL on the whole have negative values of α_{oxx} i.e., convex spectra,

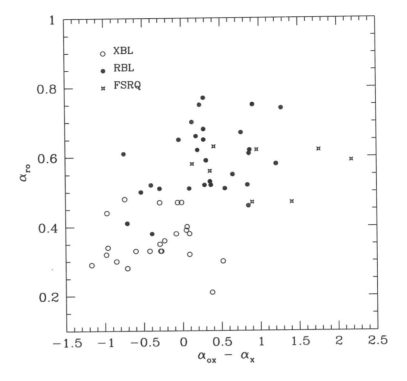

Fig. 11.8. A plot of α_{ro} against $\alpha_{oxx} = \alpha_{ox} - \alpha_x$ for X-ray selected (XBL) and radio selected (RBL) BL Lacs, and flat spectrum radio quasars (FSRQ). The data is from Sambruna *et al.* (1996).

and the FSRQ have concave spectra while the RBL are spread over a wide range of α_{oxx} with both kinds of spectra.

Sambruna *et al.* have fitted parabolic shapes of the type

$$\log[\nu L(\nu)] = A[\log L(\nu)]^2 + B \log L(\nu) + C, \tag{11.3}$$

where A, B and C are constants, to the broad band fluxes of BL Lac from the radio to the X-ray regions. The fits have been used to obtain the frequency ν_p at which $\log[\nu L(\nu)]$ has a maximum, and the bolometric luminosity between 10^9 Hz and 10^{19} Hz. They find that on average the bolometric luminosity increases from the XBL to the RBL to the FSRQ. The XBL have higher peak frequencies ν_p compared to the other two kinds, which have similar distributions of the peak frequency. Parabolic fits to the average energy distribution of each class show that for the FSRQ the spectra have a concave shape (see the discussion about Figure 11.8) with ν_p in the IR or optical range. In the case of the XBL sample the spectrum is convex, with the ν_p in the UV or soft X-ray range. The situation with the RBL is intermediate between the two.

The frequency at which the peak in the spectrum occurs can be used as a means of classification of the BL Lac population. In this scheme, objects with $v_p \lesssim 10^{14}$ Hz are called *low-frequency-peaked* BL Lacs (LBL) and those with $v_p \gtrsim 10^{14}$ Hz are called *high-frequency-peaked* BL Lacs (HBL). This division corresponds to $\alpha_{rx} > 0.8$ and $\alpha_{rx} < 0.8$ (Padovani and Giommi 1995). This is a more natural division than the RBL and XBL classes, which are dependent on the observing frequency at which a BL Lac was first discovered (radio or X-ray). A discovery-survey-dependent classification can lead to the same object's belonging to different classes depending on the depth of the surveys considered: e.g. , a radio selected BL Lac can very well appear in a sensitive X-ray survey, making it an X-ray selected BL Lac as well.

At higher energies, Sambruna *et al.* (1994) analyzed *EXOSAT* LE and ME data to obtain X-ray spectra in the 0.1—10 keV range for all blazars observed by that satellite. They find that the 24 XBL in their sample have $\langle \alpha_x \rangle \simeq 1.2$ with a 1σ range (1.05, 1.37), while five HPQs in the sample have significantly flatter indices: $\langle \alpha_x \rangle \simeq 0.62$ with a 1σ range (0.38, 1.73).

11.6   The diffuse X-ray background

We shall now review in brief the properties of the X-ray background (XRB). Our aim is not to discuss the many different theories of the origin of this background and its many interesting properties, but only to consider how much AGN and quasars contribute to it. Excellent reviews that address all the important issues may be found in Fabian and Barcons (1992) and Barcons and Fabian (1990).

The XRB, first discovered in a rocket flight by Giacconi *et al.* (1962), was also the first cosmic background to be discovered. It is highly isotropic above an energy of ~ 3 keV, which indicates an extragalactic origin. The spectrum of the XRB from 3 to ~ 50 keV was measured by the *HEAO 1* A2 experiment, while the A4 experiment on the same spacecraft measured it in the 15 keV—6 MeV range. In the lower energy range the XRB spectrum can be very well fitted by a thermal bremsstrahlung spectrum with $kT \simeq 40$ keV, which seems to imply an origin in a hot intergalactic medium. Such an origin was proposed on the basis of earlier observations by Cowsik and Kobetich (1972) and discussed in detail by Field and Perrenod (1977). There are many difficulties in this hot gas scenario, and it is necessary to look for other contributions. We will see below that AGN and quasars contribute substantially to the background, but the difficulty in the development of a detailed model has been the mismatch between the observed spectra of individual sources and the spectrum of the XRB.

11.6.1 *The X-ray background spectrum*

We shall first consider the spectrum in the 3—300 keV band where the best measurements are available. The *HEAO 1* A2 cosmic x-ray experiment measured the XRB in the 3—50 keV region with three sets of gas proportional counters, which covered different energy bands (Marshall *et al.* 1980). Each detector viewed the sky through two different apertures a few degrees square, which helped in eliminating the internal background. It was found that the observed counts could be very well fitted by a thermal bremsstrahlung spectrum with temperature $kT = 40 \pm 5\,\text{keV}$. This is shown in Figure 11.9. An analytic fit to the observed counts (Boldt 1987) in units of keV sec^{-1} steradian^{-1} cm^{-2} keV^{-1} is

$$I(E) = 5.6 \left(\frac{E}{3\,\text{keV}} \right)^{-0.29} \times \exp\left(\frac{-E}{40\,\text{keV}} \right), \quad 3\,\text{keV} < E < 50\,\text{keV}, \tag{11.4}$$

where the $E^{-0.29}$ term on the right can be taken to be the Gaunt factor of Equation 4.58. However, the interpretation of this spectrum as being emitted by an optically thin hot gas runs into trouble and the fit at this stage is to be taken merely as a convenient parameterization of the observations. Over the 3–10 keV band a good fit is provided by the simple power law form $I(E) \propto E^{-0.4}$.

The A4 experiment used scintillators that again covered different bands, spanning between them the 10 keV–10 MeV range. They had very large fields of view, which were required to obtain a large enough signal-to-noise ratio. Empirical fits, again in units of keV sec^{-1} steradian^{-1} cm^{-2} keV^{-1}, are given by

$$I(E) = 7.877 \left(\frac{E}{1\,\text{keV}} \right)^{-0.29} \times \exp\left(\frac{-E}{41.13\,\text{keV}} \right), \quad 3\,\text{keV} < E < 60\,\text{keV} \tag{11.5}$$

and

$$I(E) = 1652 \left(\frac{E}{1\,\text{keV}} \right)^{-2.00} + 1.754 \left(\frac{E}{1\,\text{keV}} \right)^{-0.70}, \quad 60\,\text{keV} < E < 6\,\text{MeV}. \tag{11.6}$$

The thermal bremsstrahlung form below 60 keV is obtained by keeping the Gaunt factor, as in Equation 11.4 and fitting the exponential to the data up to 60 keV. The first term on the right hand side of Equation 11.6 provides a continuation of the bremsstrahlung form valid at lower energies. The power law index of -0.7 in the second term was fixed at that value because it is the mean power law index for observed AGN spectra (see Section 11.1). If the index is allowed to vary in finding the best fit, the value obtained is 0.56 ± 0.20, which is not significantly different from 0.7. The flatter power law dominates towards the end of the range, but there has to be a resteepening so that the spectrum can join smoothly to the flux measured at high γ-ray energies (see Section 11.9).

We have seen in the discussion on the X-ray spectra of quasars that an excess is often observed at $\lesssim 1\,\text{keV}$ above the extrapolation of the power law spectrum at higher

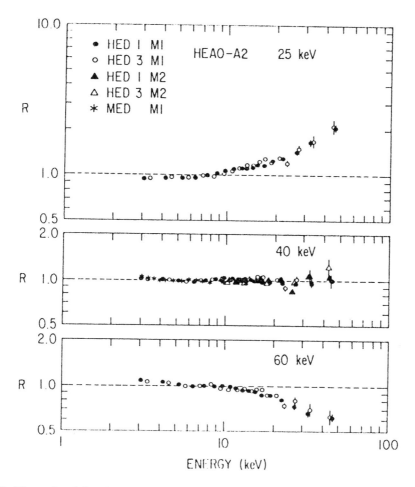

Fig. 11.9. The ratio of the observed counts from the X-ray background (XRB) to that predicted for thermal bremsstrahlung incident spectra for three different values of kT. It is clear that a temperature of 25 keV is too low while 60 keV is too high. The error bars are shown only when these are larger than the size of the symbols. Reproduced from Marshall *et al.* (1980).

energy. If these sources contribute substantially to the XRB it should show an excess too in the soft X-ray band. Wu *et al.* (1990) obtained a power law slope of ~ 0.7 for the XRB in the 0.16–3 keV range from an analysis of *EINSTEIN* IPC data. A similar slope has also been found by Hasinger *et al.* (1993) and Shanks *et al.* (1991), which points to an excess above the extrapolation of the bremsstrahlung part. Below ~ 0.5 keV most of the observed background is Galactic, with absorption in the Galaxy seriously affecting the extragalactic component. The spectrum of the X-ray background is shown in Figure 11.10.

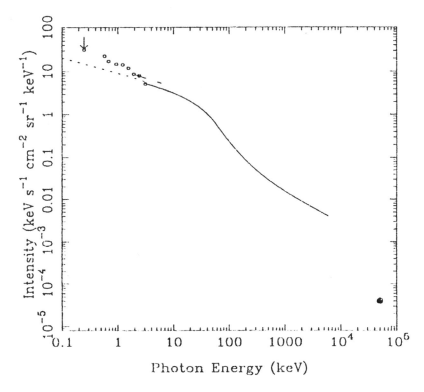

Fig. 11.10. The intensity of the X-ray background as a function of energy. The continuous line is the analytic form described above. The dotted line is the extrapolation of this fit to lower energies. The circles represent the IPC measurements due to Wu *et al.* (1990) and the upper limit is due to McCammon and Sanders (1990). Reproduced from Fabian and Barcons (1992), with permission from the *Annual Review of Astronomy and Astrophysics*, **30**, © 1992.

11.6.2 *Discrete-source contribution to the X-ray background: formulation of the problem*

Experiments that measure the XRB use a detector with a large aperture to sample the background in some energy range in different parts of the sky. The detected background can either be truly diffuse, as when it is due to a hot gas, or could include contributions from discrete sources, which are too faint to be individually resolved by the instrument measuring the background. However, the faint sources can be resolved to some limiting flux in deep surveys made with narrow beams, like the *EINSTEIN* and *ROSAT* deep surveys (see Section 10.4). If sources that have been observed as discrete objects contribute some fraction F of the background, that fraction of it is said to be resolved. There are three ways in which one can estimate the contribution of discrete sources to the XRB.

X-ray source count   If the surface density of sources as a function of the X-ray flux, $N(F_x)$, is known to some limiting flux F_x^l, the contribution of these sources to the background intensity is given by

$$I(> F_x^l) = \int_{F_x^l}^{\infty} dF_x \, F_x N(F_x). \tag{11.7}$$

For a Euclidean universe with no evolution of the comoving number density of sources with redshift, $N(F_x) \propto F_x^{-5/2}$ (see Equation 7.26) and $I(> F_x^l) \propto 1/\sqrt{F_x^l} \to \infty$ as $F_x^l \to 0$. Clearly the surface density cannot continue to increase at the Euclidean rate without exceeding the observed XRB. This is the X-ray version of the historical Olber's paradox.

This method of evaluating the contribution to the XRB is particularly useful when the surface density of sources in a deep survey is known but their identity is not established and the redshifts are unknown. If some guess can be made about the species that make up the sources, then the fractional contribution of each type can be guessed.

X-ray luminosity function   When the optical identification programme of a survey is complete to some level and the redshifts are determined, the X-ray luminosity function (see Section 10.5) can be obtained for the different types of source contributing to the source count. Let $\Phi(L_x, z)$ be the luminosity function, i.e., the comoving density of sources with X-ray luminosity L_x at redshift z, for sources of a given type (say quasars). The volume emissivity of these sources, i.e., the energy emitted by them per unit comoving volume, is $L_x \Phi(L_x, z)$ and the intensity of the background contributed by these sources is obtained by summing over the emission in different redshift shells. Applying a 'distance' factor to convert the emission from each shell to the observed background flux gives (see Equation 2.76)

$$I(E) = \frac{c}{4\pi H_0} \int_{L_{x,\min}}^{L_{x,\max}} dL_x \int_{z_{\min}}^{z_{\max}} dz \, \frac{L_x(E(1+z))\Phi(L_x(E(1+z)), z)}{(1+z)^2 \sqrt{1 + 2q_0 z}}. \tag{11.8}$$

The background is observed at some energy E (in practice it is observed in some energy band), while the X-ray luminosity on the right is at the redshifted energy $E(1+z)$. For the power law form usually assumed for X-ray spectra, $L_x(E(1+z)) = (1+z)^{-\alpha_x} L_x(E)$. The redshift dependence of the luminosity function is obtained from the results in Section 10.5. The integrals cover the observed range in X-ray luminosity and redshift.

Counts in other bands   We have seen in Section 10.6 that X-ray observations in the pointing mode have been made of a large number of quasars, Seyfert galaxies and other related objects. This has led to some idea of the distribution of α_{ox}, which is the ratio of the X-ray to the optical luminosity. Given the surface density of optical quasars, say, it is possible to estimate from it and the observed distribution of α_{ox} the contribution of quasars (or any other kind of object with similar information) to the XRB.

To see how this may be done, consider quasars (say) with optical flux in the range $F_{op}, F_{op} + dF_{op}$, at some observed frequency v_{op}. Let $N(F_{op})$ be the surface density of these quasars. From Equations 2.61 and 10.9, the X-ray flux at some observed energy E_{xs} can be written as

$$F_x = (1 + z)^{\alpha_{op} - \alpha_x} \times 10^{-k\alpha_{ox}} F_{op}, \tag{11.9}$$

where $k = \log(v_x/v_{op})$, $v_x = E_{xs}/h$ and α_x and α_{op} are the spectral indices in the optical and X-ray ranges respectively. Let $P(\alpha_{ox})d\alpha_{ox}$ be the probability of a quasar having optical-to-X-ray spectral index in the range $\alpha_{ox}, \alpha_{ox} + d\alpha_{ox}$. The number of quasars that contribute the X-ray flux given by Equation 11.9 is $P(\alpha_{ox})d\alpha_{ox}N(F_{op})dF_{op}$. For a given optical flux, the larger the value of α_{ox} the smaller will be the X-ray flux contributed by the quasar. Now suppose we want to consider the contribution to the background from quasars with X-ray flux F_x greater than some limit F_x^l. The contribution is obtained by integration over α_{ox} to some *maximum* value α_{ox}^l, which is given in terms of F_x^l through Equation 11.9. Assuming some typical value of redshift z_{typ} for all the quasars, the total background intensity contributed by quasars with optical flux greater than some limit F_{op}^l and $F_x > F_x^l$ is

$$I(> F_x^l) = (1 + z_{typ}) \int_{F_{op}^l}^{\infty} F_{op}N(F_{op}) \int_{-\infty}^{\alpha_{ox}^l(F_{op})} d\alpha_{ox} 10^{-k\alpha_{ox}} P(\alpha_{ox}). \tag{11.10}$$

In this expression we have assumed a universal distribution of α_{ox} for all quasars, but it can be easily generalized to include the dependence of the distribution on other properties of the quasars such as optical flux or redshift. It is also possible to begin with the surface density in some other wavelength region, such as the radio region, and to obtain the background in the same manner from the distribution of the appropriate spectral index connecting that band to the X-ray region.

If the surface density is expressed in terms of some optical magnitude, Equation 11.10 becomes

$$I(> F_x^l) = K(1 + z_{typ}) \int_{-\infty}^{m^l} dm 10^{-0.4m} N(m) \int_{-\infty}^{\alpha_{ox}^l(m)} d\alpha_{ox} 10^{-k\alpha_{ox}} P(\alpha_{ox}), \tag{11.11}$$

where K allows for the constants involved in passing from optical flux to magnitude and m^l is the limiting magnitude corresponding to the limiting flux F_{op}^l. If the background is evaluated at 2 keV, α_{ox} is defined between 2500 Å and 2 keV and Johnson B magnitudes are used, the constants are $k = 2.605$ and $K = (2500/4409)^{\alpha_{op}} \times 10^{-19.38}$ (Zamorani *et al.* 1981).

It must be noted that all the methods discussed above depend on X-ray surveys, which are carried out in some fixed relatively narrow band. The deepest surveys are those made with IPC on *EINSTEIN* and PSPC on *ROSAT*. The response pass band of both these detectors extends to only a few keV. An estimation of the contribution to the XRB made by discrete sources found in these surveys is limited to this range. We will see below that quasars and AGN can make up a substantial fraction of the XRB in the ~ 1–2 keV region. Extrapolating this estimate to other energies requires knowledge of the source spectrum. The difficulty here is that while the XRB has the power law form $I(E) \propto E^{-0.4}$ in the 3–10 keV range, quasars and AGN have a steeper spectrum. It is not clear how these facts are to be reconciled.

11.6.3 *Discrete-source contribution to the X-ray background: numerical estimates*

We will now consider estimates of the quasar and AGN (or, more generally, the discrete-source) contribution to the XRB.

1 Low X-ray energy We have examined the ROSAT Deep Survey (Hasinger *et al.* 1993) in subsection 10.4.4. By integrating over the surface density obtained in the survey, Hasinger *et al.* find that discrete sources contribute ~ 59 per cent of the 1–2 keV XRB, which is assumed to be 1.25×10^{-8} erg cm^{-2} sec^{-1} steradian^{-1}. The slope of the $\log N - \log F$ relation obtained in the survey from direct counts and fluctuation analysis is in the range -1.80 to -1.95 at low flux levels. This is flatter than the Euclidean value of -2.5, and extrapolating the relation to zero flux produces ~ 100 per cent of the XRB. Any resteepening below the presently observed limit would saturate the background at some non-zero flux. Hasinger *et al.* have also estimated from their fluctuation analysis that a truly diffuse component is limited to at most 25 per cent of the 1–2 keV XRB.

An early estimate of the quasar contribution to the XRB was obtained from *EINSTEIN* IPC observations of quasars by Zamorani *et al.* (1981). They combined the observed distribution of α_{ox} (taking into account upper limits) with the surface density of optical quasars to estimate that quasars with $B < 20$ contributed ~ 30 per cent of the background. This was based on the surface density $N(F_{op}) \propto F_{op}^{-2.16}$, which at that time was believed to be valid to even fainter magnitudes. Extrapolating the integration along the optical counts to $B = 21.2$ produced 100 per cent of the XRB and led to an excess at the *EINSTEIN* Deep Survey level. It was therefore suggested that the quasar counts flatten at $B \simeq 20$, which was indeed found to be the case in subsequent surveys (see Section 6.7).

In Section 10.5 we considered the X-ray luminosity function (XLF) determined by Boyle *et al.* (1993), combining data from their *ROSAT* PSPC survey survey and the extended medium sensitivity survey (EMSS). Using the XLF with parameters listed in Table 10.1, Boyle *et al.* find that quasars can contribute between 38 per cent and 78 per cent of the XRB at 2 keV for an XLF as in models **G** and **D** respectively. In this calculation the observed XRB is obtained from extrapolation of the relation in Equation 11.4 to 2 keV. These contributors here include low luminosity AGN at relatively low redshift as well as high luminosity quasars.

It is clear from the above estimates that AGN and quasars can contribute a substantial fraction of the XRB at soft X-ray energies. Taken together with the ~ 10 per cent contribution estimated to be made by clusters of galaxies as well as contributions by other kinds of source such as normal galaxies, much of the observed soft X-ray background is indeed accounted for by discrete sources. Only a small fraction at most is left to be accounted for by a truly diffuse source, which is consistent with the constraint obtained from fluctuation analysis of the *ROSAT* Deep Survey fields.

2 High X-ray energy   We have seen in subsection 10.4.1 that an X-ray luminosity function for Seyfert galaxies selected in the 2–10 keV range was derived from the *HEAO 1* All-Sky Survey. It follows from this function that Seyfert galaxies contribute ~ 20 per cent of the 2–10 keV XRB. It is more difficult to estimate the quasar contribution, since there have not been deep surveys for quasars above a few keV, and the luminosity function is not known at these energies. As a result it is not possible to estimate the quasar contribution to the XRB in this region in the manner described above. However, the contribution is expected to be rather limited, since quasar X-ray spectra in the 2–10 keV range are much steeper ($\alpha_x \sim 1$) than the X-ray background, which in this region can be approximated by a power law form $\propto E^{-0.4}$.

The spectra of Seyfert 1 galaxies in the 2–10 keV range have a power law index $\alpha_x \simeq 0.7$, and it appears that these sources too would not be able to account for the observed XRB with its flat spectrum below 40 keV. But as better data become available, Seyfert galaxy spectra are found to be rather complex (see Section 11.2), with a flattening seen above ~ 10 keV in Seyfert 1 galaxies. The emission from Seyfert 2 galaxies is heavily absorbed at low energies, while at higher energies they appear to have the same range of spectral indices as Seyfert 1 galaxies (see Section 11.3). In the unified model for Seyfert galaxies, the Seyfert 2 are obscured versions of the Seyfert 1 type, in which a molecular torus intervenes between the nucleus and observer. The optical depth of the torus is sufficient to block the low energy X-rays, but the higher energy photons get through, unless the absorbing column is unusually large, as in the case of NGC 1068. Seyfert 2 galaxies are at least twice as numerous as the Seyfert 1, and they can make important contributions to the XRB at high energy, without exceeding the observed background at low energies.

It has been suggested by several authors (see e.g. Setti and Woltjer 1989, Awaki *et al.* 1990, Morisawa *et al.* 1990) that the population of absorbed and unabsorbed Seyfert galaxies together could account for the XRB. This possibility has been examined in some detail by Madau, Ghisellini and Fabian (1994), who show that a population of the two kinds of galaxy out to a redshift of 3.5 can produce a background with spectrum matching the XRB fairly well in the 1–150 keV range. They assume that the primary spectrum has the form

$$I(E) \propto E^{-\alpha_x} \exp\left(\frac{-E}{E_c}\right), \tag{11.12}$$

with $\alpha_x = 0.9$ (see Section 11.2) and $E_c = 360$ keV. The purpose of the exponential is to provide a cutoff in the spectrum at high energy. Such a cutoff has been observed in AGN spectra by OSSE (see subsection 11.7.2), and without it the observed γ-ray background would be exceeded by the contribution from the Seyferts (see Section 11.9).

This spectrum is Compton reflected by a cold disk and the net spectrum is either directly observed, in Seyfert 1 galaxies, or after further processing in a molecular torus in the case of Seyfert 2 galaxies, which are assumed to be 2.5 times as numerous as the former. The luminosity function of the Seyfert galaxies is taken to be described by model **G** in Table 10.1, with the evolution suspended at redshift $z = 2.0$, and the integration continuing to $z = 3.5$. The background computed from this model fits the observed XRB to within 10 per cent in the 1–150 keV range. A prediction of this model is that Seyfert 2 galaxies should contribute substantially to X-ray surveys at high energy, with the observed surface density of the two kinds of Seyfert being equally numerous at a flux level of $\sim 10^{-14}$ erg cm^{-2} sec^{-1} in the 2–10 keV band. The heavily absorbed Seyfert 2 galaxies have very low flux in the soft X-ray band, and dominate the source count only below $\sim 10^{-15}$ erg cm^{-2} sec^{-1}, leading to resteepening of the source count below this level.

The model described above makes it plausible that the XRB is primarily the result of individual contributions of AGN. Such models are dependent on the unification scheme, and provide an additional consistency check. The details of the absorbed spectrum, the contribution of the different kinds of source and the evolutionary models used in obtaining the integrated spectrum are all at best tentatively known at the moment, and it cannot be expected that one kind of source provides a very good fit. A fit that is too good may in fact prove to be an embarrassment as observations become more refined and new populations of sources emerge at low flux levels. However, one can say with confidence that much of the XRB could indeed be produced by individual sources without violating the constraints imposed by the shape of the spectrum.

11.7 Gamma-ray spectra

The X-ray spectral observations described above are limited to at most a few tens of keV. The observed γ-ray band extends far beyond this range, with results from the *COMPTON* observatory (see below) available up to 30 GeV. We saw in Section 4.6 that γ-rays are produced in the Compton scattering of low energy photons by highly relativistic electrons. One may therefore expect that AGN with low optical depth to the γ-rays, i.e., with low values of the compactness parameter, are luminous γ-ray emitters. The high energy radiation can also be produced in a number of other processes, the signatures of which are expected in the γ-ray spectrum. It is also of interest to see how the X-ray spectrum connects through the few hundred keV region to the γ-ray part.

Gamma rays from discrete extragalactic sources and an extragalactic γ-ray background were observed in various satellite-based missions as well as balloon flights from the mid 1970s. The discrete sources have to be spotted against the background noise and it is difficult to determine the direction from which the high energy photons come. The number of observed sources in the pre-*COMPTON* days was therefore limited and the photon energy was $\gtrsim 100$ MeV.

A high energy γ-ray source, observed by the *COS-B* satellite and localized in a 3 deg square in the sky, was identified with the quasar 3C 273 by Swanenburg *et al.* (1978). The 50–500 MeV luminosity of the source was estimated to be $\sim 10^{46}$ erg sec^{-1}, and constituted a major part of the total radiated energy. Until the launch of *COMPTON* this was the only detected source of high energy ($\gg 1$ MeV) γ-rays . Emission above 700 keV was detected from the direction of the radio galaxy Cen A in a balloon flight by von Balmoos *et al.* (1987). The energy spectrum was found to extend to at least 8 MeV and to connect well with the power law spectrum measured at lower energies.

The Seyfert galaxy NGC 4151 was observed in the 35 keV–1.3 MeV range by the SIGMA telescope on the *GRANAT* satellite (Jourdin *et al.* 1992), and a power law continuum with energy spectral index $\alpha_\gamma \simeq 2$ was found. Comparison of this spectrum with the lower energy X-ray power law requires a break at ~ 50 keV and consistency with earlier observations calls for strong spectral variability above ~ 50–100 keV.

At lower energies merging into the X-ray region described above, the A4 experiment on *HEAO 1* measured the spectrum of 12 Seyfert 1 and broad line radio galaxies (BLRG) simultaneously in the 2–50 keV and 12–165 keV bands, resulting in spectra over the entire range covered. The mean spectrum of the sample was best fitted with a power law with spectral index $\alpha_x = 0.67 \pm 0.04$. Individual spectral indices were required to vary by just ± 0.15 about the mean. A break above 50 keV to $\alpha_x = 1.67$ could not be ruled out.

11.7.1 *The COMPTON observatory*

The aim of the *COMPTON gamma-ray observatory* (CGRO), which was launched in April 1990, is to provide broad band γ-ray observations over a wider energy range with better angular resolution and higher sensitivity than was achieved by the previous missions. The satellite has on board four major instruments (experiments), which we will now very briefly describe. An account in greater detail and further references may be found in Gehrels *et al.* (1994).

- BATSE. The burst and transient source experiment has been designed to detect intensity variations in γ-ray sources on time scales down to several μ sec over the 30 keV–1.9 MeV band. The γ-rays are detected using eight scintillation counters, which are positioned in such a way that they provide all-sky coverage. When BATSE observes a burst, a trigger signal is sent to the other instruments, which can then switch to the burst observation mode.

- OSSE. The oriented scintillation spectroscopy experiment can obtain spectra of sources or observe in another mode that provides time resolution up to 0.125 milliseconds. It has a field of view of 3.8×11.4 deg^2 in the 0.1–10 MeV band, with energy resolution of 8 per cent at 0.661 MeV.

- COMPTEL. The imaging Compton telescope has a wide field of view of ~ 1 steradian with good angular resolution for a γ-ray telescope, ~ 1 deg. It works in the 1–30 MeV band and has an energy resolution of ~ 9 per cent at 1 MeV.

- EGRET. The energetic γ-ray experiment telescope covers the broad range 20 MeV–30 GeV. It too has a wide field of view and good angular resolution, and very low background. The energy resolution is $\sim$ 15 per cent in the middle of the energy range.

EGRET and COMPTEL have been used to perform all-sky surveys in the combined 1 MeV–30 GeV range spanned by the two telescopes. These are the first surveys to be performed in the γ-ray region, and they provide maps of the diffuse γ-ray background as well as observations of discrete sources. As a result of this survey it has been found that Seyfert and radio galaxies are not strong emitters beyond a few hundred keV, while blazars often show up as very powerful γ-ray sources, the emission in the γ-ray band dominating the other bands. OSSE has provided the spectra of many AGN.

11.7.2 Seyfert galaxies

EGRET A sample of 22 Seyferts, including two Seyfert 2 galaxies, has been examined for high energy γ-ray emission in the EGRET all-sky survey (Lin *et al.* 1993). If the 2–10 keV X-ray spectrum of these objects continues to γ-ray energies, the flux expected from them at > 100 MeV is well above the detection threshold of the survey. However, there is no evidence in the EGRET data for the detection of γ-ray emission from any of the galaxies in the $\sim$ 20 MeV–30 GeV band covered by the experiment. The 2σ upper limits on the γ-ray flux from these objects are more than an order of magnitude below the limits obtained from earlier, less sensitive missions. There must therefore be a break in the spectrum beyond the region of $\sim$ 1 MeV.

OSSE The Seyfert 1.5 galaxy NGC 4151 has been observed by OSSE for two weeks (Maisack *et al.* 1993), with a sequence of two-minute observations of the source field alternated with two-minute background measurements offset from the source (this is the usual observing mode with OSSE). The total source observing time was 8.6×10^5 sec. The galaxy was detected to better than the 10σ level in the 80–150 keV range and significant flux was detected up to 300 keV. The flux was seen to vary by $\sim$ 25 per cent during the observation. A good fit to the spectrum is obtained with a broken power law, with energy index $\alpha_\gamma = 1.1 \pm 0.3$ for energies less than 103 keV and $\alpha_\gamma = 2.4^{+0.3}_{-0.4}$ above that energy. The spectrum is also well described by an exponential of the form

$$\Phi(E) = (2.33 \pm 0.05) \times 10^{-5} \exp\left[\frac{100 - E(\text{keV})}{39 \pm 2}\right] \frac{\text{photons}}{\text{cm}^2 \text{ sec keV}} \qquad (11.13)$$

It is also possible to fit the spectrum with a thermal Comptonization model (see subsection 4.3.3) with plasma temperature $kT = 37$ keV and optical depth $\tau = 2.7$. Non-

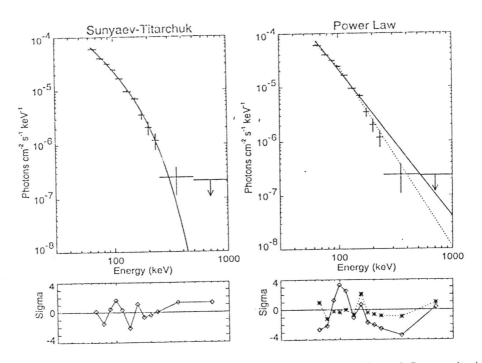

Fig. 11.11. OSSE observations of the Seyfert 1.5 galaxy NGC 4151. Thermal Comptonization and power law fits are shown. Reproduced from Maisack *et al.* (1993).

thermal pair models (see Section 4.6) do not provide a good fit. The pair annihilation line predicted by these models is not seen in the data. The spectrum with thermal Comptonization and power law fits is shown in Figure 11.11. The counts and best-fit spectra are obtained as described in the X-ray case in Section 10.3.

We shall also describe here OSSE observations of the radio galaxy Cen A (Kinzer *et al.* 1995), which has a twin-jet radio structure and is of Fanaroff–Riley class I (see Section 9.3). The galaxy was observed by OSSE in five sessions of durations five to 14 days. The spectrum was found to vary on a short time scale of 12 hr as well as a longer time scale > 4 days. Flux variations above and below 100 keV were correlated. There was no evidence of any line emission. The spectral shape at photon energy < 120 keV was found to be independent of the intensity, and could be fitted by a single power law with energy index $\alpha_\gamma \simeq 0.6$–0.7. At higher energies the spectrum can be represented by a power law with exponential cutoff at an energy that depends on the intensity. The cutoff energy varies from ~ 700 keV in the lowest intensity state to ~ 300 keV in the highest intensity state. The 50 keV–1 MeV luminosity in the high intensity state is $\sim 2 \times 10^{42} h_{100}^{-2}$ erg sec^{-1} and that in the lowest intensity state is a factor ~ 2 lower. The soft γ-ray luminosity therefore dominates the bolometric luminosity at all wavelengths, except possibly in the UV-soft X-ray range, where there is no data. In the submillimetre region too the luminosity is comparable to that in the soft γ-ray region arising from a compact source.

11.7.3 *Blazars*

EGRET In contrast to the situation with Seyfert galaxies described above, blazars are often found to be very luminous at high γ-ray energies. Von Montigny *et al.* (1995) compared the positions of (1) flat spectrum radio sources with radio flux > 1 Jy from the catalogue of Kuhr *et al.* (1981) and (2) some OVV quasars and BL Lac objects from Hewitt and Burbidge (1993) with the positions of EGRET γ-ray sources detected in the 30 MeV–30 GeV range. This comparison led to the successful detection in the γ-ray band of 33 of the sources with significance $\geq 5\sigma$. The position error boxes typically are ~ 1–2 deg and there is only a 1–2 per cent probability that an object from the catalogue would accidently fall in one such error box. From this it is estimated that there is a ~ 30 per cent probability that one of these identifications is incorrect. Von Montigny *et al.* have also detected 11 other sources with significance 4σ–5σ but the identifications in these cases are not secure.

The 33 positively identified sources are all found to be quasars or BL Lacs and have one or more properties of blazars, such as violent variability, polarization and super-luminal motion. The γ-ray spectra of these sources can be well fitted by simple power laws, with the energy index α_γ in the rather wide range from 0.4 to 2.0. Many of the 33 sources show γ-ray variability on time scales as short as a few days to a few months.

The redshifts of the sources, where these are known, are distributed over the wide range 0.03 to 2.28. The γ-ray luminosity above 100 MeV, obtained from the observed flux assuming that it is emitted isotropically, ranges from 3×10^{44} erg sec^{-1} to $> 10^{49}$ erg sec^{-1}, with typical values in the range 3×10^{47} erg sec^{-1} to 3×10^{48} erg sec^{-1}. The power in the γ-ray band is often equal to or exceeds the power in the optical and infrared bands, but the data at different wavelengths has not been taken simultaneously, except in a few cases. This can be a serious limitation because the sources are highly variable. From the optical and radio nature of the sources involved, it is to be expected that the emission is subject to relativistic beaming, and if this extends to the γ-ray band, the intrinsic luminosity of the brightest sources may be less than the isotropic estimate.

From the number of blazars detected by EGRET it seems possible that these sources as a class may be strong γ-ray emitters. To examine this possibility further, von Montigny *et al.* (1995) have looked at a sample of 35 blazars which are confirmed or suspected superluminal sources in the EGRET data, but have succeeded in deriving only γ-ray flux upper limits in all cases. The failure to detect these sources could be because only a fraction of the blazars' population is highly luminous at high γ-ray energies. Also the γ-ray emission could be more narrowly beamed than at radio wavelengths, making detection less likely, or the emission could be highly variable and above the detection threshold only for a fraction of the time.

The quasar 3C 279 was observed with EGRET for 14 days in June 1991 by Hartman *et al.* (1992). Above 70 MeV the spectrum is well fitted by a power law with energy index $\alpha_\gamma = 2.20 \pm 0.07$.[1] Assuming isotropic emission, the 100 MeV–10 GeV luminosity is $6.2 \times 10^{47} h_{100}^{-2}$ erg sec^{-1} and the energy output in the γ-ray range is larger than in any other band. The > 100 MeV flux shows variability on a time scale of a day to several days (Kniffen *et al.* 1993), with a peak flux enhancement by a factor ~ 4.

3C 279 was observed with EGRET and COMPTEL during December 1992–January 1993 for three weeks, simultaneously with observations at radio, millimetre, near-infrared, optical, ultraviolet (*IUE*) and X-ray (*ROSAT*) wavelengths (Maraschi *et al.* 1994). The quasar was found to be in a *low* state in the γ-ray region: it was only marginally detected in the EGRET observations and COMPTEL obtained an upper limit. From the multiband observations it was found that in 1993 the quasar was fainter than in 1991 at all frequencies $> 10^{14}$ Hz while at lower frequencies there was only minor variation. From the near-IR to the UV region, the quasar's spectral shape varies as the flux changes, becoming softer in the fainter state. In the low state the luminosity in the γ-ray band was comparable to that in the other bands, while in the high state the γ-ray luminosity was a factor ~ 10 higher.

The BL Lac Mrk 421, which has been detected by EGRET, has also been found to be a source of TeV photons (Punch *et al.* 1992). It was detected by the Whipple observatory ground-based γ-ray telescope, which images Cerenkov light from air showers on a two-dimensional array of photomultipliers with a pixel size of 0.25 deg. The flux above 0.3 TeV is found to be 0.3 of that from the Crab nebula. If the EGRET spectrum with power law energy index $\alpha_\gamma = 1.7 \pm 0.2$ (von Montigny *et al.* 1995) is extrapolated to TeV energy, a higher flux is expected. Joining the 100 MeV point to the flux at 0.5 TeV gives a power law index of 2. Sources brighter than Mrk 421 have not been detected at TeV energies. The reason for this could be the absorption of TeV photons by interaction with starlight and infrared photons, which becomes severe for sources at $z \gtrsim 1$.

OSSE   A sample of 17 blazars, including eight detections from EGRET, have been been observed with OSSE (McNaron-Brown *et al.* 1995) in the 50 keV–10 MeV range. Of these, 10 were positively detected, including five from the EGRET sample, and spectra were obtained for seven objects. Counts in the OSSE range were reasonably well fitted with power law spectra, with energy index α_γ ranging from 0.0 to 1.1. There is evidence for significant softening of the spectrum (i.e., steepening of the power law index) between the OSSE and EGRET bands, with $\Delta\alpha_\gamma > 0.5$ in four sources, the maximum break found being 1.7. Flux variability has been found in some sources over time scales as short as two weeks. If the emission is assumed to be isotropic, the luminosity in the 50 keV–10 MeV range extends to $\sim 3 \times 10^{48} h_{100}^{-2}$ erg sec^{-1}.

The quasar 3C 273 has been observed with OSSE on eight separate occasions over a 2 year period (Johnson *et al.* 1995). The flux in the 50–150 keV band was found to vary

---

[1] It has been stated by the authors that the flux in the 2–5 GeV range is based on 32 photons, while that in the 5–10 GeV range is based on just two photons.

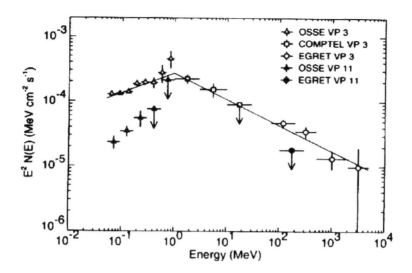

Fig. 11.12. The CGRO spectrum of the quasar 3C 273. The open symbols are contemporaneous measurements made in June 1991 by OSSE, COMPTEL and EGRET. The filled symbols are contemporaneous measurements made in October 1991 by OSSE and EGRET. The lines indicate a broken power law fit made to the June data. Reproduced from Johnson *et al.* (1995).

by a factor of ~ 3 over ~ 2 months. In the X-ray range, 3C 273 has a relatively flat spectrum with energy index $\alpha_x \simeq 0.5$. The OSSE observations to ~ 1 MeV show some steepening, with energy spectral index $\simeq 0.7$. Beyond ~ 1 MeV, the spectrum falls below the sensitivity limits of OSSE, but using contemporaneous observations by EGRET and COMPTEL it may be surmised that the spectrum steepens at ~ 1 MeV. The shape of the spectrum in the range ~ 0.5–50 MeV is not well determined, but beyond ~ 100 MeV it becomes a power law with energy index ~ 1.4. The γ-ray spectrum as measured by the various experiments on *COMPTON* is shown in Figure 11.12. The power emitted by 3C 273, as represented by $E^2N(E)$, where $N(E)$ is the differential photon spectrum, is seen to peak at ~ 1 MeV, making it similar to the steep spectrum blazars observed by EGRET.

11.8 Gamma-ray production mechanisms

EGRET has detected a number of blazars in the 30 MeV–30 GeV range and about a third of these detections exhibit superluminal motion. However, none of the Seyfert galaxies observed by EGRET have shown up as positive detections. The properties of blazars in the radio and optical band, such as rapid variability, high polarization and superluminal motion can be interpreted in terms of relativistic beaming, and the detection of a significant number of blazars makes it plausible that the beaming extends to the γ-ray band as well. In fact it is possible to show from the observed properties that if beaming were absent, the γ-rays should have been absorbed owing to photon–photon interactions.

We have seen in Section 4.6 that when the compactness parameter in a source exceeds ~ 60, the optical depth to γ-rays, due to pair production in interaction with lower energy X-ray photons, exceeds unity. The optical depth in terms of the intrinsic source luminosity and the radius of the source is given in Equation 4.66. The luminosity here pertains to the low energy photons that a γ-ray photon encounters in the source, and can be expressed in terms of the X-ray flux F at some energy, say 1 keV, using Equation 3.102. The radius can be related to the observed variability time scale using Equation 3.112, with the boosting factor that takes $\mathscr{D} \to \mathscr{D}/(1+z)$ to allow for the effect of cosmological redshift. The photon–photon interaction optical depth perceived by a γ-ray with observed energy ϵ_γ is then given by (e.g. von Montigny *et al.* 1995)

$$\tau(\epsilon_\gamma) = \left(\frac{0.2\sigma_T}{c^2}\right)\frac{(1+z)^{3+2\alpha_x}}{\mathscr{D}^{4+2\alpha_x}}D_L^2 F(1\,\text{keV})\left[\frac{\epsilon_\gamma}{2(m_ec^2)^2}\right]^{\alpha_x}\frac{1}{\Delta t_o}, \qquad (11.14)$$

where α_x is the power law spectral index in the X-ray domain. Putting $\mathscr{D} = 1$ in Equation 11.14 gives the optical depth for an unbeamed isotropic source.

The OVV radio quasar 4C 38.41(1633+289) at redshift $z = 1.814$ has been observed by EGRET to have a photon flux of $\sim 3\times10^{-6}\,\text{cm}^{-2}\,\text{sec}^{-1}$ for photon energy 100 MeV. The spectrum is of power law form with photon index 1.86 ± 0.07 and the flux is found to be variable on a time scale $\Delta t_o \simeq 2$ days. The observed flux corresponds to a 30 MeV–30 GeV luminosity of $\sim 4\times10^{48}h_{100}^{-2}\,\text{erg sec}^{-1}$ for $q_0 = 0.5$, which makes the source a factor ~ 100 brighter per decade in the γ-ray band than in any other (Mattox *et al.* 1995). The quasar has $F(1\,\text{keV}) = 0.08\,\mu\text{Jy}$ and, assuming that $\alpha_x = 0.7$, the optical depth is $\tau(\epsilon_\gamma) = 1.7\times10^4(\epsilon_\gamma/1\,\text{GeV})^{0.7}h_{100}^{-2}$, which even at the lowest energy of 30 MeV is $\sim 10^3$. The γ-rays should therefore have all been absorbed and the source should not have been detected by EGRET. The situation can be saved by introducing beaming, since a sufficiently large value of $\mathscr{D}$ on the right hand side of Equation 11.14 can reduce the optical depth to any desired value. The lower limit on $\mathscr{D}$ required to prevent significant absorption of γ-rays is obtained by setting $\tau = 1$, and in the case of 1633+289 $\mathscr{D} \geq 7.6$ for $\tau \leq 1$ (Mattox *et al.* 1995). The intrinsic γ-ray luminosity is then less by a factor $\mathscr{D}^{3.7} \simeq 2\times10^3$ than the luminosity obtained by assuming that the observed flux is being emitted in a spherically symmetric way, and the γ-ray luminosity becomes comparable to the luminosity in the other bands. In the case of 3C 279, $z = 0.54$, $\Delta t_o \simeq 2$ days, $\alpha_x = 0.7$, $F(1\,\text{keV}) = 0.8\,\text{Jy}$ and, at 3 GeV, $\tau < 1 \Rightarrow \mathscr{D} > 3.9$. Many blazars that have been observed by EGRET have remained undetected; this can be explained by assuming that the γ-ray beam is narrower than the beam in the radio domain.

If the γ-ray source is accreting in a spherically symmetric way, its intrinsic luminosity must be less than the Eddington limit, i.e., $L \leq L_{\text{Edd}} = 4\pi cGMm_p/\sigma_T$, where M is the mass of the accreting black hole. Now any intrinsic variability time scales must always be longer than the light travel time across a Schwarzschild radius, so that the lower

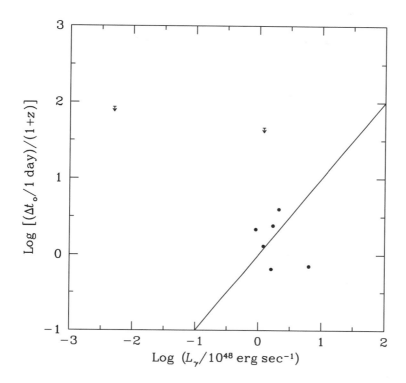

Fig. 11.13. A plot of variability time scale against observed isotropic γ-ray luminosity. The straight line is the lower limit on $\Delta t_{\mathrm{o}}/(1+z)$ given by the Elliot–Shapiro condition in Equation 11.15.

limit on the observed variability time scales is $\Delta t_{\mathrm{o}} \geq 2GM/[(1+z)c^3]$. Combining these two limits gives

$$\log\left(\frac{\Delta t_{\mathrm{o}}}{1\,\mathrm{day}}\right) \geq \log\left(\frac{L}{10^{48}\,\mathrm{erg\,sec^{-1}}}\right) - \log(1+z). \tag{11.15}$$

Such a relation was first derived by Elliot and Shapiro (1974). We show in Figure 11.13 a plot of $\log[\Delta t_{\mathrm{o}}/1+z]$ against the logarithm of the observed γ-ray luminosity, along with the line in Equation 11.15. The data is taken from von Montigny *et al.* (1995) and the figure is based on a similar figure in Schlickeiser (1996). Points that are below the line violate the Elliot–Shapiro condition. However, if beaming is taken into account, the intrinsic luminosity is a factor $\mathscr{D}^{-3}$ less than the luminosity used in Equation 11.15, while the intrinsic time scale is greater by a factor $\mathscr{D}$. If the condition is derived with these factors incorporated, the term $4\log\mathscr{D}$ is subtracted from the right hand side of Equation 11.15 and all the points are comfortably above the line.

The optical depth and variability constraints discussed above point towards relativistic beaming of the γ-ray emission, but they do not help in establishing the mechanism for the production of the high energy radiation. The γ-rays can be produced from relativistic electrons through inverse Compton scattering of low energy photons, or even through synchrotron emission or hadronic processes. We will briefly describe some models of these types. More detailed reviews and references to the literature may be found in Von Montigny *et al.* (1995), Sikora (1994) and Schlickeiser (1996).

Synchrotron and Compton processes   The γ-rays can be directly produced by synchrotron emission from electrons or pairs (Ghisellini *et al.* 1993), providing the energy density of the magnetic field is larger than the radiation energy density, so that the inverse Compton process does not dominate. However, it follows from Equation 3.13 that the production of high energy photons requires very large Lorentz factors γ: for photons with energy $> 1\,\mathrm{GeV}$, $\gamma > 2\times10^8/\sqrt{B\mathcal{D}}$. Such high energy electrons can be produced in proton–photon collisions, but the energy budget becomes prohibitively large.

The γ-rays can also result from the inverse Compton scattering of optical or UV photons by electrons in the jet. The low energy photons can be the synchrotron radiation produced by the electrons which later scatter the photons, in which case we have the synchrotron self-Compton (SSC) process. The soft photons could also be provided externally, say by the accretion disk.

Maraschi, Ghisellini and Celotti (1992) have considered a SSC model in which soft photons in the 10^{13}–10^{16} Hz range are provided by the synchrotron emission of electrons in the jet. The shape of the observed spectrum of 3C 279, to which the model was applied, cannot be produced by an SSC model with an homogeneous jet, and it is necessary to assume that (1) the jet is continuously accelerated and (2) the jet properties and the magnetic field are functions of distance along the jet. The local spectrum produced at any point in the jet is assumed to have the same slope $\alpha_l = 1$. The net observed spectrum is obtained by integrating the local spectra over the whole jet. Though the local spectra at all points have the same slope, they have different lower and upper cutoff frequencies, because of synchrotron self-absorption and the changing maximum electron energy along the jet. The X-ray and γ-ray emission is produced by inverse Compton scattering. X-ray photons in the 2–10 keV range can be produced by electrons in all part of the jets, and therefore in the summed spectrum the slope in this part of the spectrum is the same as the local slope. However, high energy γ-rays are contributed principally by the inner regions of the jet, as this is where the most energetic electrons are to be found. This makes the slope steeper than the local α_l and the variability time scales shorter. For the inverse Compton emission to dominate over the synchrotron radiation, the radiation energy density in the soft photons has to dominate over the magnetic field energy density. The Poynting flux in the jet is therefore insufficient to provide for the energy lost in the radiation, and therefore it is necessary to depend on the bulk kinetic energy of the jet, thus slowing it down as it progresses. This is, however, not consistent with the initial assumptions of the model (Sikora 1994).

Sikora, Begelman and Rees (1994) have considered a model in which soft photons for inverse Compton scattering are externally provided. These are photons from the accretion disk, which are reprocessed and scattered by (1) clouds that produce broad emission lines and the intercloud plasma or (2) an accretion disk wind or (3) absorption and thermal reradiation by dust. As the jet passes through the reprocessing region, the radiation appears to be blueshifted in the jet rest frame. This causes the radiation energy density of the diffuse field to increase by a factor $\propto \gamma_b^2$ relative to the energy density U_d as measured in the frame of the central stationary source, where γ_b is the Lorentz factor of the bulk flow. The energy density of the magnetic field in the comoving jet frame scales as γ_b^{-2} relative to its value U_B in the stationary frame, and therefore the ratio of the Compton-scattered luminosity to the synchrotron luminosity is $\propto f\gamma_b^4(U_d/U_B)$, where f is the fraction of the central region radiation that is scattered. Since γ_b^4 is large even for modest values of γ_b, the requirement that the Compton-scattered radiation dominates over the synchrotron radiation does not imply that the magnetic field has to be very low. This means that the jet need not be decelerated, as was the case with the SSC model discussed above.

The model of Sikora *et al.* can reproduce the spectrum of 3C 279 with just one component, i.e., with a homogeneous distribution of the jet properties, unlike in the SSC model. The slope of the spectrum is determined by the slope of the electron energy distribution, and the break between X-ray and γ-ray portions is produced because of the cooling of the high energy electrons above a certain energy. The model therefore predicts that the change in the power law spectral index of the continuum is $\Delta\alpha = \alpha_x - \alpha_\gamma \simeq -0.5$ (see Section 3.7). The reprocessed photons that undergo the inverse Compton scattering have a relatively narrow energy distribution. Therefore the energy ϵ_{br} where this break in the continuum occurs translates to a break in the distribution of the electron Lorentz factor at $\gamma = \sqrt{(\epsilon_{br}/\epsilon_s)}/\gamma_b$, where ϵ_s is the soft photon energy. In the SSC models, the position of the break and the change in slope at it are caused by the superposition of spectra produced in different regions of the inhomogeneous jet as well as by a break in the electron spectrum, and no specific value for the change in slope is predicted. Moreover, the synchrotron photons have a rather wide energy distribution so that the break in the continuum does not transform uniquely to a break in the electron energy distribution.

Gamma-rays with energy $> 100\,\text{GeV}$ will be absorbed in interaction with UV photons, and therefore it is necessary to modify the model in order to explain the TeV γ-ray emission from Mrk 421. The TeV radiation will remain unabsorbed if the UV density of the diffuse radiation is relatively low, and the very high energy photons are produced by the scattering of IR radiation produced by dust or SSC. The scattering of IR photons occurs in the Thomson regime, i.e., $\gamma_b\epsilon_{IR}/(m_ec^2) < 1$ (see Section 4.2). Therefore the depression in the spectrum, expected when the condition is violated and the Klein–Nishina cross-section becomes applicable, is avoided.

It has been suggested by Dermer, Schlickeiser and Mastichadis (1992) that external photons for Compton scattering by the relativistic electrons in the jet could be provided by direct radiation from the accretion disk. The production of γ-rays from the process has to take place at distances $> 10^{17}\,\text{cm}$ from the nucleus so that the γ-rays are not absorbed in interaction with soft photons. In the jet frame, photons from the disk appear to be highly redshifted, while the photons that are scattered and form a diffuse field are blueshifted. The upscattering of the diffuse field photons therefore becomes the dominant process.

Hadronic models   A model of this type has been considered by Mannheim and Biermann (1992). in this proton-induced cascade (PIC) model, electrons and protons in a relativistic jet are accelerated by shocks propagating through the jet. The process favours the acceleration of protons over electrons; because the protons see a thinner shock front due to a greater gyration radius, they can undergo resonant interactions with the plasma waves and they suffer fewer energy losses at lower energies. The protons acquire Lorentz factors γ_p in the range 10^{10}–10^{11}, which is higher than the limit of $\sim 10^8$ required for pion production. The protons interact with soft photons producing mainly pions, which decay into neutrinos, pairs and γ-ray that are further processed. The pairs and γ-rays initiate cascades in the manner described in Section 4.6. If the protons reach their maximum possible energy during the acceleration process, the spectral index in the γ-ray domain $\alpha_\gamma \leq 1$, while if the acceleration of protons is less than the maximum value, $\alpha_\gamma > 1$. The process can produce photons with energy up to the TeV range, so that the observation of such ultra-high energy photons from Mrk 421 is naturally explained.

In an alternative scenario (e.g. Eichler and Wiita 1978), highly relativistic protons close to the central source produce high energy neutrons through interactions with ambient photons. Because they are highly relativistic, the neutrons can travel to distances $\gtrsim 10^{17}$ cm before they undergo β-decay into a proton, electron and antineutrino. The ultra-relativistic protons formed in the process can produce γ-rays by interaction with ambient nucleons. The electrons produce γ-rays in the usual manner. Putting the energy into the neutrons allows it to be transported to large distances before the γ-rays are produced, which helps to avoid large photon–photon interaction opacities.

11.9   The diffuse gamma-ray background

Diffuse γ-ray flux above 50 MeV was first measured by the *OSO3* satellite, which was launched in 1968. The satellite was also responsible for the first unambiguous detection of extraterrestrial γ-rays with energy greater than a few tens of MeV (Kraushaar *et al.* 1972). More detailed observations of the diffuse background were made from the *SAS 2* satellite, which was launched in 1972. Observations in several regions of the sky with galactic latitude $|b^{II}| > 15$ deg showed a diffuse flux of γ-rays in the 35–200 MeV range. The spectrum was steep and could be represented by a power law[1] with intensity $I(E) \propto E^{-\alpha}, \alpha = 1.7^{+0.9}_{-0.7}$. The integral photon flux for $E > 100$ MeV was found to be $(2.8^{+0.9}_{-0.7}) \times 10^{-5}$ photons cm^{-2} sec^{-1} steradian^{-1}.

Fichtel *et al.* (1978) examined in detail the intensity, spectrum and spatial distribution

[1] Note that the power law index for the photon spectrum is 2.7; we will always express results in terms of the energy index, for uniformity.

of the diffuse γ-ray flux observed by *SAS 2* for $E > 35$ MeV away from the Galactic plane. They found that the emission consists of two components. (1) The first component is correlated with Galactic latitude, hydrogen column density obtained from 21 cm measurements and Galactic radio emission. This component has an energy spectrum similar to that in the Galactic plane. From these properties, this component may be taken to be of Galactic origin. (2) A second component is found to be isotropic, at least on a coarse scale ($\sim$ 5 deg) and has a steep spectrum. When extrapolated to 10 MeV the spectrum agrees well with the isotropic spectrum measured in the 0.1–10 MeV range. This component is taken to be of extragalactic origin.

Thomson and Fichtel (1982) have used galaxy counts as a tracer of galactic matter and separated the diffuse radiation observed by *SAS 2* into Galactic and extragalactic components. The latter is found to have an intensity (defined as above) of power law form with $\alpha = 2.35^{+0.4}_{-0.3}$ and a photon flux above 100 MeV of

$$(0.8-1.8) \times 10^{-5} \text{ photons cm}^{-2} \text{ sec}^{-1} \text{ steradian}^{-1}. \tag{11.16}$$

The precise value of the extragalactic background is still uncertain because of the need to separate out the Galactic component. Preliminary results from EGRET have yielded results similar to the older ones.

The origin of the extragalactic γ-ray background (EGRB) is still uncertain. On the basis of the observation of high energy γ-rays from 3C 273, it was proposed (see e.g. Bignami *et al.* 1979) that the EGRB could be generated by unresolved emission from AGN. The discovery from EGRET observations that many blazars are copious emitters of high energy γ-rays has revived this possibility.

Setti and Woltjer (1994) estimated the contribution of flat spectrum quasars and BL Lac objects to the EGRB. This was done by evaluating the ratio F_γ/F_5, of the γ-ray flux for energies greater than 100 MeV to the radio flux F_5 at 5 GHz, and multiplying it into the total 5 GHz radio surface brightness.[1] The ratio is obtained by considering the flat spectrum radio quasars and BL Lac objects from the catalogue of Véron-Cetty and Véron (1992) for which EGRET data is available. Setti and Woltjer find that the integrated radio flux from flat spectrum radio sources with $F_5 > 1$ mJy is 360 Jy steradian^{-1} and the ratio $F_\gamma/F_5 = 0.2 - 0.6$. This leads to a predicted diffuse flux for $E > 100$ MeV of

$$(0.7-1.8) \times 10^{-5} \text{ photons cm}^{-2} \text{ sec}^{-1} \text{ steradian}^{-1}. \tag{11.17}$$

Comparison with the observed range in Equation 11.16 shows that if these arguments are correct, much of the EGRB could indeed be produced by blazars. Setti and Woltjer have also estimated that normal galaxies could contribute at most $\sim$ 10 per cent of the background.

We have seen above that not all blazars are found to be luminous at high γ-ray energies. They are highly variable and are very likely to be relativistically beamed, with a beaming angle narrower than at radio wavelengths. Moreover the γ-ray spectra show

---

[1] This procedure is explained in greater detail in the context of the X-ray background in Section 11.6.

a wide distribution of spectral indices, while the high energy diffuse γ-ray emission is well fitted by a single power law. It is therefore necessary to be cautious about accepting blazars as being the source of most of the EGRB. Other sources for the EGRB have been suggested, such as the cosmic ray interactions with intergalactic gas within groups and clusters of galaxies (Dar and Shaviv 1995).

12 Unification

12.1 Introduction

We have seen that there are many kinds of quasars and AGN, with observed properties so different that they could belong to fundamentally different populations. However, seemingly different kinds of object sometimes share many properties which clearly set them apart as a class from other kinds of object. It is tempting to think of them as belonging to a single population, the differences arising because of different values taken on by one or more basic parameters.

An example of this kind of behaviour is provided by radio-quiet and radio-loud quasars. We have seen in Section 9.8 that only a small fraction of all quasars are radio-loud. The ratio of radio to optical luminosities has a bimodal distribution, with quasars termed as radio-loud having distinctly higher values of the ratio relative to the radio-quiet variety. However, the emission line properties of these two types of object are so similar that it is very difficult to tell them apart from spectroscopic criteria alone. Radio-loud quasars have higher X-ray luminosity on average than radio-quiet quasars, and flatter X-ray spectra below a few keV, but there is a great deal of overlap in the distribution of these properties. It is tempting therefore to imagine that these radio-loud and radio-quiet quasars belong to a single population. It was in fact suggested by Scheuer and Readhead (1979) that all radio-quiet quasars have relativistic radio jets in their cores, and that they appear to be luminous flat spectrum radio sources when the jet makes a small angle with line of sight to the observer. We will see below that this model is no longer considered to be tenable, but it has spawned a number of ideas which have helped in developing models that are more consistent with the data.

Emission line properties help to divide AGN into two different kinds of object, those having only narrow emission lines in their spectra, and those having broad as well as narrow lines. The former include Seyfert 2 and narrow line radio galaxies, while Seyfert 1 galaxies and broad line radio galaxies belong to the latter class. It was believed for a while that the broad-emission-line-producing regions are missing from objects that have only narrow lines, therefore making the two types of object intrinsically different from each other. However, it now seems possible that the two types belong to a single class, all members of which have narrow- as well as broad-line-producing regions. The difference between the two types is again thought to be produced by orientation.

We shall consider below different kinds of unification model in some detail, even though some of them may not be considered to be realistic in the light of data now available. The progression of ideas is important because it helps in understanding the more sophisticated models that are now being developed. We will see that even the most successful models are far from being proved to be correct; but even limited success seems to indicate that it is basically sound to unify different kinds of object. The advantage brought by unification is that differences due to extraneous factors such as orientation are separated from *intrinsic* differences, and the process of understanding the basic physics becomes somewhat simpler.

In developing unifying schemes it is necessary to keep the models simple, so that they can be tested against the available data. Elaborate constructions with a number of indeterminate parameters are not useful, as they can be fitted arbitrarily well to the data by choosing appropriate values for the parameters. In the relativistic beaming models, for example, one normally uses a single value for the bulk Lorentz factor for all the objects, even though jets in different sources are expected to move at different speeds within some range. Only when the basic nature of a model appears to be well established can one attempt to be more realistic, by e.g. introducing a distribution of Lorentz factors with one or two parameters that are to be determined from a comparison with observation.

Antonucci (1993) has summarized the simplest scheme that can be used to develop unifying models, as follows. All radio-quiet active galaxies and quasars have regions close to the nucleus that produce broad lines and featureless continuum radiation. This region is surrounded by opaque tori, approximately along the axis of which are located weak radio jets. When the torus in a given object is face-on relative to an observer, so that the line of sight reaches the nuclear region, the broad lines and continuum are seen. Otherwise only narrow lines, which are produced in a region outside the torus, are seen directly. But when the signal-to-noise ratio is sufficiently high, the nuclear region can be seen in radiation that is reflected into the observer's direction because of scattering by electrons, which also produces polarization. The torus has the same geometrical properties, such as opening angle, in all sources, and these determine the relative proportion of broad and narrow line objects. In a minority of cases, twin jets of relativistic particles are present, oriented close to the torus axis. These produce powerful radio emission through the synchrotron process and their bulk motion is relativistic, at least close to the nucleus, with the same Lorentz factor in all sources. When the axis of a radio-loud object is close to the line of sight, the observer sees a continuum superposed with broad and narrow lines, and a one-sided jet, perhaps with superluminal motion. When the orientation is very close to the line of sight, the beamed emission dominates and the object appears to be a blazar.

On the one hand, enthusiasts of beaming attempt to explain the spread of observed properties on the basis of this model, making small changes in it to accommodate sources that do not quite fit into the simple picture. The sceptics, on the other hand, use observations which are contrary to the predictions of the simplest models to argue against unification.

In the following sections we will consider different kinds of model, and provide arguments for and against unification within the framework of a model. Recent reviews that go into much greater detail have been written, amongst others, by Antonucci (1993) and Urry and Padovani (1995). Models concerning radio-loud sources have also been reviewed by Gopal-Krishna (1995).

12.2 The Scheuer–Readhead model

This model seeks to unify the populations of flat spectrum radio quasars and radio-quiet quasars. We have seen in subsection 3.5.1 that compact radio sources have flat radio spectra (i.e., a power law spectral index $\alpha \lesssim 0.5$) owing to synchrotron self-absorption. Compact flat spectrum sources are often identified with quasars and show superluminal motion. If the superluminal motion is due to relativistic motion at an angle ψ with line of sight, then the observed flux is amplified by a factor $\gamma^{-2}(1 - \beta \cos \psi)^{-m}$, where $m = 2 + \alpha$ when the moving feature is in the form of a series of knots that form a jet and $m = 3 + \alpha$ when the emitter is a single blob of matter moving relativistically.

It is expected that in a population of quasars the direction of motion is oriented at random relative to an observer. Superluminal motion occurs when $\psi \sim 1/\gamma$, where γ is the Lorentz factor. The probability that a line-of-sight angle less than this is obtained in random orientation is $P(< \psi) = 1 - \cos \psi \simeq 1/(2\gamma^2)$. For every bright flat spectrum quasar, with amplified flux and superluminal motion, there should therefore be $\sim 2\gamma^2$ quasars with motion not directed close to the line of sight and flux density closer to the intrinsic value that would be observed for $\psi = 90$ deg. If the intrinsic flux density is well below the limiting value for a survey, then only a small fraction of the quasars with amplified flux would appear to be radio-loud, while the majority would be radio-quiet.

It was proposed by Scheuer and Readhead (1979) that radio-quiet quasars and flat spectrum radio-loud quasars constitute a single population, with the latter representing the fraction with jets close to the line of sight. In their model, the intrinsic radio luminosity is proportional to the optical luminosity, which is assumed to be isotropic. Therefore the radio flux F_t of a quasar beamed transverse to the line of sight is proportional to its optical flux. At the angle ψ the radio flux is $F = F_t \gamma^{-2}(1 - \beta \cos \psi)^{-(2+\alpha)}$. To keep their model simple, Scheuer and Readhead assumed that all quasars have the same value of γ. As described above the proportion of radio-quiet quasars is $\sim 2\gamma^2$. Scheuer and Readhead concluded that since the flat spectrum radio-loud quasars constitute only a few per cent of the observed population, $\gamma \simeq 5$.

Before the VLA became available the dynamical range with which sources could be observed was rather limited and the radio flux emitted by the extended regions associated with flat spectrum sources was therefore mostly undetected. It was therefore believed that these sources were quite different from the compact cores associated with about half the quasars and a third of the radio galaxies from the 3CR survey, all of which showed extended structure. The Lorentz factor of ~ 5 obtained above cannot apply to these cores, because then the model predicts that there should be ~ 50 times as many sources without observable cores, while in the 3CR survey cores are detected in a large percentage of the sources. Scheuer and Readhead examined the ratio of the core flux to the lobe flux, in quasars with extended structures and detected cores, assuming that the emission from the extended regions was not beamed. They found a spread of $\sim 64 : 1$ in the ratio and, by attributing this to the orientation effect, concluded that a value $\gamma \simeq 2$ applied to the cores. This did not, however, mean that the intrinsic values were necessarily different in the two types of compact source. Even if the intrinsic spread of γ in the two types of sources is similar, there would be a bias towards observing higher γ values in the flat spectrum cores, because of the higher flux amplification produced. If superluminal motion with high apparent speeds were to be found in the cores of extended sources, then large values of γ would be required. Scheuer and Readhead have argued that this would be inconsistent with observation even if the high apparent speeds were found in just a few cases, and their model would be ruled out, at least in its simplest form.

The model makes a simple prediction about the distribution of radio flux density in a complete sample of optically selected quasars. These quasars are discovered without any reference to their radio emission, and if their optical emission is assumed to be isotropic, the jets in them should be randomly oriented. Consider sources with transverse flux density F_t and assume that $\beta \simeq 1$. The flux F_r from such a source with inclination ψ is $F_r = F_t(1 - \cos\psi)^{-(2+\alpha)}$. The number $N(> F_r)$ of sources with flux greater than some value F_r is equal to the number of randomly oriented sources with $\cos\psi < 1 - (F_r/F_t)^{-(2+\alpha)}$, i.e.,

$$N(> F_r) \propto (1 - \cos\psi) \propto F_r^{-1/(2+\alpha)}. \tag{12.1}$$

This relation is valid for a jet, while for a single blob the power denominator $2 + \alpha$ on the right hand side is replaced by $3 + \alpha$. If in a sample of quasars the observed range of radio flux is much greater than the range of the optical flux, the latter may be taken to be constant. In that case Equation 12.1 applies to the whole sample. The radio-loud component of such a sample therefore increases rather slowly as the flux decreases, which is in accordance with observation (see Section 9.8). We show in Figure 12.1 the integral radio flux distribution for the bright quasar survey (BQS, Kellermann *et al.* 1989), together with the lines corresponding to Equation 12.1 for $\alpha = 0$ and 1. It is clear that the observed flux distribution is not in accordance with the prediction of the Scheuer–Readhead model. The $\alpha = 1$ case provides a better fit at low flux densities, but there is an excess of bright quasars with radio flux in the range 30–500 mJy and a deficit beyond that. In fact the distribution is not consistent with a single power law. This was also shown by Strittmatter *et al.* (1980) from radio observation of 70 optically selected quasars from the Michigan Curtis–Schmidt survey.

Another argument against the Scheuer–Readhead unification scheme is the difference between the extended radio structures of radio-quiet and flat spectrum radio quasars. High dynamic range observations of a number of compact flat spectrum sources have been made with the VLA and MERLIN (Perley, Fomalont and Johnston 1979, Browne *et al.* 1982, Browne and Perley 1986, Murphy, Browne and Perley 1993). It has been

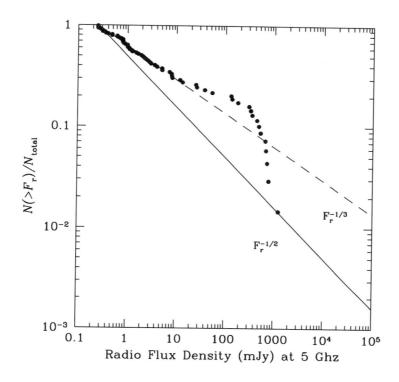

Fig. 12.1. Integral distribution of 5 GHz flux density for 69 BQS quasars brighter than 0.25 mJy. The integral distribution predicted by the Scheur-Readhead model for $\alpha = 0$ (solid line) and $\alpha = 1$ (broken line) are also shown (see text).

found from these observations that low surface brightness extended structures are associated with sources that appear to be compact in less sensitive observations. The extended structures can be one-sided, as in 3C 273 (see subsection 9.5.1), or can be in the form of a halo, which can be looked upon as emission from double lobes viewed with the line of sight close to the radio axis. Such a configuration is expected if the compact core is beamed close to the line of sight, and is approximately collinear with the overall radio axis. Murphy *et al.* (1993) have found from VLA observations of powerful core-dominated quasars that the luminosity of the extended regions is comparable to the extended luminosity of FR-II radio galaxies and quasars with prominent radio lobes. Since the flux from the extended regions is not significantly beamed, their luminosity should be independent of the orientation of the sources relative to the observer. If the Scheuer–Readhead model is correct, then quasars which are not beamed towards the observer should have the same extended luminosity as the favourably beamed flat spectrum quasars. This is clearly not the case, since the radio luminosity of a majority of the optically selected quasars is much less than that of the radio selected variety (see Section 9.8 and Figure 9.20). The luminosity of any faint extended radio component of optically selected quasars will therefore in general be very much lower than the extended luminosity of the radio selected quasars, and unification of the two kinds is not consistent. It seems more reasonable to attempt to unify core-dominated flat spectrum quasars and lobe-dominated steep spectrum quasars by attributing the core dominance in the former to the effects of relativistic beaming. Such a model will be considered in the next section.

12.3   Core-dominated and lobe-dominated quasars

The observation of faint extended regions, around compact flat spectrum quasars, having a luminosity comparable to the lobes of quasars with FR-II type structures, leads to the possibility that core- and lobe-dominated objects may be members of the same parent population. This was pointed out by Perley, Fomalont and Johnston (1979), and a detailed model was developed by Orr and Browne (1982) based on statistical comparison of the predictions of relativistic beaming with observation. We shall now describe this model, which has served as the basis for many further investigations.

It is assumed in the Orr–Browne model that the observed flux from the compact component is affected by relativistic bulk motion, while the flux from the extended structure is independent of orientation. A parameter R is defined by

$$R = \frac{F_{\mathrm{b}}(\psi)}{F_{\mathrm{ext}}}, \tag{12.2}$$

where F_{b} is the flux density of the beamed component and F_{ext} is the flux density of the unbeamed component. It is assumed that F_{b} results from symmetric jets making angles ψ and $\psi + 180$ deg with the line of sight to the observer. The angular dependence of F_{b} is given by Equation 3.105, while F_{ext} is independent of angle. We then have

$$R = \frac{R_{\mathrm{t}}}{2}(1 - \beta \cos \psi)^{-(2+\alpha)} + \frac{R_{\mathrm{t}}}{2}(1 + \beta \cos \psi)^{-(2+\alpha)}, \quad R_{\mathrm{t}} \equiv R(90 \, \mathrm{deg}), \tag{12.3}$$

where the first term is the contribution of the jet that is advancing towards the observer while the second term is from the receding jet. The minimum and maximum values of R are $R_{\min} = R(90 \, \mathrm{deg}) = R_{\mathrm{t}}$ and $R_{\max} = R(0 \, \mathrm{deg}) = R_{\mathrm{t}} \gamma^2 (2\gamma^2 - 1)$ respectively. From Equation 12.3 the *radio beaming factor* $g_{\mathrm{r}}(\beta, \psi)$ is defined as

$$g_{\mathrm{r}}(\beta, \psi) = \frac{R}{R_{\mathrm{t}}}. \tag{12.4}$$

It is convenient to use this beaming factor in place of R/R_{t} in calculations, especially when the beaming of several different wavelength regions, such as radio, optical or X-ray, is considered.

In the case where the radio spectral index α of the beamed component equals zero,

$$\cos \psi = \frac{1}{\beta} \left(\frac{2R + R_{\mathrm{t}} - R_{\mathrm{t}}^{1/2}(8R + R_{\mathrm{t}})^{1/2}}{2R} \right)^{1/2}. \tag{12.5}$$

Assuming that the sources are oriented at random, the probability of finding a source with a ratio of beamed to isotropic flux in the interval $(R, R + dR)$ is

$$P(R)dR = d(\cos \psi), \tag{12.6}$$

where R and $\cos \psi$ are related through Equation 12.5. The probability depends on R_t and the Lorentz factor γ. The probability of obtaining R in the range $R_1 < R < R_2$ is

$$P(R_1, R_2) = \cos \psi_2 - \cos \psi_1. \tag{12.7}$$

The ratio R and its transverse value R_t depend on the frequency of observation. Since a given observed frequency ν corresponds to an emitted frequency $\nu(1 + z)$ at redshift z, the same source at different redshifts will in general have different values of R_t. Defining a 'standard' value of R_t at a fiducial frequency, say 5 GHz, in the emitter's frame, one has at the frequency ν

$$R_t(\nu, z) = R_t(5\,\mathrm{GHz}) \left(\frac{\nu}{5\,\mathrm{GHz}}\right)(1 + z), \tag{12.8}$$

where the notation $R_t(\nu) \equiv R_t(\nu, 0)$ is used. Following Orr and Browne we have assumed that the radio spectral index of the beamed component $\alpha_b = 0$ and that of the extended component is $\alpha_{ext} = 1$. This considerably simplifies many of the expressions related to the model. The ratio R evaluated at the frequency ν will be denoted by R_ν.

Orr and Browne have compared the distribution of $R_{5\,\mathrm{GHz}}$ predicted by their model with the distribution of R for 32 quasars from the 3CR catalogue, which are expected to be randomly oriented. They found reasonable agreement for $R_t(5\,\mathrm{GHz}) = 0.024$ and $\gamma = 5$. The predicted distribution of the rest frame $R_{5\,\mathrm{GHz}}$ for these values, integrated over half-decade-wide bins using Equation 12.7, and the observed rest frame ratios are shown in Figure 12.2. The value of R_t is mainly constrained by the lower cutoff in the observed values of R. The shape of the distribution is sensitive to the values of γ only for large values of R, which are seldom observed. The agreement between the predicted and observed distribution can be improved by considering a distribution of R_t, but the simple model with constant parameters serves as an adequate first approximation.

Using $\alpha_b = 0$ and $\alpha_{ext} = 1$, the ratio of the total flux at the frequencies ν_1 and ν_2 is

$$\frac{F_{\nu_1}}{F_{\nu_2}} = \frac{(\nu_1/\nu_2) + R_{\nu_1}}{1 + R_{\nu_1}}, \tag{12.9}$$

where R_ν is the ratio R at the rest frequency ν. If α is the power law spectral index that connects the flux at the two frequencies, we have $F_{\nu_1}/F_{\nu_2} = (\nu_1/\nu_2)^{-\alpha}$. If, following convention, the condition for the total spectral index to be flat is taken to be $\alpha < 1/2$, then the condition in terms of R_ν is

$$R_{\nu_1} > \left(\frac{\nu_1}{\nu_2}\right)^{1/2}. \tag{12.10}$$

Consider a radio survey that is complete to a total flux density level F_0. For quasars with ratio R_ν, the corresponding completeness limit on the extended flux is $F_{ext} = F_0/(1 + R_\nu)$. Let the number of quasars with steep spectra exceeding a level F_{ext} be $n(> F_{ext})$. Let

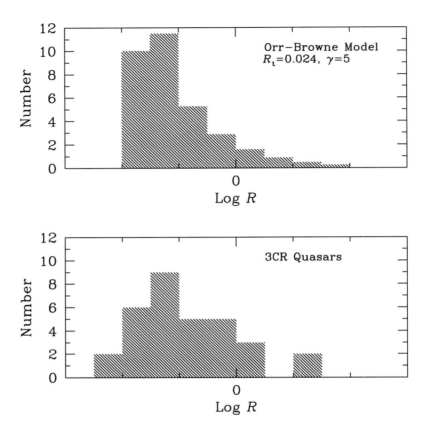

Fig. 12.2. Upper panel: the distribution of R at a rest frequency of 5 GHz obtained using the Orr–Browne model for $R_t = 0.024$ and $\gamma = 5$. Lower panel: the observed distribution of R for a sample of 32 3CR quasars from Orr and Browne.

$\rho(z)dz$ be the probability of finding a redshift in the interval $(z, z + dz)$, ρ being taken to be independent of the quasar flux density. The number of quasars above the survey flux limit as a function of R_ν and z is given by

$$N(R_\nu, z)dR_\nu dz = n\left(> \frac{F_0}{1 + R_\nu}\right)\rho(z)P(R_\nu)dR_\nu dz. \tag{12.11}$$

From the discussion following Equation 12.3 above, we find that at redshift z the values of R_ν are in the range

$$R_t(\nu, z) \leq R_\nu \leq R_t(\nu, z)\gamma^2(2\gamma^2 - 1). \tag{12.12}$$

The maximum value of $R(\nu, z)$ attained in the sample is $R_\nu^{max} = R_t(\nu, z_{max})\gamma^2(2\gamma^2 - 1)$, where z_{max} is the maximum redshift in the sample. Using Equation 12.10 the fraction of flat spectrum quasars expected in a complete sample is then given by

$$F = \frac{\int_{R_v^{\min}}^{R_v^{\max}} dR_v \int_0^{z_{\max}} dz \, N(R_v, z)}{\int_{R_t(v,0)}^{R_v^{\max}} dR_v \int_0^{z_{\max}} dz \, N(R_v, z)}. \tag{12.13}$$

With R_t reasonably well constrained using the observed distribution of R_v for 3CR quasars, the only unknown parameter in Equation 12.13 is the Lorentz factor γ. Orr and Browne have compared the predicted fraction of flat spectrum quasars with the observed fraction from seven surveys, and the weighted mean value they obtain is $\gamma = 4.7$.

A direct test of the model would be to compare the observed and predicted distributions of R_v for a complete sample. This was done in Orr and Browne (1982) for the 3CR sample, but there the number of quasars was small. Another problem was that there were not many objects at large values of R_v, although this is required to constrain γ well. Browne and Perley (1986) have considered a sample of flat spectrum ($\alpha < 0.5$) quasars with 5 GHz flux density ≥ 1 Jy for which they had VLA observations taken in the snapshot mode for estimating the extended flux density. They have obtained the model distribution of R_v using Equation 12.13 with $R_v \gg R_t(v), R_t(5\,\mathrm{GHz}) = 0.024$ and a Gaussian distribution of γ with mean 5.5 and dispersion 1.5. They find that the model and observed distributions are in good agreement for $1 \leq R_v \leq 8$, but that the observations fall well below the predicted values for $8 \leq R_v < 32$. It would be possible to achieve a better match by having a dispersion in $R_t(v)$ and a larger dispersion in γ. However, it is necessary to have high dynamic range observations for a sufficiently large complete sample of quasars in order to obtain a reliable distribution of the ratio of the beamed and extended flux densities so that a statistically meaningful comparison can be made. With this in view Murphy *et al.* (1993) obtained VLA observations of a large sample of bright core-dominated sources, and an analysis of the sample showed that the R_v distribution of the core-dominated quasars cannot be fitted by a single value of γ and a uniform distribution in the range $3 \lesssim \gamma \lesssim 11$ is required.

Kapahi and Saikia (1982) have argued that if the ratio R_v of the core to total flux is an indicator of the orientation of the source then it must be correlated with other orientation indicators. They consider three such indicators. (1) Sources oriented close to the line of sight overall should have smaller sizes and therefore there should be an anticorrelation between R_v and source size. (2) Any misalignment between the lines joining hotspots to the compact core at the centre is amplified when the axis of the source is inclined at a small angle to the line of sight. If the beaming direction is approximately collinear with the source axis, then there should be positive correlation between R_v and the misalignment angle, which is defined as the complement of the angle between the lines joining the hotspots to the core. (3) Suppose the hotspots are moving away from the core with a speed β_h, the line of motion being in the same direction as the relativistic bulk flow. Consider signals arriving from the receding and advancing hotspots and the core at some fixed time. Since the receding hotspot is moving away from the observer, the signal would have left it at an earlier time than the signal that arrives from the advancing hotspot. As a result, the advancing hotspot would be observed to be further away from the core than the receding hotspot. The ratio of the angular separation of the hotspots, which is given by $q = (1 + \beta \cos \psi)/(1 - \beta \cos \psi)$, should increase with R_v.

Kapahi and Saikia (1982) used a sample of 78 double-lobed quasars to examine these correlations, and found that in every case the trend is consistent with the Orr–Browne model for $R_t(5\,\text{GHz}) = 0.024$ and $\gamma \simeq 5$. But it is difficult to come to firm conclusions from such an exercise because evolution in the sources, as well as interaction with their environments, can change source sizes and bend angles, as well as impede motion of the source components through the host galaxy.

The main assumption of the unification scheme described in this section is that lobe-dominated quasars are the parent population of core-dominated quasars. We will see in the next section that there are arguments in favour of a more general scheme in which highly luminous *radio galaxies* form the parent population, with those galaxies beamed towards the observer appearing as quasars. In this model *all* quasars are beamed closer to the line of sight than galaxies. FR-II quasars with prominent lobes are members of the population with inclinations intermediate to those of the core-dominated quasars and the powerful radio galaxies.

12.4   Radio quasars and powerful radio galaxies

We have seen in Section 9.6 that while there are persuasive arguments in favour of relativistic beaming models, the linear size problem and the one-sidedness of large scale jets in quasars are somewhat difficult to address. In the beaming models superluminal motion is observed because fluid in the parsec scale jet moves relativistically, with the direction of motion making a small angle with the observer's line of sight. It has been pointed out that in general small scale and large scale jets are well aligned and therefore the axis of the large scale structure too should be close to the line of sight. The intrinsic sizes of superluminal quasars can be estimated by deprojecting the observed linear sizes, using the observed superluminal speeds. These intrinsic sizes are often comparable to the largest sizes found in samples that do not have an obvious bias in the orientation of their axis. In some cases the deprojected sizes are in fact much larger than those found in unbiased samples (see Figure 9.18). If these size estimates were correct, we would have been seeing many very large quasars, which is not the case. A related problem is that the largest quasars in a randomly oriented sample are found to have one-sided large scale jets. This is surprising from the point of view of the beaming models because these large quasars are expected to be oriented close to the plane of the sky. With such an orientation relative to the observer, the beaming should not be able to make an intrinsically twin-jet system appear as one-sided.

It is possible to get over the linear size problem within the framework of the beaming models by invoking, say, a misalignment of the large and small scale structures (see subsection 9.6.3). A different solution to this problem has been proposed by Barthel

(1989).[1] He makes the hypothesis that radio quasars and highly luminous radio galaxies are drawn from a single population of randomly oriented objects. When the orientation of a specific object happens to be within a certain angle to the line of sight, it has the appearance of a quasar, with superluminal motion and one-sided jets. If the orientation is not within the cone, then the object has the observed properties of a radio galaxy. This hypothesis helps in resolving the linear size and one-sidedness problems, because now *all* radio quasars are favourably oriented for the effects of relativistic beaming to be operating. Jet one-sidedness in quasars is therefore always to be expected, as well as all the effects of relativistic flows, to a greater or lesser degree. The unbeamed counterparts of the quasars are giant radio galaxies, and quasars with linear sizes much larger than those already observed are not required.

12.4.1 *Barthel's model*

Barthel has based his arguments on the quasar and radio galaxy components of the almost completely identified, low-frequency-selected sources from the 3CR catalogue (see subsection 9.6.2). The radio flux from these sources is dominated by the steep spectrum lobe emission, which is unlikely to be beamed, and therefore the orientations are expected to be distributed at random. Barthel omitted from his sample compact steep spectrum sources (CSS) identified with quasars and galaxies, on the grounds that these could be undergoing interactions with the environment that affect their sizes, making them intrinsically different from extended sources. He also restricted the redshift range to $0.5 < z < 1$ in order to avoid small-number statistics because of the paucity of quasars at $z < 0.5$ and a small number of unidentified sources at $z > 1$. The residual sample has 12 quasars and 30 radio galaxies.

The distributions of the luminosity and redshift in the two kinds of source are found to be statistically indistinguishable, which is consistent with their coming from a single population. But the radio galaxies have larger linear sizes than the quasars, as is evident from the cumulative distribution of the linear sizes shown in Figure 12.3. The linear size distribution of radio galaxies is shifted to higher values relative to the distribution of quasars. Using the Kolmogorov–Smirnov test to compare the two distributions, it is found that they are different at the 90 per cent significance level. The median linear size of the galaxies is a factor ~ 2.2 greater than the median linear size of quasars.

Quasars account for 29 per cent of the population and radio galaxies for the rest, in the combined sample. Now if the parent population is randomly oriented, the probability of having a line-of-sight angle $< \psi$ is $P(< \psi) = 1 - \cos \psi$. According to Barthel's model, a radio galaxy oriented within a certain angle of the line of sight appears like a quasar. This angle is such that the probability of a quasar-like appearance is 0.29. The angle that marks the boundary between quasars and radio

---

[1] See Peacock (1987) and Scheuer (1987) for a similar argument.

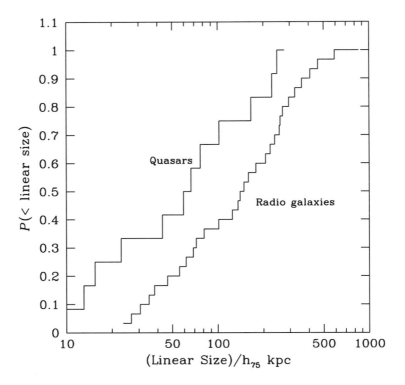

Fig. 12.3. Cumulative linear size distributions for 3CR quasars and radio galaxies with $0.5 < z < 1$. Compact steep spectrum sources have been omitted. Following Barthel (1989), $H_0 = 75\,\mathrm{km\,sec^{-1}\,Mpc^{-1}}$ has been used in computing the linear sizes.

galaxies is therefore

$$\cos^{-1}(1 - 0.29) = 44.8\,\mathrm{deg}. \tag{12.14}$$

The mean angle that each kind of object makes with the line of sight is obtained by weighting the angle with the probability density, $P(\psi)d\psi \propto \sin \psi\, d\psi$, and integrating over the allowed angles of inclination for the object. This gives for the quasars and radio galaxies the values $\langle \psi \rangle_q = 29.3\,\mathrm{deg}$ and $\langle \psi \rangle_{rg} = 68.5\,\mathrm{deg}$ respectively. The ratio of the mean projected linear sizes for the two kinds of object is therefore

$$\langle l \rangle_q / \langle l \rangle_{rg} = \sin(29.3\,\mathrm{deg}) / \sin(68.5\,\mathrm{deg}) = 0.53. \tag{12.15}$$

Radio galaxies are predicted to be, on the whole, nearly twice as large as quasars, which is consistent with the observed distributions shown in Figure 12.3. In Barthel's scheme the difference in the size distributions of quasars and radio galaxies follows from their relative numbers in a complete sample.

1 Optical polarization properties The unification of powerful radio galaxies and quasars, based on their radio properties, is consistent with the optical polarization properties of these objects, which we discussed in Chapter 8. In radio galaxies the position angle of the optical polarization is found to be either parallel or perpendicular to the radio axis (Antonucci 1984; see Antonucci 1993 for a review and references to related work). Galaxies with polarization perpendicular to the jet display a higher degree of polarization, less dominant radio cores and larger linear dimensions than galaxies with polarization parallel to the radio axis. In radio quasars there is a general alignment of the position angle of the polarization with the radio axis.

Barthel (1989) has pointed out that the optical observations of radio galaxies are consistent with the existence of an optically and geometrically thick torus that obscures the central source and the broad line region (BLR). Radiation from the obscured regions can find its way out through the opening in the torus (i.e., along the polar axis), through which also emerges the radio jet. In this picture radio galaxies, which show perpendicular polarization, are oriented with their polar axis close to the plane of the sky. The optical radiation emanating from the torus is directed away from the observer, but free electrons in the path of the radiation scatter some of it towards the observer. The scattering explains the high degree of polarization. The radio jet is also close to the sky plane, so the observer does not see an amplified flux; this keeps the radio cores from appearing to be dominant. Radio galaxies with polarization parallel to the radio axis, and those which show broad lines in their spectra, have orientations intermediate between galaxies with perpendicular polarization and radio quasars. The last have their radio axis within ~ 44.5 deg of the line of sight, which follows from the arguments given above. Quasars have a strong continuum and broad emission lines; this means that the observer is looking down the opening of the torus, which has a half-angle of ~ 44.5 deg. Narrow line luminosities in radio quasars and powerful radio galaxies are comparable, i.e., this emission is independent of the orientation of the axis of the torus. Most of the gas that emits the narrow lines should therefore be outside the torus. The unification of radio galaxies and quasars is represented schematically in Figure 12.4 (Athreya 1996).

In Barthel's model the radio axis, i.e., the jet direction, in radio galaxies is closer to the plane of the sky than in the case of quasars. It is therefore expected that relative to quasars, radio galaxies have: (1) smaller jet-to-counterjet flux ratios and jets that are relatively faint and harder to detect; (2) less common superluminal motion and smaller superluminal speeds; and (3) less pronounced Laing–Garrington polarization asymmetry. Radio galaxies should also have optical polarization perpendicular to the jet axis when the linear size is large and show polarized broad line emission. Because less of the continuum is seen in their case, narrow lines in radio galaxies should have larger equivalent widths, while the narrow line kinematics should be the same in the two kinds of object. We saw in Section 3.8 that although the statistics are not conclusive, core dominant quasars do show higher superluminal speeds than radio galaxies and in Section 9.4 that jets are indeed infrequently detected in the latter. Garrington *et al.* (1995) have shown that the fraction of radio galaxies that show stronger depolarization

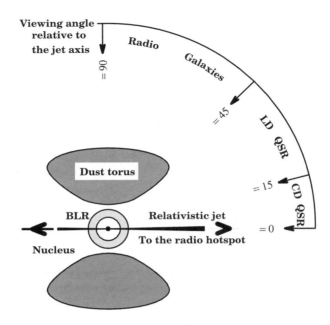

Fig. 12.4. Schematic representation of the unified model for radio galaxies and quasars. BLR: broad line region. CD: core-dominated quasars. LD: lobe-dominated quasars. Figure kindly provided by R. Athreya.

on the counterjet side is smaller than in the case of quasars; this is consistent with unification.

2 Distribution of linear sizes   Barthel has compared quasars and powerful radio galaxies in the redshift range $0.5 < z < 1$ to arrive at his model. If the two kinds of object differ only in their orientation relative to the observer, their relative number as well as the distribution of of their relative linear sizes should be independent of the redshift range examined. Singal (1993) has examined 3CR radio galaxies and quasars over the entire redshift range covered by the catalogue. He finds that the redshift distributions for the galaxies and quasars are quite dissimilar, except in the range examined by Barthel. More than a third of the galaxies in the sample are found at $z < 0.3$, where there are no quasars. The difference between the distributions remains significant at the 95 per cent confidence level even when broad line radio galaxies (BLRG) in the sample are classified as quasars, while if the BLRG are not taken to be quasars the significance of the difference is > 99 per cent. Singal has evaluated the critical angle which separates radio galaxies from quasars independently for the three redshift bins $z \leq 0.5$, $0.5 < z < 1$ and $z \geq 1$ and finds angles of 35.6 deg, 43.5 deg and 62 deg respectively. He finds that the ratios of the median values of linear sizes obtained using these angles are not in accordance with the predictions of Barthel's model. The difference is greatest at the lowest redshifts. Similar results are obtained when size ratios are evaluated for different luminosity bins. Singal therefore concludes that Barthel's unified scheme is not valid. A similar conclusion was reached by Kapahi (1990), who, however, found that 3CR quasars appear to be much more bent and misaligned, which is consistent with their being galaxies viewed at small angles.

Saikia and Kulkarni (1994) have critically examined Singal's work, and find that his conclusions about the untenability of the unified scheme may not be warranted. They find that Singal has included in his sample of powerful radio galaxies several objects that are of the FR-I type: fat doubles and sources whose structures are similar to those of the FR-I type. Such sources are not part of Barthel's scheme, which aims to unify powerful radio galaxies of the FR-II type and radio quasars. Saikia and Kulkarni also point out that the number of sources in Singal's bins is small, so that statistical fluctuations could play a major role in bringing about the perceived differences. This can be avoided by having bins with a larger number of sources and taking into account the statistical errors in derived quantities while making a comparison.

Saikia and Kulkarni have divided their sample of 44 quasars and 70 radio galaxies (in which are included compact steep spectrum sources that were excluded from earlier analysis) into two redshift bins with the boundary at $z = 0.7$. The lower redshift bin contains 32 galaxies and 17 quasars, while the higher redshift bin contains 38 galaxies and 27 quasars. Using the relative numbers and size distributions they find by making a fit to the entire data set a critical angle of 50 ± 6 deg, if the BLRG are grouped with quasars, and 45 ± 6 deg if these are grouped with galaxies. There is no evidence for a significant change in the critical angle for the low and high redshift bins, and the data are broadly consistent with the predictions of Barthel's unified scheme. Gopal-Krishna, Kulkarni and Wiita (1996) have found that the observed linear size distributions that go against the predictions of the unified model can be resolved by taking into account the available evidence for evolution with redshift in the size and luminosity of double radio sources.

12.4.2 *Beamed luminosity functions*

We have seen in Section 12.3 that there is a case for flat spectrum quasars to be considered as highly beamed versions of the lobe-dominated steep spectrum variety. Combining this with Barthel's model leads to a sequence from radio galaxies to lobe-dominated quasars to flat spectrum quasars, the beaming angle being the parameter which determines the position along this sequence. The difference between this idea and the Orr–Browne model is that in the former it is the radio galaxies which constitute the randomly oriented population, while in the latter it is the lobe-dominated quasars that play this role. Padovani and Urry (1992, see Urry and Padovani 1995 for an updated version) have tested the existence of such a sequence by constructing a luminosity function for radio galaxies, beaming it and comparing the results with the observed luminosity function of quasars.

Table 12.1. *Evolution parameter for sources in the* 2 Jy *sample (Urry and Padovani 1995). The standard deviations in* τ *correspond to deviations in* V/V_{m}.

Type	Number	Median z	$\langle V/V_{\mathrm{m}} \rangle$	τ
FR-II galaxies	61	0.15	0.55 ± 0.04	$0.26^{+0.74}_{-0.10}$
steep spectrum quasars	37	0.76	0.64 ± 0.05	$0.15^{+0.05}_{-0.02}$
flat spectrum quasars	52	0.91	0.64 ± 0.04	$0.23^{+0.07}_{-0.04}$

Urry and Padovani (1995) have used in their analysis the catalogue of Wall and Peacock (1985), which is a complete flux-limited sample of 233 sources with 2.7 GHz flux ≥ 2 Jy. The component of the catalogue relevant to the model consists of 61 FR-II radio galaxies, 37 steep spectrum and 52 flat spectrum quasars. The luminosity function is obtained by an application of the V'/V'_{m} test introduced in subsection 7.5.2. It is assumed that the evolution is of the pure luminosity type (see subsection 7.3.2), the luminosity $L(z)$ of a source at redshift z being related to its local luminosity through $L(z) = L(0) \exp[\tau(z)/\tau]$ where $\tau(z)$ is the look-back time (see Section 2.8) and τ a constant. The constant is adjusted so that the mean weighted accessible volume V'/V'_{m} becomes equal to 0.5, the goodness of fit being given by how well the distribution of V'/V'_{m} reproduces the uniform distribution. The results obtained for the three kinds of source are shown in Table 12.1. The value of $\langle V/V_{\mathrm{m}} \rangle$ for powerful radio galaxies is smaller than the value for quasars because of the smaller redshifts of the former. The evolution parameters are in agreement at the $\sim 1\sigma$ level. The local luminosity of each source is obtained using the best-fit evolution parameter and the local luminosity function for each type of source is determined from weighted accessible volumes (see subsection 7.6.2). These luminosity functions are shown in Figure 12.5.

If quasars are beamed radio galaxies, it should be possible to derive the observed quasar luminosity function from the one for the FR-II radio galaxies using the method described in subsection 3.8.6. Urry and Padovani find that good agreement between the beamed radio galaxy luminosity function and the observed function for flat spectrum quasars is obtained for a distribution of Lorentz factors $n(\gamma) \propto \gamma^{-2.3}$ with $5 \lesssim \gamma \lesssim 40$, the ratio f of intrinsic jet to unbeamed luminosity approximately equal to $\simeq 5 \times 10^{-3}$ and the angle with the line of sight restricted to $\lesssim 14$ deg. A luminosity function for steep spectrum quasars derived using the same parameters as in the case of flat spectrum quasars shows good agreement with the observed function when the beaming angle ψ is given by 14 deg $\lesssim \psi \lesssim 38$ deg. When the beaming angle of a source exceeds the upper limit of this range it appears to be a radio galaxy. The critical angle that separates quasars and radio galaxies in this model is in reasonable agreement with Barthel's result. Integration over the entire luminosity function shows that flat spectrum quasars constitute ~ 2 per cent of the population. In the integration it is assumed that the radio galaxy luminosities extend only to the lowest observed value, but the result

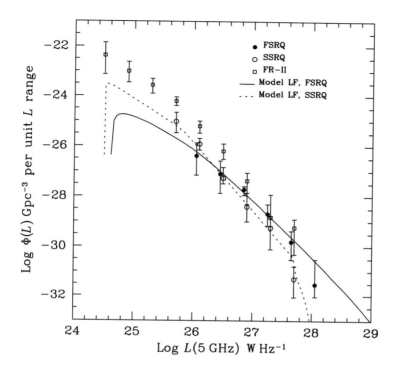

Fig. 12.5. Radio luminosity functions for flat spectrum radio quasars, steep spectrum radio quasars and FR-II galaxies. The solid line and broken lines represent luminosity functions for flat and steep spectrum quasars respectively, obtained by beaming the FR-II luminosity function. Data kindly provided by Meg Urry and Paolo Padovani.

is not sensitively dependent on where the cutoff is applied, because the luminosity function is rather flat. Urry and Padovani (1995) have derived their model parameters using $m = 3 + \alpha$ in Equation 3.117. If the value $m = 2 + \alpha$ appropriate to a jet is used, the Lorentz factor extends to higher values and the critical angle is slightly larger.

12.4.3 *Quasar host galaxies and environments*

If quasars are indeed beamed FR-II galaxies, the two classes of object should have the same extended optical morphology. We have seen in Section 9.3 that radio galaxies largely have elliptical shapes, though they do not always provide a good fit to de Vaucouleurs' law and often have disturbed morphologies indicative of mergers and interactions. We should expect radio quasars to show such structures too, but examining the situation observationally is very difficult, because of the small angular size and low surface brightness that the host galaxies are expected to have owing to their high redshift. Another difficulty is that because of the contrasting high brightness of the nucleus, any surrounding faint structure is strongly affected by it and the situation can be hopeless unless subarcsecond resolution is used. Even then it is necessary to model carefully the point spread function of the star-like nucleus, which poses various practical problems (see Dunlop *et al.* 1993 for a description of such an exercise in the near-infrared K band).

Extended optical structure around a quasar was first observed by Kristian (1973) and observations of a number of low redshift quasars have been carried out by several groups (see Dunlop *et al.* 1993 for references). One of the aims of the observations was to see whether there was a systematic difference between the hosts of radio-loud and radio-quiet quasars. Seyfert galaxies, which are often weak radio emitters, are found to be spirals, in contrast to powerful radio-loud galaxies. It therefore seems plausible that radio-quiet quasars too are embedded in spiral galaxies. In spite of a number of attempts to confirm this conjecture, the situation remains ambiguous.

Hutchings and Neff (1992) have used the Canada–France–Hawaii telescope to obtain deep optical images, in the V and I bands, of a sample of 28 radio-loud and radio-quiet quasars with redshifts in the ~ 0.1–0.5 range. They achieved a resolution (i.e., full width at half maximum of the point spread function) of ~ 0.5 arcsec in most cases, which permitted much finer detail to be observed than previously, and the examination of host galaxy luminosity profiles much closer to the nucleus. Hutchings and Neff found that the luminosity profiles range from the de Vaucouleurs' type, which are associated with ellipticals, to the exponential, which correspond to the disks of spiral galaxies. There are intermediate types of profile too. It is found that all radio-loud quasars in the sample have de Vaucouleurs' or intermediate types of profile, while the exponential profiles are always associated with radio-quiet quasars; this is consistent with the expectation about the host galaxies mentioned above. The hosts of a majority of the objects show morphological peculiarities that could be second nuclei, tidally disrupted merging galaxies, starburst knots, dust or jets in the nuclear region. Hutchings *et al.* (1994) have studied some of these features at even higher resolution with the Hubble Space Telescope.

In the sample of Hutchings and Neff, several of the radio-quiet quasars have absolute magnitudes close to the conventional dividing line between quasars and Seyfert galaxies at $M_V = -23$ for $H_0 = 50\,\mathrm{km\,sec^{-1}\,Mpc^{-1}}$.

12.5   BL Lac objects and FR-I galaxies

We have seen in Section 9.9 that several properties of BL Lac objects make it plausible that they are highly beamed versions of FR-I radio galaxies. The possibility of such a union can be examined in a formal way as was done for FR-II galaxies and quasars, by starting with a luminosity function for FR-I galaxies and beaming it to see whether the observed luminosity function for BL Lacs can be reproduced for reasonable parameter values. This has been done by Urry and Padovani (1995) following closely the treatment in subsection 12.4.2.

Urry and Padovani obtained the FR-I luminosity function from the Wall and Peacock (1985) 2 Jy catalogue. The FR-I galaxies have $\langle V/V_m \rangle = 0.42 \pm 0.05$, which shows that there is no evolution. The luminosity function derived using accessible volumes is fitted by a broken power law, which is given by

$$\Phi(L_R)\, dL_R = 6.71 \times 10^{34} L_R^{-2.02} dL_R \, \text{Gpc}^{-3}, \quad L_R < L_{R0},$$
$$\Phi(L_R)\, dL_R = 2.54 \times 10^{60} L_R^{-3.32} dL_R \, \text{Gpc}^{-3}, \quad L_R > L_{R0}, \quad (12.16)$$

where $L_{R0} = 1.56 \times 10^{25}$ W Hz^{-1}. The beamed luminosity function can be obtained from these equations using the formalism in subsection 3.8.6. The model is then fitted, by varying beaming parameters, to the observed luminosity function for BL Lac objects which was described in Section 9.9. Urry and Padovani find that it is not possible to obtain an acceptable fit with a single Lorentz factor γ for the relativistic bulk motion. But a good fit is obtained for a distribution of Lorentz factors in the range $5 \lesssim \gamma \lesssim 32$. The form of the distribution is not well constrained and for a power law distribution $N(\gamma) \propto \gamma^{-p}$, the best-fit value is $p \sim 4$. The mean value of the Lorentz factor is not very sensitively dependent on the form of the distribution and for the $p \sim 4$ power law, $\langle \gamma \rangle \sim 7$. This model implies that BL Lacs number only 2 per cent of FR-I galaxies and that the BL Lacs are aligned within ~ 12 deg of the line of sight. The ratio of the intrinsic jet to the unbeamed luminosity is $f \simeq 4.8 \times 10^{-2}$, which is ~ 10 times higher than the corresponding ratio found in the unification of powerful radio galaxies with flat spectrum quasars. The local luminosity functions of BL Lacs, FR-I radio galaxies and the fitted beaming model are shown in Figure 12.6.

A number of BL Lacs are known to have FR-II-like radio structures, and could be related to the more luminous radio galaxies (see Section 9.9). This would mean that the unifications discussed above should apply only to a subset of the BL Lacs. It is not clear whether this fraction of sources is the dominant one, with the FR-II-type objects being exceptions. The nature of the BL Lacs could depend on their luminosity and redshift, but the available statistics are not sufficient to arrive at any robust conclusion. Kollgaard *et al.* (1992) argued that only those FR-I galaxies that have intrinsically strong radio cores would have the appearance of radio selected BL Lac objects when beamed towards the observer. In this scheme, X-ray selected BL Lac objects correspond to FR-I galaxies that have weak cores, or those that have strong cores but are not beamed close to the line of sight. A bulk Lorentz factor of $\simeq 5$ is required for consistency with the data.

12.6   Seyfert galaxies

The models described above are all concerned with the unification of sources having different radio properties, and rely upon relativistic beaming to amplify the radio flux from compact regions. The degree of core dominance, and hence the observed properties of these sources, depends on the inclination of the direction of jet propagation with the line of sight. Models that depend on source orientation have also been invoked to unify

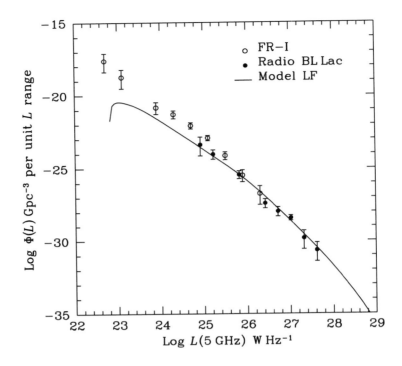

Fig. 12.6. Local radio luminosity functions for BL Lac objects, FR-I radio galaxies and the fitted beaming model for the parameters given in the text. Data kindly provided by Meg Urry and Paolo Padovani.

the populations of broad line and narrow line Seyfert galaxies. These depend not on relativistic beaming but on an opaque torus that hides the central continuum-producing source and the broad line emission region, except when the polar axis of the torus is oriented towards the observer. Broad lines are seen, and the galaxy has the properties of a Seyfert 1, only when the central region is not obscured; the fraction of cases in which this occurs depends upon opening angle. When the central region is obscured, only narrow forbidden lines are seen, since in the model the narrow-line-producing region is located outside the torus. Recent reviews of these models may be found in Antonucci (1993), Miller (1994) and Urry and Padovani (1995).

12.6.1 *Polarimetric properties*

The possibility of an obscured central region became most evident from the spectropolarimetric studies of the Seyfert 2 galaxy NGC 1068 (Miller and Antonucci 1983, Antonucci and Miller 1985) and the radio galaxy 3C 234 (Antonucci 1984).

NGC 1068 has been extensively studied because of its relative proximity (~ 22 Mpc) and brightness and has been classified as Seyfert 2 on the basis of its narrow permitted

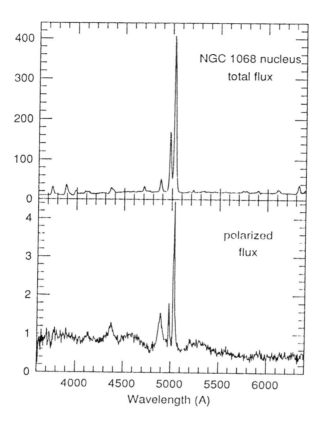

Fig. 12.7. The total flux and polarized flux spectrum of the nuclear region of NGC 1068. Reproduced from Miller *et al.* (1991).

and forbidden lines. Polarimetric observations of the galaxy have shown that, when the contribution of starlight is subtracted, the continuum is highly polarized, with the degree of polarization $P \sim 16$ per cent, independent of wavelength. The radio emission of NGC 1068 comprises a central component coincident with the optical nucleus, a linear structure of extent 13 arcsec and large scale emission that extends to ≥ 2 arcmin (Wilson and Ulvestad 1982). The polarization direction is perpendicular to the axis of the radio source. Moreover, the polarized flux spectrum shows broad Balmer lines, the degree and position angle of the polarization being similar to that of the continuum. The widths of the broad lines are similar to those found in Seyfert 1 galaxies. The total and polarized flux spectra of the nuclear region of NGC 1068 are shown in Figure 12.7.

Antonucci (1984) and Antonucci and Miller (1985) have suggested that the polarized continuum and broad component of emission lines in NGC 1068 must arise due to scattering of radiation from the nuclear region; the scatterers must be electrons, because the polarization is wavelength independent (a Rayleigh-scattering-like wavelength dependence will be present if the scattering is due to dust). In this model, the

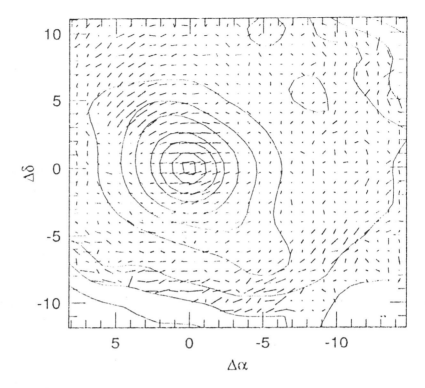

Fig. 12.8. Polarimetric image of the central $30 \times 30 \, \text{arcsec}^2$ region of NGC 1068. The contours are isophotes of the total flux, and the short lines indicate the degree and position angle of the polarization. The nucleus is at right ascension $\alpha = 0$ and declination $\delta = 0$. The knot in the north-east with offset $\Delta\alpha = 3.5 \, \text{arcsec}$ and $\Delta\delta = 4.0 \, \text{arcsec}$ is clearly seen as a region of enhanced polarization. Reproduced from Miller *et al.* (1991).

nuclear emission cannot reach the observer directly because of an obscuring torus that surrounds the continuum- and broad-line-producing regions. The narrow lines, which show low polarization, are viewed directly, and must therefore be produced outside the torus. The scattered photons mostly escape along the polar axis of the torus, which is also the axis of the radio jet. This explains the relation between the polarization position angle and the radio axis, because the polarization direction is perpendicular to the direction in which the photons last moved before being scattered into the observer's line of sight.

Further insight into the obscured broad line region of NGC 1068 has been obtained by Miller, Goodrich and Mathews (1991), using images as well as spectra of the galaxy in polarized light. Isophotes of the total continuum flux near 4260 Å, together with the distribution of the degree and direction of polarization, are shown in Figure 12.8.

The central region has isophotes elongated approximately in the north–south direction and has substantial polarization. There is a knot of high polarization at

~ 4.7 arcsec to the north-east of the nucleus. The polarization direction here and in the north and south regions is perpendicular to the radius vector from the nucleus, which, as explained above, is expected if the polarization is due to the scattering of radiation that has come from the nucleus through the opening in an obscuring torus. Polarized flux spectra show broad lines as in a Seyfert 1 spectrum, confirming the presence of a Seyfert 1-type nucleus hidden from direct view. Interestingly, the degree of polarization in the north-east knot is found to be wavelength dependent, rising rapidly towards the blue and becoming ~ 30 per cent at 3500 Å, which shows that the scattering must be due to dust, whereas the scattering from the nuclear region is due to electrons. This difference is also reflected in the width of the broad emission lines, obtained after subtraction of the narrow lines which are also present in the polarized flux. The FWHM of the broad H β line in the nucleus is 4480 km sec^{-1}, while in the north-east knot it is 2900 km sec^{-1}. These widths are consistent with the assumption that electrons are the source of the scattering in the nucleus, and slow moving dust the source in the knots.

The flux spectrum from the knot shows strong, narrow, permitted and forbidden lines with relative intensities similar to those of lines from the nuclear region. The line widths in the knots are, however, considerably narrower than nuclear line widths. The nature of the narrow lines shows that the gas in the knot is being excited by a continuum similar to the ionizing radiation that excites the lines in the nuclear region. This can be used to determine the intrinsic H β luminosity of the nucleus in the following manner. First, assuming that the nuclear flux from the forbidden line [O III] is not obscured because the narrow-line-emitting region is outside the torus, the observed flux in the line can be used to determine the [O III] luminosity. Then the nuclear H β luminosity can be determined from the observed [O III]/H β ratio in the knot, by assuming that this is the same as the ratio in the unobscured nucleus. This assumption is reasonable since the observations discussed above indicate that the knot is being excited by nuclear continuum radiation. The luminosity in the H β line, including the broad and narrow line contributions, is 1.44×10^{42} erg sec^{-1}, which is in the range of values obtained for Seyfert 1 galaxies. The obscured nucleus therefore has the spectrum as well as the luminosity of a typical Seyfert 1 nucleus.

12.6.2 *Radio galaxies*

The spectroscopic distinction between broad line radio galaxies (BLRG) and narrow line radio galaxies (NLRG) is similar to the distinction between Seyfert 1 and Seyfert 2 galaxies. It is therefore to be expected that a model that unifies the Seyferts into a single population should apply to the radio galaxies too. This point has already been made in subsection 12.4.1, where it was argued that quasars too belong to the same population, a powerful radio galaxy appearing to be a quasar when its nucleus is fully exposed to the observer and the radio emission is relativistically beamed.

An example of an NLRG with an obscured broad line region is the radio galaxy 3C 234, which is a triple radio source and has polarization properties very similar to those of NGC 1068. The starlight-subtracted continuum is polarized to $\gtrsim 14$ per cent, the position angle being oriented at 89 ± 5 deg to the radio axis. The spectrum of 3C 234 shows a relatively narrow broad Hα component, no broad Hβ and extremely strong forbidden [O III] lines. The broad component is found to be highly polarized, as in the case of NGC 1068, the position angle again being perpendicular to the radio axis. Polarized broad lines have been found in several other radio galaxies (see Urry and Padovani 1995 for references).

The highly radio luminous NLRG Cygnus A, which we have discussed in Chapter 9, has an apparent double optical nucleus, one component being an emitter of high ionization lines, and the other producing only a featureless optical continuum. The continuum-producing component has been observed at ultraviolet wavelengths with the faint object spectrograph on board the Hubble Space Telescope (Antonucci, Hurt and Kinney 1994). The spectrum shows the presence of a broad Mg II line with FWHM ~ 7500 km sec^{-1} and equivalent width ~ 25Å, which is lower than the ~ 50Å typical for quasars. Antonucci *et al.* have argued that the broad ultraviolet line is reflected radiation from the nucleus. Djorgovski *et al.* (1991) have discovered a compact, unresolved nucleus at infrared wavelengths coincident with the radio core in Cygnus A. Assuming that the infrared emission is a highly reddened extension of the radio emission, they conclude that the rest frame extinction of the nucleus is $A_V \simeq 50$ magnitudes, which puts the extinction-corrected luminosity in the quasar range. The high luminosity of the nucleus and the broad line emission mean that if the galaxy were to be viewed without obscuration, it would be classified as a quasar. The broad lines are not seen at optical wavelengths in either the total or the polarized flux, which means that the continuum at optical wavelengths must be dominated in the extended region by a component which is produced *in situ* and which swamps the scattered, polarized part. If the *in situ* extended emission is also present at ultraviolet wavelengths, the unusually low equivalent width of the broad Mg II line can be understood as being due to the presence of this extra continuum, which is not accompanied by line emission. Antonucci *et al.* have speculated that the continuum could be reddened radiation from OB stars, or free–free emission from highly photoionized gas with $10^4 < T < 10^6$ K.

12.6.3 *Emission cones*

If the presence of an obscuring torus directs the continuum emission from the nucleus to move in a beam along the polar axis, gas along the path of the beam should be excited and produce emission lines that can reveal the beam shape. Pogge (1988) has observed NGC 1068 with narrow band interference filters to isolate the emission lines Hα + [N II] $\lambda\lambda 6548, 6583$ and [O III] $\lambda 5007$. The observations revealed a cone-shaped region of ionized gas emanating from the nucleus, the cone axis being very close to the axis of a radio jet that is present. The cone has also been observed in a high resolution narrow band [O III] $\lambda 5007$ image taken with the planetary camera on board the HST, which shows that the conically distributed emitting region is divided into a number of clouds with size 0.1–0.2 arcsec (Evans *et al.* 1991).

Narrow band imaging of the Seyfert 2 galaxy NGC 5252 has shown the distribution of ionized gas as being in a double conical structure with the nucleus at the apex (Tadhunter and Tsvetanov 1989). A similar double-sided structure has been found in ∼ 0.1 arcsec narrow band imaging with the HST of the Seyfert 2 galaxy NGC 5728 (Wilson *et al.* 1993). The common apex is here coincident with the nucleus defined by a compact radio source, and the opening angle of both the cones is ∼ 55–65 deg. The cones on the two sides of the nucleus have a common axis that is within ∼ 3 deg of a one-sided nuclear radio continuum emission. The edges of the cones are sharp, which would not be the case if the cone were formed by relativistic beaming.

12.6.4 *Infrared and X-ray properties*

If Seyfert 2 galaxies are indeed obscured Seyfert 1 galaxies, then we should expect that those properties which are not affected by obscuration should be similar in the two types. Since obscuration decreases towards higher wavelengths, it should be possible to see through the torus in the infrared region. Infrared emission from the nuclear region has been found in some NLRG, but not in all the galaxies that have been observed, possibly due to large extinction. Goodrich, Veilleux and Hill (1994) have obtained infrared *J* band spectra of 15 Seyfert 2 galaxies with the aim of detecting broad Paschen lines. Three galaxies in the sample were found to have broad lines, while there were indications of such lines in a few others.

We considered the X-ray spectra of Seyfert 2 galaxies in Section 11.3 and saw how these are consistent with the unification model. To recapitulate, if the torus is present it should be opaque to X-ray photons with energy less than a few keV, but transparent to higher energy photons. Observationally, Seyfert 2 galaxies often do show the presence of strong absorption at low energies but have overall spectral shapes similar to those of Seyfert 1 galaxies. Quiet surprisingly, NGC 1068 does not show a significant absorbing column in soft X-rays. This can be explained by invoking a very large column that prevents any of the 2–10 keV X-rays from getting through, and attributing the observed X-ray flux to scattered radiation. Since only a few per cent of the radiation is scattered, the intrinsic X-ray luminosity of NGC 1068 must be comparable to the X-ray luminosity of Seyfert 1 galaxies.

12.6.5 *Unification?*

The observations described above show that some Seyfert 2 galaxies at least have hidden Seyfert 1-like central regions, and appear to be narrow line objects only because of their orientation relative to the observer. The question then is whether all Seyfert 2 galaxies are obscured and the two types may be considered to be unified into a single population. A simple unifying model, in which all the galaxies in a population are intrinsically the same, the differences arising wholly due to orientation, appears to be ruled out for several reasons.

Observations by Ulvestad and Wilson (1984a, b) appeared to show that radio properties of the two types of Seyfert galaxy were quite different, the Seyfert 2 objects having ratios of radio to non-thermal optical luminosity greater by factors in the range 40 to 100. The Seyfert 2 galaxies in their samples had larger radio sizes and the double and triple structures found had larger linear scales, as in radio galaxies (Ulvestad and Wilson 1984a, b). These difference would of course make it difficult to accept that the two types of galaxy are intrinsically the same. However, analysis of a distance-limited sample of Seyferts that includes weak Seyfert 2 galaxies has shown that the differences are much smaller than previously thought, and may not be statistically significant (Ulvestad and Wilson 1989).

Pogge (1989) has observed a sample of 20 nearby Seyfert galaxies with narrow band filters centred on H α + [N II] and [O III] emission lines. He found that only three out of eight Seyfert 1 galaxies have extended emission line regions, while eight out of 11 Seyfert 2 galaxies show extended emission. The extended emission in the type-1 objects has the same morphology in the two bands, while the morphologies are different for type-2, and four of the regions are resolved into emission cones. Pogge finds that the paucity of extended emission in Seyfert 1 galaxies cannot be attributed just to projection effects and greater brightness of the nuclear regions that could swamp the extended emission. These differences argue against unification, but again the sample size is too small for the differences to be statistically significant and the sample is biased by selection effects.

In the simple unification model, all differences between narrow and broad line galaxies are attributed to orientation. If all Seyferts have tori with the same optical depth and opening angle, independent of the luminosity of the nucleus or other properties, then the relative numbers of narrow and broad line objects too should be independent of these properties. Lawrence (1987) has considered the relative numbers of the two types of object in optical, infrared X-ray and radio selected samples. He finds that narrow line galaxies and reddened broad line galaxies occur more frequently in the lower luminosity objects, and that narrow line objects have weaker [O III] emission. These differences cannot be reconciled with the simple model but are explained if there is a distribution of cone angles, with an excess of smaller cone angles, i.e., of geometrically thick tori, at lower luminosities. It may also be that tori which produce absorption occur in only some of the objects, the probability of occurrence being higher at lower luminosities. Either of these possibilities would require a revision of the simple unification model.

13 Quasar absorption lines

13.1 Introduction

The absorption lines in quasars present a rich variety of data, which has received considerable attention both from theoreticians and observers. The absorption lines in a quasar are distinct from absorption lines in a typical normal galaxy in one important respect. A quasar with emission line redshift z_{em} may have absorption systems at several different redshifts and in general an absorption redshift z_{abs} is less than the emission redshift. There are a few exceptions to this rule, though; for example, the quasar PKS 0119–04 has $z_{em} = 1.955$ and $z_{abs} = 1.965$.

The absorption line systems are classified into three categories and it is believed that these different categories arise from different causes.

- The broad absorption line systems (BAL) are characterized by distinctive absorption troughs arising from various ions of differing excitation. These are believed to be caused by gas within the quasar itself with outflow velocity up to $\sim 60\,000$ km sec^{-1}, relative to the emission redshift. There is an analogy here with the P-Cyg profiles in stars.

- The heavy element systems such as carbon, magnesium, oxygen, etc. are characterized by sharp features. In a few cases where the z_{abs} and z_{em} are not too far apart, these features may also be intrinsic to the quasar. The more common cases, where z_{abs} is considerably less than z_{em}, are believed to arise from absorption in the galaxies or their haloes lying en route from the quasar to the observer.

- The third rather messy system is usually made up of a large number of lines representing Lyman-α absorption. Since individual lines cannot be easily distinguished in a large assembly, these systems are called Lyman-α *forests*. These are believed to arise from the tiny clouds of neutral hydrogen lying in large numbers between the observer and the quasar.

We will discuss these three types next. Our discussion is based on an excellent review by Petitjean (1995).

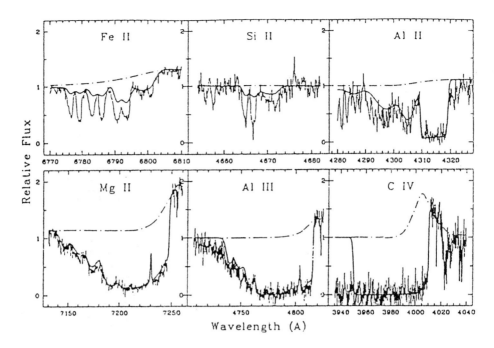

Fig. 13.1. A BAL system in the quasar Q0059-2735. Reproduced from Petitjean (1995).

13.2   Broad absorption line systems

Broad absorption lines are observed in about 12 per cent of all quasars (Weymann *et al.* 1991). As mentioned above the Doppler interpretation ascribes outflow velocities up to $60\,000\,\mathrm{km\,sec^{-1}}$ in the gas within the source. The gas is believed to lie close to the nucleus itself although the source and mechanism of its ejection are not known. The lines C IV, Al III, Al II, Si II, Si III, Fe II and Fe III show structures in the broad troughs. Wampler *et al.* (1995) have very good data showing distinct regions (corresponding to these structures) with differing velocities but similar states of ionization.

Figure 13.1 shows an excellent case where a BAL system has revealed features in the quasar Q0059-2735. There are lines of Fe II, Si II, Al II, Mg II, Al III and C IV. Studies with a 10m-class telescope will produce very valuable data on the absorption systems.

One can model the source by photoionization in the gas produced by the photons from the source. The ionization parameter is defined as the ratio of the photon density to the gas density. A satisfactory fit is obtained for a high value of this parameter and one also finds that the carbon abundance is comparable to the solar value, although the iron abundance may be enhanced by a factor 10.

A significant feature is the existence of narrow Fe II lines having $b \sim 20\,\mathrm{km\,sec^{-1}}$. These could arise from gas at low temperatures not exceeding, say, $10^4\,\mathrm{K}$, and small turbulent motions. The Fe II lines do not go the zero continuum level, which means that the corresponding clouds do not block the continuum source completely and hence are quite small. Since the absorption lines are detected from up to 4.5 eV above the ground state, one needs a high density ($n > 10^6 \, \text{cm}^{-3}$) to attain atomic levels of this order through collisional excitation. This is consistent with the compact size of the typical cloud.

It is therefore believed that there is a link between the BAL gas and the broad emission line region, and detailed studies of BAL objects could provide diagnostic tools for the investigation of the nuclear region of the QSO. Another interesting fact is that BAL quasars are only found in the radio-quiet population of quasars, whereas the associated systems of high metal abundance and high degree of ionization are seen both in radio-quiet and radio-loud quasars. This may have some bearing on the question why only 10 per cent of quasars are radio-loud.

13.3 Heavy element systems

The metal lines occur at multiple absorption redshifts in a single quasar. The absorption redshifts in this case may differ significantly from the emission redshifts, being less than the latter in most cases. In the early days of quasar astronomy the source of these redshifts was considered as being either intrinsic to the quasar or in the absorbers (galaxies or haloes) lying intermediate to the quasar. The balance of opinion has now shifted heavily towards the latter hypothesis, which we shall refer to as IGH (the intervening galaxies hypothesis).

One reason why the intrinsic hypothesis, which is invoked for BAL systems, is not considered suitable for metal systems is the rather large outflow velocities in the source that it demands. Let us take a hypothetical but generic case to illustrate this point. Suppose the emission redshift is 3 and the absorption redshift is 1. Then the Doppler redshift required to be produced by the outflowing gas is given by

$$1 + z_\text{d} = \frac{1 + z_\text{abs}}{1 + z_\text{em}} \tag{13.1}$$

which works out to be -0.5, i.e., it corresponds to an outflow velocity of $0.6c$. Such speeds are considered too high to sustain within a model. This criticism is, however, of doubtful status since, as we have seen elsewhere in this book, relativistic bulk motions have been commonly assumed in the context of beaming models. In any case, in the few cases where $z_\text{abs} > z_\text{em}$, the intrinsic model can only be invoked with the proviso that in such cases we are seeing absorption through infalling rather than outflowing gas in the QSO.

The main reason why the IGH has gained acceptance is because of observational evidence. In the 1980s this evidence was largely statistical in nature and related to the observed distribution of the redshifts of specified spectral lines such as C IV or Mg II in a sample of quasars showing absorption systems. For example, Sargent, Boksenberg and Steidel (1988) and Sargent, Steidel and Boksenberg (1988) studied the spectra of quasars, in controlled samples with a uniform signal-to-noise ratio and specified redshift range, for the C IV and Mg II absorption lines. These authors found that the number of absorbers per QSO shows a marginally significant departure from the Poisson distribution expected of randomly distributed intervening absorbers (as per the IGH).

In these studies, the authors found that the variation of the densities of the two absorption systems with redshift is epoch dependent (i.e., redshift dependent) in *different* ways. Taking the number per unit redshift range to be $N(z)$, one may express the relation as

$$N(z) = (1 + z)^\alpha \tag{13.2}$$

where $\alpha = 1.45$ with an error bar $\Delta\alpha = 0.65$ for the Mg II lines, while $\alpha = -1.2$ and $\Delta\alpha = 0.7$ for C IV . Thus evolution appears to proceed in opposite directions in the two metal line systems, which is hard to understand in terms of standard evolutionary cosmology.

A more direct approach to the problem involves a search for the absorbers themselves, if the IGH is taken as true. Although attempts to do this started in the 1970s (Weymann *et al.* 1978), actual detections became possible with the availability of more sensitive detectors. Reports of such detections at low redshifts for the Mg II system may be found in Bergeron and Boisse (1991), Steidel (1993), and Steidel *et al.* (1995). About sixty galaxy/Mg II pairs were known by 1997. From studies of the stellar contents of the galaxies and the extent of the gaseous haloes, it appears that (1) the cross-section of the halo is typically of order $35h_{100}^{-1}(L/L_*)^{0.2}$ kpc, where L_* is the constant in the Schechter luminosity function; (2) the galaxies are fairly bright with $L > 0.25L_*$; and (3) there is evidence for a mild evolution of luminosity at $z \sim 0.6$ compared with today.

The most promising case so far is that of the system at $z = 0.7913$ towards PKS 2145+06 found by Bergeron *et al.* (1993). In this a large number of lines were detected by the HST (Bahcall *et al.* 1993). The associated galaxy is detected at a projected distance of $25h_{100}^{-1}$ kpc. The metal abundance found is rather high, $\sim 0.5Z_\odot$, and this determination is claimed to be quite reliable.

Data of this kind present a question: if there are absorption features indicating a high metal abundance at such large distances from the centre of the galaxy, where the metal-generating stellar activity is taking place, what is the mechanism for it? Normally we expect metals to fizzle out at large distances into the halo, which is claimed to contain (non-baryonic?) dark matter. In the final chapter we will discuss other statistical evidence that suggests that the IGH may not fully account for the observed incidence of metal absorption systems.

The determination of metallicities in absorption line systems is complicated by the requirement of good data on the column density of H I, and on several ionization stages of the different elements in order to perform the ionization correction. To get clear data one needs to select systems in which the lines are not blended or saturated. When the H I column density is very large, as for damped systems, hydrogen can be considered neutral and the ionization correction is taken to be negligible. Since the O I, C II or Si II lines arise from singly ionized (or neutral) states, these ions can rarely be used, and since the lines are heavily saturated, the derived column densities are very uncertain.

We will not go into further details in this area, which is opening out with the new possibilities offered by large telescopes such as the Keck and the VLT. In the final chapter we will voice some reservations regarding the validity of the IGH, reservations which could be resolved by further data from complete samples.

13.4 Lyman-α systems

With the discovery of QSOs with large redshifts ($z > 2$), the possibility emerged of seeing the Lyman-α line in the visible part of the spectrum. Also, since neutral hydrogen is believed to be the most abundant element in the universe, the possibility of absorption at this wavelength by line-of-sight neutral hydrogen was also envisaged. Typically, one may imagine that the side of the continuum spectrum lying bluewards of the redshifted Ly-α would show this absorption. Gunn and Peterson (1965) estimated this effect in standard Friedmann cosmological models. They found that the optical depth in the above wavelength range at redshift z in a model with deceleration parameter q_0 is given by

$$\tau = \frac{8.3 \times 10^{10} n_{HI}}{(1 + z)(1 + 2q_0 z)^{1/2}}. \tag{13.3}$$

The early observations with 3C 9 showed no depression of the continuum within the observable limit. Setting a sensitivity limit liberally at $\tau < 0.5$ this gave, for $q_0 = 1/2$, a very low limit of $n_{HI} < 3 \times 10^{-11}$ cm^{-3} at $z = 2$. If the intergalactic medium had neutral hydrogen at a cosmological (i.e., closure) density, the number density expected would be a million times higher. Later, Davidsen *et al.* (1977) applied this method to space-based observations of 3C 273 (from space one can measure the UV continuum in low redshift quasars), and found no effect, thereby limiting the present neutral hydrogen number density to $< 6 \times 10^{-12}$ cm^{-3}.

This effect (or lack of it!) is commonly known as the Gunn–Peterson effect and was primarily responsible for drawing attention to the fact that the intergalactic medium may be more complex than the simple homogeneous medium consisting of neutral hydrogen usually assumed by the cosmological model makers.

Instead of a steady dip in the continuum, the presence of a large number of absorption lines of Lyman-α began to be noticed in the early 1970s. These lines are believed to be caused by clouds of hydrogen along the line of sight to the quasar. The Ly-α clouds were considered in early models (Sargent *et al.* 1980) to be spherical, tenuous objects of galactic dimensions that are highly ionized by the intergalactic ionizing flux due to quasars and galaxies. Thus, in a typical cloud the number density of ionized hydrogen would be $\sim 10^5$ times the number density of neutral hydrogen. The early spherical galaxy-sized models have recently been given up in favour of flat sheet-like structures extending to more than 50 kpc, largely because of the observation of large scale structures with the Keck and other telescopes and the inputs from

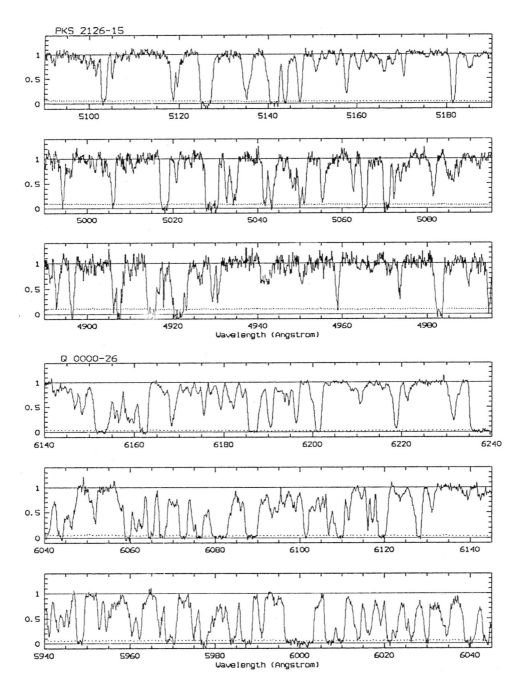

Fig. 13.2.   Examples of the Ly-α forest. Reproduced from Cristiani (1995).

structure formation theorists. It is too early to draw any definitive conclusion on these absorbers at the time of writing.

The Lyman-α forests are nowadays detected typically, by the big telescopes and big Echelle spectrographs, with resolutions of order 20 000 or more. Figure 13.2 shows two series of spectra, taken with the New Technology Telescope and the ESO multimode instrument (for details see the review by Cristiani 1995). The line profiles provide important physical information on the absorbers, such as column density, Doppler width, kinematics, clustering etc. As yet no clear picture has emerged.

We end this section with a few remarks on the confusing situation regarding the abundances of these systems at low redshifts. UV observations with the HST (Morris *et al.* 1991) showed that the number density of low redshift metal-poor Ly-α systems is five times as high as that expected from the extrapolation of the number density known at high redshifts from ground-based telescopes.

Further, Lanzetta *et al.* (1995) have claimed that these lines may be associated with galaxies. These authors have reported nine galaxies at the same redshift as the Ly-α systems near the quasar line of sight corresponding to a transverse distance from it of $150h_{100}^{-1}$ kpc. It is hard to see the galactic haloes extending to such a huge distance. Are there other classes of object much fainter than can be seen that are causing the absorption?

Future programmes with large telescopes are therefore aimed at deeper imaging of the field around a quasar and performing multi-object spectroscopy up to magnitudes comparable to those where the blue galaxy population is found, typically, around $B \sim 24$–26.

These are exciting areas, bordering on structure formation and cosmology, the absorption spectra of quasars being used as diagnostic tools of the inter-galactic medium at redshifts up to $\sim$5.

14 Gravitational lensing

14.1 The classic tests

In 1915 Einstein proposed the general theory of relativity. In the first few years afterwards, very few physicists could understand its most unusual premises, involving as they did the curvature of space–time and non-Euclidean geometry for spatial and temporal measurements. What those physicists needed was the demonstration of the physical reality of some effect not predicted by Newtonian gravity which the new theory could explain.

One such effect was the bending of light rays by gravity. In his work on light, described in the book *Opticks*, Isaac Newton had speculated whether light is affected by gravity in some way similar to the effect of gravity on matter. Unsurprisingly, he did not come to any definite conclusion. In the theory of relativity, however, there is the definite prediction that light *is* affected by gravity and a definite formula could be worked out to predict how the path of a light ray would be changed as it grazed the surface of a spherical massive object. An object of mass M and radius R would bend the path of the light ray by an angle

$$\theta = \frac{4GM}{c^2 R}. \tag{14.1}$$

For the Sun this worked out as approximately 1.75 arcsecond.

In practical terms this meant that if the Solar disk passed in front of a star, then as the star went behind or emerged past the solar surface, it would suffer an apparent change of direction by this angle. The angle, though too small for visual inspection, could be measured by photographic techniques provided of course the Sun itself was completely covered, i.e., at a total Solar eclipse. Figure 14.1 illustrates this idea.

Eddington was one of the few astronomers who understood relativity and its deep significance for physics in general. He therefore pressed for a test of this prediction during the total Solar eclipse of 1919. Two teams of astronomers carried out the measurements at Sobral in Brazil and the island of Principe in the gulf of Guinea. The observations, despite their large error-bars (which ironically were fully estimated much later) gave a positive result. It was this test that really established the general theory of relativity as a credible theory in the popular mind.

Although the test was repeated during many eclipses, the intrinsic inaccuracy of the measurements in the optical waveband, owing to refractive effects of the solar

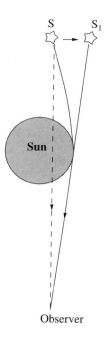

Fig. 14.1. The bending of light rays from a star S by the Sun shifts the apparent position of the star to S_1.

atmosphere and the limitations of resolution, were large. During the mid-1970s, the use of microwave technology improved the accuracy considerably. Using quasars instead of stars for this purpose, and capitalizing on the facts that the Sun is a weak emitter in microwaves and that the refractive effects at these wavelengths can be accurately estimated, these measurements were able to reduce the errors to 5 per cent (Weiler *et al.* 1975; Fomalont and Sramek 1975). The quasar 3C 279 was used as the source covered by the Sun and its deflection was measured, with 3C 273 serving as the reference position.

The bending of light rays due to the gravity of a massive object gives rise to a variety of phenomena now known as *gravitational lensing*. The lensing is caused by gravity rather than by the refraction of light passing through an inhomogeneous medium. We will discuss this topic next, since it has played a major role in quasar astronomy since 1979.

14.2 The first gravitational lens

A brief historical description of the path that led in 1979 to the discovery of the first gravitational lens involving quasars may be in order.

The first paper on the subject, entitled 'Nebulae as gravitational lenses', was published

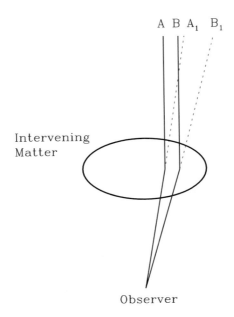

Fig. 14.2. A and B are two point sources, and A_1 and B_1 are their apparent positions produced due to the bending of light by intervening matter. The separation A_1B_1 is magnified compared to AB. This can lead to apparent superluminal motion of the sources relative to each other.

by Fritz Zwicky (1937). He clearly stressed the role of galaxies as light-deflecting objects that could produce *multiple* images of background sources. He pointed out the possibility of ring-shaped images, of flux amplification and of the use of this phenomenon for understanding the large scale structure of the universe (Zwicky 1937). Zwicky was ahead of his time in assuming that the nebulae (i.e., galaxies) would be several hundred billion times as massive as the Sun and, in another paper, he also estimated the probability for lensing to occur in extragalactic astronomy.

In the 1960s and the 1970s, S. Refsdal, J.M. Barnothy, R.K. Sachs, R. Kantowski, C.C. Dyer, R.C. Roeder, N. Sanitt and several others published papers highlighting different aspects of gravitational lensing, ranging from purely theoretical investigations in general relativity to observational predictions in astronomy. We refer the reader to the book by Schneider, Ehlers and Falco (SEF92) for further details.

In a different context, Chitre and Narlikar (1979) invoked the gravitational bending of radio waves from the VLBI components of a quasar by an intervening galaxy to explain the apparent superluminal separation of these components. If the galaxy is suitably located (i.e., close to the critical point of the lensing system) the apparent magnification of the separation between two components due to the lensing can be enormous and can convert a real subluminal speed to a superluminal one (see Figure 14.2). In Section 3.8 we showed, however, that the generally accepted interpretation of superluminal motion involves relativistic beaming.

The real stimulus to the work on gravitational lensing came from the discovery of the first lens involving the quasars 0957+561 A and B by Walsh, Carswell and Weymann (1979). The quasars A and B showed very similar spectra at a redshift of ~ 1.4. Their angular separation was ~ 6 arcsec. Although the existence of two quasars of very similar features at close separation cannot be ruled out, the circumstantial evidence pointed to a gravitational lens doubly imaging one source. The discovery of a lensing galaxy at a redshift of ~ 0.36 later lent further credibility to this scenario. The quasars and lensing galaxy are shown in Figure 14.3, while a ray diagram of the bending of light by the lens is shown in Figure 14.4.

The basic features of a gravitational lens system are described in the following section. By now there are several known lens systems and probable candidates, as listed in Table 14.1. The original expectations of Zwicky (1937) have been fully borne out.

14.3   The basic features of a gravitational lens

Figure 14.5 gives a schematic diagram of a lens system wherein S is the source, a spherical mass M provides the deflector lens d and O is the observer. (We deplore the notation of denoting the lens by d as it is a symbol normally reserved for distance, but adopt it here as it has become common in the gravitational lens literature.) The distance between the source and the observer is denoted by D_s, that between the source and the lens by D_{ds} and between the lens and the observer by D_d.

The condition that the ray from the source passing *outside* the deflector at a distance ξ reaches the observer as shown in the figure is given by the rules of projection:

$$\beta D_s = \frac{D_s}{D_d}\xi - \frac{2r_S}{\xi}D_{ds}. \tag{14.2}$$

Here $r_S = 2GM/c^2$ is the gravitational (Schwarzschild) radius of the deflector mass. We have tacitly assumed that the gravitational bending of light is small and so the angles β and α (the bending angle of the original ray) are both small compared to unity. Also, when applying this relation over cosmological distances we have to take due note of the non-Euclidean measures of redshift-related distance described in Chapter 2. Thus in general $D_{ds} \neq D_s - D_d$. The deflected ray in Figure 14.5 makes an angle θ with the line OM. Hence, with our small angle approximation, $\theta D_d = \xi$. Therefore the above equation becomes

$$\beta = \theta - \frac{2r_S D_{ds}}{D_d D_s}\theta^{-1}. \tag{14.3}$$

It is convenient to define an angle α_0 and a length ξ_0 by

$$\alpha_0 = \sqrt{\frac{2r_S D_{ds}}{D_d D_s}}, \quad \xi_0 = \alpha_0 D_d. \tag{14.4}$$

With these definitions the basic equation (14.3) reduces to the quadratic

$$\theta^2 - \beta\theta - \alpha_0^2 = 0. \tag{14.5}$$

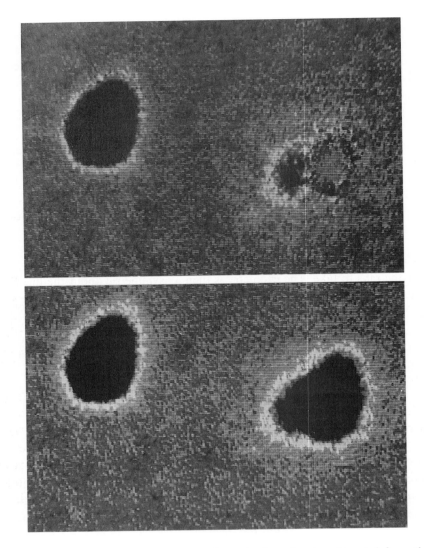

Fig. 14.3. Top panel: image with one quasar from the lensed pair subtracted from the other, revealing the presence of a lensing galaxy. Bottom panel: images of the quasar pair 0957+561 A and B. Reproduced from *The Cambridge Atlas of Astronomy*, second edition (1988).

This tells us that there are two roots, i.e., there are two possible locations of images whose angular separation is given by

$$\Delta\theta = \sqrt{4\alpha_0^2 + \beta^2}. \tag{14.6}$$

Note that the two values of the roots θ_1 and θ_2 are of opposite sign, implying that the two images are located on the opposite sides of the source. Note also that if the source, lens and the observer are collinear, then the angle $\beta = 0$ and there is no preferred

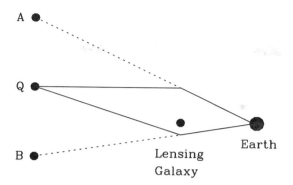

Fig. 14.4. The geometry of light bending illustrating the formation of twin images A and B from one source Q.

plane for the rays to take. The geometry is then axisymmetric about the line SO. Thus we get a ring-like distribution of images with an angular separation from the source of $\theta = \alpha_0$. What we have described is, of course, a highly symmetrical situation involving a symmetric matter distribution, a point source and a special alignment. In practice these conditions are not fully satisfied but we may still get approximately ring-shaped images of extended sources, which are called *Einstein rings* in literature. We will discuss these briefly later.

A general lens is more difficult to quantify. However, we may make a few statements that can be proved using detailed mathematics (for which the reader should consult books or monographs specializing in gravitational lensing e.g. SEF92). First, we may define two planes transverse to the line of sight. The source plane contains the source S while the image plane contains the image. Actually the images are virtual and hence there is no real plane in which they lie. But we shall project the image onto the the lens plane and use the angular coordinates β and θ to denote the source and image positions on the sky. Thus to generalize Equation 14.3 we have the relation

$$\beta = \theta - \frac{D_{ds}}{D_s}\alpha(\xi).\tag{14.7}$$

Here ξ is the vector in the transverse plane of the deflector giving the position of the light ray relative to it. α is the vectorial angle showing the change in the direction of the ray from the source after it is deflected. Its functional dependence on ξ is complicated for a general lens and gives rise to multiple solutions for θ from the above equation.

A general theorem proves that *any transparent matter distribution with a finite total mass and weak gravitational field produces an odd number of images.* In our simple case of the previous section, we got two images. In that case, if the gravitating object were

Table 14.1. *A partial list of interesting cases of gravitational lenses*

System	No. of images	Lens redshift	Image redshift	Maximum separation arcsec[a,b]
Quasar images				
0957+561	2	0.36	1.41	6.1
0142-100	2	0.49	2.72	2.2
0023+171	3	?	0.946	5.9
2016+112	3	1.01	3.27	3.8
0414+053	4	0.468[c]	2.63	3.0
1115+080	4	0.29	1.722	2.3
1413+117	4	1.4[c]	2.55	1.1
2237+0305	4	0.0394	1.695	1.8
Arcs				
Abell 370		0.374	0.725	
Abell 545		0.154	?	
Abell 963		0.206	0.77	
Abell 2390		0.231	0.913	
Abell 2218		0.171	0.702	
Cl0024+16		0.391	0.9	
Cl0302+17		0.42	0.9	
Cl0500-24		0.316	0.913	
Cl2244-02		0.331	2.237	
Rings				
MG1131+0456		?	?	2.2
0218+357		?	?	0.3
MG1549+3047		0.11	?	1.8
MG1654+1346		0.25	1.75	2.1
1830-211		?	?	1.0

[a]For arcs the maximum separation is the diameter of the corresponding Einstein ring.
[b]For rings this corresponds to the diameter of the ring.
[c]Assumed or still to be confirmed.

a point mass (black hole), the possibility exists that the incident rays would have small impact parameters, thus violating the weak field condition. If the lens were a transparent sphere of matter, the theorem would still apply.

The general problem consists of setting up Equation 14.7 and then solving it numerically to find the number and positions of images. Even though the equation yields an odd number of solutions, the result can still be reconciled with the observed even number of images by arguing that some images would be too faint to be seen. We will discuss this possibility next.

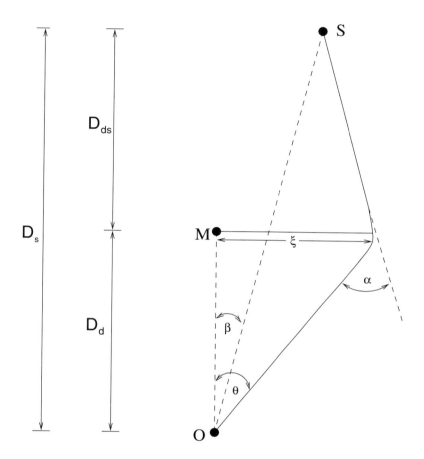

Fig. 14.5. The lensing geometry for a point source. See text for details. The lens plane is the plane through M perpendicular to OM. Likewise the source plane passes through S perpendicular to OS. After SEF92.

14.4 The magnification and amplification of images

Consider a simple example in Euclidean geometry of a spherical source of radius a and luminosity L located at a distance D. The flux of radiation received from the source is given by

$$f = \frac{L}{4\pi D^2},$$ (14.8)

while the solid angle subtended by the source at the observer is

$$\Omega = \frac{\pi a^2}{D^2}.$$ (14.9)

The surface brightness of the source is given by

$$\sigma = \frac{L}{4\pi a^2}. \qquad (14.10)$$

It is easy to verify from these results that

$$f = \Omega\sigma/\pi. \qquad (14.11)$$

In other words, the surface brightness is proportional to the ratio of the flux to the solid angle subtended. This result survives with appropriate modification (see Chapter 2) for cosmological redshift in an expanding universe. Since gravitational bending does not introduce any additional spectral shift, we may assume this result to be valid for lensed sources. Since the surface brightness of a small source is not changed by lensing, the ratio of the flux of an image to that of the source (in the absence of lensing) would simply equal the ratio of their solid angles:

$$\mu = \Omega/\Omega_0, \qquad (14.12)$$

the zero suffix standing for the unlensed situation. This is valid for the cases where the images and sources are not extended so that one may use a constant surface brightness. For extended sources one integrates over the source with suitable weighting of the surface brightness at the point.

Returning to the small source, if β, the source position, is related to θ by Equation 14.7, the angular magnification is given by the Jacobian

$$\frac{\Omega_0}{\Omega} = J[\beta;\theta] \equiv \det\left\|\frac{\partial\beta}{\partial\theta}\right\|. \qquad (14.13)$$

Thus the amplification of the flux is given by the reciprocal of the above Jacobian.

The Jacobian has great significance in the lensing calculations. The parity of the image is decided by the sign of the Jacobian for that image. If it is positive, the sense of its curves (i.e., clockwise or anticlockwise) is preserved with respect to the source. For negative parity it is reversed. Thus regions of opposite parity are separated in the lens plane. The *critical* curves separating them are those on which the factor μ diverges. This is, however, an idealization since the sources are in general extended and infinite amplification of image brightness does not take place. These critical curves in the source plane are called *caustics*. It can be shown that *the number of images changes by two if and only if the source position is changed in such a way that it crosses a caustic.* If the locations of caustics are known then, for a given source position, the number of images can be determined using this property. Another result which is useful in this respect is that for any transparent distribution of matter with finite mass the number of images of a point source sufficiently misaligned with the lens is *unity*.

Even though we do not in practice have enormously bright images (μ tending to infinity) another theorem guarantees that *for a transparent lens, there is one image with positive parity having an amplification factor not less than unity (i.e., the image is at least as bright as the source).* Thus lensing may introduce selection bias in surveys that are flux limited.

14.5 Applications to quasars and galaxies

Typical examples of point sources for gravitational lensing are quasars while for extended sources we have galaxies. As mentioned earlier there are several examples in both categories: we will discuss a few to illustrate the problems and fall-outs of such cases for the astronomy of quasars and AGN.

14.5.1 *Quasars*

Basically we may look for the following features in suspected candidates for lensing.

- There should be two or more images of quasars close together, the typical separation being no more than a few arcseconds.
- The redshifts of the members should match very accurately, say within the allowed random motions of the order of 300 km sec^{-1}.
- The flux ratios of the candidates should be invariant with respect to the waveband since gravitational lensing is wavelength invariant.
- There should be similarity in the morphology of components and in the intensity ratios of lines and continuum emission.
- A possible galaxy or cluster of galaxies should be found not too far off the line of sight to serve as the deflector lens.
- The time variations of fluxes in different members of the candidate group should be correlated, with appropriate time delays.

We describe some important configurations briefly below, dividing the cases into doubles, triples and quadruples. Reference may be made to Table 14.1 for the measured quantities of these cases.

Doubles These include the first pair, 0957+561, found in 1979 and so far the most impressive one. Walsh *et al.* (1979) found that its two components had very similar spectral features: emission lines C IV (1549 Å) and CIII] (1909 Å), both redshifted by z = 1.405, and both with an absorption feature at $z = 1.390$, most likely intrinsic to the sources. Component B was fainter than A by 0.35 mag up to 5300 Å, with a similar spectral continuum. It was redder than A at longer wavelengths ($\lambda > 5300$ Å).

The spectral similarities coupled with a small separation of less than 6 arcsecond made them suspect that here was a plausible lens case. The lensing galaxy was also found in close vicinity (see Figure 14.3) with a redshift of 0.36. It became clear that it belonged to a cluster and that the cluster also contributed to the lensing of the system. Later radio observations by the VLA revealed that the radio fluxes of B and A bore a ratio of 0.8, which was nearly the same as for the optical components. The VLA, and the later VLBI studies of the radio structure, revealed other components, C, D, E, and these suggested a more complex structure for the lens (see Figure 14.6). Detailed modelling has been carried out for this system using both the galaxy and the cluster. We refer the interested reader to the book by Schneider *et al.* (SEF92) for references to

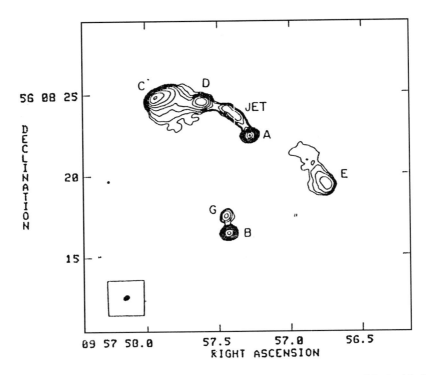

Fig. 14.6. VLA 5 GHz radio map of 0957+561. Reproduced from SEF92, with the kind permission of Springer-Verlag.

detailed work. We mention in passing a curious feature of Figure 14.6, that the centres of the components D, A, E are collinear. While it would be possible to understand this in a normal image of a radio source, here one is looking at a picture *distorted* by a lens. How did the exact alignment occur *after* lensing of the system?

A good test of whether a system is a genuine double (or multiple) system or whether it is gravitationally lensed is through observations of the time delay of a variable source. Although the images may look identical they have been formed by light rays following different paths in space and as such they will take different times to reach the observer. Thus the time variations in emission of one image should match that of the other *after allowance is made for a time delay*. This may range from a few days to a few months. 0957+561 has been monitored for such time-delay correlation analysis, but as yet without any convincing conclusion.

A second good case of pairs is that found by Surdej *et al.* (1987) after a systematic search for multiple images in a sample of very luminous quasars. Labelled UM 673 (0142-100), these images are 2.2 arcsecond apart, have the same spectra with redshift $z = 2.719$ and differ by ~ 2 magnitudes. The lensing galaxy has been found with redshift $z = 0.493$ and with a lens mass of $\sim 2.10^{11} M_\odot$. However, the data are still not sufficient to constrain the model further.

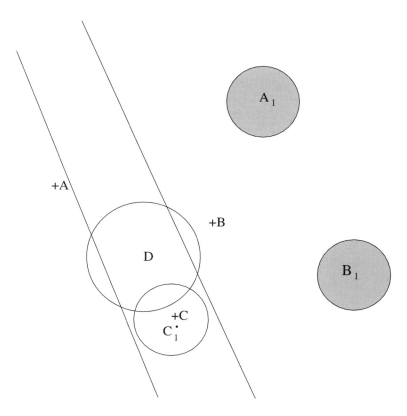

Fig. 14.7. Geometry of the triplet 2016+112. The various components are described in the text. The symbols near A, B, C and C_1 indicate positions of the corresponding optical counterpart. Reproduced from SEF92, with the kind permission of Springer-Verlag.

Triples The best documented three-component case is 2016+112, shown in Figure 14.7. It is a triple radio source with components A and B having identical spectral indices. Component C has an index that is significantly less steep but its flux is nearly three times that of either A or B (which have comparable fluxes). The common redshift of A and B is 3.27 and they are both quasars. What is C? It is probably an elliptical galaxy with an estimated redshift 0.8 and therefore not the third image. The system, however, does have a third lensed component, labelled C_1 very close to C, which was discovered later in imaging in a narrow wavelength band around the Ly-α line. The lensing galaxy D was found at redshift 1.01. A_1 and B_1 are two more components discovered in the narrow band imaging. These components are resolved and have no radio emission, and therefore are not additional lensed components of the quasar; the nature of these objects is not established.

It is difficult to make the lens models match all the features of the observed components, ensuring the right relative brightness of the three images and that there are no more images visible with the present observations. (If a lensing model predicts more images than found, it must ensure that the extra images are too faint to be seen.)

There is another triple, 0023+171, which is considered a probable three-image case but it is not yet definite enough to meet the criteria of lensing.

Quadruples   These include: PG1115+080, which was initially classified as a triple until one of its components was shown to be a double, thus making the total four; the case 2237+0305, now called the *Einstein cross*; and another, H1413+117, called the *clover leaf*. We will briefly describe the Einstein cross as it has been the centre of a controversy.

The spiral galaxy 2237+0305 with a redshift of 0.039 seemed to have a broad emission line at a redshift $z = 1.695$. This turned out to belong to a quasar that was within 0.3 arcsecond of the nucleus of a galaxy. Huchra *et al.* (1985), who found the quasar, announced it as a case of gravitational lensing wherein a faint quasar had been amplified in brightness. This was criticized by G. Burbidge (1985) on the grounds that the lensing interpretation was contrived to boost up the probability of finding a high redshift quasar near a low redshift galaxy by chance; the argument runs like this. The surface density of quasars increases rapidly with faintness. Thus a quasar that has been boosted in brightness by gravitational lensing really belongs to a fainter and hence denser population. The chance of finding one such close enough to a selected point in the sky (such as the nucleus of a galaxy) may not therefore be negligible. At the time, indeed, the system did not have any of the features needed for identifying it as a lens and the alternative would have been to admit a physical association of two objects with differing redshifts. We will discuss the implication of this latter assumption in Chapter 15.

However, subsequent studies of the system at high resolution resolved the quasar into four images, the largest angular separation between them being 1.8 arcsec. The shape of the distribution is a cross (see Figure 14.8); and hence the name 'Einstein cross'. The lensing galaxy has a very low redshift, $z = 0.039$. Arp and Crane (1992) have, however, criticized the lens interpretation on the grounds that it does not reproduce all the observed morphology of the four images. No doubt with the improved resolution of the HST this will provide further interesting food for thought.

14.5.2   *Arcs and rings*

Another more recent interest in gravitational lensing has come from the discovery of extended arc-like and ring-like structures amidst clusters of galaxies. We will briefly mention just a few examples (see Table 14.1 for a partial list); these do not really fall within the general scope of this book but are needed to complete this account of gravitational lensing.

Abell 370   In the 1970s the early work on photometry with CCD cameras began to reveal arc-like structures in some clusters including Abell 370. However, it was in

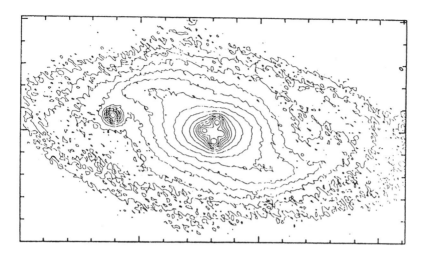

Fig. 14.8. The isophotes of 2237+0305 suggesting the name *Einstein cross*. Reproduced from SEF92, with the kind permission of Springer-Verlag.

1985 that clusters were targeted for observing such features and, within a year or two, findings of arcs in clusters were announced by two groups independently (Lynds and Petrosian 1989 and Soucail *et al.* 1987). The first such arc to be extensively studied was that in the Abell 370 cluster. Its angular length is 21 arcsecond and mean width two arcsecond. Its mean angular radius of curvature is 15 arcsecond. The arc has a rather complicated morphology, as shown in Figure 14.9, with knots, non-uniform brightness and a non-constant radius of curvature.

These features put constraints on lensing models quite different from those for point sources like quasars. Soucail *et al.* (1987) performed the spectroscopy of the arc in Abell 370 and found its redshift to be the same as that of the cluster, 0.373. At this stage it was natural to assume that the arc was part of the cluster and the early explanations were based on that hypothesis. However, in 1988 the same group (Soucail *et al.* 1988) did more careful spectroscopy and found the redshift of the parts of the arc not superposed by galaxies to be 0.724. Thus there was now a case for modelling the system as a gravitational lens. Now, assuming the arc to be non-local, the lensing model was aimed at imaging an extended source at large redshift (0.724) by the cluster. The blue colour of the arc was consistent with the blue-galaxy excess amongst high redshift clusters.

There are several other clusters including Abell clusters in which giant arcs of arc-second size are now being found and gravitational lensing is the favoured explanation for their apparent existence.

Ring MG1131+0456   Rings are more recent than arcs and the first one was found through radio astronomy. Hewitt *et al.* (1988) found what they called a possible *Einstein ring*. Their map at 5 GHz with the VLA showed the radiation coming

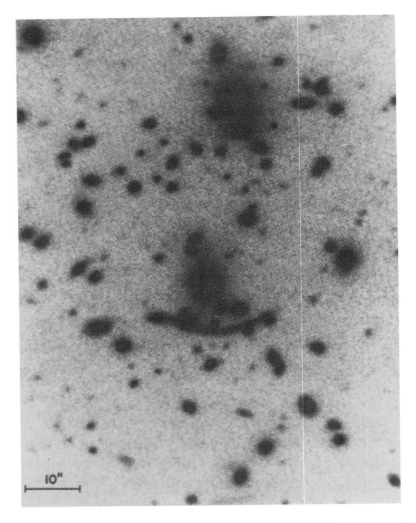

Fig. 14.9. The arc in the cluster Abell 370. Reproduced from SEF92, with the kind permission of Springer-Verlag.

from an elliptical ring with major and minor axes of angular size 2.2 arcsecond and 1.6 arcsecond respectively (see Figure 14.10). The ring has no flux at its centre, although it includes four compact sources labelled A1, A2, B and C. The optical source has an elliptical shape too, with major and minor axes of sizes 2.5 and 2.1 arcsecond respectively. However, the position angles of the radio and optical axes do not match. There are no emission lines to tell us about the redshift of the source. But the unusual morphology of the source suggests that it may be a lensed extended source. There being no redshift data, the models are not very precisely constrained.

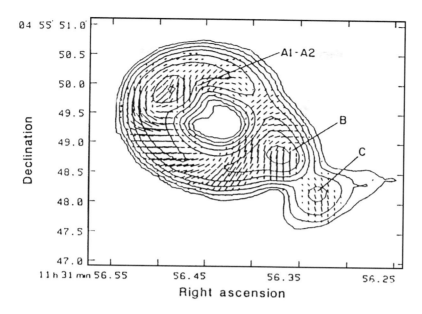

Fig. 14.10. The isophotes of MG 1131+0456, which could be an example of an Einstein ring. Reproduced from SEF92, with the kind permission of Springer-Verlag.

More rings have been found in recent times. Indeed, the gravitational lens modelling industry is flourishing with the discovery of different cases of multiple compact images as well as arcs and rings. We end this discussion with a few general remarks.

14.6 Concluding remarks

There are several difficulties that explain why the discovery of the first case of gravitational lensing came as late as 1979 and why such cases have not multiplied fast. First, the quasars must be reasonably far away (i.e., with high redshifts) so that the chance that their line of sight will be intercepted by a galaxy or a cluster is not too low. This makes them rather faint also and hence it is hard to do detailed spectroscopy and photometry.

The angular separation for a single galaxy as a deflector with a mass of $10^{12} M_\odot$ is low, being of the order of an arcsecond. This makes it necessary to have high resolution optical astronomy for probing such cases. Most cases show separations of the order of 4–6 arcsec, thereby suggesting that the lensing is caused by more massive and compact systems such as rich clusters. However, lensing systems have been identified in very few of the lensing examples (See Table 14.1).

A good example of a lensing galaxy would occur if the redshift of such a galaxy matched that of some absorption line systems of the quasar. Such a case would kill two birds with one stone, as it would not only demonstrate the credibility of the galaxy or cluster as a lensing system but would also establish a case for use of the intervening-galaxies hypothesis to explain the absorption spectra of quasars. As yet no such example exists.

The lack of a luminous lens is sometimes interpreted as an indirect proof of the existence of dark matter. For, one could argue that if lensing explains the observed features but no lens is visible then it must be invisible. That is, the light rays are bent by intervening masses of dark matter. Thus lensing is sometimes proposed as a probe for dark matter.

A clear signature for lensing would be the observation of a time delay. That is, if the quasar is undergoing flux variation, it would be noticed that its images were not synchronous but had a relative time delay. As mentioned earlier, searches in this respect have been inconclusive so far, the pair studied most systematically being the first one, 0957+561 A and B. Two problems arise.

First, the theoretical modelling has to be very precise and unique for the experiment to mean anything. An order-of-magnitude estimate of the time delay for an image separation of $\delta\theta$ measured in arcsecond is $\sim 15(\delta\theta)^2$ days. Secondly, these observations may be confused by so-called *microlensing*, in which a star or a Jupiter- or brown-dwarf-like object may cross the line of sight and cause a rise in luminosity that lasts for a few days to months.

Any calculation of time delay is based on an assumption on the extragalactic distance scale, i.e., on the value of the Hubble constant. One could turn the calculation around and argue that if one is sure of the lens model and of the measured time delay then one can measure Hubble's constant. Going by the present status of the subject this hope is not likely to be realized soon.

Gravitational lensing has also been put forward as a cause for the observed associations of quasars with bright galaxies. The general argument (see Chapter 15) is based on statistical estimates of finding such quasar–galaxy associations by chance. If one can show that owing to lensing the background quasars are amplified in intensity (the possibility discussed earlier that $\mu \gg 1$), then the probability of finding relatively bright quasars near a galaxy might go up enough to make such phenomena of no great significance. However, detailed calculations show that there is an insufficient increase in the background number density of quasars around galaxies to resolve the observed anomaly (see, for example, Hogan *et al.* 1989).

15 Problems and controversies

15.1 Introduction

In the earlier chapters of this book we have covered the present-day understanding of quasars and AGN, based on certain paradigms. We may refocus on these to begin with.

- Quasars and AGN are extragalactic phenomena that form natural steps in the overall scenario of galactic evolution.
- The redshifts of these objects are of cosmological origin.
- The primary source of production of the energy emitted by these objects is the gravity of a highly collapsed supermassive object, which is idealized as a spinning black hole with an accretion disk.
- The ejection of matter from the central region (in the form of jets) is relativistic.
- Relativistic beaming and orientation effects play a key role in explaining certain crucial observed features of these objects.

Astronomy has developed through paradigms, some of which were correct right from the beginning whereas others have had to be corrected, modified or abandoned. From the early days of the geocentric theory to the colliding galaxies hypothesis of radio sources, astronomy also has a history in which a majority have enthusiastically subscribed to a mistaken paradigm, which has thus become dogma. It is against this background that we now look at the above assumptions with the eyes of a sceptic.

15.2 Quasars, AGN and galaxies

Unlike stellar evolution, astronomers are still far from piecing together a scenario of galactic evolution in which all the different types of galaxy would find a natural place. Do spirals and ellipticals come in a sequence along a unique evolutionary track? Or, are they stages in two different and alternative tracks? Where do the irregulars fit, in all this? Are quasars (which are comparatively short duration phenomena) expected to occur in the beginning, or towards the end of galactic evolution – or do they occur at random at any time?

We have described the considerable work in relating the activity of quasars and AGN to star formation regions. It is still too early to say at present how these ideas will stand up to greater multiwavelength scrutiny. Even in the case of galactic evolution the relationship to stellar populations is still not understood.

15.3   The black hole paradigm

The black hole and accretion disk picture has been extensively studied by theoreticians. It is the logical outcome of the early speculations about quasar energy sources by Hoyle and Fowler (1963). There are, however, certain problems with this paradigm that have not been adequately addressed. We mention them below.

No evidence for inward flow?   The presence of a black hole, or of dark matter in general, is inferred from the study of the dynamics of the environment. We expect that a black hole would pull in matter which will fall *towards* it.

In reality, we have as yet no spectroscopic evidence of such an *inward* flow. The increase in velocity dispersion as we approach the central region shows increasing dynamical activity at the centre. *But it could be either way, inward or outward.*

The evidence, *prima facie*, seems to suggest an explosion: matter and radiation are both being ejected from the central region, with the jets, where seen, believed to be directed outward. The arguments suggesting how inward motion (not seen) can be converted to outward motion (seen) are at best qualitative and lack conviction.

The same criticism is applicable to the accretion disk. With so many examples in high energy astrophysics where the accretion disk is used, do we have a single case where it is actually observed? Let us briefly review the much publicized example of the disk around M87, observed by the Hubble Space Telescope in May 1994. Radial velocity measurements show that the disk has a radius of 20 pc and that it is rotating.

The mass of the central black hole is estimated at about $3 \times 10^9 M_\odot$. A simple calculation shows that the Schwarzschild radius of such a black hole would be about 10^{10} km. Let us assume that the accretion disk is some hundred times larger. Its radius will then be $\sim 10^{12}$ km, i.e., about 0.03 pc. Thus the claimed observed accretion disk is 600 times larger than the predicted radius!

The existence of the disk in M87 cannot be denied; but disks have been postulated and observed in astrophysical scenarios that have nothing to do with a black hole, e.g., the proto-planetary disks around stars. We therefore advocate caution in interpreting the observation of the M87 disk as confirming the black hole accretion disk paradigm.

The energy problem   The supermassive black hole scenario demands a very high level of efficiency if it is to deliver the estimated high luminosity of a quasar or an AGN. This can be seen in the following way.

The maximum luminosity that can be sustained by an accreting source is the Eddington luminosity given in Equation 5.26. The typical variability time scale τ for the emitting region is the light travel time across the region. If the size of the emitting region is taken to be a hundred times the Schwarzschild radius, we can combine the expression for Eddington luminosity with Equation 5.35 to give

$$L_{\rm Edd} = 10^{41} \left(\frac{\tau}{1\,{\rm sec}} \right) {\rm erg\,sec^{-1}}.$$ (15.1)

For time variations observed on the scale of hours we should set τ at the order of 10^3 and thus arrive at a maximum luminosity of order $10^{44}\,{\rm erg\,sec^{-1}}$. The black hole in this case has to have a mass $\sim 10^6 M_\odot$. The most luminous quasars, however, are at least a hundred times as powerful. Thus, if the above scenario is correct, then we should not expect the time variability of the order of hours that is *actually observed* in some quasars.

Theorems in general relativity tell us that only about 30 per cent of the rest-mass energy of a spinning black hole is extractible. Thus from a black hole of the above mass we can draw at most a total of $\sim 6\times 10^{59}\,{\rm erg}$. With this reservoir the source would be able to shine with the given luminosity for 2×10^8 yr. Again we should remind ourselves that these time scales have been obtained with the maximum efficiency allowed by the laws of physics. With a more realistic efficiency, a hundred times less, the lifetime is reduced to about 10^6 yr.

The inverse-Compton catastrophe Hoyle, Burbidge and Sargent (1966) pointed out another difficulty arising from the compact high density storage of radiation in a quasar. We have already considered this in Section 4.4, but will reconsider it here in a slightly different form. Suppose that the radius R of a source is limited by the time scale τ of variability:

$$R < c\tau.$$ (15.2)

Suppose, further, that the volume of this size is filled with magnetic flux, optical and infrared radiation and relativistic electrons, which are the sources for synchrotron radiation. The photons so generated can scatter off the electrons and lose their energy by the Compton effect. For the source to be sustained the synchrotron losses (which pump energy into the photons) must dominate the Compton losses. For an isotropic radiation field to be so maintained the following condition must be satisfied:

$$8\pi U_{\rm rad} < B^2,$$ (15.3)

where $U_{\rm rad}$ is the energy density of radiation and B is the magnetic field. If L is the luminosity of the source,

$$U_{\rm rad} = \frac{L}{\pi R^2 c}.$$ (15.4)

Thus the condition required to hold within the source is

$$B^2 > \frac{8L}{R^2 c}.$$ (15.5)

Taking $L = 10^{47}\,{\rm erg\,sec^{-1}}$ and $R < 10^{17}\,{\rm cm}$, based on a value for τ of a month or so for 3C 273, we find

$$U_{\text{rad}} > 100 \, \text{erg cm}^{-3}, \quad B > 15 \, \text{gauss}. \tag{15.6}$$

The corresponding photon density is as high as $10^{16} \, \text{cm}^{-3}$. The lifetime of an electron, emitting synchrotron radiation of frequency v, is

$$\tau_e = 10^{12} v^{-1/2} B^{-3/2} \, \text{sec}, \tag{15.7}$$

during which it will travel a distance $\sim c\tau_e$. At a characteristic frequency $v = 10^{13} \, \text{Hz}$, and using the lower bound on the magnetic field calculated above, we arrive at an electron travel range of

$$r_e < 1.5 \times 10^{14} \, \text{cm} \ll R. \tag{15.8}$$

Thus we are forced to conclude that to sustain a coherent source of linear dimension R, the injection of high speed electrons must be made at a number

$$\left(\frac{R}{r_e} \right)^3 \sim 10^9 \tag{15.9}$$

of local centres, an enormously large value! This feedback effect of Compton scattering, known as the *Compton catastrophe*, imposes a severe constraint on the energy producing mechanism.

Relativistic beaming   It will be seen that the energy problem as described above becomes greatly constrained by the short time scales of the luminosity variation of the object. We also saw that the short time scales of VLBI components land us with the task of explaining apparent superluminal motions.

Relativistic beaming offers an ingenious solution out of these problem, as we saw in Chapter 3. However, there are problems there too.

Take for example, the observation that a typical quasar shows only one jet. A counterjet expected from the principle of the conservation of linear momentum is never seen. One possible explanation is that the jet seen is beamed at us and, being Doppler blueshifted, it is enhanced in luminosity whereas the counterjet is weakened by redshifting. The same argument is used for the VLBI jets. Now if the large jets and VLBI jets are both beamed at us, they should be aligned with each other. Yet, there are cases where this is not so.

Similarly, the larger the superluminal factor, the smaller is the critical beaming angle θ. So, such cases should be seen on a random-observer basis with a very small probability. For the abnormally bright quasar 3C 273 with a superluminal speed of $10c$, the probability comes out as low as 10^{-5}. One is therefore appealing to a relatively rare combination of events in order to understand the phenomenon.

15.4 Are quasars at their cosmological distances?

Cosmological-hypothesis constraints The above considerations tell us how constrained the standard quasar model becomes with an interpretation of quasar distance according to the cosmological hypothesis. Because a typical quasar has a large redshift, it becomes a distant source and, for its observed flux of radiation, its luminosity becomes very high, exceeding that of a typical galaxy by two or three orders of magnitude. Consequently, its energy budget becomes high also, and one is forced to postulate a very efficient energy machine and a large energy reservoir at its core. Matters are not helped by the short time scales of quasar variability, which make the reservoir very compact. The black hole accretion disk paradigm is considered the best buy theory for a quasar powerhouse, despite the various problems outlined above.

These difficulties, however, are alleviated if we give up the assumption that quasars are very far away. For example, if a typical quasar is a hundred times closer, say at a distance of 30 Mpc instead of 3000 Mpc, its luminosity is reduced by at least a factor 10^{-4}, and the energy budget is considerably eased. The VLBI components are then no longer in superluminal motion. The short time scales do not present problems of compactness. The Compton catastrophe also does not exist, as can be easily verified by the application of the above calculation in the *local* context.

The redshift–magnitude diagram Theoretical considerations apart, one may ask what is the hard evidence that a quasar is actually at the distance implied by its redshift, as per Hubble's law. For galaxies the applicability of Hubble's law is justified by the relatively scatter-free redshift–magnitude relation. Figure 15.1 shows the relation for first-ranked cluster member galaxies. Contrast it with the redshift–magnitude diagram for all the ~ 7000 quasars in the latest Hewitt–Burbidge catalogue, shown in Figure 15.2. The diagram is a scatter diagram with no Hubble-type relation clearly apparent.

Of course, it may be argued that the scatter arises because of the variation in luminosity from quasar to quasar. In that case, one should try to identify a 'standard candle' class of quasars for which the scatter in luminosity is minimal. The identification of such a class for galaxies (viz. the brightest member of a cluster) led Allan Sandage and his coworkers to a tight Hubble relation such as that in Figure 15.1. Such an exercise has not so far been successful for quasars. So, as far as the redshift–magnitude relation is concerned the validity of Hubble's law is more in the nature of a paradigm than a confirmed fact.

Evidence from absorption lines The absorption lines of quasars could arise either in the quasar itself or in the intervening absorbing material. It is generally assumed that the broad absorption line regions with redshifts very close to the emission redshift of the quasar arise in the immediate local environment of the quasar. A large fraction of metal absorption lines having redshifts often considerably less than the corresponding emission redshifts are believed to arise in the intervening galaxies or haloes of galaxies. In some cases, galaxies with redshifts matching those in the quasar

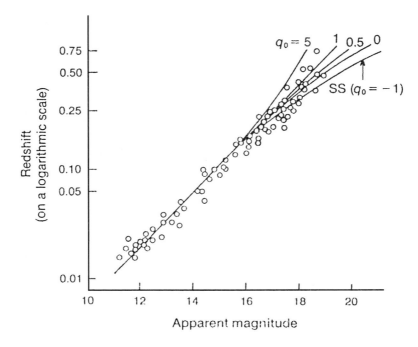

Fig. 15.1. The relatively tight redshift–magnitude plot for galaxies. The fitted curves represent different cosmological models characterized by the deceleration parameter q_0. Reproduced from N93.

absorption spectra are found close to the lines of sight to the quasars. These results support the intervening galaxies hypothesis (IGH) and suggest that the quasars are at least farther away than these intervening galaxies. A similar though less definitive argument is made for the Lyman-α forest in the absorption spectra of large redshift quasars.

Nevertheless, statistical studies aiming to match the occurrence of absorption lines in the quasar spectra as predicted by the IGH with the known distributions of galaxies and large haloes have not been entirely successful. The early work of Burbidge *et al.* (1977) was repeated more recently by Duari and Narlikar (1995), who found that the frequency of multiple absorption systems is far greater than could be accounted for from chance interception by absorbers as large as 80–100 kpc in radius. Despite the allowances made for incomplete samples, the argument remains significant.

Evidence of gravitational lensing   The discovery of multiple and closely spaced quasar systems with near-identical members has led in the past few years to the modelling of these observations in terms of the gravitational lensing of a single quasar by an intervening galaxy or cluster of galaxies. The fact that viable lensing models

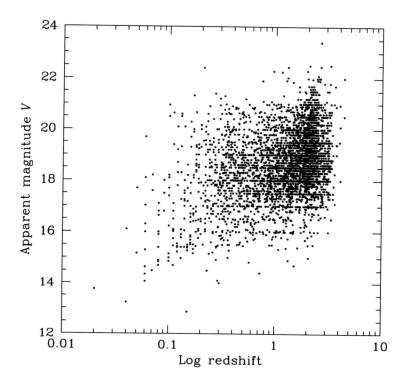

Fig. 15.2. The redshift–magnitude plot for quasars from the Hewitt and Burbidge catalogue. The points are widely scattered, unlike in the case of galaxies in Figure 15.1.

can be made again suggests that the quasars in question must be behind the lensing galaxies, i.e., very far away.

A more convincing argument would be the discovery of relative time delays in the light variations in the different images. These, if in agreement with the theoretical models, would provide a distance scale to the quasar. Also, a model may be further verified if absorption lines at redshifts matching the lensing galaxy redshifts appeared in the quasar spectra. So far these observations have not yielded any definitive results of this kind.

To sum up, while a consistent case can be built upon these issues for the cosmological hypothesis, none of the arguments above tell us in unequivocal terms that the redshifts of quasars conform to the Hubble law. Thus it may still be possible to look for alternative explanations that contribute at least a significant part to the total redshift z_t. Thus we will write

$$(1 + z_t) = (1 + z_c)(1 + z_i), \tag{15.10}$$

where the right hand side has two components to the redshift: the cosmological part z_c

which is as per Hubble's law, and an *intrinsic* part z_i that may arise from some other causes.

In Chapter 1 we discussed two aspects that make the cosmological hypothesis the most acceptable interpretation of the redshifts of quasars. Based on this hypothesis we interpreted the data and worked out the detailed models. From what has been said above, despite various difficulties it may appear that it is still possible to make the cosmological hypothesis survive, i.e., to assume that $z_i = 0$.

However, in the following section we will briefly review evidence that, at face value, is hard to accommodate within this framework.

15.5   Nonconformist evidence

The purpose of this section is not so much to advocate the case for non-cosmological redshifts as to draw the reader's attention to the evidence claimed for them that has steadily accumulated over the last three decades. The evidence outlined here is rarely discussed in standard textbooks, although it is found in conference proceedings and review articles. We will, however, be brief and refer the reader to more detailed sources for further information (see e.g. A87, Narlikar 1989).

Basically, the violation of the cosmological hypothesis (CH hereafter) can be demonstrated by showing the existence of two extragalactic sources in close proximity but with different redshifts. The CH requires a unique relation between redshift z and distance D,

$$z = f(D), \tag{15.11}$$

say, and hence if close neighbours have the same value of D, it is expected that their redshifts will be the same. We now look at a few examples that have a bearing on this expectation.

Quasar–galaxy association   Assuming that galaxies have only cosmological redshifts, one may use them as benchmarks for determining whether the quasars in their proximity also share this property. With this assumption, Stockton (1978) selected a sample of 27 quasars with $z \gtrsim 0.45$ and looked for galaxies within 45 arcsec of each member of the sample. The search was a *success* if a galaxy with redshift within $(1000/c)$ km sec^{-1} could be found. (The difference in redshift of this order could be attributed to peculiar velocities.) Stockton registered eight successes. The probability of obtaining these by chance, given the surface density of quasars, worked out to be as low as 1.5×10^{-6}. This probability is small enough to reject the hypothesis that the quasars happened to be projected near galaxies by chance and thus to conclude that these eight quasars do indeed have redshifts given by their respective distances.

While this argument appears convincing, it leaves out the so-called 'failures', of which there were twelve, galaxies that were within 45 arcsec of the respective quasars but with redshift differences beyond the specified limit. Why should we ignore them? Do they not refute the CH?

A systematic computer-aided search was made by Burbidge *et al.* (1990) for close pairs of quasars from the then available Hewitt–Burbidge catalogue and bright galaxies from the catalogue of Sulentic and Tifft (ST73). The separation limit was set at 600 arcsec. They found over 500 close pairs of quasars and galaxies of which 28 low redshift quasars were associated with 42 galaxies with the same redshifts (this is what the CH predicts), while for the remainder the quasar redshift considerably exceeded the galaxy redshift.

The conservative reaction to the finding is to ignore, as Stockton did, the latter set, which contradicts the CH, with the argument that these are chance projections. However, statistical computations of the chance projection hypothesis yield probabilities $< 10^{-2}$, which are low enough to make us doubt or reject the hypothesis.

There have been some attempts to explain such associations as arising from gravitational lensing by the foreground galaxy, which brightens the quasar in the background. These, however, do not succeed where angular separations are more than, say, 10 arcsec.

Close pairs of quasars   Quasars being rarer than galaxies, close pairs of quasars are even more rare. Before looking for such pairs we may set limits of a priori probability by specifying the magnitude limit m for quasars and estimate the number of such pairs with separation less than a preassigned value θ. This number is

$$\langle n \rangle = 2.4 \times 10^{-7} N \Gamma(< m) \theta^2, \tag{15.12}$$

where $\Gamma(< m)$ is the surface density (per deg^2) of quasars brighter than magnitude m, θ is in arcsec and N is the number of galaxies searched for the effect.

The number $\langle n \rangle$ is usually small and one may use Poissonian statistics to estimate the probability that the observed number of pairs could have come from a chance projection of quasars in the same direction. All such pairs, excluding those ascribed to multiple images of the same source through gravitational lensing, have members with discrepant redshifts.

Studies of this kind have shown that the Poissonian probabilities are of the order of 10^{-4}, again rejecting the null hypothesis of chance projection.

Alignments and groupings   Another way, mentioned above, of discrediting the CH is by demonstrating close groupings of objects of discrepant redshifts on the sky. Arp and others have found close groupings of high redshift quasars near a typically bright NGC galaxy. Given the low surface densities of quasars, such groupings stand out as statistically significant.

Additionally, such groupings also show alignments of quasars across the galaxy, with the redshifts of the aligned objects close in value (but higher than that of the galaxy). Alignments of triplets of quasars with different redshifts have also been found. These alignments are so precise that their happening by chance projection has a probability $< 10^{-4}$. We have illustrated such observations in Figures 15.3 to 15.6. Because so much hinges on the outcome of these arguments, one should naturally exercise caution in accepting these conclusions. Nevertheless, it cannot be denied that such 'anomalous' findings have become numerous enough that they should not be ignored.

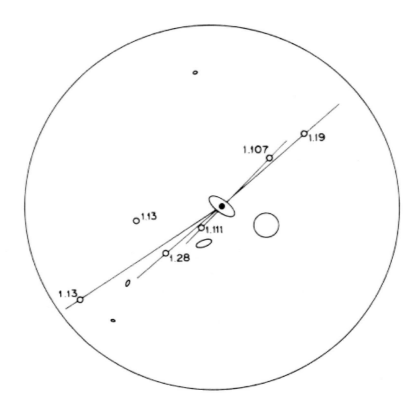

Fig. 15.3.   Six quasars bunched across NGC 3384. Reproduced from Arp *et al.* (1979).

Redshift periodicities   We end this section with a finding which, though somewhat unrelated to the studies just described, is nonetheless hard to understand within
the CH framework.

In 1968, with about 70 emission redshifts of quasars then available, Geoffrey Burbidge found a periodic peaking of the distribution with a period of ~ 0.06 in the
redshift. Since then the result has been scrutinized from time to time as the observed
quasar population has grown. The most recent study was conducted by Duari *et al.*
(1992), who found that with over 2000 quasar redshifts, chosen after avoiding any selection bias, the periodicity is still there (more precisely it is 0.0565). Duari *et al.* found
this effect by several different statistical techniques and concluded that its significance
increases if one goes over to the galacto-centric frame of reference (see Figure 15.7).

Karlsson (1977) had earlier examined the redshift distribution for a larger period,
and found one on a logarithmic scale:

$$\Delta \left[\log(1 + z) \right] = 0.089. \tag{15.13}$$

This has been subject to controversy, with some analyses refuting any such effect and
others confirming it, while a few others are inconclusive.

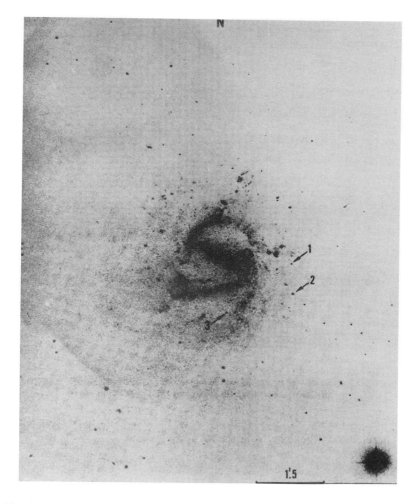

Fig. 15.4. Three quasars within 2 arcsec of NGC 1073. Reproduced from A87.

If we take the Burbidge effect seriously, then it implies, as per the CH, that there is a cellular structure in the universal distribution of quasars with a typical linear dimension of $\sim 180 h_{100}^{-1}$ Mpc. (We remind the reader that we express the value of the Hubble constant as $100 h_{100}$ km sec^{-1} Mpc^{-1}.) No conventional cosmological theory or theory of quasar formation has come up with an explanation for this cellular type of structure.

15.6 Non-cosmological alternatives

Having aired the above evidence, which cast doubt on the universality of the CH, we may now go back to the relation 15.10 between z_t, z_i and z_c. If in a total redshift of 3.5,

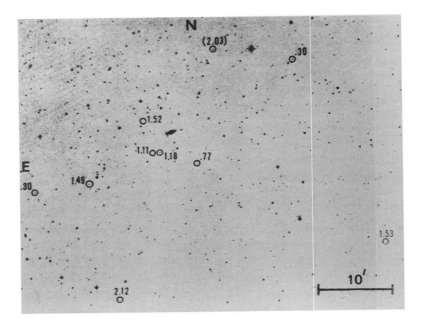

Fig. 15.5. Ten quasars surrounding the galaxy companion to NGC 2639. Reproduced from Arp A87.

say, the cosmological contribution is only 0.5, then the intrinsic component required is $z_i = 2$. To what could this contribution be due?

We have already discussed the Doppler and gravitational alternatives. With a large intrinsic component of this type, one would expect no correlation between the redshift and apparent magnitude and could thus understand the scatter diagram. But are these alternatives able to explain the anomalous phenomena described in the previous section? Let us discuss them briefly.

The Doppler option   We introduced this in Chapter 1 and noted there the major drawback that it not only allows for intrinsic blueshifts but that these also dominate over the redshifts. Why are no blueshifts found? There are some selection effects that might go against the detection of blueshifts.

- For example, the continuum gets enhanced in intensity thus making it relatively difficult to pick out lines. Such objects may be classified as a separate category of lineless objects (e.g. BL Lac objects?) or as highly evolved stars.
- The observers as a rule look for line ratios in the redshifted part rather than in the blueshifted part. Thus there has been no systematic search for lines that may have been blueshifted to the observed wavelengths.
- The quasars are often detected from radio surveys and it is likely that, when blueshifted, the peak of the synchrotron radiation maximum could move to such

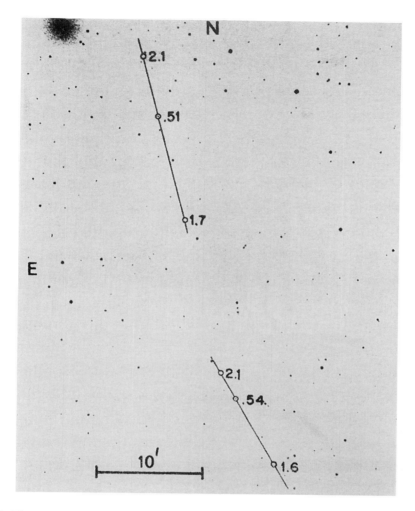

Fig. 15.6. The two triplets of quasars found by Arp and Hazard on the same photographic plate. Reproduced from Arp and Hazard (1980).

high frequencies that at the frequency of the survey the source is too faint to be detected.

In addition, Strittmatter in the 1960s suggested that if the quasar selectively radiates in the backward direction as it moves, we would see only those quasars which are receding from us. The effect of relativistic aberration as calculated by Fred Hoyle in 1980 is to magnify this effect. Thus, if in the rest frame of the quasar the emission is confined to a backward cone of semi-vertical angle θ_{H} given by

$$\theta_{\mathrm{H}} = \cos^{-1}\left(\frac{c - \sqrt{c^2 - v^2}}{v}\right), \qquad (15.14)$$

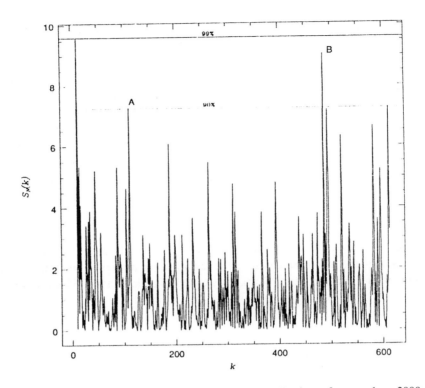

Fig. 15.7. A power spectrum analysis of the redshift distribution of more than 2000 quasars shows a significant peak at $z = 0.0565$, corresponding to a redshift periodicity at this value of z. Reproduced from Duari *et al.* (1992).

where v is the speed of the quasar with respect to the rest frame of the observer, then no quasar will be seen by the observer as blueshifted.

Why should a quasar moving rapidly in the intergalactic medium (IGM) radiate in the backward cone? The answer was provided by K. Subramaniam and one of the present authors (Narlikar and Subramaniam 1982) in terms of the ram pressure of the IGM on the jets issuing from the central region of the quasar. The pressure is strong enough to block the forward jet and does not affect the backward one. This hypothesis, incidentally, also explains why quasars (as opposed to radio galaxies) appear to be single-jet systems.

The alignments mentioned in the previous section also get explained, as ejection effects, and we build up a theory wherein quasars are the ejecta of explosions in big galactic nuclei. An observational confirmation of this picture, however, requires the detection of proper motions of substantial magnitudes. So far no case has been reported with a positive claim for proper motions. Pending such findings the Doppler option has not gained much backing.

Gravitational redshift It is somewhat ironical that this option has remained relatively sterile, considering that the energy machine powering a quasar or an AGN is popularly believed to be a massive black hole. Apart from the work described in Chapter 1 there has been only one attempt, by Fred Hoyle in the mid-1980s, to revive the idea, using further variations on the Hoyle–Fowler model of 1967 described in Chapter 1.

Gravitational redshift thus does not offer any radical resolution of the anomalous redshift problem. It could possibly have an explanation of redshift periodicity if some quantum effect in the quasar were able to produce quantized gravitational redshifts. Quantum gravity so far proving intractable, this idea will have to await some future breakthrough.

The variable mass hypothesis We now discuss in somewhat greater detail a new idea that goes beyond standard relativity. Its origin broadly lies in Mach's principle, which states that the inertia of matter arises from other matter in the universe. To put the statement in a mathematical form, Hoyle and Narlikar (1964, 1966) assumed that the space–time geometry is Riemannian with a metric

$$ds^2 = g_{ik}dx^i dx^k \qquad (15.15)$$

for coordinates x^i ($i = 0, 1, 2, 3$); x^0 is time-like, and the signature is $(+ - - -)$.

Now imagine particles of matter labelled $a, b, c, \ldots$, with x_a^i the coordinates of the ath particle, whose world line will be denoted by Γ_a. Then the 'mass-function' $m(X)$ at a world point X is defined as the contribution to the inertia at X of all particles $a, b, \ldots$:

$$m(X) = \sum_a m^{(a)}(X) \qquad (15.16)$$

where

$$m^{(a)}(X) = \int_{\Gamma_a} G(X, A)ds_a. \qquad (15.17)$$

Here the inertia at X due to particle a is communicated by the propagator $G(X, A)$, which satisfies a conformally invariant wave equation. The simplest form of such an equation is

$$\Box m^{(a)} + \tfrac{1}{6}Rm^{(a)} = N^{(a)} \qquad (15.18)$$

where $\Box$ is the wave operator, R the scalar curvature and $N^{(a)}$ the number density function of particle a at point X.

The dynamical equations of this theory are derived from the variation of a simple action:

$$\sum_a \int m_a(A)ds_a, \qquad (15.19)$$

where

$$m_a(A) = \sum_{b \neq a} m^{(b)}(A). \tag{15.20}$$

The action given by Equation 15.19 may be varied with respect to g_{ik} to get the field equations and with respect to particle world lines to get the equations of motion. The former gives, in the many particle approximation,

$$\tfrac{1}{2}m^2(R_{ik} - \tfrac{1}{2}g_{ik}R) = -3T_{ik} - m(g_{ik}\Box m - m_{;ik}) - 2(m_{,i}m_{,k} - \tfrac{1}{4}g_{ik}m^{,l}m_{,l}). \tag{15.21}$$

These equations allow us to talk of a variable inertial mass. Since the equations are conformally invariant, we may be able to choose a conformal frame in which $m = $ constant. In such a frame Equations 15.21 become

$$R_{ik} - \frac{g_{ik}R}{2} = -\frac{6}{m^2}T_{ik}, \tag{15.22}$$

These equations are identical with the field equations of general relativity if we set

$$\frac{8\pi G}{c^4} = \frac{6}{m^2}, \tag{15.23}$$

G being the Newtonian gravitational constant.

However, this transformation breaks down if we a choose part of space–time which has $m = 0$. Indeed, one can show that the space–time singularities of general relativity are due to the 'forcing' of equations such as 15.21 into the more compact form of 15.22 even when $m = 0$ hypersurfaces exist (Kembhavi 1978). It is at such hypersurfaces that relativistic singularity is found. As we shall see later, one can avoid referring to Equations 15.22 and their singular solutions and instead use the non-singular Equations 15.21.

This is when we encounter a new interpretation for redshift that applies equally well to regular as well as anomalous situations.

We illustrate this statement with the flat space–time solution of Equations 15.21. It can be easily verified that the solution of these equations is given by the Minkowski metric

$$ds^2 = c^2 dt^2 - dr^2 - r^2(d\theta^2 + \sin^2 d\phi^2), \tag{15.24}$$

with the mass function

$$m = at^2, \quad a = \text{constant}, \tag{15.25}$$

the number density of particles being constant in the comoving reference frame (r, θ, ϕ).

We have here a flat space–time cosmology in which light waves travel without spectral shift. How then do we explain redshift? Consider a galaxy G at a given radial coordinate r, the observer being at $r = 0$. A light ray leaving the galaxy at $t_0 - r/c$ reaches the observer at time t_0. Since the masses of all subatomic particles scale as t^2, the emitted wavelengths scale as $m^{-1} \propto t^{-2}$. Hence we get the factor

$$1 + z = \frac{t_0^2}{\left(t_0 - r/c\right)^2} \tag{15.26}$$

as the ratio of the wavelength *actually emitted* by the galaxy to the wavelength emitted in the laboratory of the observer. As such the observed cosmological redshift is a consequence of the systematic increase in particle masses with the t-epoch.

This solution is observationally *no different* from the Einstein–de Sitter model of standard relativistic cosmology because we can effect a conformal transformation that makes the mass function constant, by choosing a conformal function $\propto t^2$. Thus, writing

$$ds_R \propto t^2 ds \tag{15.27}$$

the line element in the *relativistic frame*, ds_R^2, becomes the familiar Einstein–de Sitter line element if we make a coordinate transformation to the new time τ given by

$$t \propto \tau^{1/3}, \qquad t_0 = 3\tau_0. \tag{15.28}$$

The present value of the Hubble constant in the model is $H_0 = 2/t_0$.

Notice that in a well-behaved conformal transformation the conformal function should not vanish or become infinite. Here we have to pay the price of choosing a conformal function that vanishes at $t = 0$, for in the relativistic frame the $\tau = 0$, $t = 0$ hypersurface has the (Big Bang) singularity.

The flat space–time cosmology admits anomalous redshifts in a natural way, as was shown by Narlikar (1977). Suppose the zero mass hypersurface has a kink as shown in Figure 15.8. The world line of a quasar Q intersects it at an epoch $t_Q > 0$. As shown in Narlikar (1977), the particle mass function in Q starts ticking from this epoch; thus at an epoch $t > t_Q$ it will be $\propto (t - t_Q)^2$. The interpretation of this result is simple; the particle receives all inertial contributions of type $1/r$ from a past light cone extending from t to t_Q.

In Figure 15.8 we see a quasar Q and a galaxy G_1 which are close neighbours, but the world line of Q passes through the kink while that of G_1 does not. For particles in G the mass function is $\propto t^2$ at epoch t. If both Q and G_1 are at a distance r from the observer, the formula in Equation 15.26 gives the respective redshifts as

$$1 + z_Q = \frac{t_0^2}{(t_0 - r/c - t_Q)^2}, \qquad 1 + z_G = \frac{t_0^2}{(t_0 - r/c)^2}. \tag{15.29}$$

So we have $z_Q > z_G$ and thus an anomalous redshift for the quasar! Narlikar and Das (1980) considered such pairs.

As illustrated in Figure 15.8 the world lines of Q and G continue on both sides of the zero mass hypersurface. However, the appearance of $m = 0$ corresponds in the relativistic frame to the space–time singularity, thus giving an incomplete (and erroneous) view of a universe 'beginning' at $\tau = 0$. In practice we may interpret

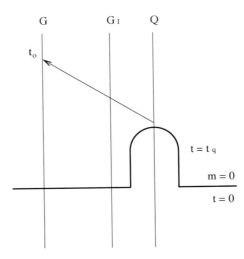

Fig. 15.8. A kink in the zero mass surface through which the world line of quasar Q passes implies the delayed ejection of Q from its neighbouring galaxy G_1. Both the observer's galaxy and the galaxy G_1 have their world lines passing through the zero mass epoch much earlier than Q.

Figure 15.8 as describing a quasar ejected from the neighbour galaxy. Narlikar and Das (1980) gave a detailed dynamical study of such pairs.

How could a static, matter-filled universe remain stable? Would it not collapse, as Einstein (and even earlier Newton) found? The answer is that stability is guaranteed by the mass-dependent terms on the right hand side of Equation 15.21. Small perturbations of the flat Minkowski space–time would lead to small oscillations about the line element given in Equation 15.24 rather than to a collapse.

Quantized redshifts Redshifts that arise from a difference in age, however, could solve the quantization problem in a natural way. Creation processes that produce galaxies at different times must originate at a zero mass surface. Close to the zero mass surface the classical action is very small and hence the physics is dictated by quantum considerations. Thus one could argue that material emerges from the zero mass surface within a quantum mechanical realm and may do so in discrete bursts spaced at discrete intervals instead of continuously. This could lead to a quantized distribution of redshift intervals.

This explanation must assume that there are no other 'contaminating' redshift contributions such as the Doppler or gravitational ones that would spoil the observed exact periodicity. This remains a serious difficulty of the present explanation as, indeed, of any other explanation of this effect.

The idea of redshifts arising from variable particle masses is, however, novel and uses an extension of known standard physics. It could be criticized on the grounds of being speculative, as are the present discussions of the very early universe (quantum cosmology, inflation, cosmic strings, etc.). Nevertheless, it makes certain testable predictions that may inspire observational probes. These are as follows.

- Particles in anomalously redshifted objects have smaller masses than their counterparts in normally redshifted objects. If there are common regions where the two kinds of matter mix, e.g., in filaments connecting the two kinds of object, one may be able to work out the high energy astrophysics of their interaction.
- The claim that anomalously redshifted objects are younger can be tested by looking at their stellar populations, if such are found in these objects. If the stars are systematically younger, this may provide some proof.
- The Narlikar–Das work describes the dynamics of the ejection of quasars by galaxies. It may therefore be compared to the claimed examples of quasars ejected by galaxies.

15.7 Concluding remarks

To sum up the contents of this chapter, we first note that if we accept the cosmological hypothesis (CH) *in toto* we have energy and time scale problems on our hands, problems which are not entirely resolved by the relativistic beaming hypothesis. We are expected to believe in the massive black hole plus accretion disk paradigm, despite any direct observational evidence for it. Nor do we have any direct demonstration of a redshift–magnitude relation for quasars.

The CH fails to account for a variety of phenomena observed for quasars in relation to galaxies, for their alignments and special configurations as well as for their periodically distributed redshifts. Even granting that a fraction of the claimed results may be statistically insignificant, the growing body of such phenomena will soon have to be addressed within an acceptable theoretical framework.

It is also true that no acceptable framework exists as an alternative to the CH. The Doppler and the gravitational redshift options have their own difficulties while the variable mass hypothesis is speculative and needs further probing.

It is here that we should express a word of caution. Against the well-known dictum 'Believe no theory unless it is supported by observation' there is another, less well known: 'Believe no observations unless they are explained by a theory'. The latter can err on the excessively safe side and miss new discoveries. The examples of Fraunhofer lines and stellar energy sources tell us that astronomy can point to new physics if one is adventurous enough to stick one's neck out.

In this book we have by and large taken the 'safe and hence sound' approach in describing the properties and models of quasars and AGN. But these are remarkable objects and it is optimistic to believe that we can understand them fully with the physics we know. The authors would not be surprised if behind the host of (what many may call peculiar and crazy) anomalous phenomena lurks some new physics.

References

Anderson, S.F. and Margon, B. (1987). The X-ray properties of high-redshift quasi-stellar objects, *Astrophys. J.* **314**, 111.

Angel, J.R.P. and Stockman, H.S. (1980). Optical and infrared polarization of active extragalactic objects, *Ann. Rev. Astron. Astrophys.* **18**, 321.

Antonucci, R. (1984). Optical spectropolarimetry of radio galaxies, *Astrophys. J.* **278**, 499.

Antonucci, R. (1988). Polarization of active galactic nuclei, in K88, p. 26.

Antonucci, R. (1993). Unified models for active galactic nuclei and quasars, *Ann. Rev. Astron. Astrophys.* **31**, 473.

Antonucci, R., Hurt, T. and Kinney, A. (1994). Evidence for a quasar in the radio galaxy Cygnus A from observation of broad-line emission, *Nature* **371**, 313.

Antonucci, R., Kinney, A.L. and Ford, H.C. (1989). The Lyman edge test of the quasar emission mechanism, *Astrophys. J.* **342**, 64.

Antonucci, R. and Miller, J.S. (1985). Spectropolarimetry and the nature of NGC 1068, *Astrophys. J.* **297**, 621.

Arnaud, K.A., Branduardi-Raymont, G., Culhane, J.L., Fabian, A.C. and Hazard, C. (1985). *EXOSAT* observation of a strong soft X-ray excess in Mkn 841, *Mon. Not. Roy. Astr. Soc.* **217**, 105.

Arp, H.C. and Crane, P. (1992). Testing the gravitational lens hypothesis in 2237+0305, *Phys. Lett. A* **168**, 6.

Arp, H.C. and Hazard, C. (1980). Peculiar configuration of quasars in two adjacent areas of the sky, *Astrophys. J.* **240**, 726.

Arp, H.C., Sulentic, J.W. and Di Tullis, G. (1979). Quasars aligned across NGC 3384, *Astrophys. J.* **229**, 489.

Athreya, R.M. (1996). Multifrequency studies of high redshift radio galaxies, Ph.D. thesis, unpublished, 104.

Avni, Y. and Bahcall, J.N. (1980). On the simultaneous analysis of several complete samples. The V/V_{max} and V_e/V_a variables, with applications to quasars, *Astrophys. J.* **235**, 694.

Avni, Y., Sottan, A., Tananbaum, H. and Zamozari, G. (1980). A method for determining luminosity functions incorporating both flux measurements and flux upper limits. With applications to the average X-ray optical luminosity ratio of quasars, *Astrophys. J.* **238**, 800.

Avni, Y. and Tananbaum, H. (1986). X-ray properties of optically selected QSOs, *Astrophys. J.* **305**, 83.

Awaki, H., Koyama, K., Kunieda, H. and Tawara, Y. (1990). X-ray evidence of an obscured nucleus in the type 2 Seyfert galaxy Mkn 3., *Nature* **346**, 544.

Baade, W. and Minkowski, R. (1954). On the identification of radio sources, *Astrophys. J.* **119**, 215.

Baath, L.B., Padin, S., Woody, D., Rogers, A.E.E., Wright, M.C.H., Zensus, A. *et al.* (1991). The microarcsecond structure of 3C 273 at 3 mm, *Astron. Astrophys.* **241**, L1.

Bahcall, J.N., Bergeron, J., Boksenberg, A., Hartig, G.F., Januzzi, B.T., Kirhkos, S. *et al.* (1993). The Hubble Space Telescope quasar absorption line key project. I. First observational results including Lyman-alpha and Lyman-limit systems, *Astrophys. J. Suppl.* **87**, 1.

Baker, J.C., Hunstead, R.W. and Brinkmann, W. (1995). Radio and X-ray beaming in steep-spectrum quasars, *Mon. Not. Roy. Astr. Soc.* **277**, 553.

Bambynek, W., Crasemann, B., Fink, R.W., Freund, H.-U., Mark, H., Swift, C.D., Price, R.E. and Rao, P.V. (1972). X-ray fluorescence yields, Auger, and Koster-Kronig transition probabilities, *Rev. Mod. Phys.* **44**, 716.

Barcons, X., Branduardi-Raymont, G., Warwick, R.S., Fabian, A.C., Mason, O., Hardy, I.M. and Rowan-Robinson, M. (1994). Deep X-ray source counts from a fluctuation analysis of *ROSAT* PSPC images, *Mon. Not. Roy. Astr. Soc.* **268**, 833.

Barcons, X. and Fabian, A.C. (1990). A fluctuation analysis of the X-ray background in the *EINSTEIN* Observatory IPC, *Mon. Not. Roy. Astr. Soc.* **243**, 366.

Bardeen, J.M. (1973). Timelike and null geodesics in the Kerr metric, in DD73, p. 215.

Barr, P., White, N.E., Sanford, P. and Ives, J.C. (1977). An increase in the X-ray absorption of NGC 4151, *Mon. Not. Roy. Astr. Soc.* **181**, 43p.

Barthel, P.D. (1987). Feeling uncomfortable, in ZP87, p. 148.

Barthel, P.D. (1989). Is every quasar beamed?, *Astrophys. J.* **336**, 606.

Barthel, P.D., Hooimeyer, J.R., Schilizzi, R.T., Miley, G.K. and Preuss, E. (1989). Superluminal motion in the giant quasar 4C 34.47, *Astrophys. J.* **336**, 601.

Barthel, P.D., Pearson, T.J., Readhead, A.C.S. and Canzian, B.J. (1986). 0851 + 581: another superluminal radio source, *Astrophys. J.* **310**, L7.

Barvainis, R. (1993). Free–free emission and the big blue bump in active galactic nuclei, *Astrophys. J.* **412**, 513.

Baum, S.A., Zirbel, E.L. and O'Dea, C.P. (1995). Toward understanding the Fanaroff–Riley dichotomy in radio source morphology and power, *Astrophys. J.* **451**, 88.

Becker, R.H., White, R.L. and Helfand, D.J. (1995). The FIRST survey: faint images of the radio sky at twenty centimeters, *Astrophys. J.* **450**, 559.

Begelman, M.C., Blandford, R.D. and Rees, M.J. (1984). Extragalactic radio sources, *Rev. Mod. Phys.* **56**, 265.

Bergeron, J. and Boisse, P. (1991). A sample of galaxies giving rise to MG II quasar absorption systems, *Astron. Astrophys.* **243**, 344.

Bergeron, J., Petitjean, P., Sargent, W.L.W., Bahcall, J.N., Boksenberg, A. (1993). The Hubble Space Telescope quasar absorption line key project. VI. Properties of the metal-rich systems, *Astrophys. J.* **436**, 33.

Bignami, G.F., Fichtel, C.E., Hartman, R.C. and Thompson, D.J. (1979). Galaxies and gamma-ray astronomy, *Astrophys. J.* **232**, 649.

Biretta, J.A. (1993). The M87 jet, in BLO93, p. 263.

Biretta, J.A. and Meisenheimer, K. (1993). The jet of M87, in RM93, p. 159.

Black, A.R.S. (1993). Jets and related issues, in RM93, p. 159.

Blandford, R.D. (1990). Physical processes in active galactic nuclei, in BNW90, p. 161.

Blandford, R.D., McKee, C.F. and Rees, M.J. (1977). Super-luminal expansion in extragalactic radio sources, *Nature* **267**, 211.

Blandford, R.D. and Rees, M.J. (1974). A 'twin-exhaust' model for double radio sources, *Mon. Not. Roy. Astr. Soc.* **169**, 395.

Blandford, R.D. and Rees, M.J. (1978). Some comments on radiation mechanisms in Lacertids, in W78, p. 328.

Blandford, R.D. and Znajek, R. (1977). Electromagnetic extraction of energy from Kerr black holes, *Mon. Not. Roy. Astr. Soc.* **179**, 433.

Blumenthal, G.R. and Gould, R.J. (1970). Bremsstrahlung, synchrotron radiation, and Compton scattering of high-energy electrons traversing dilute gases, *Rev. Mod. Phys.* **42**, 237.

Boldt, E. (1987). The cosmic X-ray background, *Phys. Rep.* **146**, 215.

Bondi, H. (1952). On spherically symmetric accretion, *Mon. Not. Roy. Astr. Soc.* **112**, 195.

Bondi, H. (1964). Massive spheres in general relativity, *Proc. Roy. Soc. Lond. A* **282**, 303.

Bowyer, S., Lampton, M. and Mack, J. (1970). Detection of X-ray emission from 3C273 and NGC5128, *Astrophys. J.* **161**, L1.

Boyle, B.J. (1990). A composite QSO spectrum, *Mon. Not. Roy. Astr. Soc.* **243**, 231.

Boyle, B.J., Fong, R., Shanks, T. and Peterson, B.A. (1987). The evolution of optically selected QSOs, *Mon. Not. Roy. Astr. Soc.* **227**, 717.

Boyle, B.J., Griffiths, R.E., Shanks, T., Stewart, G.C. and Georgantopoulos, I. (1993). A deep *ROSAT* survey – I. The QSO X-ray luminosity function, *Mon. Not. Roy. Astr. Soc.* **260**, 49.

Boyle, B.J., Jones, L.R., Shanks, T., Marano, B., Zitelli, V. and Zamorani, G. (1991). QSO evolution and clustering at $z < 2.9$, in C91, p. 191.

Boyle, B.J., Shanks, T. and Peterson, B.A. (1988). The evolution of optically selected QSOs – II, *Mon. Not. Roy. Astr. Soc.* **235**, 935.

Bracessi, A., Zitelli, V., Bonoli, F. and Formiggini, L. (1980). A sample of very faint ultraviolet excess objects in the $13\,h + 36$ deg field, *Astron. Astrophys.* **85**, 80.

Bradt, H.V.D., Ohashi, T. and Pounds, K.A. (1992). X-ray astronomy missions, *Ann. Rev. Astron. Astrophys.* **30**, 391.

Branduardi-Raymont, G., Mason, K.O., Warwick, R.S., Carrera, F.J., Graffagnino, V.G., Mittaz *et al.* (1994). The UK deep and medium surveys with *ROSAT*, *Mon. Not. Astr. Soc.* **270**, 947.

Bregman, J.N. (1990). Continuum radiation from active galactic nuclei, *Astron. Astrophys. Rev* **2**, 125.

Bregman, J.N. (1994). The origin of continuum emission in active galactic nuclei, in CB94, p. 5.

Bregman, J.N., Glassgold, A.E., Huggins, P.J., Neugebauer, G., Soifer, B.T., Matthew, K. *et al.* (1990). Multifrequency observations of BL Lacertae, *Astrophys. J.* **352**, 574.

Bregman, J.N., Glassgold, A.E., McHardy, I., *et al.* (1988). Multifrequency observation of the optically violent variable quasar 3C 446, *Astrophys. J.* **331**, 746.

Bregman, J.N. and Hufnagel, B.R. (1989). Radio and optical variability in blazars, in MMU89, p. 159.

Bridle, A.H. (1990). The dark side of radio jets, in ZP90, p. 186.

Bridle, A.H., Hough, D.H., Lonsdale, C J., Burns, J.O. and Laing, R.A. (1994). Deep VLA imaging of twelve extended 3CR quasars, *Astron. J.* **108**, 766.

Bridle, A.H. and Perley, R.A. (1984). Extragalactic radio jets, *Ann. Rev. Astron. Astrophys.* **22**, 319.

Browne, I.W.A. (1987). Extended structure of superluminal radio sources, in ZP87, p. 129.

Browne, I.W.A. and Murphy, D.W. (1987). Beaming and the X-ray, optical and radio properties of quasars, *Mon. Not. Roy. Astr. Soc.* **226**, 601.

Browne, I.W.A., Orr, M.J.L., Davis, R.J., Foley, A., Muxlow, T.W.B. and Thompson, P. (1982). Jodrell Bank MTRLI observations of nine core-dominated sources at 408 MHz, *Mon. Not. Roy. Astr. Soc.* **198**, 673.

Browne, I.W.A. and Perley, R.A. (1986). Extended radio emission around core-dominated quasars – constraints on relativistic beaming models, *Mon. Not. Roy. Astr. Soc.* **222**, 149.

Brunner, H., Lamer, G., Worrall, D.M. and Staubert, R. (1994). X-ray spectra of a complete sample of core-dominated radio sources, *Astron. Astrophys.* **287**, 436.

Burbidge, G.R. (1959). Estimates of the total energy in particles and magnetic field in non-thermal radio sources, *Astrophys. J.* **129**, 849.

Burbidge, G.R. (1968). The distribution of redshifts in quasi-stellar objects, N-systems, and some radio and compact objects, *Astrophys. J.* **154**, L41.

Burbidge, G.R. (1985). A comment on the discovery of the QSO and related galaxy 2237+0305, *Astron. J.* **90**, 1399.

Burbidge, G.R., Crowne, A.H. and Smith, H.E. (1977). An optical catalogue of quasi-stellar objects, *Astrophys. J. Suppl.* **33**, 113.

Burbidge, G.R. and Hewitt, A. (1987). An updated list of BL Lac objects, and their relation to galaxies and quasistellar objects, *Astron. J.* **92**, 1.

Burbidge, G.R., Hewitt, A., Narlikar, J.V. and Das Gupta, P. (1990). Associations between quasi-stellar objects and galaxies, *Astrophys. J. Suppl.* **74**, 675.

Burn, B.J. (1966). On the depolarization of discrete radio sources by Faraday dispersion, *Mon. Not. Roy. Astr. Soc.* **133**, 67.

Burstein, D. and Heiles, C. (1982). Reddenings derived from H I and galaxy counts: accuracy and maps, *Astron. J.* **87**, 1165.

Caillaut, J. and Helfand, D.J. (1985). The *EINSTEIN* soft X-ray survey of the Pleiades, *Astrophys. J.* **289**, 279.

Camenzind, M. (1993). Formation of jets in extragalactic radio sources, in RM93, p. 109.

Canizares, C.R. and White, J.L. (1989). The X-ray spectra of high-redshift quasars, *Astrophys. J.* **339**, 27.

Carilli, C.L. and Barthel, P.D. (1996). Cygnus A, *Astron. Astrophys. Rev* **7**, 1.

Carleton, N.P., Elvis, M., Fabbiano, G. and Willner, S.P. (1987). The continuum of type 1 Seyfert galaxies II. Separating thermal and nonthermal components, *Astrophys. J.* **318**, 595.

Cavaliere, A. and Padovani, P. (1988). Eddington ratios as a constraint on the activity patterns of the active galactic nuclei, *Astrophys. J.* **333**, L33.

Cavaliere, A. and Padovani, P. (1989). The connection between active and normal galaxies, *Astrophys. J.* **340**, L5.

Cawthorne, T.V., Scheuer, P.A.G., Morrison, I. and Muxlow, T.W.B. (1986a). A sample of powerful radio sources for VLBI studies, *Mon. Not. Roy. Astr. Soc.* **219**, 883.

Cawthorne, T.V., Scheuer, P.A.G., Morrison, I. and Muxlow, T.W.B. (1986b). A sample of powerful radio sources for VLBI studies, *Erratum,, Mon. Not. Roy. Astr. Soc.* **222**, 895.

Celotti, A., Maraschi, L., Ghisellini, G., Caccianiga, A. and Maccacaro, T. (1993). Unified model for X-ray- and radio-selected BL Lacertae objects, *Astrophys. J.* **416**, 118.

Cheng, F., Danese, L., De Zotti, G. and Franceschini A. (1985). The optical luminosity function of Seyfert galaxies, *Mon. Not. Roy. Astr. Soc.* **212**, 857.

Chitre, S.M. and Narlikar, J.V. (1979). On the apparent superluminal separation of radio source components, *Mon. Not. Roy. Astr. Soc.* **187**, 655.

Clavel, J. (1994). Multifrequency variability of AGNs, in CB94, p. 131.

Clavel, J., Nandra, K., Makino, F., Pounds, K.A., Reichert, G.A., Urry, C.M. *et al.* (1992). Correlated hard X-ray and ultraviolet variability in NGC 5548, *Astrophys. J.* **393**, 113.

Clavel, J., Reichert, G.A., Alloin, D., Crenshaw, D.M., Kriss, G., Krolik, J.H. *et al.* (1991). Steps towards determination of the size and structure of the broad-line region in active galactic nuclei. I. An 8 month campaign of monitoring NGC 5548 with IUE, *Astrophys. J.* **366**, 64.

Cohen, M.H. (1990). Significance of small angles, in ZP90, p. 317.

Cohen, M.H., Barthel, P.D., Pearson, T.J. and Zensus, J.A. (1988). Expanding quasars and the expansion of the universe, *Astrophys. J.* **329**, 1.

Cohen, M.H., Cannon, W., Purcell, G.H., Shaffer, D.B., Broderick, J.J., Kellermann, K.I. and Jauncey, D.L. (1971). The small scale structure of radio galaxies and quasi-stellar sources at 3.9 centimeters, *Astrophys. J.* **170**, 207.

Cohen, M.H., Kellermann, K.I., Shaffer, D.B., Linfield, R.P., Moffet, A.T., Romney, J.D. *et al.* (1977). Radio sources with superluminal velocities, *Nature* **268**, 405.

Coleman, H.H. and Shields, G.A. (1990). The polarization of thermal radiation from supermassive accretion disks, *Astrophys. J.* **363**, 415.

Comastri, A., Molendi, S. and Ghisellini, G. (1995). *ROSAT* observations of radio selected BL Lac objects, *Mon. Not. Roy. Astr. Soc.* **277**, 297.

Comastri, A., Setti, G., Zamorani, G., Elvis, M., Giommi, P., Wilkes B.J. and McDowell, J.C.M. (1992). *EXOSAT* X-ray spectra of quasars, *Astrophys. J.* **384**, 62.

Condon, J.J. (1974). Confusion and flux density error distributions, *Astrophys. J.* **188**, 279.

Condon, J.J., Condon, M.A., Jauncey, D.L., Smith, M.G., Turtle, A.J. and Wright, A.E. (1981). Multifrequency radio observations of optically selected quasars, *Astrophys. J.* **244**, 5.

Condon, J.J., Condon, M.A., Mitchell, K.J. and Usher, P.D. (1980). Radio observations of a new class of optically selected quasi-stellar objects, *Astrophys. J.* **242**, 486.

Conway, J.E., Pearson, T.J., Readhead, A.C.S, Unwin, S.C., Xu, W. and Mutel, R.L. (1992). The compact triples $0710 + 439$ and $2352 + 495$: a new morphology of radio galaxy nuclei, *Astrophys. J.* **396**, 62.

Cowsik, R. and Kobetich, E.J. (1972). Comment on inverse Compton models for the isotropic X-ray background and possible thermal emission from a hot intergalactic gas, *Astrophys. J.* **177**, 585.

Crampton, D., Cowley, A.P. and Hartwick, F. (1989). The CFHT/MMT survey: further evidence for isolated groups, *Astrophys. J.* **345**, 59.

Crawford, D.F., Jauncey, D.L. and Murdoch, H.S. (1970). Maximum-likelihood estimation of the slope from number-flux-density counts of radio sources, *Astrophys. J.* **162**, 405.

Cristiani, S. (1996). Cosmological adventures in the Lyman forest, in *Proceedings of the International School of Physics 'Enrico Fermi'*, eds. S. Bonometto, J. Primach and A. Provenzale, IOS publishers, Amsterdam, p. 137.

Cristiani, S., Barbieri, C., Iovino, A., La-Franca, F. and Nota, A. (1989). Quasars in the field of SA 94. III. A colour survey, *Astron. Astrophys. Suppl.* **215**, 409.

Curtis, H.D. (1918). Descriptions of 762 nebulae and clusters photographed with the Crossley reflector, *Pub. Lick Obs.* **13**, 9.

Dallacasa, D., Fanti, C. and Fanti, R. (1993). Compact steep spectrum radio sources: a progress report, in RM93, p. 27.

Damle, S.V., Kunte, P.K., Naranan, S., Sreekantan, B.V. and Venkatesan, D. (1987). Hard X-ray observations of the quasar 3C 273, *Astron. Astrophys.* **182**, L1.

Dar, A. and Shaviv, N.J. (1995). Origin of the high energy extragalactic diffuse gamma ray background, *Phys. Rev. Lett.* **75**, 3052.

Das, P.K. (1977). Physical properties of collapsed objects with large central gravitational redshifts, *Mon. Not. Roy. Astr. Soc.* **177**, 391.

Datt, B. (1938). Über eine Klasse von Lösungen der Gravitationsgleichungen der Relativität, *Zeit. Phys.* **108**, 314.

Davidsen, A., Hartig, G.F. and Fastie, W.G. (1977). Ultraviolet spectrum of quasistellar object 3C 273, *Nature* **269**, 203.

Davidson, K. and Netzer, H. (1979). The emission lines of quasars and similar objects, *Rev. Mod. Phys.* **51**, 715.

Davis, R.J., Muxlow, T.W.B. and Conway, R.G. (1985). Radio emission from the jet and lobe of 3C 273, *Nature* **318**, 343.

De Kool, M. & Begelman, M.C. (1989). Production of self-absorbed synchrotron spectra steeper than $v^5/2$, *Nature* **338**, 484.

De Sitter, W. (1917). On the relativity of inertia: remarks concerning *EINSTEIN*'s latest hypothesis, *Proc. Akad. Weteusch. Amsterdam* 19, 1217.

De Young, D. (1993). On the relation between Fanaroff-Riley types I and II radio galaxies, *Astrophys. J.* **405**, L13.

Della Ceca, R., Maccacaro, T., Gioia, I., Wolter, A. and Stocke, J. (1992). The properties of X-ray selected active galactic nuclei. II. A deeper look at the cosmological evolution, *Astrophys. J.* **389**, 491.

Dermer, C.D., Schlickeiser, R. and Mastichadis, A. (1992). High-energy gamma radiation from extragalactic radio sources, *Astron. Astrophys.* **256**, L27.

Djorgovski, S., Weir, N., Matthews, K. and Graham, J.R. (1991). Discovery of an infrared nucleus in Cygnus A: an obscured quasar revealed?, *Astrophys. J.* **372**, L67.

Done, C. and Fabian, A.C. (1989). The behaviour of compact non-thermal sources with pair production, *Mon. Not. Roy. Astr. Soc.* **240**, 81.

Done, C., Pounds, K.A., Nandra, K. and Fabian, A.C. (1995). The complex variable soft X-ray spectrum of NGC 5548, *Mon. Not. Roy. Astr. Soc.* **275**, 417.

Draine, B.T. (1990). Mass determinations from far-infrared observations, in *The Interstellar Medium in Galaxies*, eds. H.A. Thornson, Jr. and J.M. Shull, Kluwer, Dordrecht, p. 483.

Dreher, J.W., Carilli, C.L. and Perley, R.A. (1987). The Faraday rotation of Cygnus A: magnetic fields in cluster gas, *Astrophys. J.* **316**, 611.

Duari, D., Das Gupta, P. and Narlikar, J.V. (1992). Statistical tests of peaks and periodicities in the observed redshift distribution of quasi-stellar objects, *Astrophys. J.* **384**, 35.

Duari, D. and Narlikar, J.V. (1995). The intervening galaxies hypothesis of the absorption spectra of quasi-stellar objects: some statistical studies, *Int. J. Mod. Phy. D* **4**, 367.

Dunlop, J.S, Taylor, G.L., Hughes, D.H. and Robson, E.I. (1993). Infra-red imaging of the host galaxies of radio-loud and radio-quiet quasars, *Mon. Not. Roy. Astr. Soc.* **264**, 455.

Eddington, A.S. (1940). The correction of statistics for accidental error, *Mon. Not. Roy. Astr. Soc.* **100**, 354.

Edelson, R.A. (1986). Far-infrared properties of optically selected quasars, *Astrophys. J.* **309**, L69.

Edelson, R.A. (1992). A survey of ultraviolet variability in blazars, *Astrophys. J.* **401**, 516.

Edelson, R.A., Krolik, J.H. and Pike G.F. (1990). Broad-band properties of the CfA Seyfert galaxies III. Ultraviolet variability, *Astrophys. J.* **359**, 86.

Edelson, R.A. and Malkan, M.A. (1986). Spectral energy distributions of active galactic nuclei between 0.1 and 100 microns, *Astrophys. J.* **308**, 59.

Edelson, R.A. and Malkan, M.A. (1987). Far infrared variability in active galactic nuclei, *Astrophys. J.* **323**, 535.

Edelson, R.A., Malkan, M.A. and Rieke, G.H. (1987). Broad band properties of the CfA sample. II. Infrared to millimeter properties, *Astrophys. J.* **321**, 233.

Eichler, D. and Wiita, P.J. (1978). Neutron beams in active galactic nuclei, *Nature* **274**, 38.

Eilek, J.A., Burns, J.O., O'Dea, C.P. and Owen, F.N. (1984). What bends 3C 465?, *Astrophys. J.* **278**, 37.

Einstein, A. and de Sitter, W. (1932). On the relation between the expansion and the mean density of the universe, *Proc. Nat. Acad. Sci. USA* **18**, 213.

Elliot, J.L. and Shapiro, S.L. (1974). On the variability of the compact nonthermal sources, *Astrophys. J.* **192**, L3.

Elvis, M. (1990). The ultraviolet continua of active galactic nuclei, Center for Astrophysics preprint, 2846.

Elvis, M., Fiore, F., Wilkes, B. and McDowell, J. (1994). Absorption in X-ray spectra of high-redshift quasars, *Astrophys. J.* **422**, 60.

Elvis, M., Maccacaro, T., Wilson, A.S., Ward, M.J. and Penston. M.V. (1978). Seyfert galaxies as X-ray sources, *Mon. Not. Roy. Astr. Soc.* **183**, 129.

Elvis, M., Matsuoka, M., Siemiginowska, A., Fiore, F., Mihara, T. and Brinkmann, W. (1994). An *ASCA* GIS spectrum of S5 0014 + 813 at $z = 3.384$, *Astrophys. J.* **436**, L55.

Evans, I.N., Ford, H.C., Kinney, A.L., Antonucci, R.R.J., Armus, L. and Caganoff, S. (1991). HST imaging of the inner 3 arcseconds of NGC 1068 in the light of [O III] λ5007, *Astrophys. J.* **369**, L27.

Fabian, A.C. (1979). Theories of the nuclei of active galaxies, *Proc. Roy. Soc. Lond. A* **366**, 449.

Fabian, A.C. (1994). Hot plasmas and the generation of gamma rays, *Astrophys. J. Suppl.* **92**, 555.

Fabian, A.C. and Barcons, X. (1992). The origin of the X-ray background, *Ann. Rev. Astron. Astrophys.* **30**, 429.

Fabian, A.C., Blandford, R.D., Guilbert, P.W., Phinney, E.S. and Cueller, L. (1986). Pair-induced spectral changes and variability in compact X-ray sources, *Mon. Not. Roy. Astr. Soc.* **221**, 931.

Fabian, A.C., George, I.M., Miyoshi, S. and Rees, M.J. (1990). Reflection-dominated hard X-ray sources and the X-ray background, *Mon. Not. Roy. Astr. Soc.* **242**, 14p.

Fabian, A.C., Shioya, Y., Iwasawa, K., Nandra, K., Crawford, C., Johnstone, R. *et al.* (1994). Fe K emission from the hidden quasar IRAS P 09104 + 4109, *Astrophys. J.* **436**, L51.

Falomo, R., Pesce, J.E. and Treves, A. (1995). Host galaxy and environment of the BL Lacertae object PKS 0548–322: observations with subarcsecond resolution, *Astrophys. J.* **438**, L9.

Fanaroff, B.L. and Riley, J.M. (1974). The morphology of extragalactic radio sources of high and low luminosity, *Mon. Not. Roy. Astr. Soc.* **167**, 31p.

Fanti, R., Fanti, C., Schilizzi, R.T., Spencer, R.E., Nan Rendong, Parma, P. *et al.* (1990). On the nature of compact steep spectrum radio sources., *Astron. Astrophys.* **231**, 333.

Fasano, A. and Franceschini, A. (1987). A multidimensional version of the Kolmogorov–Smirnov test, *Mon. Not. Roy. Astr. Soc.* **225**, 155.

Fayyad, U. and Roden, J. (1995). The discovery of five quasars at $z > 4$ using the second Palomar sky survey, *Astron. J.* **110**, 78.

Feigelson, E.D. (1992). Censoring in astronomical data due to nondetections, in FB92, p. 221.

Feigelson, E.D., Isobe, T. and Kembhavi, A.K. (1984). Radio and X-ray emission in radio-selected quasars, *Astron. J.* **89**, 1464.

Felten, J.E. (1976). On Schmidt's V_m estimator and other estimators of luminosity functions, *Astrophys. J.* **207**, 700.

Felten, J.E. and Morrison, P. (1966). Omnidirectional inverse Compton and synchrotron radiation from cosmic distributions of fast electrons and thermal photons, *Astrophys. J.* **146**, 686.

Fernini, I. and Burns, J.O., Bridle, A.H. and Perley, R.A. (1993). Very large array imaging of five Fanaroff–Riley II 3CR radio galaxies, *Astron. J.* **105**, 1690.

Fichtel, C.E., Kniffen, D.A. and Hartman, R.C. (1973). Celestial diffuse gamma radiation above 30 MeV observed by SAS-2, *Astrophys. J.* **186**, L99.

Fichtel, C.E., Simpson, G.A. and Thompson, D.J. (1978). Diffuse gamma radiation, *Astrophys. J.* **222**, 833.

Field, G.B. and Perrenod, S.C. (1977). Constraints on a dense hot intergalactic gas, *Astrophys. J.* **215**, 717.

Fiore, F., Elvis, M., McDowell, J.C., Siemiginowska, A. and Wilkes, B.J. (1994). The complex optical to soft X-ray spectra of the low-redshift radio-quiet quasars. I. The X-ray data, *Astrophys. J.* **431**, 515.

Fomalont, E.B. and Sramek, R.A. (1975). A confirmation of Einstein's general theory of relativity by measuring the bending of microwave radiation in the gravitational field of the Sun, *Astrophys. J.* **199**, 749.

Francis, P.J., Hewett, P.C., Foltz, C.B., Chaffee, F.H., Weymann, R.J. and Morris, S.L. (1991). A high signal-to-noise ratio composite quasar spectrum, *Astrophys. J.* **373**, 465.

Friedmann, A. (1924). Über die Krummung des Reumes, *Zeit. Phys.* **10**, 377.

Gabuzda, D.C., Cawthorne, T.V., Roberts, D.H. and Wardle J.F.C. (1992). A survey of the milliarcsecond polarization properties of BL Lacertae objects at 5 GHz, *Astrophys. J.* **388**, 40.

Gabuzda, D.C., Mullan, C.M., Cawthorne, T.V., Wardle, J.F.C. and Roberts, D.H. (1994). Evolution of the milliarcsecond total intensity and polarization structures of BL Lacertae objects, *Astrophys. J.* **435**, 140.

Garrington, S.T., Conway, R.G. and Leahy, J.P. (1991). Asymmetric depolarization in double radio sources with one-sided jets, *Mon. Not. Roy. Astr. Soc.* **250**, 171.

Garrington, S.T., Holmes, G.F. and Saikia, D.J. (1995). Depolarization studies of radio sources and the unified scheme, in EFP95, 111.

Garrington, S.T., Leahy, J.P., Conway, R.G. and Laing, R.A. (1988). A systematic asymmetry in the polarization properties of double radio sources with one jet, *Nature* **331**, 147.

Gehrels, N., Chipman, E. and Kniffen, D. (1994). The Compton gamma ray observatory, *Astrophys. J. Suppl.* **92**, 351.

Georgantopoulos, I., Stewart, G.C., Shanks, T., Griffiths, R.E. and Boyle, B.J. (1993). *ROSAT* survey – II. Observations of the isotropy of the 1–2 keV X-ray background, *Mon. Not. Roy. Astr. Soc.* **262**, 619.

George, I.M. and Fabian, A.C. (1991). X-ray reflection from cold matter in active galactic nuclei and X-ray binaries, *Mon. Not. Roy. Astr. Soc.* **249**, 352.

Ghisellini, G., George, I.M., Fabian, A.C. and Done, C. (1991). Anisotropic inverse Compton scattering, *Mon. Not. Roy. Astr. Soc.* **248**, 14.

Ghisellini, G., Guilbert, P. and Svensson, R. (1988). The synchrotron boiler, *Astrophys. J.* **334**, L5.

Ghisellini, G. and Maraschi, L. (1989). Bulk acceleration in relativistic jets, *Astrophys. J.* **340**, 181.

Ghisellini, G., Padovani, P., Celotti, A. and Maraschi, L. (1993). Relativistic bulk motion in active galactic nuclei, *Astrophys. J.* **407**, 65.

Ghosh, K.K. and Soundararajaperumal, S. (1992). X-ray spectra of eight Seyfert galaxies, *Mon. Not. Roy. Astr. Soc.* **259**, 175.

Giacconi, R., Branduardi, G., Briel, U., Epstein, A., Fabricant, D., Feigelson, E. *et al.*

(1979). The *EINSTEIN* (*HEAO 2*) X-ray observatory, *Astrophys. J.* **230**, 540.

Giacconi, R., Gursky, H., Paolini, F. and Rossi, B. (1962). Evidence for X-rays from sources outside the solar system, *Phys. Rev. Lett.* **9**, 439.

Giacconi, R., Murray, S., Gursky, H., Kellogg, E. and Schreier, E. (1974). The third Uhuru catalogue of X-ray sources, *Astrophys. J.* **188**, 667.

Gioia, I.M., Maccacaro, T., Schild, R.E., Wolter, A., Stocke, J.T., Morris, S.L. and Henry, J.P. (1990). The *EINSTEIN* observatory extended medium-sensitivity survey. I. X-ray data and analysis., *Astrophys. J. Suppl.* **72**, 567.

Giommi, P., Barr, P., Garilli, B., Maccagni, D. and Pollock, A.M.T. (1990). A study of BL Lacertae-type objects with *EXOSAT*. I. Flux correlations, luminosity variability, and spectral variability, *Astrophys. J.* **356**, 432.

Gnedin, Y.N., and Silant'ev, N.A. (1978). Polarization effects in the emission of a disk of accreting matter, *Soviet Astron.* **22**, 325.

Goldschmidt, P., Miller, L., La Franca, F. and Cristiani, S. (1992). The high surface density of bright ultraviolet-excess quasars, *Mon. Not. Roy. Astr. Soc.* **256**, 65p.

Goodrich, R.W., Veilleux, S. and Hill, G.J. (1994). Infrared spectroscopy of Seyfert 2 galaxies: a look through the obscuring torus?, *Astrophys. J.* **422**, 521.

Gopal-Krishna (1995). The case for unification, *Proc. Nat. Acad. Sci. USA* **92**, 11399.

Gopal-Krishna, Kulkarni, V.K. and Wiita, P. (1996). The linear sizes of quasars and radio galaxies in the unified scheme, *Astrophys. J.* **463**, L1.

Grandi, P., Tagliaferri, G., Giommi, P., Barr, P. and Palumbo, G.G.C (1992). X-ray luminosity and spectral variability of hard X-ray-selected active galactic nuclei, *Astrophys. J. Suppl.* **82**, 93.

Green, P.J., Schartel, N., Anderson, S.F., Hewett, P.C., Foltz, C.B., Brinkmann, W. *et al.* (1995). The soft X-ray properties of a large optical QSO sample: *ROSAT* observations of the Large Bright Quasar Survey, *Astrophys. J.* **450**, 51.

Greenstein, J.L. and Schmidt, M. (1964). The quasi-stellar radio sources 3C 48 and 3C 273, *Astrophys. J.* **140**, 1.

Gregg, M.D., Becker, R.H., White, R.L., Helfand, D.J., McMahon, R.G. and Hook, I.M. (1996). The FIRST bright QSO survey, *Astron. J.* **112**, 407.

Griffiths, R.E., Murray, S.S., Giacconi, R., Bechtold, J., Murdin, P., Smith, M. *et al.* (1983). The optical identification content of the *EINSTEIN* observatory deep survey of a region in Pavo, *Astrophys. J.* **269**, 375.

Grindlay, J.E., Steiner, J.E., Forman, W.R., Canizares, C.R. and McClintock, J.E. (1980). Low-redshift X-ray selected quasars: new clues to the QSO phenomenon, *Astrophys. J.* **239**, L43.

Guilbert, P.W., Fabian, A.C. and Rees, M.J. (1983). Spectral and variability constraints on compact sources, *Mon. Not. Roy. Astr. Soc.* **205**, 593.

Guilbert, P.W. and Rees, M.J. (1988). 'Cold' material in non-thermal sources, *Mon. Not. Roy. Astr. Soc.* **233**, 475.

Gunn, J.E. and Peterson, B.A. (1965). On the density of neutral hydrogen in intergalactic space, *Astrophys. J.* **142**, 1633.

Gurvits, L.I. (1993). Milliarcsecond structures of AGN at different redshifts, in DB93, p. 380.

Haenhelt, M.G. and Rees, M.J. (1993). Formation of nuclei in newly-formed galaxies and the evolution of the quasar population, *Mon. Not. Roy. Astr. Soc.* **263**, 168.

Hamilton, T.T. and Helfand, D.J. (1987). The origin of the diffuse X-ray background, *Astrophys. J.* **318**, 93.

Hargrave, P.J. and Ryle, M. (1974). Observations of Cygnus A with 5 km radio telescope, *Mon. Not. Roy. Astr. Soc.* **166**, 305.

Hartman, R.C., Bertsch, D.L., Fichtel, C.E., Hunter, S.D., Kanbach, G., Kniffen, D.A. *et al.* (1992). Detection of high-energy gamma radiation from quasar 3C 279 by the *EGRET* telescope on the Compton Gamma Ray observatory, *Astrophys. J.* **385**, L1.

Hartwick, F.D.A. and Schade, D. (1990). The space distribution of quasars, *Ann. Rev. Astron. Astrophys.* **28**, 437.

Hasinger, G., Burg, R., Giacconi, R., Hartner, G., Schmidt, M., Trumper, J. and Zamorani, G. (1993). A deep X-ray survey in the Lockman hole and the soft X-ray $\log N - \log S$, *Astron. Astrophys.* **238**, 1.

Hasinger, G., Burg, R., Giacconi, R., Hartner, G., Schmidt, M., Trumper, J. and Zamorani, G. (1994). A deep X-ray survey in the Lockman hole and the soft X-ray $\log N - \log S$, *Erratum, Astron. Astrophys.* **291**, 348.

Hazard, C., Mackey, M.B. and Shimmins, A.J. (1963). Investigations of the radio source 3C 273 by the method of lunar occultations, *Nature* **197**, 1037.

Hazard, C. and McMahon, R.G. (1985). New quasars with $z = 3.4$ and 3.7 and the surface density of very high redshift quasars, *Nature* **314**, 21.

Hazard, C., McMahon, R.G. and Sargent W.L.W. (1986). A QSO with redshift 3.8 found on a UK Schmidt telescope IIIa–F prism plate, *Nature* **322**, 38.

Henriksen, M.J., Marshall, F.E. and Mushotzky, R.F. (1984). An X-ray survey of variable radio bright quasars, *Astrophys. J.* **284**, 49.

Herterich, K. (1974). Absorption of gamma rays in intense X-ray sources, *Nature* **250**, 311.

Hewett, P.C., Foltz, C.B. and Chaffee, F.H. (1993). The evolution of bright, optically selected QSOs, *Astrophys. J.* **406**, L43.

Hewett, P.C., Foltz, C.B. and Chaffee, F.H. (1995). The large bright quasar survey. VI. Quasar catalog and survey parameters, *Astron. J.* **109**, 1498.

Hewitt, A. and Burbidge, G.R. (1991). An optical catalog of emission-line objects similar to quasi-stellar objects, *Astrophys. J. Suppl.* **75**, 297.

Hewitt, A. and Burbidge, G.R. (1993). A revised and updated catalog of quasi-stellar objects, *Astrophys. J. Suppl.* **87**, 451.

Hewitt, J.N., Turner, E.L., Schneider, D.P., Burje, B.F., Langston, G.I. and Lawrence, C.R. (1988). Unusual radio source MG 1131+0456: a possible Einstein ring, *Nature* **333**, 537.

Hill, G.J. and Lilly, S.J. (1991). A change in the cluster environments of radio galaxies with cosmic epoch, *Astrophys. J.* **367**, 1.

Hogan, C.J., Narayan, R. and White, S.D.M (1989). Quasar lensing by galaxies, *Nature* **339**, 106.

Hook, I.M., McMahon, R.G., Boyle, B.J, and Irwin, M.J. (1994). The variability of optically selected quasars, *Mon. Not. Roy. Astr. Soc.* **268**, 305.

Hough, D.H., Vermeulen, R.C. and Readhead, A.C.S. (1993). The search for superluminal motion in a complete sample of lobe-dominated quasars, in DB93, p. 193.

Hough, D.S. and Readhead, A.C.S (1989). Complete sample of double lobed radio quasars for VLBI tests of source models: definitions and statistics, *Astron. J.* **98**, 1208.

Hoyle, F. and Burbidge, G.R. (1966). On the nature of quasi-stellar objects, *Astrophys. J.* **144**, 534.

Hoyle, F., Burbidge, G.R. and Sargent, W.L.W. (1966). On the nature of quasi-stellar sources, *Nature* **209**, 751.

Hoyle, F. and Fowler, W.A. (1962). On the nature of strong radio sources, *Mon. Not. Roy. Astr. Soc.* **125**, 169.

Hoyle, F. and Fowler, W.A. (1963). Nature of strong radio sources, *Nature* **197**, 533.

Hoyle, F. and Fowler, W.A. (1967). Gravitational redshifts in quasi-stellar objects, *Nature* **213**, 373.

Hoyle, F. and Narlikar, J.V. (1964). A new theory of gravitation, *Proc. Roy. Soc. Lond. A* **282**, 191.

Hoyle, F. and Narlikar, J.V. (1966). A conformal theory of gravitation, *Proc. Roy. Soc. Lond. A* **294**, 138.

Hubble, E. (1929). A relation between distance and radial velocity among extragalactic nebulae, *Proc. Nat. Acad. Sci. USA* **15**, 168.

Huchra, J. and Burg, R. (1992). The spatial distribution of active galactic nuclei. I. The density of Seyfert galaxies and liners, *Astrophys. J.* **393**, 90.

Huchra, J.P., Gorenstein, M., Kent, S., Shapiro, I., Smith, G., Horine, E. and Perley, R. (1985). 2227+0305: a new and unusual gravitational lens, *Astron. J.* **90**, 691.

Hughes, D.H., Robson, E.I., Dunlop, J.S. and Gear, W.K. (1993). Thermal dust emission from quasars – I. Submillimeter spectral indices, *Mon. Not. Roy. Astr. Soc.* **263**, 607.

Hutchings, D.J., Holtzman, J., Sparks, W.B., Morris, S.C., Hanisch, R.J. and Mo. J. (1994). HST imaging of quasi-stellar objects with WFPC 2, *Astrophys. J.* **429**, L1.

Hutchings, J.B. and Neff, S.G. (1992). The optical imaging of QSOs with 0.5 arcsec resolution, *Astron. J.* **104**, 1.

Impey, C.D. and Neugebauer, G. (1988). Energy distributions of blazars, *Astron. J.* **95**, 307.

Inoue, H. (1989). *GINGA* observations of the X-ray spectra of AGN, in HB89, 783.

Irwin, M., McMahon, R.G. and Hazard, C. (1991). APM optical surveys for high redshift quasars, in C91, p. 117.

Jackson, N., Browne, I.W.A. and Warwick, R.S. (1993). The soft X-ray spectra of quasars and X-ray beaming models, *Astron. Astrophys.* **274**, 79.

Jennison, R.C. and Das Gupta, M.K. (1953). Fine structure of the extragalactic radio source Cygnus I, *Nature* **172**, 996.

Johnson, W.N., Dermer, C.D., Kinzer, R.L., Kurfess, J.D., Strickman, M.S., McNaron-Brown, K., Jourdain, E., Jung, G.V., Grabelsky, D.A., Purcell, W.R. and Ulmer, M.P. (1995). *OSSE* observations of 3C 273, *Astrophys. J.* **445**, 182.

Jourdin, E., Bassani, L., Bouchet, L., Mandrou, P., Ballet, J., Lebrun, F. *et al.* (1992). *SIGMA* observation of a steep spectral shape in NGC 4151 above 35 keV, *Astron. Astrophys.* **256**, L38.

Königl, A. (1996). Active galactic nuclei in the extreme ultraviolet, in BM96, p. 27.

Kafka, P. (1967). How to count quasars, *Nature* **213**, 346.

Kapahi, V.K. (1990). On the unification of quasars and radio galaxies, in ZP90, p. 304.

Kapahi, V.K. and Murphy, D.W. (1990). Beaming, unification, and H_0 parsec-scale radio jets, in ZP90, p. 313.

Kapahi,, V.K. and Saikia, D.J. (1982). Relativistic beaming in the central components of double radio quasars, *J. Astrophys. Astron.*, 465.

Karlsson, K.G. (1977). On the existence of significant peaks in quasar redshift distribution, *Astron. Astrophys.* **58**, 237.

Katz, J.I. (1976). Non-relativistic Compton scattering and models of quasars, *Astrophys. J.* **206**, 910.

Kellermann, K.I. (1966). On the interpretation of radio-source spectra and the evolution of radio galaxies and quasi-stellar sources, *Astrophys. J.* **146**, 621.

Kellermann, K.I. and Pauliny-Toth, I.I.K. (1969). The spectra of opaque radio sources, *Astrophys. J.* **155**, L71.

Kellermann, K.I. and Pauliny-Toth, I.I.K. (1981). Compact radio sources, *Ann. Rev. Astron. Astrophys.* **19**, 373.

Kellermann, K.I., Sramek, R.A., Schmidt, M., Green, R.F. and Shaffer, D.B. (1994). The radio structure of radio loud and radio quiet quasars in the Palomar Bright Quasar Survey, *Astron. J.* **108**, 1163.

Kellermann, K.I., Sramek, R.A., Schmidt, M., Shaffer, D.B, and Green, R. (1989). VLA observations of the objects in the Palomar bright quasar survey, *Astron. J.* **98**, 1195.

Kembhavi, A.K. (1978). Zero mass surfaces and cosmological singularities, *Mon. Not. Roy. Astr. Soc.* **185**, 807.

Kembhavi, A.K. (1986). Radio induced X-ray emission in radio quasars, in SK86, p. 239.

Kembhavi, A.K. (1993). X-ray beaming in radio quasars, *Mon. Not. Roy. Astr. Soc.* **264**, 683.

Kembhavi, A.K. and Fabian, A.C. (1982). X-ray quasars and the X-ray background, *Mon. Not. Roy. Astr. Soc.* **198**, 921.

Kembhavi, A.K., Feigelson, E.D. and Singh, K.P. (1986). X-ray and radio core emission in radio quasars, *Mon. Not. Roy. Astr. Soc.* **220**, 51.

Kembhavi, A.K., Wagh, S.M. and Narasimha, D. (1989). Relativistic beaming of X-ray quasars, in OM89, p. 209.

Kennefick, J.D., Djorgovski, S.G. and De Carvalho, R.R. (1995). The luminosity function of $z > 4$ quasars from the second Palomar sky survey, *Astron. J.* **110**, 2553.

Kii, T., Williams, O.R., Ohashi, T., Awaki, H., Hayashida, K., Inoue, H. *et al.* (1991). X-ray continuum and evidence for an iron emission line from the quasar E 1821+643, *Astrophys. J.* **367**, 455.

Killeen, N.E.B., Bicknell, G.V. and Ekers, R.D. (1986). The radio galaxy IC 4296 (PKS 1333–33). I. Multifrequency VLA observations, *Astrophys. J.* **302**, 306.

Kinzer, R.L., Johnson, W.N., Dermer, C.D., Kurfess, J.D., Strickman, M.S., Grove, J.E. and Kroeger, R.A. (1995). *OSSE* observations of gamma-ray emission from Centaurus A, *Astrophys. J.* **449**, 105.

Kniffen, D.A., Bertsch, D.L., Fichtel, C.E., Hartman, R.C., Hunter, S.D., Kanbach, G. *et al.* (1993). Time variability in the gamma-ray emission of 3C 279, *Astrophys. J.* **411**, 133.

Köhler, T., Groote, D., Reimers, D. and Wisotzki, L. (1997). The local luminosity function of QSOs and Seyfert 1 nuclei, *Astron. Astrophys.* **325**, 502.

Kollgaard, R.I., Wardle, J.F.C., Roberts, D.H. and Gabuzda, D.C. (1992). Radio constraints on the nature of BL Lacertae objects and their parent population, *Astron. J.* **104**, 1687.

Kompaneets, A.S. (1957). The establishment of thermal equilibrium between quanta and electrons, *Soviet Phys. JETP* **4**, 730.

Koo, D.C. and Kron, R.G. (1982). QSO counts: a complete survey of stellar objects to $B = 23$, *Astron. J.* **105**, 107.

Koo, D.C. and Kron, R.G. (1988). Spectroscopic survey of QSOs to $B = 22.5$: the luminosity function, *Astrophys. J.* **325**, 92.

Koo, D.C., Kron, R.G. and Cudworth, K.M. (1986). Quasars to $B > 22.5$ in selected area 57: a catalog of multicolor photometry, variability, and astrometry, *Proc. Astron. Soc. Pacific* **98**, 285.

Koratkav, A.P., Kinney, A.L. and Bohlin, R.C. (1989). Search for partial systematic line edges in nearby quasars, *Astrophys. J.* **400**, 435.

Koyama, K. (1992). *GINGA* results of type 2 AGNs and their contribution to the cosmic X-ray background, in *X-ray Emission from Active Galactic Nuclei and the Cosmic X-ray Background*, eds. W. Brinkmann and J. Trümper, MPE–Report 235, p. 74.

Koyama, K., Inoue, H., Tanaka, Y., Awaki, H., Takano, S., Ohashi, T. and Matsuoka, M. (1989). An intense iron line emission from NGC 1068, *Proc. Astron. Soc. Japan* **41**, 731.

Kraushaar, W.L., Clark, G.W., Garmire, G.P., Borken, R., Higbie, P., Leong, V., *et al.* (1972). High energy cosmic gamma-ray observations from the *OSO 3* satellite, *Astrophys. J.* **177**, 341.

Krichbaum, T.P., Quirrenbach, A. and Witzel, A. (1992). Intraday variability and compact extragalactic radio sources, in VV92 , p. 331.

Krichbaum, T.P., Witzel, A., Graham, A., Standke, K.J., Schwartz, R., Lochner, R. *et al.* (1993). First 43 GHz VLBI-observations with the 30 m radio telescope at Pico Veleta, *Astron. Astrophys.* **275**, 375.

Kriss, G.A. and Canizares, C.R. (1982). Optical and X-ray properties of X-ray selected active galactic nuclei, *Astrophys. J.* **261**, 51.

Kriss, G.A. and Canizares, C.R. (1985). X-ray properties of quasars and results from a deep X-ray survey of optically selected objects, *Astrophys. J.* **297**, 177.

Kriss, G.A., Canizares, C.R. and Ricker, G.R. (1980). X-ray observations of Seyfert galaxies

with the *EINSTEIN* observatory, *Astrophys. J.* **242**, 492.

Kristian, J. (1973). Quasars as events in the nuclei of galaxies: the evidence from direct photographs, *Astrophys. J.* **179**, L61.

Krolik, J.H. and Kallman, T.R. (1987). Fe K features as probes of the nuclear reflection region in Seyfert galaxies, *Astrophys. J.* **320**, L5.

Kron, R.G. and Chiu, L.-T.G. (1981). Stars with zero proper motion and the number of faint objects, *Proc. Astron. Soc. Pacific* **93**, 397.

Kruiper, J.S., Urry, C.M. and Canizares, C.R. (1990). Soft X-ray properties of Seyfert galaxies, I. Spectra, *Astrophys. J. Suppl.* **74**, 347.

Ku, W.H.-M., Helfand, D.J. and Lucy, L.B. (1980). The X-ray properties of quasars, *Nature* **288**, 323.

Kuhn, O., Bechtold, J., Cutri, R., Elvis, M. and Rieke, M. (1995). The spectral energy distribution of the $z = 3$ quasar: HS 1946 + 7658, *Astrophys. J.* **438**, 643.

Kuhr, H., Pauliny-Toth, I.I.K., Witzel, A. and Schmidt, J. (1981). The 5 GHz strong source surveys. V. Survey of the area between declinations 70° and 90°, *Astron. J.* **86**, 854.

Laing, R.A. (1984). Interpretation of radio polarization data, in BE84, p. 90.

Laing, R.A. (1988). The sidedness of jets and depolarization in powerful extragalactic radio sources, *Nature* **331**, 149.

Laing, R.A. (1993). Radio observations of jets: large scales, in BLO93, p. 95.

Lampton, M., Margon, B. and Bowyer, S. (1976). Parameter estimation in X-ray astronomy, *Astrophys. J.* **208**, 177.

Landau, R., Golisch, B., Jones, T.J., Jones, T.W., Pedelty, J., Rudnick, L. (1986). Active extragalactic sources: nearly simultaneous observations from 20 centimeters to 1400 Å, *Astrophys. J.* **308**, 78.

Lanzetta, K.M., Bowen, D.V., Tytler, D. and Webb, J.K. (1995). The gaseous extent of galaxies and the origin of the Lyman-alpha absorption systems: a survey of galaxies in the field of Hubble Space Telescope spectroscopic target QSOs, *Astrophys. J.* **442**, 538.

Laor, A. (1990). Massive accretion disks–III. Comparision with the observations, *Mon. Not. Roy. Astr. Soc.* **246**, 369.

Laor, A., Fiore, F., Elvis, M., Wilkes, B.J. and McDowell, J.C. (1994). The soft X-ray properties of a complete sample of optically selected quasars I. First results, *Astrophys. J.* **435**, 611.

Laor, A., Netzer, H. and Piran, T. (1990). Massive thin accretion disks – II. Polarization, *Mon. Not. Roy. Astr. Soc.* **242**, 560.

Laurent-Muehleisen, S.A., Kollgaard, R.I., Moellenbrock, G.A. and Feigelson, E.D. (1993). Radio morphology and parent population of X-ray selected BL Lacertae objects, *Astron. J.* **106**, 875.

Lawrence, A. (1987). Classification of active galaxies and the prospect of a unified phenomenology, *Proc. Astron. Soc. Pacific* **99**, 309.

Lawson, A.J., Turner, M.J.L., Williams, O.R., Stewart, G.C. and Saxton, R.D. (1992). Radio-loud and radio-quiet quasars observed by *EXOSAT, Mon. Not. Roy. Astr. Soc.* **259**, 743.

Ledden, J.E. and O'Dell, S.L. (1985). The radio–optical–X-ray spectral flux distribution of blazars, *Astrophys. J.* **298**, 630.

Lemaitre, Abbé G. (1931). A homogeneous universe of constant mass and increasing radius accounting for the radial velocity of extragalactic nebulae (translation of original paper published in French published in 1927), *Mon. Not. Roy. Astr. Soc.* **91**, 483.

Liedahl, D.A., Paerels, F., Hur, M.Y., Kahn, S.M., Fruscione, A. and Bowyer, S. (1996). Extreme ultraviolet spectroscopy of the Seyfert 1 galaxy Markarian 478, in BM96, p. 57.

Lightman, A.P. and White, T.E. (1988). Effects of cold matter in active galactic nuclei: a broad hump in the X-ray spectra, *Astrophys. J.* **335**, 57.

Lightman, A.P. and Zdziarski, A.A. (1987). Pair production and Compton scattering in compact sources and comparison to observations of active galactic nuclei, *Astrophys. J.* **319**, 643.

Lin, Y.C., Bertsch, D.L., Dingus, B.L., Fichtel, C.E., Hartman, R.C., Hunter, S.D. *et al.* (1993). *EGRET* limits on high-energy gamma-ray emission from X-ray- and low-energy gamma-ray-selected Seyfert galaxies, *Astrophys. J.* **416**, L53.

Liu, F.K. and Xie, G.Z. (1992). A finding list of extragalactic radio jets and statistical results, *Astron. Astrophys. Suppl.* **95**, 249.

Longair, M.S. (1979). Radio astronomy and cosmology, in GLR78, p. 127.

Longair, M.S. and Scheuer, P.A.G. (1970). The luminosity–volume test for quasi-stellar objects, *Mon. Not. Roy. Astr. Soc.* **151**, 45.

Lynden-Bell, D. (1971). A method of allowing for known observational selection in small samples applied to 3CR quasars, *Mon. Not. Roy. Astr. Soc.* **155**, 95.

Lynds, R. and Petrosian, V. (1972). On the ability of the luminosity–volume test to reveal the statistical evolution of the luminosity of quasi-stellar sources, *Astrophys. J.* **175**, 591.

Lynds, R. and Petrosian, V. (1989). Luminous arcs in clusters of galaxies, *Astrophys. J.* **336**, 1.

Maccacaro, T., Della Ceca, R., Gioia, I.M., Morris. S.L. and Stocke, J.T. (1991). The properties of X-ray selected active galactic nuclei. I. Luminosity function, cosmological evolution and contribution to the diffuse X-ray background, *Astrophys. J. Suppl.* **374**, 117.

Madau, P., Ghisellini, G. and Fabian, A.C. (1994). The unified Seyfert scheme and the origin of the cosmic X-ray background, *Mon. Not. Roy. Astr. Soc.* **270**, L17.

Madejski, G.M., Done, C., Turner, T.J., *et al.* (1994). Solving the mystery of the periodicity of in the Seyfert galaxy NGC 6814 , in CB94, p. 127.

Madejski, G.M. and Schwarz, D.A. (1983). X-ray studies of BL Lacertae objects with the *EINSTEIN* observatory: confrontation with the synchrotron self-Compton predictions., *Astrophys. J.* **275**, 467.

Mahabal, A.A. (1998). The optical and near infrared morphology of radio galaxies, Ph. D. thesis (unpublished).

Maisack, M., Johnson, W.N., Kinzer, R.L., Strickman, M.S., Kurfess, J.D., Cameron, R.A. *et al.* (1993). *OSSE* observations of NGC 4151, *Astrophys. J.* **407**, L61.

Majewski, S., Munn, J.A., Kron, R.G., Bershady, J.J., Smetanka, J.J. and Koo, D.C. (1991). A proper motion and variability survey to $B = 22.5$, in C91, p. 55.

Malkan, M.A. (1983). The ultraviolet excess of luminous quasars II. Evidence for massive accretion disks, *Astrophys. J.* **268**, 582.

Malkan, M.A. and Sargent, W.L.W. (1982). The ultraviolet excess of Seyfert 1 galaxies and quasars, *Astrophys. J.* **254**, 22.

Mannheim, K. and Biermann, P.L. (1992). Gamma-ray flaring of 3C 279: a proton-initiated cascade in the jet?, *Astron. Astrophys.* **253**, L21.

Marano, B., Zamorani, G. and Zitelli, V. (1988). A new sample of quasars to $B = 22.0$, *Mon. Not. Roy. Astr. Soc.* **232**, 111.

Maraschi, L., Ghisellini, G. and Celotti, A. (1992). A jet model for the gamma-ray emitting blazar 3C 279, *Astrophys. J.* **397**, L5.

Maraschi, L., Grandi, P., Urry, C.M., Wehrle, A.E., Madejski, G.M. and Fink, H.H. (1994). The 1993 multiwavelength campaign on 3C 279: the radio to gamma-ray energy distribution in low state, *Astrophys. J.* **435**, L91.

Marscher, A.P. (1987). Synchro–Compton emission from superluminal sources, in ZP87, p. 280.

Marscher, A.P. (1993). Compact extragalactic radio jets, in BLO93, p. 73.

Marshall, F.E., Boldt, E.A., Holt, S.S., Miller, R.B., Mushotzky, R.F., Rose, L.A. *et al.* (1980). The diffuse X-ray background spectrum from 3 to 50 keV, *Astrophys. J.* **235**, 4.

Marshall, F.E., Holt, S.S. and Mushotzky, R.F. (1983). Rapid X-ray variability from the

Seyfert 1 galaxy NGC 4051, *Astrophys. J.* **269**, L31.

Marshall, H.L. (1985). The evolution of optically selected quasars with $z < 2.2$ and $B < 20$, *Astrophys. J.* **299**, 109.

Marshall, H.L. (1987). The radio luminosity function of optically selected quasars, *Astrophys. J.* **316**, 84.

Marshall, H.L. (1996). Variability and spectra of AGN in the EUV and the relation to other bands, in BM96, p. 63.

Marshall, H.L., Avni, Y., Braccesi, A., Huchra, J.P., Tananbaum, H., Zamorani, G. and Zitelli, V. (1984). A complete sample of quasars at $B = 19.80$, *Astrophys. J.* **283**, 50.

Marshall, H.L., Avni, Y., Tananbaum, H. and Zamorani, G. (1983). Analysis of complete quasar samples to obtain parameters of luminosity and evolution functions, *Astrophys. J.* **269**, 35.

Marshall, H.L., Tananbaum, H., Zamorani, G. and Huchra, J.P. (1983). Optical and X-ray observations of faint quasars in an optically selected sample, *Astrophys. J.* **269**, 42.

Marshall, N., Warwick, R.S. and Pounds, K.A. (1981). The variability of X-ray emission from active galaxies, *Mon. Not. Roy. Astr. Soc.* **194**, 987.

Masnou, J.-L., Wilkes, B.J., Elvis, M., McDowell, J.C. and Arnaud, K.A. (1992). The soft X-ray excess in *EINSTEIN* quasar spectra, *Astron. Astrophys.* **253**, 35.

Mathez, G. (1976). Evolution of the luminosity function of quasars. A model with constant density and luminosity evolution, *Astron. Astrophys.* **53**, 15.

Mathez, G. (1978). Evolution of the luminosity function of quasars, *Astron. Astrophys.* **68**, 17.

Mathur, S., Elvis, M. and Wilkes, B. (1995). Testing unified X-ray/ultraviolet absorber models with NGC 5548, *Astrophys. J.* **452**, 230.

Matsuoka, M., Ikegami, T., Inoue, H. and Koyama, K. (1986). Detection of an intense iron line at 6.4 keV in the X-ray spectrum of NGC 4151, *Proc. Astron. Soc. Pacific* **38**, 285.

Matsuoka, M., Yamauchi, M., Piro, L. and Murakami, T. (1990). X-ray spectral variability and complex absorption in the Seyfert 1 galaxies NGC 4051 and MCG-6-30-15, *Astrophys. J.* **361**, 440.

Matthews, T.A. and Sandage, A.R. (1963). Optical identification of 3C 48, 3C 196, and 3C 286 with stellar objects, *Astrophys. J.* **138**, 30.

Mattox, J.R., Bertsch, D.L., Chiang J., Fichtel, C.E., Hartman, R.C., Hunter, S.D. *et al.* (1995). The *EGRET* detection of quasar 1633+382, *Astrophys. J.* **410**, 609.

McAlary, C.W. and Rieke, G.H. (1988). A near-infrared and optical study of X-ray selected Seyfert galaxies. II. Models and interpretation, *Astrophys. J.* **333**, 1.

McCammon, D. and Sanders, W.T. (1990). The soft X-ray background and its origins, *Ann. Rev. Astron. Astrophys.* **28**, 657.

McDowell, J.C., Elvis, M., Wilkes, B.J., Willner, S.P., Oey, M.S., Polomski, E. *et al.* (1989). Weak bump quasars, *Astrophys. J.* **345**, L13.

McHardy, I.M. (1989). X-ray variability of AGN, in HB89, 1111.

McMahon, R.G. (1991). Radio loud quasars with high redshift, in C91, p. 129.

McMahon, R.G. and Irwin, M.J. (1992). APM surveys for high redshift quasars, in MT92, p. 147.

McNaron-Brown, K., Johnson, W.N., Jung, G.V., Kinzer, R.L., Kurfess, J.D., Strickman, M.S. *et al.* (1995). *OSSE* observations of blazars, *Astrophys. J.* **451**, 575.

Miley, G.K. (1980). The structure of extended extragalactic radio sources, *Ann. Rev. Astron. Astrophys.* **18**, 165.

Miley, G.K., Hogg, D.E. and Basart, J. (1970). The fine scale structure of Virgo A, *Astrophys. J.* **159**, 219.

Miller J.S. (1994). The unification of active galaxies: Seyferts and beyond, in BDQ94, p. 149.

Miller, J.S. and Antonucci, R.R.J (1983). Evidence for a highly polarized continuum in the nucleus of NGC 1068, *Astrophys. J.* **271**, L7.

Miller, J.S., Goodrich, R.W. and Mathews, W. (1991). Multidirectional views of the active nucleus of NGC 1068, *Astrophys. J.* **378**, 47.

Miller, L., Peacock, J.A. and Mead, A.R.G. (1990). The bimodal radio luminosity function of quasars, *Mon. Not. Roy. Astr. Soc.* **244**, 207.

Mitchell, K.J., Warnock III, A. and Usher, P.D. (1984). A medium bright quasar sample: new quasar surface densities in the magnitude range $16.4 < B < 17.65$, *Astrophys. J.* **287**, L3.

Mitchell, P.S., Miller, L. and Boyle, B.J. (1990). Luminous quasars at high redshift, *Mon. Not. Roy. Astr. Soc.* **244**, 1.

Mittaz, J.P.D., Lieu, R., Bowyer, S., Hwang, C.-Y. and Lewis, J. (1996). EUV and soft X-ray evidence for partially ionized gas in active galactic nuclei, in BM96, p. 45.

Moffett, A.T. (1968). Strong non-thermal radio emission from galaxies, in S68, p. 211.

Moore, R.L., Schmidt, G.D. and West S.C (1987). The hiss of BL Lacertae, *Astrophys. J.* **314**, 176.

Morgan, P., Swings, J.-P., Borgeest, U., Courvoisier, T.J.-L., Kayser, R. *et al.* (1987). A new case of gravitational lensing, *Nature* **329**, 695.

Morisawa, K., Matsuoka, M., Takahara, F. and Piro, L. (1990). Effect of X-ray spectra of Seyfert galaxies on the cosmic X-ray background, *Astron. Astrophys.* **236**, 299.

Morris, S.L., Weymann, R.J., Savage, B.D., Gilliland, R.L. (1991). First results from the Goddard high resolution spectrograph: the Galactic halo and the Lyman-alpha forest at low redshift in 3C 273, *Astrophys. J.* **377**, L21.

Morrison, P. (1969). Are quasi-stellar radio sources giant pulsars?, *Astrophys. J.* **157**, 273.

Morrison, R. and McCammon, D. (1983). Interstellar photoelectric absorption cross sections, 0.03–10 keV, *Astrophys. J.* **270**, 119.

Mulchaey, J.S., Mushotzky, R.F. and Weaver, K.A. (1992). Hard X-ray tests of the unified model for an ultraviolet-detected sample of Seyfert 2 galaxies, *Astrophys. J.* **390**, L69.

Murphy, D.W., Browne, I.W.A. and Perley, R.A. (1993). VLA observations of a complete sample of core-dominated radio sources, *Mon. Not. Roy. Astr. Soc.* **264**, 298.

Mushotzky, R.F. (1982). The X-ray spectrum and time variability of narrow emission line galaxies, *Astrophys. J.* **256**, 92.

Mushotzky, R.F. (1984). X-ray spectra and time variability of active galactic nuclei, *Adv. Space Res.* **3**, 157.

Mushotzky, R.F., Done, C. and Pounds, K.A. (1993). X-ray spectra and time variability of active galactic nuclei, *Ann. Rev. Astron. Astrophys.* **31**, 717.

Mushotzky, R.F., Holt, S.S. and Serlemitsos, P.J. (1978a). X-ray observation of a flare in NGC 4151 from OSO 8, *Astrophys. J.* **225**, L115.

Mushotzky, R.F., Marshall, F.E., Boldt, E.A., Holt, S.S. and Serlemitsos, P.J. (1980). *HEAO 1* spectra of X-ray emitting Seyfert 1 galaxies, *Astrophys. J.* **235**, 377.

Mushotzky, R.F., Serlemitsos, P.J., Becker, R.H., Boldt, E.A. and Holt, S.S (1978b). The X-ray emitting galaxy Centaurus A, *Astrophys. J.* **220**, 790.

Muxlow, T.W.B. and Garrington, S.T. (1991). Observations of large scale extragalactic jets, in H91, p. 52.

Nandra, K., Fabian, A.C., Brandt, W.N., Kunieda, H., Matsuoka, M., Mihara, T. *et al.* (1995). X-ray spectra of two quasars at $z > 1$, *Mon. Not. Roy. Astr. Soc.* **276**, 1.

Nandra, K. and Pounds, K.A. (1992). Highly ionized gas in the nucleus of the active galaxy MCG-6-30-15, *Nature* **359**, 215.

Nandra, K. and Pounds, K.A. (1994). *GINGA* observations of the X-ray spectra of Seyfert galaxies, *Mon. Not. Roy. Astr. Soc.* **268**, 405.

Nandra, K., Pounds, K.A. and Stewart, G.C. (1990). The X-ray spectrum of MCG-6-30-15 and its temporal variability, *Mon. Not. Roy. Astr. Soc.* **242**, 660.

Nandra, K., Pounds, K.A., Fabian, A.C. and Rees, M.J. (1989). Detection of iron features in the X-ray spectrum of the Seyfert 1 galaxy

MCG-6-30-15, *Mon. Not. Roy. Astr. Soc.* **236**, 39p.

Narlikar, J.V. (1977). Two astrophysical applications of conformal gravity, *Ann. Phys.* **107**, 325.

Narlikar, J.V. (1989). Noncosmological redshifts, *Space Sci. Rev.* **50**, 523.

Narlikar, J.V. and Das, P.K. (1980). Anomalous redshift of QSOs, in AP80, p. 127.

Narlikar, J.V. and Subramaniam, K. (1982). Observational limitation of the Doppler theory of gravitation, *Astrophys. J.* **260**, 469.

Netzer, H. (1990). AGN emission lines, in BNW90, p. 57.

Neugebauer, B.T., Soifer, K., Matthews, K. and Elias, J.H. (1989). The near infra-red variability of optically selected quasars, *Astron. J.* **97**, 957.

Neugebauer, G., Miley, G.K., Soifer, B.T. and Klegg, P.E. (1986). Quasars measured by the infrared astronomical satellite, *Astrophys. J.* **308**, 815.

Northover, K.J.E. (1973). The radio galaxy 3C 66, *Mon. Not. Roy. Astr. Soc.* **165**, 369.

Novikov, I.D. and Thorne, K.S. (1973). Black hole astrophysics, in DD73, p. 343.

O'Brien, P.T., Gondhalekar, P.M. and Wilson, R. (1988). The ultraviolet continuum of quasars – I. The shape of the continuum, continuum reddening and intervening absorption, *Mon. Not. Roy. Astr. Soc.* **233**, 801.

O'Dea, C.P. and Owen, F.N. (1986). Multi-frequency VLA observations of the prototypical narrow-angle tail radio source NGC1265, *Astrophys. J.* **301**, 841.

Oke, J.B. (1963). Absolute energy distribution in the optical spectrum of 3C 273, *Nature* **197**, 1040.

Oppenheimer, J.R. and Snyder, H. (1939). On continued gravitational contraction, *Phys. Rev.* **56**, 455.

Orr, M.J.L. and Browne, I.W.A. (1982). Relativistic beaming and quasar statistics, *Mon. Not. Roy. Astr. Soc.* **200**, 1067.

Osmer, P.S., and Smith, M.G. (1980). Discovery and spectrophotometry of the quasars in the −40 deg zone of the CTIO Curtis Schmidt survey, *Astrophys. J. Suppl.* **42**, 333.

Oster, L. (1961). Emission, absorption and conductivity of a fully ionized gas at radio frequencies, *Rev. Mod. Phys.* **33**, 525.

Ostriker, J.P. and Vietri, M. (1990). Are some BL Lacs artifacts of gravitational lensing?, *Nature* **344**, 45.

Owen, F.N. (1986). The extended radio structure of quasars, in SK86, p. 173.

Owen, F.N., Helfand, D.J. and Spangler, S.R. (1983). The correlation of X-ray emission with strong millimeter activity in extragalactic sources, *Astrophys. J.* **280**, L55.

Owen, F.N. and Laing, R.A. (1989). CCD surface photometry of radio galaxies – FR class I and II sources, *Mon. Not. Roy. Astr. Soc.* **238**, 357.

Owen, F.N. and Ledlow, M.J. (1994). The FRI/II break and the bivariate luminosity function in Abell clusters of galaxies, in BDQ94, p. 319.

Padovani, P. and Giommi, P (1995). The connection between X-ray and radio selected BL Lacertae objects, *Astrophys. J.* **444**, 567.

Padovani, P. and Urry C.M. (1992). Luminosity functions, relativistic beaming, and unified theories of high-luminosity radio sources, *Astrophys. J.* **387**, 449.

Papaloizou, J.C.B. and Pringle, J.E. (1984). The dynamical stability of differentially rotating discs with constant specific specific angular momentum, *Mon. Not. Roy. Astr. Soc.* **208**, 721.

Peacock, J.A. (1983). Two dimensional goodness of fit testing in astronomy, *Mon. Not. Roy. Astr. Soc.* **202**, 615.

Peacock, J.A. (1985). The high-redshift evolution of radio galaxies and quasars, *Mon. Not. Roy. Astr. Soc.* **217**, 601.

Peacock, J.A. (1987). Unified beaming models and compact radio sources, in K87, p. 171.

Pearson, T.J. (1990). Parsec-scale radio jets, in ZP90, p. 1.

Pearson, T.J., Blundell, K.M., Riley, J.M. and Warner, P.J. (1992). A jet in the nucleus of the giant quasar 4C 74.26, *Mon. Not. Roy. Astr. Soc.* **259**, 13p.

Pearson, T.J. and Readhead, A.C.S. (1987). The milli-arcsec structure of a complete sample of radio sources. I. VLBI maps of seven sources, *Astrophys. J.* **248**, 61.

Pearson, T.J. and Readhead, A.C.S. (1988). The milli-arcsec structure of a complete sample of radio sources. II. First epoch maps at 5GHz, *Astrophys. J.* **328**, 114.

Pearson, T.J., Unwin, S.C., Cohen, M.H., Linfield, R.P., Readhead, A.C.S., Seielstad, G.A. (1981). Superluminal expansion of quasar 3C 270, *Nature* **290**, 365.

Pearson. T.J, and Zensus, J.A. (1987). Superluminal radio sources: introduction, in ZP87, p. 1.

Pelletier, G. and Roland, J. (1989). Two-fluid model of superluminal radio sources: application to cosmology, *Astron. Astrophys.* **224**, 24.

Penrose, R. (1969). Gravitational collapse: the role of general relativity, *Riv. Nuovo Cimento* **1**, 252.

Perley, R.A. (1989). Hotspot radio galaxies – an introduction, in MR89, p. 1.

Perley, R.A., Dreher, J.W. and Cowan, J.J. (1984). The jet and filaments of Cygnus A, *Astrophys. J.* **285**, L35.

Perley, R.A., Fomalont, E.B. and Johnston, K.J. (1979). Compact radio sources with faint components, *Astron. J.* **85**, 649.

Perley, R.A., Willis, A.G. and Scott, J.S. (1979). The structure of the radio jets in 3C 449, *Nature* **281**, 437.

Perlman, E.S. and Stocke, J.T. (1993). The radio structure of X-ray-selected BL Lacertae objects, *Astrophys. J.* **406**, 430.

Perry, J.J. (1994). Activity in galactic nuclei: starbursts and black holes, in BDQ94, p. 417.

Peterson, B.A. (1988). Selection effects in QSO surveys, in OPGF88, p. 23.

Petitjean, P. (1995). QSO absorption line systems, in WD95, p. 339.

Petre, R., Mushotzky, R.F., Krolik, J.H. and Holt, S.S. (1984). Soft X-ray spectral observations of quasars and high X-ray luminosity Seyfert galaxies, *Astrophys. J.* **280**, 499.

Petrosian, V. (1973). The luminosity function of quasars and its evolution: a comparison of optically selected quasars and quasars found in radio catalogs, *Astrophys. J.* **183**, 359.

Phillips, R.B. and Mutel, R.L. (1982). On symmetric structure in compact radio sources, *Astron. Astrophys.* **106**, 21.

Piccinotti, G., Mushotzky, R.F., Boldt, E.A., Holt, S.S., Marshall, F.E., Serlemitsos, P.J. and Shafer, R.A. (1982). A complete X-ray sample of the high-latitude ($|b| > 20$ deg) sky from *HEAO1* A2: log N – log S and luminosity functions, *Astrophys. J.* **253**, 485.

Pogge, R.W. (1988). An extended ionizing radiation cone from the nucleus of the Seyfert 2 galaxy NGC 1068, *Astrophys. J.* **328**, 519.

Pogge, R.W. (1989). The circumnuclear environment of nearby, noninteracting Seyfert galaxies, *Astrophys. J.* **345**, 730.

Pounds, K.A. (1990). The X-ray spectrum of the broad-line radio galaxy 3C 445, *Mon. Not. Roy. Astr. Soc.* **242**, 20p.

Pounds, K.A., Nandra, K., Fink, H.H. and Makino, F. (1994). Constraining the complexities in Seyfert X-ray spectra – an analysis of simultaneous observations with *GINGA* and *ROSAT*, *Mon. Not. Roy. Astr. Soc.* **267**, 193.

Pounds, K.A., Nandra, K., Stewart, G.C., George, I.M. and Fabian, A.C. (1990). X-ray reflection from cold matter in the nuclei of active galaxies, *Nature* **344**, 132.

Pounds, K.A., Nandra, K., Stewart, G.C., and Leighly, K. (1989). Iron features in the X-ray spectra of three Seyfert galaxies, *Mon. Not. Roy. Astr. Soc.* **240**, 769.

Prestage, R.M. and Peacock, J.A. (1988). The cluster environments of powerful radio galaxies, *Mon. Not. Roy. Astr. Soc.* **230**, 131.

Primini, F.A., Murray, S.S., Huchra, J., Schild, R., Burg, R. and Giacconi, R. (1991). The CfA *EINSTEIN* observatory extended deep X-ray survey, *Astrophys. J.* **374**, 440.

Pringle, J.E. (1981). Accretion discs in astrophysics, *Ann. Rev. Astron. Astrophys.* **19**, 137.

Punch, M., Akerlof, C.W., Cawley, M.F., Chantell, M., Fegan, D.J., Fennell, S. *et al.* (1992). Detection of TeV photons from the active galaxy Markarian 421, *Nature* **358**, 477.

Qian, S.J., Quirrenbach, A., Witzel, A., Krichbaum, T.P., Hummel, C.A. and Zensus T.A. (1991). A model for the rapid radio variability in the quasar 0917 + 624, *Astron. Astrophys.* **241**, 15.

Readhead, A.C.S. (1993). Parsec-scale jets, in DB93, p. 173.

Rees, M.J. (1966). Appearance of relativistically expanding radio sources, *Nature* **211**, 468.

Rees, M.J. (1971). New interpretation of extragalactic radio sources, *Nature* **229**, 312.

Rees, M.J. (1984). Black hole models for active galactic nuclei, *Ann. Rev. Astron. Astrophys.* **22**, 471.

Rees, M.J., Begelman, M.C., Blandford, R.D. and Phinney, E.S. (1982). Ion-supported tori and the origin of radio jets, *Nature* **295**, 17.

Rees, M.J. and Schmidt, M. (1971). On the V/V_m test applied to quasi-stellar radio sources, *Mon. Not. Roy. Astr. Soc.* **154**, 1.

Rees, M.J., Silk, J.I., Werner, M.W. and Wickramasinghe, N.C. (1969). Infrared radiation from dust in Seyfert galaxies, *Nature* **223**, 788.

Richstone, D.O. and Schmidt, M. (1980). The spectral properties of a large sample of quasars, *Astrophys. J.* **235**, 361.

Rieke, G.H. and Lebofsky, M.J. (1979). Infrared emission of extragalactic sources, *Ann. Rev. Astron. Astrophys.* **17**, 477.

Robertson, H.P. (1935). Kinematics and world structure, *Astrophys. J.* **82**, 248.

Rocca-Volmerange, B. and Guiderdoni, B. (1988). An atlas of synthetic spectra of galaxies, *Astron. Astrophys. Suppl.* **75**, 93.

Rudnick, L. (1987). A different perspective on superluminal sources, in ZP87, p. 217.

Ryle, M. and Longair, M.S. (1967). A possible method for investigating the evolution of radio galaxies, *Mon. Not. Roy. Astr. Soc.* **136**, 123.

Saikia, D.J., Junor, W., Cornwell, T.J., Muxlow, T.W.B. and Shastri, P. (1990). A VLA and MERLIN study of extragalactic radio sources with one-sided structure, *Mon. Not. Roy. Astr. Soc.* **245**, 408.

Saikia, D.J. and Kulkarni, V.K. (1994). On the evidence against the unified scheme, *Mon. Not. Roy. Astr. Soc.* **270**, 897.

Saikia, D.J. and Salter, C.J. (1988). Polarization properties of extragalactic radio sources, *Ann. Rev. Astron. Astrophys.* **26**, 93.

Sambruna, R.M., Barr, P., Giommi, P., Maraschi, L., Tagliaferri, G. and Treves, A. (1994). The X-ray spectra of blazars observed with *EXOSAT*, *Astrophys. J.* **434**, 468.

Sambruna, R.M., Maraschi, L. and Urry, C.M. (1996). On the spectral energy distributions of blazars, *Astrophys. J.* **463**, 444.

Sandage, A. (1965). The existence of a major new constituent of the universe: the quasi-stellar galaxies, *Astrophys. J.* **141**, 1560.

Sandage, A. and Lyuten, W.J. (1967). On the nature of faint blue objects in high galactic latitudes. I. Photometry, proper motions, and spectra in PHL field $1:36+6^0$ and Richter field M3, II, *Astrophys. J.* **148**, 767.

Sanders, D.B., Phinney, E.S., Neugebauer, G., Soifer, B.T. and Matthews, K. (1989). Continuum energy distributions of quasars: shapes and origins, *Astrophys. J.* **347**, 29.

Sargent, W.L.W., Boksenberg, A. and Steidel, C.S. (1988). C IV absorption in a new sample of 55 QSOs: evolution and clustering of the heavy-element absorption redshifts, *Astrophys. J. Suppl.* **68**, 539.

Sargent, W.L.W., Steidel, C.S. and Boksenberg, A. (1988). MG II absorption in the spectra of high redshift and low redshift QSOs, *Astrophys. J.* **334**, 22.

Sargent, W.L.W., Young, P.J., Boksenberg, A. and Tytler, D. (1980). The distribution of Lyman-alpha absorption lines in six QSOs: evidence for an intergalactic origin, *Astrophys. J. Suppl.* **42**, 41.

Savage, A., Jauncey, D.L., White, G.L., Peterson, B.A. and Peters, W.L. (1988). A complete sample of flat spectrum radio sources from the Parkes 2.7 GHz survey, in OPGF88, p. 204.

Scheuer, P.A.G. (1960). The absorption coefficient of a plasma at radio frequencies, *Mon. Not. Roy. Astr. Soc.* **120**, 231.

Scheuer, P.A.G. (1974a). Models of extragalactic radio sources with a continuous energy supply, *Mon. Not. Roy. Astr. Soc.* **166**, 513.

Scheuer, P.A.G. (1974b). Fluctuations in the X-ray background, *Mon. Not. Roy. Astr. Soc.* **166**, 329.

Scheuer, P.A.G. (1987). Tests of beaming models, in ZP87, p. 104.

Scheuer, P.A.G. (1995). Lobe asymmetry and the expansion speeds of radio sources, *Mon. Not. Roy. Astr. Soc.* **277**, 331.

Scheuer, P.A.G. and Readhead A.C.S. (1979). Superluminally expanding radio sources and the radio-quiet QSOs, *Nature* **277**, 182.

Schilizzi, R.T. (1992). VLBI-scale radio jets, in RSP92, p. 92.

Schlickeiser, R. (1996). Models of high-energy emission from active galactic nuclei, *Astron. Astrophys. Suppl.* **120**, 481.

Schlickeiser, R., Biermann, P.L. and Crusius-Wätzel, A. (1991). On a nonthermal origin of steep far-infrared turnovers in radio-quiet active galactic nuclei, *Astron. Astrophys.* **247**, 283.

Schmidt, M. (1963). 3C273: a star-like object with large redshift, *Nature* **197**, 1040.

Schmidt, M. (1968). Space distribution and luminosity function of quasi-stellar radio sources, *Astrophys. J.* **151**, 393.

Schmidt, M. (1970). Space distribution and luminosity function of quasars, *Astrophys. J.* **162**, 371.

Schmidt, M. and Green, R.F. (1983). Quasar evolution derived from the Palomar bright quasar survey and other complete quasar surveys, *Astrophys. J.* **269**, 352.

Schmidt, M., Schneider, D.P. and Gunn, J.E. (1995). Spectroscopic CCD surveys for quasars at large redshift IV. Evolution of the luminosity function from quasars detected by their Lyman-alpha emission, *Astron. J.* **110**, 68.

Schneider, D.P., Schmidt, M. and Gunn, J.E. (1994). Spectroscopic CCD surveys for quasars at large redshift III. The Palomar transit grism survey catalog, *Astron. J.* **107**, 1245.

Schonfelder, V. (1994). Gamma-ray properties of active galactic nuclei, *Astrophys. J. Suppl.* **92**, 593.

Serlemitsos, P., Yaqoob, T., Ricker, G., Woo, J., Kunieda, H., Terashima, Y. and Iwasawa, K. (1994). The complex X-ray spectra of two high redshift quasars observed with *ASCA*, *Proc. Astron. Soc. Japan* **46**, L43.

Setti, G. and Woltjer, L. (1989). Active galactic nuclei and the spectrum of the X-ray background, *Astron. Astrophys.* **224**, L21.

Setti, G. and Woltjer, L. (1994). The gamma-ray background, *Astrophys. J. Suppl.* **92**, 629.

Seyfert, C.K. (1943). Nuclear emission in spiral nebulae, *Astrophys. J.* **97**, 28.

Shakura, N.I. and Sunyaev, R.A. (1973). Black holes in binary systems: observational appearance, *Astron. Astrophys.* **24**, 337.

Shanks, T., Georgantopoulos, I., Stewart, G.C., Pounds, K.A., Boyle, B.J. and Griffiths, R.E. (1991). The origin of the cosmic X-ray background, *Nature* **353**, 315.

Shapiro, S.I., Lightman, A.P. and Eardley, D. (1976). A two-temperature accretion disc model for Cygnus X-1: structure and spectrum, *Astrophys. J.* **204**, 187.

Shashtri, P. (1990). A relativistically beamed X-ray component in quasars?, *Mon. Not. Roy. Astr. Soc.* **249**, 640.

Shastri, P., Wilkes, B.J., Elvis, M. and McDowell, J. (1993). Quasar X-ray spectra revisited, *Astrophys. J.* **410**, 29.

Shields, G.A. (1978). Thermal continuum from accretion disks in quasars, *Nature* **272**, 706.

Sikora, M. (1994). High energy radiation from active galactic nuclei, *Astrophys. J. Suppl.* **90**, 923.

Sikora, M., Begelman, M.C. and Rees, M.J. (1994). Comptonization of diffuse ambient radiation by a relativistic jet: the source of gamma-rays from blazars?, *Astrophys. J.* **421**, 153.

Sikora, M. and Wilson D.B. (1981). The collimation of particle beams from thick accretion tori, *Mon. Not. Roy. Astr. Soc.* **197**, 529.

Simpson, C., Ward, M.J. and Wilson, A.S. (1995). Evidence for an obscured quasar in the giant radio galaxy PKS 0634–205, *Astrophys. J.* **454**, 683.

Singal, A.K. (1993). Evidence against the unified scheme for powerful radio galaxies and quasars, *Mon. Not. Roy. Astr. Soc.* **262**, L27.

Singh, K.P., Garmire, G.P. and Nousek, J. (1985). Observation of soft X-ray spectra from a Seyfert 1 and a narrow emission-line galaxy, *Astrophys. J.* **297**, 633.

Singh, K.P., Rao, A.R. and Vahia, M.N. (1990). *EXOSAT* observations of the blazar PKS 1510–089, *Astrophys. J.* **365**, 455.

Singh, K.P., Westergaard, N.J. and Schnopper, H.W. (1987). *EXOSAT* observations of a broad absorption-line quasar: PHL 5200, *Astron. Astrophys.* **172**, L11.

Singh, K.P., Westergaard, N.J., Schnopper, H.W., Awaki, H. and Tawara, Y. (1990). *GINGA* observations of X-rays from the Seyfert I galaxy Markarian 509, *Astrophys. J.* **363**, 131.

Skibo, J.G., Dermer, C.D. and Kinzer, R.L. (1994). Is the high-energy emission from Centaurus A Compton-scattered jet radiation?, *Astrophys. J.* **426**, L23.

Small, T.A. and Blandford R.D. (1992). Quasar evolution and the growth of black holes, *Mon. Not. Roy. Astr. Soc.* **259**, 725.

Smith, M.G. and Wright, A.E. (1980). A radio study of optically selected QSOs, *Mon. Not. Roy. Astr. Soc.* **191**, 871.

Soucail, G., Mellier, Y., Fort, B., Hammer, F. and Mathez, G. (1987). Further data on the ring-like structure in A 370, *Astron. Astrophys.* **184**, L7.

Soucail, G., Mellier, Y., Fort, B., Mathez, G. and Cailloux, M. (1988). The giant arc in A 370: spectroscopic evidence for gravitational lensing from a source at $z = 0.724$, *Astron. Astrophys.* **191**, L19.

Sparks, W.B., Fraix-Burnet, D., Macchetto, F. and Owen, F.N. (1992). A counterjet in the elliptical galaxy M87, *Nature* **355**, 804.

Spencer, R.E. and Akujor, C.E. (1992). Merlin and VLBI observations of two-sided sources, in RSP92, p. 119.

Sramek, R.A. and Weedman, D.W. (1980). The radio properties of optically selected quasars, *Astrophys. J.* **238**, 435.

Stark, A.A., Gammie, C.F., Wilson, R.W., Bally, J., Linke, R.A., Heiles, C. and Hurwitz, M. (1992). The Bell Laboratories H I survey, *Astrophys. J. Suppl.* **79**, 77.

Steidel C.S. (1993). The environment and evolution of galaxies, in ST93, p. 263.

Steidel, C.S., Bowen, D.V., Blades, J.C., and Dickinson, M. (1995). The $z = 0.8596$ damped Ly α absorbing galaxy towards PKS 0454+039, *Astrophys. J.* **440**, L45.

Stein, W.A., O'Dell, S.L. and Strittmatter, P.A. (1976). The BL Lacertae objects, *Ann. Rev. Astron. Astrophys.* **14**, 173.

Stiavelli, M., Biretta, J., Moller, P. and Zeilinger, W.W. (1992). Optical counterpart of the east radio lobe of M 87, *Nature* **355**, 802.

Stickel, M., Padovani, P., Urry, C.M., Fried, J.W. and Kuhr, H. (1991). The complete sample of 1 Jansky BL Lacertae objects I. Summary of properties, *Astrophys. J.* **374**, 431.

Stocke J.T., Morris, S.L., Gioia, I.M., Maccacaro, T., Schild, R. and Wolter, A. (1991). The *EINSTEIN* observatory extended medium-sensitivity survey. II. The optical identifications, *Astrophys. J. Suppl.* **76**, 813.

Stocke, J.T., Morris, S.L., Weymann, R.J. and Foltz, C.B. (1992). The radio properties of broad absorption line QSOs, *Astrophys. J.* **396**, 487.

Stockton, A. (1978). The nature of QSO redshifts, *Astrophys. J.* **273**, 747.

Stritmatter, P.A., Hill, P., Pauliny-Toth, I.I.K., Steppe, H. and Witzel, A. (1980). Radio observations of optically selected quasars, *Astron. Astrophys.* **88**, L12.

Stritmatter, P.A., Serkowski, K., Carswell, R., Stein, W.A., Merrill, K.M. and Burbidge, E.M. (1972). Compact extragalactic nonthermal sources, *Astrophys. J.* **175**, L7.

Stubbs, P. (1971). Redshift without reason, *New Scientist* **50**, 254.

Sun, W.-H. and Malkan, M.A. (1989). Fitting improved accretion disk models to the multiwavelength continua of quasars and active galactic nuclei, *Astrophys. J.* **346**, 68.

Surdej, T., Magin, P., Swings, J.-P. *et al.* (1987). A new case of gravitational lensing, *Nature* **329**, 695.

Svensson, R. (1986). Physical processes in active galactic nuclei, in MW89, p. 325.

Svensson, R. (1987). Non-thermal pair production in compact X-ray sources: first order Compton cascades in soft radiation fields, *Mon. Not. Roy. Astr. Soc.* **227**, 403.

Svensson, R. (1990). An introduction to relativistic plasmas in astrophysics, in BFG90, p. 357.

Svensson, R. (1994). The nonthermal pair model for the X-ray and gamma-ray spectra from active galactic nuclei, *Astrophys. J. Suppl.* **92**, 585.

Swanenburg, B.N. *et al.* (1978). *COS-B* observation of high energy radiation from 3C 273, *Nature* **275**, 298.

Tadhunter, C. and Tsvetanov, Z. (1989). Anisotropic ionizing radiation in NGC5252, *Nature* **341**, 422.

Tanaka, Y., Inoue, H. and Holt, S.S. (1994). The X-ray astronomy satellite *ASCA*, *Proc. Astron. Soc. Japan* **46**, L37.

Tananbaum, H., Avni, Y., Branduardi, G., Elvis, M., Fabbiano, G., Feigeson, E. *et al.* (1979). X-ray studies of quasars with the *EINSTEIN* observatory, *Astrophys. J.* **234**, L9.

Tananbaum, H., Avni, Y., Green, R.F., Schmidt, M. and Zamorani, G. (1986). X-ray observations of the bright quasar survey, *Astrophys. J.* **305**, 57.

Tananbaum, H., Peters, G., Forman, W., Giacconi, R. and Jones, C. (1978). *UHURU* observations of X-ray emission from Seyfert galaxies, *Astrophys. J.* **223**, 74.

Tananbaum, H., Wardle, J.F.C., Zamorani, G. and Avni, Y. (1983). X-ray studies of quasars with the *EINSTEIN* observatory. III. The 3CR sample, *Astrophys. J.* **268**, 60.

Tennant, A.F. and Mushotzky, R.F. (1983). The absence of rapid X-ray variability in active galaxies, *Astrophys. J.* **264**, 92.

Terlevich, R. (1992). The starburst models for AGN, in F92, p. 133.

Terrell, N.J. (1966). Quasi-stellar objects: possible local origin, *Science* **154**, 1281.

Thomson, D.J. and Fichtel, C.E. (1982). Extragalactic gamma radiation: use of galactic counts as a galactic tracer, *Astron. Astrophys.* **109**, 352.

Tucker, W., Kellogg, E., Gursky, H., Giacconi, R. and Tananbaum, H. (1973). X-ray observations of NGC 5128 (Centaurus A), *Astrophys. J.* **180**, 715.

Turland, B.D. (1975). Observations of M 87 at 5GHz with the 5 km telescope, *Mon. Not. Roy. Astr. Soc.* **170**, 281.

Turner, M.J.L., Thomas, H., Patchett, B., Reading, D. and Makishima, K. (1989). The Large Area Counter on *GINGA*, *Proc. Astron. Soc. Japan* **41**, 345.

Turner, M.J.L., Williams, O.R., Courvoisier, T.J.-L., Stewart, G.C., Nandra, K., Pounds, K.A. *et al.* (1990). The X-ray emission of 3C 273, *Mon. Not. Roy. Astr. Soc.* **244**, 310.

Turner, M.J.L., Williams, O.R., Saxton, R., Stewart, G.C., Courvoisier, T.J-L., Ohashi, T. *et al.* (1989). The spectra (2–10 keV) of high luminosity quasars, in HB89, p. 769.

Turner, T.J. and Pounds, K.A. (1989). The *EXOSAT* spectral survey of AGN, *Mon. Not. Roy. Astr. Soc.* **240**, 833.

Ulrich, M.-H., Kinman, T.D., Lynds, C.R., Rieke, G.H. and Ekers, R.D. (1975).

Nonthermal continuum radiation in three elliptical galaxies, *Astrophys. J.* **198**, 261.

Ulrich, M.-H. (1989). The host galaxy of BL Lac objects, in MMU89, p. 45.

Ulvestad, J.L. and Wilson, A.S. (1984a). Radio structures of Seyfert galaxies. V. A flux limited sample of Markarian galaxies, *Astrophys. J.* **278**, 544.

Ulvestad, J.L. and Wilson, A.S. (1984b). Radio structures of Seyfert galaxies. VI. VLA observations of a nearby sample, *Astrophys. J.* **285**, 439.

Ulvestad, J.L. and Wilson, A.S. (1989). Radio structures of Seyfert galaxies. VII. Extension of a distance limited sample , *Astrophys. J.* **343**, 659.

Unwin, S.C. and Davis, R.J. (1988). VLBI observations of the radio jet in 3C 273, in RM88, p. 127.

Unwin., S.C. and Wehrle, A.E. (1992). Kinematics of the parsec-scale relativistic jet in 3C 343, *Astrophys. J.* **398**, 74.

Urry, C.M. and Padovani, P. (1995). Unified schemes for radio-loud active galactic nuclei, *Proc. Astron. Soc. Pacific* **107**, 803.

Urry, C.M., Padovani, P. and Stickel, M. (1991). Fanaroff-Riley I galaxies as the parent population BL Lacertae objects III. Radio constraints, *Astrophys. J.* **382**, 501.

Urry, C.M., Sambruna, R.M., Worrall, D.M., Kollgaard, R.I., Feigelson, E.D., Perlman, E.S. and Stocke, J.T. (1996). Soft X-ray properties of a complete sample of radio-selected BL Lacertae objects, *Astrophys. J.* **463**, 424.

Urry, C.M. and Shafer, R.A. (1984). Luminosity enhancement in relativistic jets and altered luminosity functions for beamed objects, *Astrophys. J.* **280**, 569.

Van der Hulst, J.M. (1991). Emission from the nuclei of normal galaxies, in DC91, p. 215.

Vermeulen, R.C. and Cohen, M.H. (1994). Superluminal motion statistics and cosmology, *Astrophys. J.* **430**, 467.

Vermeulen, R.C., Readhead, A.C.S. and Backer, D.C. (1994). Discovery of a nuclear counterjet in NGC 1275: a new way to probe the parsec-scale environment, *Astrophys. J.* **430**, L41.

Véron-Cetty, M.P. and Véron, P. (1993). Catalogue of quasars and active nuclei, *ESO Sci. Rep.* **13**, 1.

Visnovsky, K.L., Impey, C.D., Foltz, C.B., Hewett, P.C., Weymann, R.J. and Morris S.L. (1992). Radio properties of optically selected quasars, *Astrophys. J.* **391**, 560.

Visvanathan, N. (1969). The continuum of BL Lac, *Astrophys. J.* **155**, L133.

Von Balmoos, P., Diehel, R. and Schönfelder, V. (1987). Centaurus A observation at MeV-gamma-ray energy, *Astrophys. J.* **312**, 134.

Von Montigny, C., Bertsch, D.L., Chiang, J., Dingus, B.L., Esposito, J.A., Fichtel, C.E. *et al.* (1995). High-energy gamma-ray emission from active galaxies: *EGRET* observations and their implications, *Astrophys. J.* **440**, 525.

Wagh, S.M. and Dadhich, N. (1989). The energetics of black holes in electromagnetic fields by the Penrose process, *Phys. Rep.* **183**, 137.

Wagh, S.M., Dhurandhar, S.V. and Dadhich, N. (1985). Revival of Penrose process for astrophysical applications, *Astrophys. J.* **290**, 12.

Wagner, S.J. and Witzel, A. (1992). Intraday variability of active galactic nuclei: radio and optical observations, in RSP92, p. 59.

Walker, A.G. (1936). On Milne's theory of world structure, *Proc. Lond. Math. Soc. (2)* **42**, 90.

Walker, R.C., Benson, J.M. and Unwin, S.C. (1987). The radio morphology of 3C 120 on scales from 0.5 parsecs to 400 kiloparsecs, *Astrophys. J.* **316**, 546.

Wall, J.V. and Peacock, J.A. (1985). Bright extragalactic radio sources at 2.7 GHz – III. The all-sky catalogue, *Mon. Not. Roy. Astr. Soc.* **216**, 713.

Wallinder, F.H., Kato, S. and Abramowicz, M.A. (1992). Variability of the central region in active galactic nuclei, *Astron. Astrophys. Rev* **4**, 79.

Walsh, D., Carswell, R.F. and Weymann, R.J. (1979). 0957+561 A,B: twin quasistellar objects or gravitational lens?, *Nature* **279**, 381.

Walter, R. and Fink, H.H. (1993). The ultraviolet to soft X-ray bump of Seyfert 1 type active galactic nuclei, *Astron. Astrophys.* **274**, 105.

Wampler, E.J., Chugai, N.N. and Petitjean, P. (1995). The absorption spectrum of nuclear gas in Q 0059–2735, *Astrophys. J.* **443**, 586.

Wampler, E.J. and Ponze D. (1985). Optical selection effects that bias quasar evolution studies, *Astrophys. J.* **298**, 448.

Wang, B., Inoue, H., Koyama, K., Tanaka, Y., Hirano, T. and Nagase, F. (1986). X-ray observations of Centaurus A from Tenma, *Proc. Astron. Soc. Pacific* **38**, 685.

Wardle, J.F.C. and Potash, R.I. (1994). Observations of large scale jets in quasars and the sidedness problem, in BE94, p. 30.

Warren, S.J. and Hewett, P.C. (1991). The detection of high redshift quasars, *Rep. Prog. Phy.* **54**, 243.

Warren, S.J., Hewett, P.C., Irwin, M.J. and Osmer, P.S. (1991). A wide-field multicolor survey for high-redshift quasars, $z \geq 2.2$. I. Photometric catalog and survey selection function, *Astrophys. J. Suppl.* **76**, 1.

Warren, S.J., Hewett, P.C. and Osmer, P.S. (1991). A wide-field multicolor survey for high-redshift quasars, $z \geq 2.2$. II. The quasar catalog, *Astrophys. J. Suppl.* **76**, 1.

Warren, S.J., Hewett, P.C. and Osmer, P.S. (1994). A wide-field multicolor survey for high-redshift quasars, $z \leq 2.2$. III. The luminosity function, *Astrophys. J.* **421**, 412.

Warwick, R.S., Barstow, M.A. and Yaqoob, T. (1989). The X-ray spectrum of QSO 1821+643, *Mon. Not. Roy. Astr. Soc.* **238**, 917.

Weiler, K.W., Ekeres, R.D., Raimond, E. and Wellington, K.S. (1975). Dual-frequency measurement of the Solar gravitational microwave deflection, *Phys. Rev. Lett.* **35**, 134.

Westfold, A.C. (1959). The polarization properties of synchrotron radiation, *Astrophys. J.* **130**, 241.

Weymann, R.J., Boroson, T.A., Peterson, B.M. and Butcher, H.R. (1978). An attempt to detect faint objects near quasi-stellar objects with low-redshift absorption systems, *Astrophys. J.* **226**, 603.

Weymann, R.J., Morris, S.L., Foltz, C.B. and Hewett, P.C. (1991). Comparison of the emission line and continuum properties of broad absorption line and normal quasi-stellar objects, *Astrophys. J.* **373**, 23.

White, T.R., Lightman, A.P. and Zdziarski, A.A. (1988). Compton reflection of gamma rays by cold electrons, *Astrophys. J.* **331**, 939.

White, T.R., Lightman, A.P., Zdziarski, A.A. (1988). Compton reflection of gamma rays by cold electrons, *Astrophys. J.* **331**, 939.

Whitney, A.R., Shapiro, I.I., Rogers, A.E.E., Robertson, D.S., Knight, C.A., Clark, T.A. *et al.* (1971). Quasars revisited: rapid time variations observed via very long baseline interferometry, *Science* **173**, 225.

Wiita, P.J. (1991). The production of jets and their relation to active galactic nuclei, in H91, p. 379.

Wilkes, B.J. and Elvis, M. (1987). Quasar energy distributions. I. Soft X-ray spectra of quasars, *Astrophys. J.* **323**, 243.

Wilkes, B.J., Tananbaum, H., Worrall, D.M., Avni, Y., Oey, S. and Flanagan, J. (1994). The *EINSTEIN* database of X-ray observations of optically selected and radio selected quasars. I., *Astrophys. J. Suppl.* **92**, 53.

Williams, O.R., Bennet, K., Bloemen, H., Collmar, W., Diehl, R., and Hermsen, W. (1995). COMPTEL observations of the quasars 3C273 and 3C279, *Astron. Astrophys.* **298**, 33.

Williams, O.R., Turner, M.J.L., Stewart, G.C., Saxton, R.D., Ohashi, T., Makishima, K. *et al.* (1992). The X-ray spectra of high-luminosity active galactic nuclei observed by *GINGA*, *Astrophys. J.* **389**, 157.

Williams, P.J.S. (1963). Absorption in radio sources of high brightness temperature, *Nature* **200**, 56.

Wills, B.J. and Browne, I.W.A. (1986). Relativistic beaming and quasar emission lines, *Astrophys. J.* **302**, 56.

Wilson, A.S., Braatz, J.A., Heckman, T.M., Krolik, J.H. and Miley, G.K. (1993). The ionizing cones in the Seyfert galaxy NGC 5728, *Astrophys. J.* **419**, L61.

Wilson, A.S. and Ulvestad, J.S.. (1982). Radio structures of Seyfert galaxies. IV. Jets (?) in NGC 1068 and NGC 4151, *Astrophys. J.* **263**, 576.

Winkler, P. and White, A. (1975). A sudden increase in the flux of Centaurus A, *Astrophys. J.* **199**, L139.

Witzel, A., Schalinski, C.J., Johnston, K.J., Biermann, P.L., Krichbaum, T.P., Hummel, C.A. and Eckart, A. (1988). The occurrence of bulk relativistic motion in compact radio sources, *Astron. Astrophys.* **206**, 245.

Wolter, A., Caccianiga, A., Della Ceca, R.D. and Maccacaro, T. (1994). Luminosity functions of BL Lacertae objects, *Astrophys. J.* **433**, 29.

Woltjer, L. (1990). Phenomenology of active galactic nuclei, in BNW90, p. 1.

Worrall, D.M (1987). Superluminal radio sources: what does X-ray emission tell us?, in ZP87, p. 251.

Worrall, D.M., Bololt, E.A., Holt, S.S. and Serlemitsos, P.J. (1980). The X-ray spectrum of Q50 0241+622, *Astrophys. J.* **240**, 421.

Worrall, D.M., Giommi, P., Tananbaum, H. and Zamorani, G. (1987). X-ray studies of quasars with the *EINSTEIN* observatory. IV. X-ray dependence on radio emission, *Astrophys. J.* **313**, 596.

Worrall, D.M., Mushotzky, R.F., Boldt, E.A., Holt, S.S. and Serlemitsos, P.J. (1979). The X-ray spectrum of 3C273, *Astrophys. J.* **232**, 683.

Worrall, D.M. and Wilkes, B.J. (1990). X-ray spectra of compact extragalactic radio sources, *Astrophys. J.* **360**, 396.

Wrobel, J.M. (1987). VLA polarimetry of the active galaxy 3C371, in ZP87, p. 186.

Wrobel, J.M. (1991). Active nuclei and star formation in ES0 galaxies, in DC91, p. 197.

Wu, X., Hamilton, T., Helfand, D.J. and Wang, Q. (1990). The intensity and the spectrum of the diffuse X-ray background, *Astrophys. J.* **379**, 564.

Zamorani, G., Henry, J.P., Maccacaro, T., Tananbaum, H., Soltan, A. and Avni, Y. (1981). X-ray studies of quasars with the *EINSTEIN* observatory (II), *Astrophys. J.* **245**, 357.

Zdziarski, A.A., Ghisellini, G., George, I.M., Svensson, R., Fabian, A.C. and Done, C. (1990). Electron–positron pairs, Compton reflection, and the X-ray spectra of active galactic nuclei, *Astrophys. J.* **363**, L1.

Zdziarski, A.A., Johnson, W.N., Done, C., Smith, D. and McNaron-Brown, K. (1995). The average X-ray/gamma-ray spectra of Seyfert galaxies from *GINGA* and *OSSE* and the origin of the cosmic X-ray background, *Astrophys. J.* **438**, L63.

Zitelli, V., Mignoli, M., Zamorani, G., Marano, B. and Boyle, B.J. (1992). A spectroscopically complete sample of quasars with $B_J \geq 22.0$, *Mon. Not. Roy. Astr. Soc.* **256**, 349.

Zwicky, F. (1937). Nebulae as gravitational lenses, *Phys. Rev.* **51**, 290.

Books, reviews and proceedings

[A87] Arp, H. 1987. *Quasars, Redshifts and Controversies* (Berkeley, Interstellar Media).

[AP80] Abell, G.O. and Peebles, P.J.E. 1980. *Objects of High Redshifts, Proc. International Astronomical Symposium No. 92* (Ridel, Dordrecht).

[B52] Burhope, E.H.S. 1952. *The Auger Effect and Other Radiationless Transitions* (Cambridge University Press, Cambridge).

[B69] Bevington, P.R. 1969. *Data Reduction and Error Analysis for the Physical Sciences* (McGraw-Hill, New York).

[BDQ94] Bicknell, G.V., Dopita, M.A. and Quinn, P.J. 1994. *The First Stromlo Symposium: The Physics of Active Galaxies* (Astronomical Society of the Pacific, San Francisco).

[BE84] Bridle A.H. and Eilek J.A. 1984. *Physics of Energy Transport in Extragalactic Radio Sources. Proc. of NRAO Workshop No. 9* (National Radio Astronomy Observatories, Greenbank).

[BF92] Barcons, X. and Fabian, A.C. 1992. *The X-ray Background* (Cambridge University Press, Cambridge).

[BFG90] Brinkmann, W., Fabian, A.C. and Giovannelli, F. 1990. *Physical Processes in Hot Cosmic Plasmas* (Kluwer Academic Publishers, Dordrecht).

[BLO93] Burgarella, D., Livio, M. and O'Dea, C.P. 1993. *Astrophysical Jets* (Cambridge University Press, Cambridge).

[BM96] Bowyer, S. and Malina, F. 1996. *Astrophysics in the Extreme Ultraviolet* (Kluwer Academic Publishers, Dordrecht).

[BNW90] Blandford, R.D., Netzer, H. and Woltjer, L. 1990. *Active Galactic Nuclei* (Springer-Verlag, Berlin).

[C91] Crampton, D. 1991. *The Space Distribution of Quasars* (Astronomical Society of the Pacific, San Francisco).

[CB94] Courvoisier.T.J.-L. and Blecha, A. 1994. *Multi-wavelength Continuum Emission of AGN, Proc. International Astronomical Union Symp. 159* (Kluwer, Dordrecht).

[CS95] Charles, P.A. and Seward, F.D. 1995. *Exploring the X-ray Universe* (Cambridge University Press, Cambridge).

[CW97] Clarke, D.A. and West, M.J. 1997. *Computational Astrophysics. Proc. 12th 'Kingston Meeting' on Theoretical Astrophysics* (Astronomical Society of the Pacific, San Francisco).

[D90] Dyson, N.A. 1990. *X-rays in Atomic and Nuclear Physics* (Cambridge University Press, Cambridge).

[DB93] Davis, R.J. and Booth, R.S. 1993. *Sub-arcsecond Radio Astronomy* (Cambridge University Press, Cambridge).

[DC91] Duric, N. and Crane, P.C. 1991. *The Interpretation of Modern Synthesis Observations of Spiral Galaxies* (Astronomical Society of the Pacific, San Francisco).

[DD73] Dewitt, C. and Dewitt, B.S. 1973. *Black Holes* (Gordon and Breach, New York).

[DW92] Duschl, W.J. and Wagner, S.J. 1992. *Physics of Active Galactic Nuclei* (Springer-Verlag, Berlin).

[EFP95] Ekers, R., Fanti, C. and Padrielli, L. 1995. *Extragalactic Radio Sources, Proc.*

International Astronomical Union Symp. 175 (Kluwer, Dordrecht).

[F92] Fillipenko, A.V. 1992. *Relationships Between Active Galactic Nuclei and Starburst Galaxies, Proc. Taipei Astrophysics Workshop* (Astronomical Society of the Pacific, San Francisco).

[FB92] Feigelson, E.D. and Babu, J.D. 1992. *Statistical Challenges in Modern Astronomy* (Springer-Verlag, New York).

[FKR92] Frank, J., King, A. and Raine, D. 1992. *Accretion Power in Astrophysics* (Cambridge University Press, Cambridge).

[GLR78] Gunn, J.E., Longair, M.S. and Rees, M.J. 1978. *Observational Cosmology* (Geneva Observatory, Sauverny).

[GS64] Ginzburg, V.L. and Syrovatskii, S.I. 1964. *The Origin of Cosmic Rays* (Pergamon Press, Oxford).

[H91] Hughes, P.A. 1991. *Beams and Jets in Astrophysics* (Cambridge University Press, Cambridge).

[HB89] Hunt, J. and Battrick, B. 1989. *Two Topics in X-ray Astronomy, Proc. 23rd ESLAB Symp.* (European Space Agency, Munich).

[HC89] Hogg, R.V. and Craig, A.T. 1989. *Introduction to Mathematical Statistics*, fourth edition (Macmillan, New York).

[HNU92] Holt, S.S., Neff, S.G. and Urry, C.M. 1992. *Testing the AGN Paradigm* (American Institute of Physics, New York).

[J75] Jackson, J.D. 1975. *Classical Electrodynamics*, second edition (John Wiley, New York).

[K87] Katz, J.I. 1987. *High Energy Astrophysics* (Addison-Wesley, California).

[K88] Kafatos, M. 1988. *Supermassive Black Holes* (Cambridge University Press, Cambridge)

[MB81] Mihalas, D. and Binney, J. 1981. *Galactic Astronomy* (W.H. Freeman and Company, New York).

[MMU89] Maraschi, L. , Maccacaro, T. and Ulrich, M. H. 1989. *BL Lac Objects* (Springer-Verlag, Berlin).

[MR89] Meisenheimer, K. and Röser, H. J. 1989. *Hotspots in Extragalactic Radio Sources* (Springer-Verlag, Berlin).

[MT92] MacGillivray, H.T. and E.B. Thomson, E.B. 1992. *Digitised Optical Sky Surveys* (Kluwer, Dordrecht).

[MTW73] Misner, C.W., Thorne, K.S. and Wheeler, J.A. 1973. *Gravitation* (W.H. Freeman & Co., San Francisco).

[MW89] Mihalas, D. and Winkler, K.-H. 1989. *Radiation Hydrodynamics in Stars and Compact Objects, Proc. International Astronomical Union Colloq. 89* (Ridel, Dordrecht).

[MW91] Miller, H.R. and Wiita, P.J. 1991. *Variability of Active Galactic Nuclei* (Cambridge University Press, Cambridge).

[N93] Narlikar, J.V. 1993. *Introduction to Cosmology*, second edition (Cambridge University Press, Cambridge).

[N96] Narlikar J.V. 1996. *The Lighter Side of Gravity*, second edition (Cambridge University Press, Cambridge).

[O89] Osterbrock, D.E. 1989. *Astrophysics of Gaseous Nebulae and Active Galactic Nuclei* (University Science Books, Mill Valley).

[OM89] Osterbrock, D.E. and Miller, J.S. 1989. *Active Galactic Nuclei, Proc. International Astronomical Union Symp. 134* (Kluwer, Dordrecht).

[OPGF88] Osmer, S., Porter, A.C., Green, R.F. and Foltz, C.B. 1988. *Proc. Workshop on Optical Surveys for Quasars* (Astronomical Society of the Pacific, San Francisco).

[PTVF92] Press, W.H., Teukolsky, S.A., Vetterling, W.T. and Flannery, B.P. 1992. *Numerical Recipes* (Cambridge University Press, Cambridge).

[RL79] Rybicki, G.B. and Lightman, A.P. 1979. *Radiative Processes in Astrophysics* (John Wiley, New York).

[RM88] Reid, M.J. and Moran, J.M. 1988. *The Impact of VLBI on Astrophysics and Geophysics, Proc. International Astronomical Union Symp. 129* (Kluwer, Dordrecht).

[RM93] Röser, H. J. and Meisenheimer, K. 1993. *Jets in Extragalactic Radio Sources* (Springer-Verlag, Berlin).

[RSP92] Roland, J., Sol, H. and Pelletier, G. 1992. *Extragalactic Radio Sources – From Beams to Jets* (Cambridge University Press, Cambridge).

[RT94] Robinson, A. and Terlevich, R.J. 1994. *The Nature of Compact Objects in Active Galactic Nuclei* (Cambridge University Press, Cambridge).

[S68] Sandage, A.R. 1968. *Galaxies and the Universe: Stars and Stellar Systems Volume 9* (The University of Chicago Press, Chicago).

[S88] Siegel, S. 1988. *Nonparametric Statistics for the Behavioural Sciences* (McGraw-Hill, New York).

[S89] Sellwood, J.A. 1989. *Dynamics of Astrophysical Discs* (Cambridge University Press, Cambridge).

[SEF92] Schneider, P., Ehlers, J., and Falco, E.E. 1992. *Gravitational Lenses* (Springer-Verlag, Berlin).

[SK86] Swarup, G. and Kapahi, V.K. 1986. *Quasars, International Astronomical Union Symposium 119* (Kluwer, Dordrecht).

[ST73] Sulentic, J.W. and Tifft, W.G. 1973. *The Revised General Catalogue of Nonstellar Astronomical Objects* (University of Arizona Press, Tucson).

[ST83] Shapiro, S.L., Teukolsky, S.A. 1983. *Black holes, White Dwarfs, and Neutron Stars: The Physics of Compact Objects* (John Wiley & Sons, New York).

[ST93] Shull J.M. and Thornson, H.A. 1993. *Proc. 3rd Tetons School* (Kluwer, Dordrecht).

[VV92] Valtoja, E. and Valtonen, M. 1992. *Variability of Blazars* (Cambridge University Press, Cambridge).

[W78] Wolfe, A.M. 1978. *Proc. Pittsburgh Conference on BL Lac Objects* (University of Pittsburgh, Pittsburgh).

[W86] Weedman, D.W. 1986. *Quasar Astronomy* (Cambridge University Press, Cambridge).

[WD95] Walsh, J.R. and Danziger, I.J. 1995. *Science with the VLT, Proc. ESO Workshop* (Springer-Verlag, Berlin).

[ZP87] Zensus, J.A. and Pearson, T. J. 1987. *Superluminal Radio Sources* (Cambridge University Press, Cambridge).

[ZP90] Zensus, J.A. and Pearson, T. J. 1990. *Parsec-scale Radio Jets* (Cambridge University Press, Cambridge).

Author index

Subject index

Pages where the topic is first introduced or discussed are shown in italics. The italic capitals refer to satellites.